Impacts of Climate Change and Economic and Health Crises on the Agriculture and Food Sectors

Vítor João Pereira Domingues Martinho
Polytechnic Institute of Viseu, Portugal

IGI Global
PUBLISHER of TIMELY KNOWLEDGE

A volume in the Practice, Progress, and Proficiency in Sustainability (PPPS) Book Series

Published in the United States of America by
IGI Global
Engineering Science Reference (an imprint of IGI Global)
701 E. Chocolate Avenue
Hershey PA, USA 17033
Tel: 717-533-8845
Fax: 717-533-8661
E-mail: cust@igi-global.com
Web site: http://www.igi-global.com

Library of Congress Cataloging-in-Publication Data

Names: Martinho, Vítor João Pereira Domingues, editor.
Title: Impacts of climate change and economic and health crises on the agriculture and food sectors / Vitor Martinho, editor.
Description: Hershey PA : Engineering Science Reference, [2022] | Includes bibliographical references and index. | Summary: "This book discusses new trends in the agricultural and food sectors with themes that propose to make it possible to approximate the various and current dimensions related to food production adding new insights to bring relevant value added for stakeholders"-- Provided by publisher.
Identifiers: LCCN 2021045747 (print) | LCCN 2021045748 (ebook) | ISBN 9781799895572 (hardcover) | ISBN 9781799895589 (paperback) | ISBN 9781799895596 (ebook)
Subjects: LCSH: Agricultural industries. | Food industry and trade. | Food security. | Crops and climate. | Crisis management.
Classification: LCC HD9000.5 .I455 2022 (print) | LCC HD9000.5 (ebook) | DDC 338.1/9--dc23/eng/20211028
LC record available at https://lccn.loc.gov/2021045747
LC ebook record available at https://lccn.loc.gov/2021045748

This book is published in the IGI Global book series Practice, Progress, and Proficiency in Sustainability (PPPS) (ISSN: 2330-3271; eISSN: 2330-328X)

British Cataloguing in Publication Data
A Cataloguing in Publication record for this book is available from the British Library.

All work contributed to this book is new, previously-unpublished material. The views expressed in this book are those of the authors, but not necessarily of the publisher.

For electronic access to this publication, please contact: eresources@igi-global.com.

Practice, Progress, and Proficiency in Sustainability (PPPS) Book Series

Ayman Batisha
International Sustainability Institute, Egypt

ISSN:2330-3271
EISSN:2330-328X

Mission

In a world where traditional business practices are reconsidered and economic activity is performed in a global context, new areas of economic developments are recognized as the key enablers of wealth and income production. This knowledge of information technologies provides infrastructures, systems, and services towards sustainable development.

The **Practices, Progress, and Proficiency in Sustainability (PPPS) Book Series** focuses on the local and global challenges, business opportunities, and societal needs surrounding international collaboration and sustainable development of technology. This series brings together academics, researchers, entrepreneurs, policy makers and government officers aiming to contribute to the progress and proficiency in sustainability.

Coverage

- Global Business
- Knowledge clusters
- Eco-Innovation
- Socio-Economic
- Environmental informatics
- E-Development
- Global Content and Knowledge Repositories
- Outsourcing
- Innovation Networks
- ICT and knowledge for development

IGI Global is currently accepting manuscripts for publication within this series. To submit a proposal for a volume in this series, please contact our Acquisition Editors at Acquisitions@igi-global.com or visit: http://www.igi-global.com/publish/.

The Practice, Progress, and Proficiency in Sustainability (PPPS) Book Series (ISSN 2330-3271) is published by IGI Global, 701 E. Chocolate Avenue, Hershey, PA 17033-1240, USA, www.igi-global.com. This series is composed of titles available for purchase individually; each title is edited to be contextually exclusive from any other title within the series. For pricing and ordering information please visit http://www.igi-global.com/book-series/practice-progress-proficiency-sustainability/73810. Postmaster: Send all address changes to above address.

Titles in this Series

For a list of additional titles in this series, please visit: http://www.igi-global.com/book-series/practice-progress-proficiency-sustainability/73810

Achieving Sustainability Using Creativity, Innovation, and Education A Multidisciplinar Approach
Ziska Fields (University of Johannesburg, South Africa)
Information Science Reference • © 2022 • 260pp • H/C (ISBN: 9781799879633) • US $215.00

Handbook of Research on Sustainable Development Goals, Climate Change, and Digitalization
Rui Alexandre Castanho (WSB University, Poland)
Engineering Science Reference • © 2022 • 684pp • H/C (ISBN: 9781799884828) • US $315.00

Sociological Perspectives on Sustainable Development and Poverty Reduction in Rural Populations
Ladislaus M. Semali (Pennsylvania State University, USA)
Information Science Reference • © 2022 • 303pp • H/C (ISBN: 9781799846468) • US $195.00

Handbook of Research on Novel Practices and Current Successes in Achieving the Sustainable Development Goals
Cristina Raluca Gh. Popescu (University of Bucharest, Romania & The Bucharest University of Economic Studies, Romania & University of Craiova, Romania & The National Institute for Research and Development in Environmental Protection (I.N.C.D.P.M.), Romania & National Research and Development Institute for Gas Turbines COMOTI, Romania)
Engineering Science Reference • © 2021 • 461pp • H/C (ISBN: 9781799884262) • US $295.00

Advancements in Sustainable Architecture and Energy Efficiency
Roberto A. González-Lezcano (Universidad CEU San Pablo, Spain)
Engineering Science Reference • © 2021 • 337pp • H/C (ISBN: 9781799870234) • US $195.00

Management and Conservation of Mediterranean Environments
Rui Alexandre Castanho (WSB University, Poland) and José Martín Gallardo (Universidad de Extremadura, Spain)
Engineering Science Reference • © 2021 • 336pp • H/C (ISBN: 9781799873914) • US $195.00

Handbook of Research on Institution Development for Sustainable and Inclusive Economic Growth in Africa
Evans S. Osabuohien (Covenant University, Nigeria) Emmanuel A. Oduntan (Covenant University, Nigeria) Obindah Gershon (Covenant University, Nigeria) Olaronke Onanuga (Covenant University, Nigeria) and Oluyomi Ola-David (Covenant University, Nigeria)
Business Science Reference • © 2021 • 507pp • H/C (ISBN: 9781799848172) • US $235.00

IGI Global
PUBLISHER of TIMELY KNOWLEDGE

701 East Chocolate Avenue, Hershey, PA 17033, USA
Tel: 717-533-8845 x100 • Fax: 717-533-8661
E-Mail: cust@igi-global.com • www.igi-global.com

Table of Contents

Detailed Table of Contents

Since 2016, STP has been funding the implementation of greenhouses, viewed as a viable way to guarantee, increase, and diversify production; supply the market; improve farmers' incomes; and mitigate climate change impacts. The greenhouses in selected districts were based on farmers' experiences in horticultural production, available agricultural area, and capacity of rural communities to organize themselves into small farmers' cooperatives. There are also private greenhouse initiatives. This chapter analyzed the current situation of the STP greenhouse project and its socioeconomic contribution to rural communities, proposing actions for its improvement, addressing climate changes and poverty reduction. Despite several weaknesses, mainly linked to lack of knowledge and mastery of technology, greenhouse production represents a viable alternative for horticulture development. Greenhouses, properly exploited, are a mechanism to mitigate climate change effects and ensure an increase in income and consequently reduce poverty and improve individual and collective living conditions.

In India, the enormous contribution of mass media in the agricultural sector helps to develop agriculture as well as the socio-economic structure of the nation. The chapter discusses the contribution of mass media in agricultural sector of India and how it supports the socio-economic development of the nation. Agriculture and economy are two essential factors of modern society. Agriculture helps in the development of economy whereas mass media gives a pace. Mass media is attracting the attention of the farmer, giving them information related to agronomy and creating needs and demands of agriculture through marketing techniques, which ultimately enhance the productivity and economy of the nation. Mass media is ultimately helping the overall sustainable developmental process of the agricultural sector of the nation.

This chapter aims to study the trends in sustainable agricultural development in China. The deterioration of the world's climate and environment can be counteracted with the trend of development in internet and technological spheres. How to create value out of scientifically advanced agricultural techniques in China so as to achieve sustainable development in the future is the subject of the author's discussion. This chapter will investigate the development of smart and sustainable agricultural techniques that are being employed in China in integration with the internet and information industry. Further, the sustainable ecological development of agriculture in China in the past two decades will be analyzed. Finally, this chapter will provide decision-makers with analysis and suggestions on the way forward and direction with respect to sustainable agricultural development in China in the future.

The production of bioenergy and biofertilizers based on animal and plant biomass is a crucial pillar in circular economy (CE). CE conceptual model and main aims are closely related to the 3 "R" (reduce, reuse, and recycle) rule, which is to improve the use of resources, minimize waste, and assure sustainability. Although bioenergy offers many opportunities and could be an alternative to fossil fuels use, the path for a broader implementation of this type of activity is still long. This study marks the starting point or direction of research to be taken, ensuring the existence of benefits from plant and animal biomass for the production of bioenergy and biofertilizer, as well as the contributions of this type of production to the circular economy and the mitigation of the climate change impacts.

COVID-19 is a pandemic of the 21st century, a disease that shook the world and altered the lives of entire communities. Due to the enormous negative influence on the economy, it permanently alters the way an organization operates, leading businesses to develop crisis management techniques and implement new innovative practices. Agriculture is no exception. Given the sector's constant growth, which is not only due to population growth but also to continuous lifestyle changes, it is critical to implement recovery plans at the organizational and government levels. Thus, this chapter provides an overview of crisis management, including its key characteristics and framework; analyzes the importance of innovation in the agricultural sector; provides an overview of the agricultural sector; examines the impact of the pandemic on this sector and some recovery strategies; and examines the attitude of agricultural professionals toward the COVID-19 crisis.

Agricultural production is influenced by environmental factors such as temperature, air humidity, soil water, light intensity, and CO2 concentration. However, climate change has influenced the values of average temperature, precipitation, global atmospheric CO2 concentration, or ozone level. Thus, climate change could lead to different situations on plants and consequently influence agricultural production. With this chapter, the authors intend to research how climate change influences some plant metabolisms (such as photosynthesis, photorespiration, transpiration, among others) and therefore agricultural production.

This chapter will aim to explicate the challenges posed by global warming or the climate change conditions on food security especially from the point of view of India. The negative impact of global warming has been seen, especially in developing economies, on the agricultural yields leading towards food insecurity. The four pillars of food security (i.e., availability, accessibility, utilization, and stability) are having an impact on climate change. The present study will begin by highlighting the concept of global warming. It will further provide an overview of the Indian food security system followed by the impact of global warming on the food security level in India. The study will also highlight the global warming and food security scenario in the present situation of the ongoing COVID-19 pandemic in India.

The social role of the farms is, especially, relevant in the rural areas where the socioeconomic problems are, often, more visible. In this perspective, this study aims to investigate the interrelationships of the labour input with other variables inside the farms and assess how the sector may create more employment in a sustainable way. For that, the labour input was, first, correlated with other farm variables and after analysed through factor analysis approaches and cross-section econometric methodologies, considering as basis the Cobb-Douglas and Verdoorn-Kaldor models. The main findings highlight relevant insights to improve the social dimension of the European Union farms. The labour input growth rate is positively influenced by the total output growth rates and negatively impacted by the total productivity growth. The effects from the investment and from the subsidies are residual or not significant.

A deeper assessment of the main determinants associated with the use of fertilisers, for example, in the European Union farms may bring relevant insights about the respective frameworks and support to find more sustainable solutions. In this context, the main objective of this study is to identify factors that influence the use of fertilisers in the agricultural sector of the European Union regions. To achieve this objective, statistical information, at farm level, from the European Farm Accountancy Data Network was considered. These data were first analysed through exploratory approaches and after assessed with classification and regression tree methodologies. The results obtained provide interesting insights to promote a more sustainable development in the European farms, namely supporting the policymakers to design more adjusted measures and instruments. In addition, the fertilisers costs on the European Union farms are mainly explained by crop output, costs with inputs, current subsidies, utilised agricultural area, and gross investment.

Despite the recognized advantages of Mediterranean Diet (MD), the adherence to it decreased with modern lifestyle, where the time dedicated to acquisition, preparation/confection of food and meals diminished. At the same time, Mediterranean regions face a growth in the levels of non-communicable diseases, such as obesity, diabetes, and hypertension, sometimes together with undernutrition that affects other parts of the population. This chapter make a presentation about MD as a sustainable food system, essential to promote food security, at the same time that the methods of food production and consumption must respect the environment, maintain biodiversity, and economic society valorisation. Also, it shows MD associated with several factors such as gender, marital status, education level, lifestyle, and body weight. Maintaining the traditional MD pattern is crucial for public health, particularly in pandemic contexts such as COVID-19 where it shows the opportunity and relevance of adopt and promote MD as a healthy and sustainable diet.

As living standards change with the development of modern industry and social encounters, people tend to change their lifestyle and environment exposure along with their psychophysiological factors, leading to an imbalance of homeostasis and increasing the risk for chronic diseases. In addition to ingredients, methods, and food conditions storage and processing, the use of additives and certain new foods have facilitated the increased occurrence of chronic diseases in children or adults. The interaction of some components of the food system with enzymes that metabolize different types of drugs can affect the body's clearance and therapeutic index.The objective of this chapter was to present the general principles of food development for special nutritional conditions, also the adjuvants used for chronic disease status improvement, under the condition of nutritional nutrivigilence and food safety standards, and specific to introduce an adjuvant food for atopic dermatitis management.

Agriculture is one of India's most significant industries, serving half of the nation. Climate is a big factor for agricultural production. Agriculture is primarily rain-fed and subjected by small-scale farmers. Rain-fed crops are 48% of the overall area under food crops and 68% on non-food crops. The complete worldwide agricultural production in India accounts for 7.39%. India's total emissions from all GHGs in 2014, according to the World Resource Institutes, total about 3,200 MTCO2 eq out of 48,892 MTCO2s eq worldwide in 2014. Agriculture is ascribed 626.86 2 MTCO2 eq of the Indian greenhouse gas emissions averaged in 2014 of 3,200 MTCO2 eq. This chapter reviews the concept of sustainable agriculture, establishes the link between climate change and agriculture, the origination of climate-smart agriculture, and relevant practical approaches, case studies, and geospatial assessment methods responding to climate-smart agriculture.

The objective of this study is to determine the suitability for the cultivation of emerging fruit crops in the Beira Baixa region. The suitability was examined for the present time and in the face of two future emission scenarios (RCP 4.5 and 8.5). For this purpose, the biophysical criteria determining the cultivation of pistachio tree and almond tree were processed using a G. The analysis was performed by the AHP. After dividing the problem into hierarchical levels of decision making, a pairwise comparison of criteria was performed to evaluate the weights of these criteria, based on a scale of importance. In the present conditions, about 16.4% of the study area is classified as highly suitable for almond tree and 15.9% to pistachio tree. For the future scenarios, the area with high suitability will increase both for almond tree and pistachio tree. The AHP was adequate in the evaluation of the emerging fruit tree species suitability, since it allowed the integration of the several criteria studied, being a useful tool, which allows the decision making and the resolution of problems.

Preface

This book entitled *Impacts of Climate Change and Economic and Health Crises on the Agriculture and Food Sectors* intends to be a relevant contribution to the current challenges worldwide, namely those associated with global warming and pandemics and their consequent impacts on different dimensions sustainability (economic, social and environmental). The focus of this book is to analyze, assess and discuss these implications specifically in the agricultural and food sectors.

The subjects covered here are interconnected with other issues, such as those associated with new technologies and advances in the various fields of science. In these cases, nanotechnologies and integrated-smart agriculture are of particular importance. The idea with these new approaches is to find a compromise between the requirement for more agricultural and food production to meet the food needs of the world population and the importance of maintaining sustainable development. The integrated-smart agriculture approach aims to allow improvements in agricultural productivity without compromising sustainability, namely the environmental one. This is only possible with new technologies, specifically those related to, for example, IoT (Internet of Things) approaches. These new perspectives can be interesting contributions to more sustainable activities throughout the entire agrifood chains.

Considering the objectives outlined for the book and the topics covered, this publication can be an interesting contribution to the various stakeholders present in the different stages of the agrifood chains, including students, researchers, policy makers and international operators.

The book received contributions, organized into 16 chapters, from several authors around the world, thus giving a broad global perspective on the topics covered. A brief description of the importance of each chapter will be presented following.

CHAPTER 1. AGRICULTURAL GREENHOUSES IN SÃO TOMÉ AND PRÍNCIPE: A WAY TO MITIGATE CLIMATE CHANGES, PROMOTE FOOD SECURITY, AND REDUCE FARMER POVERTY

In this chapter the present condition of the São Tomé and Príncipe greenhouse project and its socioeconomic impact to rural communities were analyzed, suggesting proposals for its improvement, covering global warming and poverty mitigation.

CHAPTER 2. AGRICULTURE, MASS MEDIA, AND THE ECONOMIC DEVELOPMENT OF INDIA

This chapter broadly debates the involvement of mass media in farming sector of India and how it contributes in the socioeconomic performance of the country.

CHAPTER 3. ANALYSIS ON THE PRESENT SITUATION OF CHINA'S AGRICULTURAL DEVELOPMENT AND SUSTAINABLE DEVELOPMENT OF AGRICULTURAL ECONOMY

The improvement of smart and sustainable farming practices that are being applied in China, in combination with Internet and information industry, was explored. In addition, the sustainable and ecological evolution of the agricultural sector in China in the past two decades was analyzed. Finally, this chapter provides suggestions and directions with respect to sustainable farming development in China in the future.

CHAPTER 4. CONTRIBUTIONS OF SUSTAINABLE BIOMASS AND BIOENERGY IN AGRICULTURE TRANSITIONS TOWARDS A CIRCULAR ECONOMY

This chapter highlights the existence of advantages from plant and animal biomass for the generation of bioenergy and production of biofertilizer, as well as the contributions of this type of production to the circular economy and the reduction of the global warming impacts.

CHAPTER 5. CRISES ON THE AGRICULTURE: CONCEPTS, RESPONSES, AND RECOVERY STRATEGIES

An overview of agricultural crisis management, including its key characteristics and framework, was presented. In addition, this chapter analyzes the relevance of innovation in the farming sector, presents an overview of agriculture, analyzes the influence of the pandemic on this sector and investigates the attitude of farmers toward the COVID-19 crisis.

CHAPTER 6. EFFECT OF CLIMATE CHANGE ON AGRICULTURAL PRODUCTION

This chapter intends to investigate how global warming influences some plant metabolisms (such as photosynthesis, photorespiration, transpiration, among others) and, consequently, the agricultural outputs.

CHAPTER 7. EFFECT OF GLOBAL WARMING ON FOOD SECURITY: AN INDIAN PERSPECTIVE

The challenges created by the climate change circumstances on food security particularly from the point of view of India were explained. Negative implications of the climate change have been seen, principally in developing economies, on the agricultural yields conducting towards food insecurity.

CHAPTER 8. LABOUR DRIVERS IN THE AGRICULTURAL SECTOR OF THE EUROPEAN UNION: THE SOCIAL ROLE OF FARMS

The social role of the farms is, especially, relevant in the rural areas where the socioeconomic problems are, often, more visible. In this perspective, this study aims to investigate the interrelationships of the labour input with other variables inside the farms and assess how the sector may create more employment in a sustainable way.

CHAPTER 9. MAIN FACTORS THAT EXPLAIN THE USE OF FERTILISERS ON FARMS IN THE EUROPEAN UNION: CONTRIBUTIONS TO A MORE SUSTAINABLE DEVELOPMENT

A deeper assessment of the main determinants associated with the use of fertilisers, for example, in the European Union farms may bring relevant insights about the respective frameworks and support to find more sustainable solutions. In this context, the main objective of this study is to identify factors that influence the use of fertilisers in the agricultural sector of the European Union regions.

CHAPTER 10. MEDITERRANEAN DIET AS A HEALTHY, SUSTAINABLE, AND SECURE FOOD PATTERN

This chapter discusses about Mediterranean Diet (MD) as a sustainable food structure, indispensable to stimulate food security, at the same time that the approaches of food production and consumption should be compatible with the environment, preserving biodiversity and socioeconomic dimensions. In addition, presents MD associated with numerous factors such as gender, marital status, education level, lifestyles and body weight. Maintaining the traditional MD configuration is crucial for public health, principally in pandemic frameworks such as Cobvid-19 where it presents the opportunity and importance of implement MD healthy and sustainable diet.

CHAPTER 11. NEW NUTRITIONAL PERSPECTIVES IN THE CONTEXT OF CHRONIC DISEASE PATIENT MANAGEMENT

The overall assumptions of food development for special nutritional circumstances were presented. In addition, the adjuvants used for chronic disease's status improvement were highlighted.

CHAPTER 12. PERSPECTIVES ON CLIMATE SMART AGRICULTURE: A REVIEW

This chapter reviews the notion of sustainable agriculture, analyzes the link among global warming and farms, the beginning of Climate-Smart Agriculture (CSA), and pertinent applied methods, case studies, and geospatial valuation approaches answering to CSA.

CHAPTER 13. PRESENT AND FUTURE LAND SUITABILITY ANALYSIS FOR ALMOND AND PISTACHIO CROPS IN THE BEIRA BAIXA REGION USING SPATIAL MULTICRITERIA DECISION SYSTEMS

The aim of this chapter is to investigate the appropriateness for the production of emerging fruit crops (namely, almond and pistachio) in the Beira Baixa region (Portugal), considering spatial multicriteria decision systems.

CHAPTER 14. SUSTAINABILITY OF RICE PRODUCTION AT BAIXO MONDEGO, PORTUGAL: DRIVERS, RISKS, AND SYSTEMS IMPROVEMENT

The rice production structure at the Baixo Mondego Valley, to understand the principal worries, was analyzed. Field investigation and field trials were performed to examine rice production, marketing strategies and different irrigation options.

CHAPTER 15. SUSTAINABLE COCOA VALUE CHAIN: A REVIEW AND CRITICAL ANALYSIS OF "TRIPLE BOTTOM LINE" SCENARIOS

This chapter conducts a systematic review of the literature on the dimensions of sustainability associated with the Cocoa Value Chain. At the end, suggestions for future research on the topic were suggested.

CHAPTER 16. SUSTAINABLE DEVELOPMENT IN THE AGRIFOOD RICE CHAIN IN TIMOR-LESTE

The rice agri-food chain sustainable development was assessed, highlighting the main stakeholders involved and understanding their viewpoints. Interviews, questionnaires, observation and focus group have been considered to understand how sustainable development may be achieved.
Vítor João Pereira Domingues Martinho

Acknowledgment

I would like to thank everyone who has contributed in any way to this book.

Chapter 1
Agricultural Greenhouses in São Tomé and Príncipe:
A Way to Mitigate Climate Change, Promote Food Security, and Reduce Farmer Poverty

Kiakisiki Quaresma Nascimento
CEGAFE, University of Évora, Portugal & CETRAD, University of Trás-os-Montes and Alto Douro, Portugal

Maria Raquel Raquel Lucas
Universidade de Évora, Portugal & CEFAGE, University of Algarve, Portuga

Pedro Damião Henriques
Universidade de Évora, Portugal & MED, Portugal

ABSTRACT

Since 2016, STP has been funding the implementation of greenhouses, viewed as a viable way to guarantee, increase, and diversify production; supply the market; improve farmers' incomes; and mitigate climate change impacts. The greenhouses in selected districts were based on farmers' experiences in horticultural production, available agricultural area, and capacity of rural communities to organize themselves into small farmers' cooperatives. There are also private greenhouse initiatives. This chapter analyzed the current situation of the STP greenhouse project and its socioeconomic contribution to rural communities, proposing actions for its improvement, addressing climate changes and poverty reduction. Despite several weaknesses, mainly linked to lack of knowledge and mastery of technology, greenhouse production represents a viable alternative for horticulture development. Greenhouses, properly exploited, are a mechanism to mitigate climate change effects and ensure an increase in income and consequently reduce poverty and improve individual and collective living conditions.

DOI: 10.4018/978-1-7998-9557-2.ch001

INTRODUCTION

São Tomé e Príncipe economy is predominantly supported by agriculture and services (Arteta and Kirby, 2019) and agriculture aggregates 33% of the population, it is only responsible for 5% of GDP (gross domestic product) (BM, 2018). Cocoa is the main agricultural production and main source of income for rural families, generating 70% of rural employment and 80% of export earnings (UNDP, 2016). Agriculture has a very low level of productivity, mainly due to traditional agricultural practices, lack of infrastructure (irrigation systems, markets and rural roads), insufficient technical support and resources, low quality products, and lack of organization and strategy in the the sector (Espírito Santo, 2008). Additionally, there is a lack of supportive agricultural research and investments to ensure food security (UNDP, 2016).

As a result, domestic production has been decreasing, increasing the economic and social constraints of rural populations, as well as farmers' abandonment of agricultural activity and the replacement of sustainable agricultural practices by others more harmful to natural resources and the environment. If added the effects of climate change is added the situation of agriculture in the country problematic (UNDP, 2016).

The latest estimates point to about a third of the population living on less than US$1.9 a day, and more than two thirds being poor and on the poverty line of 3.2 US dollars per day (World Bank, 2019). Urban areas and southern districts in STP such as Caué and Lembá are the ones with the highest levels of poverty incidence. Thus, although GDP growth has been relatively stable since 2009 (it has grown at an average rate of 4.0% between 2009 and 2016 and has decelerated moderately since 2014), the increase in government expenditure growth has not contributed to poverty alleviation (Gomes et al., 2018). Still, STP is above average in the UNDP Human Development Index indicator compared to the average for sub-Saharan African countries and has been making significant progress in improving other social indicators. For example, the gross enrolment rate in primary education is 110%; life expectancy is 66 years; the mortality rate for children under five is 51 per 1000 live births, access to an improved water source is 97% for the population and access to electricity is 60% (World Bank, 2019).

Similarly, to other small island states, STP also faces constraints related to the high cost of imports, territorial limitations that make economies of scale unfeasible, the non-diversification of the economy and the high public expenditure on the provision of services to the population, all this affect the country's ability to deal with external shocks and achieve a balanced budget (Gomes et al., 2018).

Alongside agriculture and services, tourism tends to be an important economic activity as STP has a natural potential, although there is still much to do in this particular area (World Bank, 2019). The commercial exploitation of oil, a resource considered promising for the country's economy, has been delayed for almost a decade, making the country heavily dependent on imports, including oil for energy generation and food production.

Although the country's resources and edaphoclimatic conditions are adequate for agricultural practice and rainfall is, on average, abundant, constraints on productivity have been increased by the increasingly frequent occurrence of drought episodes. This situation has contributed to the increased vulnerability of agricultural communities, particularly in the districts of Caué, Me-Zochi, Príncipe, Lembá, Cantagalo and Lobata.

In recent decades, there have been significant changes in the weather pattern in STP, mainly the decrease in rainfall at a rate of 1.7mm/year between 1951 and 2010. This reduction in rainfall disturbs the hydrologic cycle, altering the precipitation/runoff ratio and disrupting the supply of groundwater, as well as the reduction of the water table (UNDP, 2016). This study referrs that the annual temperatures have increased by approximately 0.4°C between 1960 and 2006. It is predicted that by 2060 there will be a temperature increase in the country between 0.8°C and 2.4°C. The country has registered an average temperature increase of 0.1°C, per decade, and a significant decrease in rainfall (5.2% per decade) – in the months of March, April and May, as well as October, November and December. The volubility of climate change will have negative impacts on the primary sector (agriculture, fisheries and forestry), which by the way, are the most vulnerable sectors of STP that have already suffered negative impacts from climate change, such as reduced agricultural production and uncontrolled use of forests and soils.

The UNDP (2016) report estimates that in the medium and long term, the following natural phenomena may be observed: i) worsening food insecurity; ii) increase in the level of poverty; iii) increased dependence on food imports and, consequently, an increase in the chronic trade balance deficit. Indeed, climate change could aggravate the poverty rate, reducing access to basic needs (such as food and water), and consequently, the population's health. It is estimated, however, that the cyclicality of this problem and its effects will progressively make the population more vulnerable to the impacts of adverse climatic conditions. This Report concludes that climate change poses a threat to agricultural production and can compromise the country's increased productivity and income, as well as the food security of small farmers (who constitute the most vulnerable strata of rural communities) whose livelihoods are highly dependent on natural resources, namely, rudimentary agriculture (which in itself is poorly diversified).

Faced with the country's challenges to overcome insularity, the small size of the market, limited human capital and scarce tradable resources to generate inclusive sustainable growth and reduce poverty and vulnerability to natural shocks and climate change, a multifaceted answer was proposed that combines: i) rural development and knowledge of the capacities of key institutions and livelihoods; ii) the development of decision-making structures that improve the livelihoods of agricultural communities; iii) the spread of climate-resilient livelihoods in the most vulnerable communities; and, iv) the promotion of investments to increase the livelihoods of communities against climate risks (UNDP, 2016).

Thus, since 2016, the government of São Tomé and Príncipe (STP) with its development partners, in particular the United Nations Development Program (UNDP) and the Global Environment Fund (GEF), has been funding the acquisition and implementation of greenhouses for production of vegetables, in particular peppers and tomatoes, among other species. In their view, the implementation of greenhouses is a viable and adequate way to guarantee production during the year, which is greatly reduced during the rainy season and, increase quality and price stabilization in the domestic market. Also, is considering a way to increase and diversify production, supply the market, improve farmers' incomes and rural community's resilience, as well as, mitigate climate changes impacts.

The country currently has 10 greenhouses in selected districts and localities based on farmers' experience in terms of horticultural production, available agricultural area adjusted for this type of crop, and the capacity of rural communities to organize themselves into small farmers' cooperatives, being the main criterion considered in the selection of the communities, the difficulty in the production of vegetables, due to their climatic conditions. There are also private initiatives in the greenhouse implementation framework that also receive occasional support from the State and its partners.

The purpose of this chapter is to analyze the current situation of the STP greenhouse project and its socioeconomic contribution to rural communities, proposing actions for its improvement in addressing climate changes and poverty reduction.

Apart the introduction, the second section reviews the concept of rural development and the role of greenhouses to achieve it and to address climate challenges, then the methodology used to analyze the greenhouses project in STP is offered, the characterization of the current situation of greenhouses and its socioeconomic contribution to rural communities, are presented in the section of the results and the chapter ends with the conclusion which included some propositions for greenhouses improvement towards climate changes mitigation and, the additional literature reading suggestions.

GREENHOUSES, RURAL DEVELOPMENT AND CLIMATE CHANGES

Rural Development and Climate Changes

Climate change have impact in the livelihoods of local communities and on people's welfare (Sørensen, 2018) and can be reflected on rural development (Biegańska et al., 2018; Xie, 2020) and on the economic development of a region (Barbier & Hochard, 2018; Tanure et al., 2020; Stern, 2018; Moreno-Cruz & Smulders, 2018). This condition can also be exacerbated when rural population is far from the centre of decision (Nørgaard and Thuesen, 2021).

Despite rural development being widely discussed in the academic, politic and NGOs field, there is no comprehensive and universally acknowledged definition of it, taking account of the diverse disciplines, it involves and the divergent contexts in which it is embedded (Lu and de Vries, 2021). According to the World Bank Sector Paper on Rural Development (World Bank, 2019), rural development is a strategy designed to improve the economic and social life of a specific group of people, the rural poor. It could include small-scale farmers, tenants and the landless (Chambers, 1983), been identified as a policy goal (Copp, 1972), refer to processes of change in rural societies, not all of which involve action by the government (Harriss, 1982) or be recognized as a multi-level, multi-actor and multi-faceted process (Van der Ploeg et al., 2000).

In other perspective, rural development can be defined as efforts to improve changes in the quality of life and for the greatest welfare of rural communities (Nogueira, 2014). Thus, the development goals in rural areas means improving the welfare of the rural community and the quality of human life and poverty reduction through meeting basic needs, construction of rural facilities and infrastructure, development of local economic potential, and use of natural resources and environmentally sustainable (Faradiba and Zet, 2020).

Concerning the research in rural development, it became more multidisciplinary and interdisciplinary, reflecting the dynamics and complexity of rural contexts worldwide and the differences in the socio-economic, environmental, and political development between different regions at different times (Lu and de Vries, 2021). These authors suggest that it is more feasible to analyse the pattern of rural development when focusing on one region with a consistent context.

Apart others, the factors that can become obstacles to rural development are natural disasters that often occur in a certain area, induced by climate changes (Zhang & Managi, 2020; Sawada & Takasaki, 2017; Straub et al., 2020; Jackson et al., 2020; Jiang et al., 2016).

Climate change, changes the social conditions as is reflected in changes in global livelihoods (Arronson & Schöb, 2018) and particularly in the most affected sector, agriculture (Faradiba, 2018; Javed et al., 2020, Soubry et al., 2020; Sousa et al., 2018; Falco et al., 2019). When climate change impacts can be anticipated, controlled and mitigate properly, it can maintain rural development performance. In this condition, the impact on rural development can go according to planned and produce prosperity and comfort for the community (Austin et al., 2020). Particularly in developing countries, mitigate climate changes is crucial to assure rural development as a step to increase socio-economic conditions and attenuate rural community's limitations (Uthes et al., 2017).

Rural development as to be understood as a broad concept, embodied in the time factor (long-term trajectory), in the space factor (the territory and its resources) and in the social structures present in each case (Nogueira, 2014). It also entails the development of policies focused on increasing production, improving infrastructure, training farmers and modernizing production systems.

From the perspective of the International Labor Organization (ILO, 2014), rural development is the key to Africa's prosperity, despite being underestimated by governments, international funders and policy advisers. For its Deputy Director General, Gilbert Houngbo, boosting agriculture in Africa and building a strong rural economy is crucial in order to create jobs, wealth, social inclusion, food security, resilience to crises and climate changes, and social peace. It is also to recognize the potential of rural communities for sustainability and make the necessary investment to empower them through integrated and integrative approaches. Therefore, it advocates investing in basic physical and social infrastructure, such as roads, energy, education and health centres, as well as the development of skills and entrepreneurship, individually or through cooperatives, and innovative financial mechanisms (ILO, 2014).

Houngbo also warned that the failure to recognize the value of rural areas and to invest in rural development has resulted in per capita food production that has barely grown over the past 50 years and an incipient agricultural sector in terms of its weight in the gross domestic product of Sub-Saharan Africa countries (only 17%) and on the decline of its already low productivity.

Thus, becoming aware of this issue is one of the prerequisites for its resolution and for the agricultural transformation in Africa, which also necessarily involves enabling integrated approaches that establish links between public and private agents, leading to the training of rural workers, including young people and women, and the development of entrepreneurial structures and encouragement of dialogue between stakeholders.

Greenhouses, Climate Changes Mitigation and Poverty Reduction

There seems to be a consensus among scientist, researchers and all other stakeholders that climate change is already happening and primarily caused by human activities while some disagree with this view and explain climate change as a natural phenomenon (Nda et al., 2018). Based on the current state of climate change, immediate development of viable mitigation and adaptation mechanisms is of extreme importance, particularly in developing countries where the poor smallholder face simultaneously, short-term socioeconomic objectives (subsistence food production and income) and long-term objectives (natural resources preservation, particularly soil fertility) (Rakotovao et al., 2021). However, there is no ultimate solution to tackle climate change and all technologies and techniques that are technically and economically being viable should be deployed (Fawzy et al., 2020).

Due to the decarbonisation efforts alone are not sufficient to meet the targets stipulated by the Paris agreement, alternatives approach to mitigate climate changes are inevitable. Capturing CO2 through photosynthesis is a straightforward and solid process that needs to be effectively integrated within a technological framework (Fawzy et al., 2020). In the perspective of Santos (2018), for the context of STP, one of these approaches are the greenhouses, which have potential and economic viability, especially for the production of vegetables addressing also climate changes. However, Santos also says that is necessary to invest in the training and qualification of managers of these protected structures, helping horticulturists learn and improve how to handle these structures and taking better and greater advantage of them.

The seasonality in the production of vegetables in STP is mainly caused by the presence of rain during eight months of the year, so the production of most vegetables is very difficult during this season, which severely affects availability, quality and the price of these products throughout the year. The use of greenhouses is a way to minimize this problem of seasonality, especially the two most economically and socially important vegetables most affected by seasonality, tomatoes and peppers. Although the edaphoclimatic characteristics are more favorable for outdoor vegetable production during the dry season (June to September), greenhouse production can be done throughout the year, especially during the rainy season (October to May) when it is more favorable the market price. In these months, it is often the only production of farmers and the source of income for the family (Santos, 2018).

Neto (2015) states that agricultural production in a greenhouse, in addition to representing one of the ways of modernizing agriculture in developing countries, ensures livelihoods for farmers that allow them to minimize their poverty and feed their families. By providing favorable microclimates that guarantee the stabilization of production throughout the year, regardless of adverse weather conditions, the greenhouse production technology also presents the advantages, in relation to outdoor production, of increasing the quantity and quality of production, responding more closely to the needs and demands of the consumer in the marketplace (Marques, 2017).

Greenhouse production is one of the most demanding and intense forms of all agricultural enterprises once is high technology and capital intensive (Jensen, 1997). The main distinction between protected and outdoor crops is that greenhouses create a microclimate that protects crops from wind, precipitation, weeds, pests, diseases and animals, allowing the producer to condition the environment using air conditioning systems (heating, cooling, increasing the concentration of carbon dioxide or effective application of crop protection measures, chemical or biological) (Perdigones et al., 2015).

Globally greenhouses have two major contributions. One is the modernization of agriculture with the increment in food production and farmer income due the benefits from market higher prices and quality (Santos, 2018). The other is to help mitigate the damage caused by climate changes and adverse weather conditions (Louro, 2011; Rebouças et al., 2015).

In order to get the most out of the greenhouses, the producer must know how to use this technology and comply with some fundamental principles such as: i) proper control of the environment, especially temperature and humidity; ii) the choice of selected varieties and cultivars adjusted to greenhouse conditions; iii) the adequacy of cultural techniques (watering, fertilization, pest and disease control). In addition to the demand for high investments, there is also the need for specialization of the farmer or the

Despite several studies found a positive effect of greenhouse investments on poverty reduction (e.g. Soumare, 2015; Agarwal et al., 2017, Magombeyi and Odhiambo, 2017; Ganié, 2019), others (e.g., Ali et al., 2010) found an inverse relationship, while Ogunniyi and Igberi (2014) found no beneficial effect of foreign direct investment in reducing poverty especially in developing countries which has continually record increase in poverty level despite the massive inflow of foreign direct investment.

More recently, Dada and Akinlo (2021) sustain that both, the financial and rural development, contribute to poverty reduction implying that more effort is required from the policymakers to increase the level of financial deepening to have for more impact of financial and rural development on poverty reduction. Hence, policies makers should increase the access of poor farmers to credit facilities and other financial services which can significantly reduce their level of poverty (Moulton, 2015).

Along with increasing attention on climate change and food security, the implementation of adaptation-mitigation investment projects and the adoption of agricultural systems that reduce poverty and benefit farmers and the society should be taken (Dada et al., 2021). This is especially important in STP, where a high proportion of the people work in the agricultural sector, and where agriculture is the primary activity and source of income for its households. Additionally, the agricultural production of the country relies almost entirely upon smallholder farms which are poor and vulnerable to climate change.

MATERIALS AND METHODS

Study Area

The study area was the ten communities selected for the implementation of the national greenhouses project, eight in São Tomé island and two in Príncipe island (see Table 1). These communities are located in the Districts of Caué, Mé-Zochi (Saudade and Bem Posta regions), Lembá (Canavial and Fernão Dias regions), Cantagalo (Solidade, Uba Budo and Bom Sucesso regions) and Lobata located on the south, central and north regions of São Tomé island, respectively, and the Principe Island (Nova Estrela and Santa Rita regions). The criteria used to select the communities for the implementation of the greenhouse project were: the isolation that characterizes these communities; the state of poverty in which they live; the difficulty in accessing agricultural information, the absence of modern agricultural practices (few inputs and poor quality), the poor state of infrastructure to support agriculture (irrigation systems, rural markets, rural trails.), the absence of advice and strategies to supply farmers with good quality inputs, to channel agricultural products to the market and to create a commercial network that allows access to production factors and facilitates for sale of agricultural products.

Table 1. Selected communities to the greenhouses project by region and district

Region	District	Community	Quantity
South	Cantagalo	Soledade	1
Centre-South		Uba-Budo	1
Centre	Mé-Zochi	Bom Sucesso	1
		Saudade	1
		Bem Posta	1
Centre-North	Lobata	Canavial	1
		Fernão Dias	1
North	Lembá	Dependência de Lembá	1
Região Autónoma do Príncipe	South	Nova Estrela	1
	North	Santa Rita	1
	Total		10

Source: PNUD (2016)

Regarding STP climate, is wet tropical with two distinct seasons: a dry, cool season from June to September and a warm, humid one from October to March. The average annual air temperature and rainfall are 25.6 °C and about 2000 to 7000 mm yr^{-1}, respectively. The country, with almost 1001 km2, is formed by two volcanic islands, São Tomé with 859 km2 and Principe with 142 km2, in which 33% of the population of the 187,739 inhabitants, work in the agriculture sector.

Lowlands are dedicated to production of vegetables, while uplands are dedicated to rainfed crops that include cocoa, coffee, banana, cassava, kills candy, bread fruit, coconut, corn, and sugar cane. About 84% of the arable land is cultivated and exploited, mainly by familiar smallholders. The national average of cropland per household average 1.5 ha and only 4.8% of farmers own more than 4 ha (INE, 2019). The typical farmer uses none or low levels of external inputs, and exercises little control over water supply in cropping systems. Because about 60% of the cultivated land receives no fertilization agricultural productivity is low. The proximity of São Tomé city in the main island, allows producers of the study area to access the market for selling their production in the capital's markets, despite the non-structured production channels.

Data Collection and Analysis

The study used a mixed (quantitative and qualitative) methodology, supported by an exploratory research approach to collect primary data, and an extensive analysis of the literature and deep analysis of the Greenhouses STP project titled "Reinforcement of the capacities of the rural communities for adaptation to the effects of climate change in STP in the districts of Caué, Mé-Zochi, Autonomous Region of the Príncipe, Cantagalo and Lobata", to gather secondary data. The exploratory research was based on an inductive approach, in which secondary sources, observation, the researcher's point of view and qualitative interviews were fundamental for the design of the second phase, which consisted of a conclusive-descriptive research, for which the deductive approach and the quantitative research methods were considered. The interviews were conducted with the ten main players in the process of implementing greenhouses in STP, farmers using greenhouses, cooperatives of agricultural producers with greenhouses, technicians of the different institutions dedicated to this production and political decision-makers. Then, a structured questionnaire targeted twenty-five farmers, ten of them were using greenhouses from the STP project, two have own greenhouses under private investment in Lobata and Mé-Zochi districts and, other ten farmers practiced horticulture in the outdoor agriculture model. The reason behind employing farmers with and without greenhouses, cultivating the same type of crops, in the same regions and with equivalent dimension, was its comparability. Field observation was done during the entire period when empirical data collection took place, in order to complement the information collected and, help adding information and triangulate the data generated by the other research methods in place.

The information treatment included content analysis of the interviews, productivity calculation, costs and feasibility of the implementation of greenhouses and, analysis of socio-economic impact, through the logical landmark (MML) matrix, a tool suggested by Toni (2019) to take into account the impact, objectives, results and activities of one specific project.

RESULTS AND DISCUSSION

This section presents the results of the work. It starts with the greenhouses project description, which includes greenhouses localization, characterization of the current situation and implementation process and, results in terms of productivity, costs, feasibility and, socio-economic contribution to rural communities.

The STP Greenhouses Project

The greenhouses built in STP were conceived as part of the Climate Change Adaptation Project. In the understanding of the government and its development partners, greenhouses are a solution for the production of vegetables, particularly in communities with significant socio-economic restrictions. In each community, cooperatives were created to manage these structures. The project aimed the vegetables greenhouse production as a reinforcement of the resilience of existing vulnerable communities in STP (five districts, including both, São Tomé and Principe islands) by reducing the vulnerability of rural livelihoods to climate risks through infrastructure and climate risk management mechanisms and the design and transfer of adaptation strategies to strengthen the climate resilience.

Recognizing the difficulties faced by these communities, the project was designed based on a social and associative strategy, taking advantage of the existing agricultural and community dynamics, in which men and women traditionally form working groups for the most varied activities. It focused mainly on horticulture because this activity is carried out in almost all communities and districts in the country. Also, because the income obtained is managed by the farmers, which usually results in a greater application of this income for the benefit of the household through education and food.

Within the scope of the project, the greenhouse technology represents "an environment, where the plant and the soil are protected from torrential rains, pests and diseases and adverse temperatures, allowing for a certain control of climatic variables such as temperature, air humidity, solar radiation and wind (UNDP, 2016).

The ten greenhouses of the project, are characterized by multi-chapels. Each greenhouse consists of four chapels of 9.2 m wide and 20 m in length. The height of the trough is 3 m and the usable area of 736 m2 (Figure 1). Table 2 presents the overall greenhouses characteristics.

Figure 1. Greenhouses multi-chapels
Source: PNUD (2016)

Table 2. Multi-chapel greenhouses overall characteristics

Characteristics	Values I Specificities
Chapel number	4
Height in the Guteau	3m
Width in the chapel	9,2 m
Overall width	36,8 m
Length	20m
Greenhouse area	736 m2
Roof cover	3-layer polyethylene plastic 200 μm
Front and lateral closing cap	10x20 insect fillet with raffia plastic base
Ventilation	Zenital fixed opener with insect protection 6x9
Fixation System: Doors: Truss System:	Profile of steel with PVC male and female profile 1 galvanized steel sliding door Included with 12 lines of support / chapel
Distance from outer pillars	2,5m
Distance from interior pillars	5 m
Pillars	80x80x2 mm
Arches: Round Tube	60x1,5 mm
Separation of the arcs	2,5 m
Gutter	410x2 mm
Connection bars: round tube	40x1,5 mm
Reinforcement of the vertical arch	3 round tube units 32 mm
Failure	Longitudinal profile M and C 30x35x1.5 mm
Steel	Galvanized by Sendzimer
Fixing	Bi chromate against corrosion
Watering system	Droplet

Source: PNUD (2019)

Production Quantity and Diversity

Horticulture is carried out mainly by small farmers, throughout the country and with greater intensity in the districts of Água Grande, Mé-Zochi and Lobata. The main products are peppers, tomatoes, green beans, onions, carrots and cabbage and others with reduced expression. According to Gomes et al. (2018), in the year 2016 were produced in the open field model approximately 3,869 ton of vegetables, having been imported in the same period around 827 ton. According to the same source, the average production and imports in the period 2010 at 2016 were 4,000 ton and 670 ton, respectively, the per capita consumption during the same period was 22.2 kg (Table 3).

According to the data collected in 2018, in the different greenhouses of the project just over 4.714.5 ton were of pimento and 1.585.5 ton of tomatoes (Figure 3 and Table 4) were produced. Although some farmers produce other varieties of crops, tomatoes and especially peppers are the primordial ones to provide the highest income. According to the new production strategy, all greenhouses will produce

Table 3. Horticultural production diversity by year (Ton)

Cultures	2008	2009	2010	2011	2012	2013	2014	2015	2016
Tomato	715	922	991	1065	1145	1230	1292	1202	1021
Green bean	1014	866	687	545	432	342	360	497	487
Carrot	1082	771	837	909	987	1072	1126	1024	1015
Onion	100	115	190	315	522	864	907	588	500
Lettuce	40	46	54	63	74	86	90	77	78
Cabbage	1193	901	838	780	726	676	710	784	768

Source: Gomes et al. (2018)

Figure 2. Proportion of crops in greenhouses (2018)

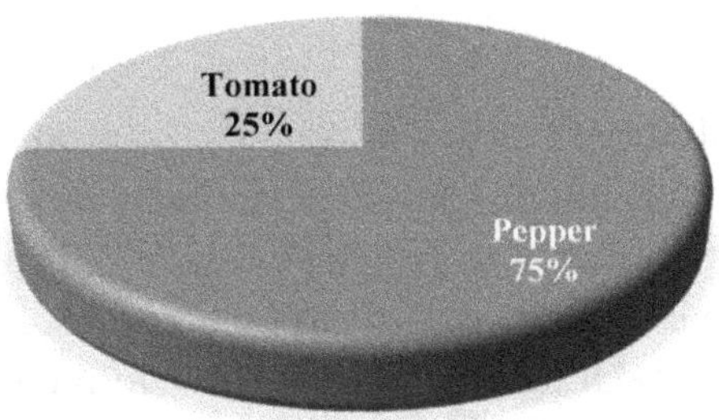

twelve tomato beds and twelve pimento beds, and a flowerbed was reserved for a special type of tomato (Cherry) for the months traditionally of larger scarcity (October, November, December, February, March and April). In the two private greenhouses, their owners invested in the production of pimento due to their commercial value, especially in the rainy season, when price reaches their maximum. The definition of the price is marked essentially by the supply and demand law, regardless of the product to come from greenhouses or outdoor.

Table 4. Production of horticultural products in 2018

Products	Jan	Feb	Mar	Apr	May	Jun	Total
Pepper (kg)	374.5	487	597	1386	1280	590	4.714.5
Tomato (Kg)	443	739	403.5	-	-	-	1.585.5

By community, Bom Sucesso were the one that most contributed to the production of pepper while Uba-Budo and Lembá for tomatoes (Table 5). In 2019 the production cycles of all greenhouses did not come to an end due to pests and diseases that plagued the plantation and the year the total production of peppers and tomatoes in the 10 greenhouses did not exceed 2,500 kg and 1,500 kg, respectively.

Table 5. Total production of horticultural products in cooperatives in 2018

Products	Cooperatives						
	Canavial	Uba Budo	Saudade	Bem Posta	Lembá	Bom Sucesso	Fernão Dias
Pimento	1.389	331	2.013	109	41	820	180
Tomate		226			338.5		

For the private greenhouses, despite their owners not recording the production, the data from the interviews and questionnaire estimates an average annual production of 5,000 kilos of pepper. The same situation occurs with conventional open-air horticulturists, where the annual production, for equivalent cultivation area, does not exceed 1,300 kilos, for peppers and tomatoes.

Production Costs

In the year 2019, more specifically from January to August, the production cost of project greenhouses, private greenhouses, as well as conventional open-air agriculture varied differently from one greenhouse to another and from one production model to another (Table 6). The project greenhouses, the private greenhouses and conventional model had an average production cost of around 23,675, 28,145 and 18,660 thousand Dobras (Dbs), respectively. It should be noted that in the open-air model, the average costs with Phytopharmaceuticals and fertilizers are slightly higher than in greenhouses modes of production. This is due to the need to use pesticides for the unfavorable weather conditions for horticulture. Also in this model, higher production costs combined with reduced sales revenue due to weak production, particularly in the rainy months, leads to inferr the positive effects of the use of greenhouses in mitigating the adverse effects of the climate on the results of the activity. In the dry season, where production in the conventional model is even lower than production in a greenhouse, the profitability of this mode is even weaker.

Table 6. Production cost in the project greenhouses, private greenhouses and in conventional model

Cost Assessment Criteria		Phytopharmaceuticals and Fertilizers	Fuel	Administrative Costs	Maintenance Cost	Other Costs	Total Cost
Project Greenhouses	Maximum Cost	20.000	18.000	2.400	2.000	800	43.200
	Minimum Cost	2.000	3.600	200	600	100	6.500
	Average Cost	14.000	7.882	680	985	128	23.675
Private Greenhouses	Average Cost	4.000	5620	275	1250	17000	28.145
Conventional Production (open air model)	Average Cost	16.250	1550	300 (estimated value)	300 (estimated value)	125	18.660,5

It can be concluded that the project's greenhouses, due to poor seed quality, the lack of technical knowledge of horticulturists and the problem of pests and diseases, are producing an average of 785.75 kilos of pepper and 528.5 kilos of tomatoes, equivalent to 31,430 and 10,564,000 Dobras (Total of 41,994 Dbs), with an average production cost estimated at 23,675 thousand Dbs. Even so, the economic viability of greenhouses is a reality when compared to conventional production in the open-air model.

Greenhouses Production Value

The production value of a greenhouse is influenced by several factors, from adequate soil preparation, seed quality and compliance with technical guidelines, to the prevention and fight against pests and diseases. The estimated production value of a greenhouse with maximum yield, with the pepper crop supported on good quality seeds, is shown in Table 7.

Table 7. Production value of pepper considering a package of one thousand seeds of good quality

Maximum production capacity for each improved seed plan			=	7 kl
7 kl	X	1000 plants	=	7.000 Ton
7.000 Ton	X	40 STN	=	280.000 STN
280.000 STN	-	50.000 (Amortization Value)	=	230.000 STN
230.000 STN	/	2 (Production cycles)	=	115.000 STN
115.000 STN	-	10.000 STN (other costs)	=	105.000 STN
105.000 STN	/	8 cooperative members	=	13.125 SNT (by cooperative member)

As can be seen, in a package of a thousand seeds it would be possible to produce a thousand plants, with each plant having a maximum production capacity of seven kilos of pepper. If each plant produces 7 kilos, the total production of the greenhouse would be 7,000 kilos of pepper, that is, seven tons of pepper to be sold in the national market for a standard value of 40 Dbs. This would ensure a gross yield of 280,000 Dbs which, deducting the production cost and amortization corresponding to 50,000 STN, would remain at 230,000 Dbs, divided into two production cycles per year. From the income of 115,000 Dbs, the amount of 10,000 Dbs should also be subtracted for other expenses, leaving 105,000 Dbs, to be divided by 8 members of a given greenhouse management cooperative. This would mean that each member would have an income of 13,125 Dobras in each production cycle, making a total annual value in the order of 26,250 Dbs. This value can be compared with the current reality in which horticulturists use, in general, the "Mongal" seed, whose maximum production of pepper per plant is approximately 2.5 kg (Table 8). In this situation, each member would have an income of 2,500 Dbs in each production cycle, making a total annual value in the order of 5,000 Dbs.

Considering the cost of acquisition and assembly of each greenhouse structure in STP in the amount of 60,000 USD (equivalent to 1,290,000.00 Dbs at the exchange rate of 21.5) and the amortization value equivalent to 50,000.00 Dbs per year, it can be inferred an amortization period of around 25.8 years. On the other hand, considering the current reality of the production and marketing capacity of greenhouses, within the standards of cooperatives, the amortization period would be 69.5 years. Given the current

Table 8. Production scheme for a package of one thousand "Mongal"seeds

Maximum production capacity for each MONGAL seed plants			=	2,5kl
2,5 kl	X	1000 plants	=	2.500 Ton
2.500 Ton	X	40 STN	=	100.000 STN
100.000 STN	-	50.000 (Amortization Annual Value)	=	50.000 STN
50.000 STN	/	2 (Production Cycles)	=	25.000 STN
25.000 STN	-	5.000 (small expenses)	=	20.000 STN
20.000 STN	/	8 cooperative members	=	2.500 SNT (by cooperative member)

context of pepper production in protected structures, it can be said that even the greenhouse with an above average yield in 2018 (such as Canavial) was only operated at 28.8% of its maximum capacity. As for the greenhouse with the worst yield (Lembá) it was operated at 0.5% of its maximum capacity.

In view of the above, if the amortization period of the amount invested by each structure is taken into account the situation of the project's greenhouses presents increased risks of not being economically viable. This statement is corroborated by the way greenhouses have been managed by cooperatives, that is, a large part of the cooperative members does not demonstrate a serious commitment to the project's financial investment. Therefore, in order to ensure the financial sustainability of these structures there is the need to create greater dynamics and efficiency in the management of greenhouses with improvements of the method of production and marketing of crops. At the level of the private greenhouse, taking into account the case of the farmer Abel Bom Jesus located in Mesquita, whose size of the greenhouse is the same as the project, applying the same model and criteria, the results obtained are shown in Table 7.

Table 9. Production scheme for a thousand seeds package – private greenhouse

Production capacity for each Mongal seed plant – private greenhouse			=	5 kl
5 kl	X	1000 plants	=	5.000 Ton
5.000 Ton	X	40 STN	=	200.000 STN
200.000 STN	-	50.000 (Amortization Annual Value)	=	150.000 STN
150.000 STN	/	2 (Production Cycles)	=	75.000 STN
75.000 STN	-	5.000 (small expenses)	=	70.000 STN
70.000 STN	/	1 owner	=	70.000 SNT SNT (by cooperative member)

In summary, it can be concluded that Mesquita's private greenhouse has currently better yields than the project's greenhouses, both in production and financial terms.

Greenhouses Benefits

As mentioned, the implementation of greenhouses in STP had as its main purpose to mitigate the impacts of climate change effects on some vulnerable rural communities. These protected structures for agriculture are a new technology in STP for which a small-scale pilot experiment has not been carried out to assess their functionality and feasibility. This study sought to identify some advantages and disadvantages of the use of greenhouses perceived by stakeholders, the relationship of greenhouse technology in the production of healthy food in areas where traditionally this is difficult, the way in which they have helped to drive the growth of the vegetables market and how they have contributed to increasing the income of benefited farmers and those who, on their own initiative, invested in the construction of a greenhouse.

Regarding the benefits, all the interviewees, including the Director of the Scientific Department of CIAT (Center for Agricultural and Technological Research) stated that the agricultural greenhouses allowed the creation of favorable microenvironments for the production of vegetables. Furthermore, the control over the climate inside the structures generated, in São Tomé, benefits for the producer and the consumer. The evaluation of the advantages of greenhouses are based on the following factors: off-season cultivation; protection against pests, saving irrigation water; product quality increase; pesticide reduction and sustainability.

With regard to off-season cultivation, the implementation of greenhouses in STP has allowed the vegetable growers of the benefited cooperatives, as well as those who have structures protected by private initiative, to no longer be hostage to seasonality. Therefore, the cooperatives managed to cultivate and harvest, mainly peppers and tomatoes in times of greatest scarcity, which coincides with the rainy season, namely the months of September, October, November, December, January, February, March and April. This situation made the production of these two products meet the demand for consumption, offering food with higher quality and stabilizing market prices. On the other hand, the greenhouses also allowed the cultivation of vegetables (pepper and tomato) in regions that, naturally, do not provide adequate conditions, as in the case of the south zone (Caué) and north zone (Lembá). Consequently, the production of these two products in greenhouses has increased the diversity of supply, boosting local and national trade of this product.

Pest Protection is another benefit. In the conventional agricultural model (outdoor), even with the proper handling, the plantation is much more susceptible to attack by diseases and pests. All of the surveyed conventional production horticulturists were unanimous in stating that the main constraint on horticultural production was due to the constant rain for nine months, which favored the appearance of pests and diseases in their plantations, causing them to use pesticides reducing the quality of the final products. The cultivation of peppers and tomatoes in protected structures allowed them to reduce the attack of these pathogens, although this reduction is not yet the most desired. According to information from the Scientific Department of CIAT, this entity monitored the problem of pests and diseases in greenhouses on a weekly basis and when pathogens were detected in certain greenhouses, they drew up technical guidelines for technicians to travel to cooperatives in compliance with the guidelines for treatment of the plants. However, this procedure did not always happen. From the perspective of the interviewed horticulturists, the lack of financial means to purchase chemical products and the momentary scarcity of these products in the national market, were some of the causes pointed out for this failure.

The economy in irrigation is another benefit, particularly important considering the effects of climate change, which in STP are increasingly evident. The increasingly long periods of Gravana constitute a problem of water scarcity at certain times and regions of the country. The greenhouses, with their drop-by-drop irrigation system and the respective dosing of the amount of water for watering the plants, which also allows the use of the fertigation method, generates a great savings in water use and greater efficiency in supplying the necessary nutrients to the plants.

Despite the increase in product quality being one of the benefits attributed to greenhouse production, around 60% of respondents said that the quality of the product is still not the best and point to the poor quality of seeds used by cooperative gardeners. For example, the price of a packet of seed for Mongal tomato costs in the national market 335 Dobras (13.4 Euros) and the price of seed for lamatropic pepper costs 352 Dobras (14.1 Euros) and California pepper 39 Dobras (1,56 Euros), respectively. According to Angel Oliva, a greenhouse specialist who monitors the greenhouse process in STP, the price of a quality seed unit costs 400 Euros and the quality of the product in the country is medium-low. However, the product produced in greenhouses in STP, compared to that produced in the open field, has superior quality in terms of presentation and longevity. This factor has contributed to the increased demand for products from greenhouses in national supermarkets and hotels, since in the vegetable sector there is concern about color, taste, smell, texture and other visual characteristics (not just the product quantity). Therefore, greenhouse cultivars are able to maximize plant performance, however, the quality of products from greenhouses in STP may still improve, if cooperative gardeners comply with CIAT technical guidelines, if the three is availability of chemical product to be applied to the plants in the national market and access to higher quality seeds.

In the context of STP, the effects of climate change are more harmful in certain regions of the country than in others. In the former, they induce greater difficulty in conceptual production, which leads horticulturists to resort, very often, to chemical products to mitigate the effects of pests and diseases. In protected crops, not only are the destructive effects of pests and diseases less, but they also allow for more effective and efficient control in combating these pathogens and a closer and more correct follow-up to CIAT guidelines. Agricultural greenhouse production in STP provides an environmentally much cleaner production system, which drastically reduces the use of pesticides. It should be noted that the population's growing dissatisfaction with the use of agrochemicals in conventional crops means that protected cultivation plays an important role and meets the demand of more conscientious and demanding consumers.

The lower environmental impacts associated with greenhouses, when compared to conventional agriculture, due to the more rational use of resources (water and fertilizers) and a reduction in pesticides, is a guarantee of the sustainability of this production model. Although there are differences between some greenhouses in relation to others, in the rational use of these resources, production is unanimously considered cleaner and more sustainable.

Greenhouses Disadvantages

The disadvantages of greenhouses pointed out by interviewees were the following: construction and maintenance work, financial cost, access to adequate seeds, lack of qualified personnel for planting in greenhouses and difficulty in fighting pests or diseases.

Construction and maintenance work is a disadvantage to the adoption of greenhouses. These are extremely difficult to install and maintain, requiring constant care and attention. A proper location for the greenhouse should be chosen to ensure that the plants receive the right amount of sunlight. Another important factor is that the greenhouses must be installed on an area of fertile soil enough for the plants to grow well. In the case of STP, these criteria were considered when building the greenhouses, however, the biggest problem is related to the maintenance of these structures. According to information from technicians connected to the project, the lack of technical and specific knowledge of the greenhouse managers has conditioned the better maintenance of these structures. The cooperatives, as managers of the greenhouses, claim that they do not have sufficient financial means to replace the damage and wear that the structure and its accessories have had over these nearly 3 years of project life. This situation should have been taken care of when designing and implementing the project, since it was foreseeable that this difficulty would happen as it happened in other projects financed by external entities.

Other disadvantage is the financial costs. Within the framework of the project, the acquisition, transport and assembly of the greenhouses in general were made based on an international public tender, and the unit cost of each greenhouse and its respective accessories were in the order of 1,500,000.00 (one million and five hundred Dobras), equivalent to 60 thousand Dollars (USD). Regarding the reality of the country, as well as the reduced productive capacity of the benefited cooperatives associated with several factors, namely, lack of technical knowledge and handling of the new technology, the market's own consumption capacity, it is understood that the acquisition value is far above the average, the financial capacity of the members. In addition to this factor, cooperatives have not had a desirable amortization capacity according to the value of the investment. On the other hand, the beneficiaries still do not feel that they truly own the greenhouses, that is, they think that the structures are property of the State and its partners. This factor inhibits them from making financial efforts for their proper maintenance. This situation reports to the need to lower the cost of installation through non-repayable subsidies and attribution of property title to collective and/or individual users and managers.

According to the literature on agricultural greenhouses, and confirmed by the Scientific Department of CIAT, the environment inside the greenhouses requires the use of appropriate seeds. In the specific case of STP, the use of adequate seeds has not happened for two reasons: the absence of adequate seeds in the national market and the high price of these seeds in the international market. These constraints make horticulturists resort to seeds, generally used in open-air horticulture, which by the way are those available on the national market.

The lack of qualified staff for planting in greenhouses is another disadvantage, especially given that greenhouse structures are a new technology in STP, as is the associated drip irrigation system. This situation represents important challenges for national technicians and for the benefited horticulturists, whose experience is exclusive to the conventional model. Given this situation, it is necessary to invest a lot in training and capacity building for technicians and for gardeners in order to obtain greater income from production in greenhouses. In the specific case of the project, it has the assistance of an international specialist in protected agriculture to provide training and follow up on the greenhouses.

Finally, the disadvantage linked to the difficulty of fighting pests or diseases. CIAT is the body attached to the Ministry of Agriculture that has the role of controlling and monitoring the situation of pests and diseases in greenhouses. This control, carried out every week, allows to detect the presence of pests and diseases in the greenhouses with some regularity. It also allows, and depending on this detection, to prescribe technical fighting guidelines so that technicians, together with the cooperative gardeners, can put these guidelines into practice, in an attempt to combat such pathogens. This control and technical guidance did not always take place in a convenient way, as alleged by the cooperative members who claim that they do not have financial conditions to purchase phytopharmaceuticals or because these, at certain periods, do not exist in the country. It should be noted that most of the diseases and pests caused by pathogens in plants are due to the misuse of water and the lack of technical knowledge on the part of the cooperative members.

Greenhouses Viability, Performance and Socioeconomic Impact

As for the assessment of the feasibility of implementing the study, all respondents were unanimous in stating that this is a viable alternative for the context of horticulture in STP. However, they also consider that the criterion for the implementation and distribution of greenhouses at national level should take into account the district areas depending on the climatic conditions (humidity or drought), as well as the product chain market. With regard to the market factor, it should be noted that all production is oriented towards the domestic market, which can represent a problem when the greenhouses were fully producing, given the small market and lack of capacity to absorb the product. As a result of a predictable oversupply, the price may fall, causing some instability in the market and reduced profitability. The suggested alternative is to specialize each greenhouse in a type of crop and find the respective outlet, not forgetting the process of transforming the surplus or even outflowing.

Within the framework of the project, the model found for managing the greenhouses was the creation of cooperatives and, in the selected communities, meetings were held in order to identify and select the members. On the other hand, still according to the project's philosophy, the income from the production in greenhouses should be applied in the realization of social actions in the communities; amortization of the investment and payment of income to the cooperative gardeners so that, in the medium term, they could invest in the construction of other greenhouses or in the application of investments, such as the construction of a poultry house, guaranteeing the sustainability of the project.

However, the cooperatives created under the project have presented difficulties and constraints in their functioning, namely the lack of understanding among the members, the little involvement and engagement of some members, the individual interests of some members above the collective interests, some cooperative members do not identify themselves as horticulturists, and lack of motivation and feeling of dispossession of investment, since none of the members made their own financial effort. In other words, 99% of the surveyed cooperatives present the aforementioned problems, with emphasis on human relations, behaviors and attitudes that do not contribute to the project's success. Therefore, this issue constitutes a real obstacle to the good performance of cooperatives. It is also important to emphasize that in the communities, it is verified that the population, including the cooperative members, have weak characteristics and profiles, due to the high levels of poverty, the low level of education, little scientific knowledge about protected agriculture, resistance to the acquisition of new knowledge and the high rates of alcohol consumption. In short, these constraints have contributed to the poor performance of cooperatives and, consequently, to the reduction in the success of greenhouse management.

The socioeconomic impact, analyzed through the Logical Framework Matrix (MML), a tool used to establish the logic of social projects, allows us to infer the project's success in its various stages, considering its impact, objectives, results and activities (Table 8). The matrix must be interpreted in the logic that if the activities are carried out, the results will be produced and if they exist, the objectives are met and if the objectives are met, the impact will be generated. All of this must be properly measured, so that we can have a real sense of the impact and a better understanding if the objectives are not met.

The information obtained in the interviews, as well as the findings made through observation, allow us to state that, within the framework of the project, the activities were carried out. In other words, there was a set of actions that were materialized that allowed the generation of results. In this sense, these results were in line with the objectives advocated by the project, which, in turn, denote a socioeconomic impact at the level of rural communities in view of the currently verifiable indicators. Therefore, the presence of protected structures at the country level is already a reality, with consequences for the production and marketing of vegetables, mainly tomatoes and peppers. On the other hand, from an economic point of view, the data currently available do not allow measuring the effects on the region and country's economy and the contribution to GDP.

Table 10. Logical framework matrix

	Description	Indicators	Verification Sources	Assumption
Impact	Cooperatives already regularly produce crops in greenhouses. (pepper and tomato)	1) Cooperatives already make regular payments to members. 2) Regions that traditionally did not produce horticulture (Lembá and Caué) already do. 3) The products from the greenhouses are already available in markets, fairs, commercial surfaces, hotels and restaurants.	1) Analysis of the content of the interviews. 2) Observation and observation in the field.	1) Improvement of the region's and country's economy. 2) Improved HDI 3) Poverty reduction. 4) The population of the communities already has a cleaner product. 5) Cooperatives already carry out social works in the communities.
Objectives	Strengthen the resilience of gardeners in rural communities vulnerable to the effects of climate change.	1) Ensure livelihoods for over 80 horticulturists. 2) Ensure cleaner and more sustainable production.	1) Analysis of the content of the interviews.	1) Through the income from production in greenhouses, horticulturists are already able to satisfy some basic needs, such as building their own house, purchasing household appliances and food.
Results	There are already 10 greenhouse production cooperatives in STP.	1) A little more than 80 gardeners nationwide are already employed in the communities. 2) More than 320 people in different communities benefit directly/indirectly from the greenhouse income.	1)Evaluation Report. 2) Analysis of the content of the interviews.	1) Increase in the income of benefited greenhouses and poverty reduction. 2) More and better social dignity. 3) Improved quality of life (more financial capacity).
Activities	1) Construction of 10 greenhouses and distribution of accessories and agricultural equipment to 10 rural communities in STP. 2) Soil preparation and delivery of seeds and inputs.	1) In each community, cooperatives composed of 6 to 9 people were created to manage the greenhouses. 2) Technical training was given to more than 80 horticulturists from cooperatives and 55 technicians from the Ministry of Agriculture and Rural Development.	1)Project document. 2) Field Observation.	1) Horticulturists from the 10 cooperatives already produce pepper and tomato crops in greenhouses throughout the year. 2) CIAT and CADR technicians provide technical guidance to beneficiaries.

The recommendations for the optimization of the greenhouses, took into account the cooperatives, the phytotechnical improvement and the management of the greenhouses. Regarding cooperatives, the following measures are recommended: i) more and better inclusion of all cooperative members so that they feel committed and engaged with the project, since only a commitment attitude will add value and increase and diversify the production; ii) greater participation of all members in the production process at the same level, avoiding imbalances between some who make an effort and commit and others do little or nothing to improve the situation; iii) encourage collaborative, solidary, democratically participatory and shared responsibility work; iv) invest, as a priority, in training and continuous training of members of the cooperative, with a view to providing them with information and knowledge about agriculture in protected structures; v) acquisition of fertilizers and seeds made jointly by greenhouse producers in order to achieve better negotiations and, consequently, lower prices. vi) promote the sharing of experience and the creation of business networks among the cooperative members; vii) create a confederation of producers, considering it as a body capable of defending the interests of members in different domains.

viii) generate and distribute income equitably within cooperatives in order to promote the human and economic development of communities and reduce poverty, ix) generate employment and income for cooperative members, improving their living conditions.

Regarding phytotechnical improvements, it is recommended that technicians and horticulturists have more and better knowledge in the following areas: i) irrigation and adequate plant nutrition, in order to avoid shortages and excess water that can cause imbalances, generating possible pest problems and from diseases; ii) nutritional balance to obtain more resistant and productive plants; iii) control of amounts of water and fertilizers to be applied to the plants, with a view to optimizing production; iv) handling the biological control of the plant to allow the producer to replace or eliminate synthetic pesticides; v) correct use of biological control agents in order to achieve high efficiency in protecting against pests and diseases; vi) effective and effective use of natural insecticides and biological agents in the production system; vii) correct soil preparation using organic compounds as a way to maximize the nutrients to be absorbed by the plant; viii) use nebulizers, shading screens, curtains in order to minimize the destructive effects of heat on the plant, whenever heating the greenhouse on sunny days is excessive. On the other hand, they can also integrate in the greenhouse a manual or automatic opening system for the roof or windows in the greenhouse wall, facilitating quick aeration.

With regard to improvements in the management of greenhouses, it is recommended that technicians and horticulturists have more and better knowledge about: i) model of greenhouse used, with a view to making the most of their structure; ii) correct handling of the crop practiced within the possibilities of the greenhouse, to obtain the best result; iii) business feasibility, sustainability and management and know-how to analyze the economic feasibility of the greenhouse business. that is, learn to plan according to the cultural calendar so that the harvest period coincides with the time of great demand in the market; iv) analysis of production costs, most important factors and knowing how decisions about daily management can influence the project's financial result; for this purpose, it is important that the greenhouse manager and his team carry out an adequate and rigorous record of the harvests in order to better plan the future investment; v) how to add value to the production and increase profitability in the market; vi) how to invest in the image and reputation of the brand, in order to obtain recognition from consumers; vii) product promotion and customer acquisition through partnerships, including the establishment of strategic partnerships with other companies, namely hotels and restaurants; and, viii) adoption of a marketing strategy that prioritizes participation in all fairs, markets, workshops in order to publicize the product's brand and increase awareness levels.

FUTURE RESEARCH AGENDA

The knowledge and experience gained from this study allow us to draw useful inferences for the structuring and development of future investigations. In this same theme, the importance of carrying out a parallel study on the reality of production in greenhouses in STP with another country with similar characteristics is highlighted. It would also be entirely pertinent to extend the sample to final consumers, in order to understand how the greenhouse product is accepted by the São Tomé society.

In the field of greenhouses, it would be important to find a model of greenhouses that is smaller or of different dimensions and with ventilation, which could be used individually by small producers for their own production, aimed at the market and the family's self-consumption. Interconnected with the above, it would be very important to analyze the economic and financial viability of the different dimensions of greenhouses, to evaluate the forms of support to individual farmers to carry out the investment, such as subsidized credit and subsidies to non-repayable investments. Finally, another suggestion would be to investigate, in a particular way, the constraints of existing cooperatives in order to extract the cause of the problems and find solutions for improvement, which may involve other forms of organization such as the formation of companies.

CONCLUSION

Climate change mitigation, agricultural technological innovation and community's behaviour are important factors in rural development. Climate is a natural phenomenon, in which the changes has been felt by all countries around the world, including STP, with a negative impact on rural communities' welfare if it is not accompanied by good mitigating measured that assure properly community respond and survive without aggravates the state of the environment.

The study highlighted the greenhouses production potential to improve farmer's income and reduce their poverty level, as well as, mitigate climate changes effects. Nevertheless, improvements in the assessments, technologies and production systems, are still needed as well as funding for technology research and development.

Various constraints slow private farmers in adopting greenhouses vegetables production: lack of resources and funding, lack of secure land tenure and the investment risks. These constraints are specific of the local context.

Future research may focus on assess the temporal and spatial variability of greenhouses production adoption to mitigate climate changes effects and to have economic benefits over time, and on the impact of greenhouses on the resilience of agricultural production facing climate change. Models to predict agricultural production under climate change according to specific practices have to be include in future assessments to clarify the potential impact of greenhouses adoption in a changing climate.

While economic benefits and food production are good arguments for encouraging and building policies to support greenhouses adoption, they are not the only ones. Other criteria have to be included in the assessments of the farmer sustainable livelihoods such as labour demand, labour productivity, income stability, farmer incorporation of agroecological approaches, access to the market, influence on decision making, and recognition of transition costs (D'Annolfo et al., 2017). Evaluating all of these criteria still constitutes work prospects and calls for multidisciplinary and participatory research efforts.

To conclude, greenhouses adoption, such as most of the climate mitigation actions and poverty reduction, has to be done on large scales to have a significant impact although it depends on and is implemented at more local scales by farmers. An adoption of greenhouses is necessary to achieve climate change mitigation, and provides other sustainable development goals as food security, poverty reduction and farmer well-being. The results of the present study, even being exploratory, showed that greenhouses vegetable production is a promising way to achieve these goals.

ACKNOWLEDGMENT

This work is financed by national funds through the FCT – Foundation for Science and Technology, I. P., under the projects UID/04007/2020 and UIDB/05183/2020.

REFERENCES

Agarwal, M., Atri, P., & Kundu, S. (2017). Foreign direct investment and poverty reduction: India in regional context. *South Asia Economic Journal, 18*(2), 135–157. doi:10.1177/1391561417713129

Ali, M., Nishat, M., & Anwar, T. (2010). Do foreign inflows benefit Pakistan poor? *Pakistan Development Review, 48*(4), 715–738. Doi.org/10.30541/v48i4IIpp.715-738

Aronsson, T., & Schöb, R. (2018). Climate change and psychological adaptation: A behavioral environmental economics approach. *Journal of Behavioral and Experimental Economics, 74*, 79–84. doi:10.1016/j.socec.2018.03.005

Arteta, C., & Kirby, P. (2019). *Country Economic Memorandum for Sao Tome and Principe - Background Note 2: Is it Sustainable for Sao Tome and Príncipe to Have a Large Current Account Deficit and a Fixed Exchange Rate?* World Bank. Retrieved from: https://openknowledge.worldbank.org/handle/10986/32141

Austin, E. K., Rich, J. L., Kiem, A. S., Handley, T., Perkins, D., & Kelly, B. J. (2020). Concerns about climate change among rural residents in Australia. *Journal of Rural Studies, 75*, 98–109. doi:10.1016/j.jrurstud.2020.01.010

Barbier, E. B., & Hochard, J. P. (2018). The Impacts of Climate Change on the Poor in Disadvantaged Regions. *Review of Environmental Economics and Policy, 12*(1), 26–47. doi:10.1093/reep/rex023

Biegańska, J., Środa-Murawska, S., Kruzmetra, Z., & Swiaczny, F. (2018). Peri-Urban Development as a Significant Rural Development Trend. *Quaestiones Geographicae, 37*(2), 125–140. doi:10.2478/quageo-2018-0019

Chambers, R. (1983). *Rural Development: Putting the Last First*. Routledge.

Copp, J. H. (1972). Rural sociology and rural development. *Rural Sociology, 37*, 515–533.

Dada, J. T., & Akinlo, T. (2021). Foreign direct investment and poverty reduction in sub-Saharan Africa: Does environmental degradation matter? *Future Bussiness Journal, 7*(1), 21. doi:10.118643093-021-00068-7

Espírito Santo, S. N. (2008). *Programas de ajustamento estrutural, produção agrícola e segurança alimentar na África Sub-saariana: caso específico de S. Tomé e Príncipe* [Unpublished doctoral dissertation]. Instituto Superior de Agronomia, Universidade Técnica de Lisboa, Lisboa, Portugal.

Falco, C., Galeotti, M., & Olper, A. (2019). Climate change and migration: Is agriculture the main channel? *Global Environmental Change, 59*, 101995. doi:10.1016/j.gloenvcha.2019.101995

Faradiba, F. (2018). Forecasting Rainfall and Area of Plant Pest Disturbance in Bogor Regency. *Pro Life, 5*(3), 688-699. http://ejournal.uki.ac.id/index.php/prolife/article/view/847/685

Faradiba, F., & Zet, L. (2020). The Impact of Climate Factors, Disaster, and Social Community in Rural Development. *Journal of Asian Finance, Economics and Business, 7*(9), 707–717. doi:10.13106/jafeb.2020.vol7.no9.707

Fawzy, S., Osman, A. I., Doran, J., & Rooney, D. W. (2020). Strategies for mitigation of climate change: A review. *Environmental Chemistry Letters*, *18*(6), 2069–2094. doi:10.100710311-020-01059-w

Ganić, M. (2019). Does foreign direct investment (FDI) contribute to poverty reduction? Empirical evidence from Central European and Western Balkan countries. *Scientific Ann Econ Bus*, *66*(1), 15–27. doi:10.2478aeb-2019-0003

Gomes, C. B., Tenjua, E., Paquete, H., & Dória, A. (2018). *Relatório Revisão Estratégica "Fome Zero"- Horizonte 2030. Programa Alimentar Mundial*. PAM.

Harriss, J. (1982). *Rural Development: Theories of Peasant Economy and Agrarian Change*. Routledge.

INE-Instituto Nacional de Estatística. (2019). *Estatísticas Económicas*. www.ine.st/economia.html

Jackson, G., Mcnamara, K. E., & Witt, B. (2020). "System of hunger": Understanding causal disaster vulnerability of indigenous food systems. *Journal of Rural Studies*, *7*(3), 163–175. doi:10.1016/j.jrurstud.2019.10.042

Javed, S. A., Haider, A., & Nawaz, M. (2020). How agricultural practices managing market risk get attributed to climate change? Quasi-experiment evidence. *Journal of Rural Studies*, *73*, 46–55. doi:10.1016/j.jrurstud.2019.11.020

Jensen, M. H. (1997). Food Production in Greenhouses. In E. Goto, K. Kurata, M. Hayashi, & S. Sase (Eds.), *Plant Production in Closed Ecosystems*. Springer., doi:10.1007/978-94-015-8889-8_1

Jiang, W., Deng, Y., Tang, Z., Cao, R., Chen, Z., & Jia, K. (2016). Adaptive capacity of mountainous rural communities under restructuring to geological disasters: The case of Yunnan Province. *Journal of Rural Studies*, *47*, 622–629. doi:10.1016/j.jrurstud.2016.05.002

Louro, M. J. S. (2011). *Modelação do Clima em Estufa Climatizada para Produção e Enraizamento de Estacas de Eucalyptus Globulus Labill* [Unpublished master thesis]. Instituto Politécnico de Viana do Castelo - Escola Superior Agrária de Ponte de Lima.

Lu, Y., & de Vries, W. T. (2021). A Bibliometric and Visual Analysis of Rural Development Research. *Sustainability*, *13*(11), 6136. doi:10.3390u13116136

Magombeyi, M. T., & Odhiambo, N. M. (2017). Causal relationship between FDI and poverty reduction in South Africa. *Cogent Economic Finance*, *5*(1), 1357901. doi:10.1080/23322039.2017.1357901

Marques, F. C. (2017). *Metodologia TRIZ Aplicada ao Desenvolvimento de uma Estufa "Inteligente" Energeticamente Sustentável* [Unpublished master thesis]. Faculdade em Ciência e Tecnologia, Universidade Nova de Lisboa.

Moreno-Cruz, J. B., & Smulders, S. (2017). Revisiting the economics of climate change: The role of geoengineering. *Research in Economics*, *71*(2), 212–224. doi:10.1016/j.rie.2016.12.001

Moulton, A. (2015). *Greenhouse Governmentality: Discourses of Rural Development and the Negotiation of Farmer Subjectivity in Jamaica* [Unpublished master thesis]. East Carolina University.

Nda, M., Adnan, M. S., Ahmad, K. A., Usman, N., Razi, M. A. M., & Daud, Z. (2018). A Review on the Causes, Effects and Mitigation of Climate Changes on the Environmental Aspects. *International Journal of Integrated Engineering*, *10*(4), 169–175. doi:10.30880/ijie.2018.10.04.027

Nogueira, L. R. T. (2014). *Ações voltadas para a Agricultura Familiar que Contribuem com o Desenvolvimento de Municípios do Sul de Minas Gerais* [Unpublished doctoral dissertation]. Minas Gerais. Brazil.

Nørgaard, H., & Thuesen, A. A. (2021). Rural community development through competitions, prizes, and campaigns: The villagers' perspective. *Journal of Rural Studies*, *87*, 465–473. doi:10.1016/j.jrurstud.2020.03.006

Ogunniyi, M. B., & Igberi, C. O. (2014). The impact of foreign direct investment on poverty reduction in Nigeria. *Journal of Economics and Sustainable Development*, *5*(14), 73–89.

Perdigones, A., Benedicto, S., & García, J. L. (2015). *Automatização da estufa: a evolução da tecnologia.* Universidade Técnica de Madrid.

Rakotovao, N. H., Chevallier, T., Chapuis-Lardy, L., Deffontaines, S., Mathé, S., Ramarofidy, M. A., Lepage, A., Masso, C., Albrech, A., & Razafimbelo, T. M. (2021). Impacts on greenhouse gas balance and rural economy after agroecology development in Itasy Madagascar. *Journal of Cleaner Production*, *291*, 125220. doi:10.1016/j.jclepro.2020.125220

Rebouças, P. M., Dias, I. F., Alves, M. A., & Barbosa Filho, J. A. D. (2015). Radiação solar e temperatura do ar em ambiente protegido. *Revista Agrogeoambiental, Porto Alegre*, *7*(2), 115–125.

Santos, M. A. M. (2018). *Potencial e Viabilidade económica para Utilização de Estufas Para produção de Hortaliças em STP* [Unpublished master thesis]. Instituto Superior de Agronomia, Universidade de Lisboa.

Sawada, Y., & Takasaki, Y. (2017). Natural Disaster, Poverty, and Development: An Introduction. *World Development*, *94*, 2–15. doi:10.1016/j.worlddev.2016.12.035

Sørensen, J. F. (2018). The importance of place-based, internal resources for the population development in small rural communities. *Journal of Rural Studies*, *59*, 78–87. doi:10.1016/j.jrurstud.2018.01.011

Soubry, B., Sherren, K., & Thornton, T. F. (2020). Are we taking farmers seriously? A review of the literature on farmer perceptions and climate change, 2007–2018. *Journal of Rural Studies*, *74*, 210–222. doi:10.1016/j.jrurstud.2019.09.005

Soumare, I. (2015). *Does Foreign Direct Investment Improve Welfare in North Africa?* Working Paper, North Africa Policy Series. Africa Development Bank.

Sousa, K. D., Casanoves, F., Sellare, J., Ospina, A., Suchini, J. G., Aguilar, A., & Mercado, L. (2018). How climate awareness influences farmers' adaptation decisions in Central America? *Journal of Rural Studies*, *64*, 11–19. doi:10.1016/j.jrurstud.2018.09.018

Stern, N. (2018). Public economics as if time matters: Climate change and the dynamics of policy. *Journal of Public Economics*, *162*, 4–17. doi:10.1016/j.jpubeco.2018.03.006

Straub, A. M., Gray, B. J., Ritchie, L. A., & Gill, D. A. (2020). Cultivating disaster resilience in rural Oklahoma: Community disenfranchisement and relational aspects of social capital. *Journal of Rural Studies*, *73*, 105–113. doi:10.1016/j.jrurstud.2019.12.010

Tanure, T. M., Miyajima, D. N., Magalhães, A. S., Domingues, E. P., & Carvalho, T. S. (2020). The Impacts of Climate Change on Agricultural Production, Land Use and Economy of the Legal Amazon Region between 2030 and 2049. *Economía*, *21*(1), 73–90. doi:10.1016/j.econ.2020.04.001

Toni, J. (2019). *Curso de ZOPP e Método do Quadro Lógico*. Escola Nacional de Administração Pública.

UNDP - The United Nations Development Program. (2016). *Reforço das capacidades das comunidades rurais para a adaptação aos efeitos às mudanças climáticas em São Tomé e Príncipe nos distritos de Cauê, Me-Zochi, Príncipe, Lembá, Cantagalo e Lobata (CMPLCL). Projeto Adaptação (CMPLCL) - PNUD/GEF (2015 - 2016)*. Programa Das Nações Unidas Para Desenvolvimento – STP.

Van der Ploeg, J. D., Renting, H., Brunori, G., Knickel, K., Mannion, J., Marsden, T., De Roest, K., Sevilla-Guzmán, E., & Ventura, F. (2000). Rural development: From practices and policies towards theory. *Sociologia Ruralis*, *40*(4), 391–408. doi:10.1111/1467-9523.00156

Vaz, J. (2018). *Plano de Gestão de Resíduos - 2018 – República Democrática de São Tomé e Príncipe*. Ecogestus Lda. doi:10.13140/RG.2.2.30218.62400

WHO. (2019). *Strategic Plan to Reduce Malnutrition in Africa Adopted*. WHO.

World Bank. (2018). *Turning Smallness into Uniqueness: Six Key Challenges to Unlock São Tomé and Príncipe Growth's Potential. Country Economic Memorandum for São Tomé and Príncipe Summary Report*. Author.

World Bank. (2019). *World Bank Group Report – WBG (2019). The Obstacles And Opportunities for STP's Agriculture*. Value Chains.

World Bank. (2019). *Rural Development: Sector Policy Paper*. http://documents.worldbank.org/curated/en/688121468168853933 /Rural-development-sector-policy-paper

Xie, Z., Zhang, F., Lun, F., Gao, Y., Ao, J., & Zhou, J. (2020). Research on a diagnostic system of rural vitalization based on development elements in China. *Land Use Policy*, *92*, 104421. doi:10.1016/j.landusepol.2019.104421

Zhang, D., & Managi, S. (2020). Financial development, natural disasters, and economics of the Pacific small island states. *Economic Analysis and Policy*, *66*, 168–181. doi:10.1016/j.eap.2020.04.003

ADDITIONAL READING

Akrofi-Atitianti, F., Ifejika Speranza, C., Bockel, L., & Asare, R. (2018). Assessing climate smart agriculture and its determinants of practice in Ghana: A case of the cocoa production system. *Land (Basel)*, *7*(1), 30. doi:10.3390/land7010030

Rakotovao, H. N. (2017). *Impacts environnementaux de l'adoption des pratiques agroécologiques: empreinte carbone d'exploitation agricole et stocks de carbone, Région Itasy, Madagascar* [Unpublished PhD thesis]. Ecole Supérieure des Sciences Agronomiques, Université d'Antananarivo, Madagascar.

Sadovska, V. (2018, July-August). *Sustainable value creation in the agricultural sector. A literature review* [Paper presentation]. International Association of Agricultural Economists (IAAE), Vancouver. . doi:10.22004/ag.econ.276984

KEY TERMS AND DEFINITIONS

Climate Changes: Refers to long-term shifts in temperatures and weather patterns (warming and precipitation) which are altering the geographical ranges of many plant and animal species and the timing of their life cycles. Although these shifts may be natural, are mainly caused by human activities primarily due to burning fossil fuels like coal, oil, and gas.

Food Security: Is a basic human right and the measure of an individual's ability to access food that is nutritious and sufficient in quantity. Additionally, that food must also meet an individual's food preferences and dietary needs for active and healthy lifestyles.

Poverty: Is a state or condition in which a person or community lacks the financial resources and essentials for a minimum standard of living. Poverty means that the income level is so low that basic human needs can't be met. Poverty-stricken people and families might go without proper housing, clean water, healthy food, and medical attention.

Rural Development: Is the process of improving the quality of life and economic well-being of people living in rural areas, often relatively isolated and sparsely populated areas.

Sustainable Development: Is development that meets the needs of the present without compromising the ability of future generations to meet their own needs.

United Nations Development Program (UNDP): Is a United Nations organization tasked with helping countries eliminate poverty and achieve sustainable economic growth and human development. Headquartered in New York City, it is the largest UN development aid agency, with offices in 170 countries.

Chapter 2
Agriculture, Mass Media, and the Economic Development of India

Subir Sinha
https://orcid.org/0000-0001-7578-9809
Dum Dum Motijheel College, India

ABSTRACT

In India, the enormous contribution of mass media in the agricultural sector helps to develop agriculture as well as the socio-economic structure of the nation. The chapter discusses the contribution of mass media in agricultural sector of India and how it supports the socio-economic development of the nation. Agriculture and economy are two essential factors of modern society. Agriculture helps in the development of economy whereas mass media gives a pace. Mass media is attracting the attention of the farmer, giving them information related to agronomy and creating needs and demands of agriculture through marketing techniques, which ultimately enhance the productivity and economy of the nation. Mass media is ultimately helping the overall sustainable developmental process of the agricultural sector of the nation.

INTRODUCTION

In India, the significance of agriculture and its allied sectors are highly essential and significant. It plays a major role in maintaining the livelihood for the majority of population. It provides foods, shelter and economy to around 1.2 billion people of India. M. Sharma in the opening of the article with entitled *'The Future of Indian Agriculture'* mentioned about the significance of Indian agriculture as "Agriculture in India is livelihood for a majority of the population and can never be underestimated" – (Sharma, 2021). The Indian agriculture has shown a wider impact not only on the Indian society but also across the globe. It is satisfying the need of the Indian citizen and exports a wide variety of crops across the globe. In India, the agricultural sector from the beginning remains as a backbone of the Indian economy and served the society in several ways. India produces variety of crops in huge quantity from across the nation and strengthens the economic development. The huge production is due to its favourable climate

DOI: 10.4018/978-1-7998-9557-2.ch002

but the climatic changes in the recent few decades fluctuates the rate of production to a certain percentage. However, the activities of mass media support in the promotion and production of agriculture and ultimately help to boost the economy of India. Mass media propagates several information and messages among the farmers, traders and common public whereas on the other it helps to promotes various agricultural projects across the nation for the welfare of Indian agricultural system. The roles of mass media in the development of Indian agriculture are enormous and significant.

OBJECTIVES OF THE RESEARCH

Indian agriculture and its allied sectors are highly essential for the growth and development of India. It shows a steady rapid growth since the arrival of mass media in the post-independence era and strengthens the sustainable economic development of the nation. The immensity and the essentiality of the agriculture and its allied sectors along with the involvement of mass media make it significant and turn it as a vital subject of research. The recent development after the involvement of mass media make it essential to find out various impacts of mass media on the Indian agriculture and its allied sectors. However, the article entitled ***'Agriculture, Mass Media and the Economic Development of India'*** shows a wide scope of research and was conducted to find out the correct explanation while focusing the following objectives:

- Significance of Indian agriculture and its allied sectors in the era of mass media.
- How agriculture is helping to enhance the economic development of India.
- How mass media are helping to promote the field of agriculture.
- How Indian government and several Indian institutions are using mass media to enhance the growth of the Indian agriculture.

METHODOLOGIES OF THE RESEARCH

The article entitled ***'Agriculture, Mass Media and the Economic Development of India'*** is based on a thorough analysis of various contents and documents related to agriculture, its allied sectors and the involvement of mass media in the field of Indian agricultural system. It analyses several articles and chapter published in journals and book, along with several statements and contents published and projected in various websites and web portals of Government of India, various educational institutes, agricultural organisations and international agencies. The article also reflects several case studies of agricultural projects that run across the India or the Indian states for the welfare of the farmers as well as for the development of Indian agriculture.

DISCUSSION

Agriculture is the art or the technique of cultivating soil, growing crops and raising livestock. It is the process that helps to reduce poverty, raise income and improve food security. In the - *'Dictionary of Agriculture'* the term agriculture is defining as *"cultivation of the land, including horticulture, fruit growing, crop and seed growing, dairy farming and livestock breeding"- (Stephen, 2015)*. In the - *'Ox-*

ford Dictionary of Agriculture and Land Management'- Agriculture is defined as *"A very broad term covering the production of food, fibre, energy medicines and other products primarily from plants and animal...."- (Manley, Foot and Davis, 2019).* In a nutshell, the term agriculture can be referring as the cultivation of variety of crops and livestock. However, the significance of agriculture is colossal from the beginning of human civilization. It not only provides foods and livestock to the human beings but also helps to develop the socio-economic infrastructure of the nation. Agriculture has a close attachment with the economic development as it is diaphanous that increased agricultural output and productivity tend to contribute substantially to an overall economic development of a nation.

The terms 'Agriculture' and 'Economic Development' are closely related to each other and both the fields are considered as the backbone of human civilization. The significance of agriculture and economy are vital for both developed and developing countries. Almost every nation of our globe is giving huge attention on the agricultural sector to increase the productivity and the sustainable economic development of the nation. However, in the development of the agricultural sector, mass media and communication system is playing a significant role. Mass media giving a lot of information to the farmers related to farming and helping them to understand the modern farming process on the one hand while on the other it is creating a steady demand for the agricultural products globally.

In India, the enormous contribution of mass media in the agricultural sector helps to develop agricultural productivity as well as the socio-economic infrastructure of the nation. Mass media attracting the attention of the farmer, giving them, latest information related to agronomy, new technique of farming and creating needs and demands of agriculture through marketing technique which ultimately enhancing the productivity and economy of the nation. The support of mass media has given Indian agriculture sector a new dimension. It is gradually changing the farming pattern and the life style of the rural India. Farmers are getting well informed about various farming pattern, seeds, fertilizers, chemicals, pesticides, weather forecast etc. Mass media is ultimately helping in the overall agricultural developmental process of the nation. In the report of the National Round Table Conference on *"Role of Media in Agriculture"* of *Indian Council of Food and Agriculture* the significance of mass media in Indian agriculture was mentioned as:

Mass media plays a significant role in dissemination of agricultural technologies. The success of agricultural development programmes largely depends on the nature and extent of use of mass media in mobilization of people for development. Moreover, it can be decisive in helping farmers access the information that they need and transmitting their concerns. Radio, Television has been acclaimed to be most effective media for diffusing the scientific knowledge to the masses. - (Report, National Round Table Conference of Indian Council of Food and Agriculture, 2017)

Along with the traditional broadcasting media like radio and television, new media is also playing a vital role in the development of agricultural sector. Kishan Call Centre system of The Department of Agriculture and Cooperation under Ministry of Agriculture is one of the renowned rural projects of recent days. The system is functioning from January 2004 with the help of Telecommunication Consultants India Ltd (TCIL). Under this scheme, farmers can call from any part of India to a toll-free number for necessary information about farming and agriculture - (Mkisan. About KCC). 'Kissan Kerala' is another project where various data about agriculture and farming were disseminated among the farmers by a central processing unit with the help of internet, telephone landline and mobile telephony. These project and technological system are ultimately helping the farmers and rural people to promote their economy.

Karshaka Information Systems Service And Networking (KISSAN) is an integrated, multi-modal delivery of agricultural information system, which provides several dynamic and useful information and advisory services for the farming community across Kerala. It is one of the leading citizen centric e-governance projects of the Department of Agriculture, Government of Kerala. The project was conceived, developed and managed by the Indian Institute of information Technology and Management, Kerala for the Department of Agriculture, Government of Kerala – (Kissan Kerala).

SIGNIFICANCE OF AGRICULTURE IN INDIA:

India shows a steady development in the field of agriculture and rises globally as a country with surplus of crops and livestock. Robert E. Evenson, Carl E. Pray and Mark W. Rosegrant in the opening chapter of the book - *'Agricultural Research and Productivity Growth in India'*- which was published as a report of *International Food Policy Research Institute* of Washington .D.C. narrated as *" India's agriculture has grown rapidly enough in recent decades to move the country from the severe food crises of the early 1960's to the food surpluses of the early 1990's even though the population grew by 424 million people between 1963 and 1993"- (Evenson, Pray and Rosegrant, 1999).* India rises predominantly as an agricultural country. In India, the significance and the impact of the agriculture are enormous from the early days of Indian civilization but in the age of globalization its role in the supply of food and in the socio-economic development of the nation is extraordinary. In the introductory part of the book *'Fundamentals of Agriculture'* it was mentioned as:

Today, India ranks second worldwide in farm output. Agriculture and its allied sectors like forestry and logging accounted for 16.6% of the GDP in 2007, employed 52% of the total workforce and despite a steady decline of its share in the GDP, is still the largest economic sector and played a significant role in the overall socio-economic development of India. (Arya. Et.al., 2020)

India expertise in agriculture and served several essential roles globally, few essential roles of agriculture are as - source of food and livestock, production of raw material, maintain international trade and support in the economic development, generate employment, maintain agricultural marketing and continue the study flow of market.

Source of Food and Livestock

India with around 1.3 billion of population considered as the second largest populated nation in the world. The surplus of population does not create food crisis or hunger within the nation rather it shows food security to every citizen. In the post-independence era, India focuses on various agricultural developmental policies with objectives of agricultural development, eradication of food crisis and providing food security to every citizen. Agriculture make India self-sufficient in the production and supply of food crops and livestock. Major food crops of India are rice, wheat, pulse, jowar, bajra, millet, maize etc whereas few cash crops which also used as food and livestock are sugarcane, groundnut, sunflower seed, soya, tea and coffee.

In rice cultivation, India is the second largest producer of rice in the world after China. Wheat is another major food crop of India which most of the Indian use regularly. The production of wheat in the country has increased significantly from 75.81 million MT in 2006-07 to an all-time record high of 94.88 million MT in 2011-2012. The productivity of wheat which was 2602 kg/hectare in 2004-05 has increased to 3140 kg/hectare in 2011-12. - (Farmer' Portal, 2021). Food and Agriculture Organization of United Nation has mentioned in a specific section 'India at a Glance' in its web portal as "India is the world's largest producer of milk, pulse and jute and rank as the second largest producer of rice, wheat, sugarcane, groundnut, vegetables, fruit and cotton. It is also one of the leading producers of spices, fish, poultry, livestock and plantation crops" – (FOA in India, 2021)

Production of Raw Material

The growing population of India has shown a sharp rise in the graph of requirement of various products and goods. Raw material is essential to manufacture products and goods. The sufficient raw material helps to produce valuable products and goods on the one hand while on the other it helps to reduce the prize of the products. In India, the growth in agricultural output over the past few decades have been sufficient enough to meet the demand of the growing population. Agriculture is the main source of raw material for the agro based industries. Agriculture provides sufficient cash crops to produce raw material. Most of the raw materials for the agro based industries are mainly cash crops like groundnuts, sunflower seeds, coconuts, soya, sugarcane, tea, coffee, cocoa, cotton, jute, rubber, resin, etc.

Groundnuts are mainly eaten raw or roasted but they have some industrial use also. They are used to extract cooking oil, making peanut butter, margarine, and body lotions. Sunflower seed are used to extract cooking oil, coconuts is highly useful for various purpose. Coconut water is highly nutritious, dry coconuts are used for bakery and confectionery, coconut oil and margarine are used for cooking purpose, dried fibre of coconut are used for making ropes, coir and mattress, leaves of the coconut tree are also used to make baskets, brooms and brushes. Soya or the soya bean is another cash crop which is used in a multipurpose way. Soya bean are highly nutritious and are mainly used as a vegetable but along with this soya bean are also used for making vegetable oil, soya sauce, milk powder, soya pops, etc. Sugarcane is another highly useful cash crop used to make sugarcane juice, sugar and jiggery. After extracting the juice, the residual part is mainly used as bio fuel in several parts of India. Tea, coffee and cocoa are mainly used for beverage but cocoa is also used to make chocolate, cocoa butter, soft drink, cake, cookies, etc. India is well known for its fine cotton from the early days. Cotton is used mainly to produce cloths, pillow, sleeping mattress, etc. Next to cotton is the jute, which is also a fine cash crop used to produce variety of products. Jute is mainly used to make cloths, sack bags, bags, rope, carpet, etc. Rubber is another form of raw material used to make various products like waterproofing and insulating good, elastic products, eraser, rubber gum, tyre and tube of cars and bicycle, etc. Rosin which is also known as colophony is the solid form of resin obtained mainly from pine trees which also served as a raw material used to polish wood, varnish, paints, art works, adhesives, etc

Maintain International Trade and Support in the Economic Development

Agriculture is the backbone of Indian economy. Agricultural sector provides a valuable contribution in the development of socio-economic structure of the nation. It contributes 16 percent of the overall GDP – (Farmer' Portal, 2021). The contribution of agriculture in Indian economy has been steadily increasing over the few years. According to an economic survey, the share of agriculture in gross domestic product (GDP) has reached almost 20% for the first time in last 17 years making a solo bright spot in GDP performance during financial year 2020-2021 – (Kapil,2021). India is also maintaining the international trade of agricultural products which supporting in the development of national economy. In 2013, India exported nearly 39 billion US dollar of agricultural product globally, which makes India in 2013 as the seventh largest agricultural exporter in the world. India enlisted itself among the fastest growing exporters of the agricultural product – (USDA. Foreign Agricultural Service. International Agricultural Trade Report, 2014). India has become one of the largest suppliers of rice, cotton, sugar and wheat. Most of its agricultural export serve developing nation. In 2011, it exported around 2 million metric tonnes of wheat and 2.1 million metric tonnes of rice to Africa, Nepal, Bangladesh and several other regions around the world – (Parija and Mishra, 2011). During the period of March- June 2020 the total agricultural commodities exported was US$3.50 billion which shows an increase of 23.24% from the same period of 2019. In the recent days India exported its agricultural product to more than 120 countries across the globe in which a major part exported to Japan, South East Asian countries and SAARC countries, countries of European Union and the United State. During the financial year 2020, India exported basmati rice of worth 3.88 billion US dollar, non-basmati rice of worth 1.84 billion US dollar and other processed food worth 2.71 billion US dollar – (IBEF, 2021). India has shown a steady growth in the field of agriculture, which not only helps to develop the economic infrastructure of the nation but also helps to raise India as a global provider of food and livestock.

Generate Employment

In India, agriculture is one of the main sources of employment. It plays an essential role in generating employment opportunities for the rural workforce and support in the process of rural development. It not only provides employment and economy to the rural sector but also played a significant role in the employment and economic development of the urban India. The huge global market of agriculture involves not only the farmers but also various employees at various stages. In the 21st century traditional farming and ploughing the land is not the only option of employment in the agricultural sector rather it shows a wide variety of scope like agricultural research, agricultural engineering, horticulture, dairy farming, fishery, agronomy, etc. All these field are the part of modern agriculture which involved every year a major workforce from both rural and urban India.

However, in the recent days with the rise of technologies and modernisation, dependency on agricultural sector for employment show a sharp decline but still it gives a major contribution in the field of employment and opportunity for the workforce. As per the Farmers' Portal of Government of India, agriculture accounts for employment of approximately 52% of the Indian population – (Farmer'Portal, 2021). Whereas a report on Statista, indicated that in 2020, 41.49 percent of the workforces in India were employed in agriculture, while the other half was almost evenly distributed among the two other sectors, industry and service – (Neill, 2020). The percentage of the workforce in agricultural sector still shows a majority compare to the other fields.

Agricultural Marketing and the Steady Flow of Market

Agriculture promotes and support in the development of agricultural marketing. Agricultural Marketing is a complex process which mainly deals with the business activities related to the agriculture. It is the agricultural marketing which helps to maintain a steady flow of agricultural market. The process begins with decision to produce crops, or a livestock commodity which involves various support from market system, development of farming and ultimately the post farming process including assembling of crops, stocking and storing, transportation, distribution and finally selling in the market. During the process, agricultural products moves from farmers land to the stockist, dealer and finally reached the common public. However, in several cases farmers directly sell their product in the open market but in most cases farmer sell their agricultural product to the stockist.

In India, the system of agricultural marketing is playing a significant role in maintaining the national and international agricultural markets. It helps the entire nation to provide food and livestock. Recently the government of India launches online agricultural marketing portal like e-NAM to promote uniformity in agriculture marketing by streamlining of procedures across the integrated markets, removing information asymmetry between buyers and sellers and promote real time price discovery based on actual demand and supply – (Govt. Of India. eNAM). In India, there are also several central government organisations who are involved in agricultural marketing such as Commission of Agricultural Costs and Prices, Food Corporation of India, Cotton Corporation of India, Jute Corporation of India, etc. There are also some specialised body for rubber, tea, coffee, tobacco, spices and vegetables – (TNAU AGRITECH PORTAL). Each organisation has their own website which provides enormous data and information to the farmers and the general public for the benefit of the society and for the continuous flow of the market.

THE SIGNIFICANCES OF MASS MEDIA IN AGRICULTURAL DEVELOPMENT

In India, the rapid growth of population and the rise of agricultural studies make the farmers more focused on the productivity of the land. Agriculturalist or the farmers has change the need pattern of information. They are searching for information related to the result of various new research based on agriculture, modern agronomy and intensive farming. Mass media bring a perfect solution to them. Mass media such as newspaper, radio, television, along with various forms of new media helps them by providing various message and information regarding farming and agriculture. They are regularly updating the farmers with new information regarding modern agronomy and brief them about various new innovative trends. The information provided by mass media has changed the agricultural pattern and which pushes the farmers toward greater benefits that ultimately help to raise the GDP and the overall economy of the nation.

Mass media also promotes various products of agro based industries and attract the consumers to buy those products. Mass media helps to maintain a steady demand of various products in the markets and for making those products industries need raw material which the farmer is farming and supplying to the industries. There is an interrelationship between agriculture, mass media and industries. In India, it has been found that whenever there is a sluggish or low agricultural growth, the industrial sector suffer due to lack of raw industrial products. The increase in agricultural productivity shows growth in the home market for manufactured goods and services which support to speeds up the rate of economy. According to World Development Report of the year 1979, "*a stagnant rural economy with low purchasing power holds back industrial growth in many developing countries*." – (http://www.economicsdiscussion.net)

Print Media and the Agricultural Development

Print media, the oldest from of media which includes newspaper, magazine, journal, brochures, books, etc. They are promoting the agricultural journalism in its own traditional way. However, in India, it plays a significant role in the development of agriculture and in the progress of national economy. The Indian farmers and agricultural communities highly prefer various forms of print media like newspaper, magazine, journals, etc for accessing agricultural and farming news. The print media shows various positives sides to the farmers such as they are highly information rich, available at a very low cost and they are easy to access. Another facility of the print media such as newspaper and magazine is that, it published mainly in a permanent form which any one can store and show anywhere if required. The newspaper and agricultural magazine are serving a key role in information dissemination regarding agriculture. Several Rural newspapers and magazine published various news articles related to agriculture, new pattern of farming, effect of new fertilizers along with various stories of farmers and farming which revitalised the concept of the farmers and create new hope among them. The contribution of print media in the development of agricultural sector in India is enormous. They not only provide the information and data to the farming communities but also give moral support to them.

Radio and the Agricultural Development

India has shown a huge rural and semi urban sector, which need a proper infrastructure for sustainable development. Information and effective communication plays a significant role in the socio economic developmental process and radio has an active part in it. Radio is the early form of electronic media serving the mankind since its arrival. The main objectives of radio are to disseminate information, education and entertainment. The Dominant Paradigm also accepted its importance and signifying as tool of modernization. In India, radio arrived during the British period but started to develop in the post-independence era. Radio is a mass media which beautifully disseminates the message through its clear audio and various sound effects. It is the medium through which various government agencies and non- governmental organisation disseminating various message and information among the masses for agricultural and socio-economic development.

In India, radio has played a major role in various agricultural and rural developmental programmes. One of the vital radio projects that India has seen is 'Radio Farm Forum' in 1956 which started in Bombay. The project was financed by UNESCO and promoted with an active co-operation of Government of Maharashtra. The broadcast was successfully carried out by Poona AIR station. The project use radio for disseminating a wide range of information in around 145 villages spread over five Marathi dominated state of Maharashtra. However, on November 1959 the Community Development and Co-operation minister Shri. S.K.Dey inaugurated the All India Radio's 'Radio Rural Forum' project. It is also known as 'Charcha Mandals'. It was an ambitious project and an extension of Radio Farm Forum of Maharashtra – (Bhatt and Krishnamoorthy, 1965)

All India Radio (AIR) broadcast several programmes on agriculture which easily reach every corner of the nation. Farmers of India easily access that information broadcasted by AIR and use them in farming and cultivation. Several agricultural plan and policy presented by the government are also broadcasted by All India Radio. In '*Pradhan Mantri Fasal Bima Yojna*' All India radio was involved by the government of India to make publicity and awareness campaign on the policy – (Prasar Bharati. National Programme, Farm and Home Unit, 2016). All India Radio has various radio stations under the project 'Kisanvani'. The project shows collaboration between All India Radio and Department of Agriculture & Cooperation, Ministry of Agriculture – (Indian Council of Food and Agriculture, 2017). The purpose of the project is to provide media support in agriculture extension by making the local farmer aware about the daily market rates, giving them accurate weather report, briefing current scenario of the agriculture development in India.

Television and the Agricultural Development

Agriculture is considered as the back bone of Indian economy and television is playing a significant role in it. Television support and telecast rural communication and agricultural development programme from the very beginning of its arrival. It widely propagates information and inductive technique related agriculture and various new farming processes in all over India. During 1967 All India Radio with the help of Department of Atomic Energy, Indian Agricultural Research Institute, Delhi administration and the State Government of Haryana and Uttar Pradesh launched a weekly television programme known as 'Krishi Darshan' in an experimental basis. In the initial stage, it covers around 80 villages near Delhi, later few villages of Haryana and Uttar Pradesh also come under range. The project mainly focuses on agricultural development – (Tripathy, 2011). Other than this television also telecast various governmental advertisements for the welfare of rural agricultural development. SITE also broadcast agricultural programme for agricultural development. Recently Doordarshan has launched a new channel known as 'DD Kisan Channel' especially for the farmers. DD Kisan is an Indian agricultural based 24-hour television channel, which is owned by Doordarshan and was launched on 26 May 2015 – (Prasar Bharati, 2021). The channel has been dedicated to agriculture and related sectors, which disseminates real-time inputs to farmers on new farming techniques, water conservation and organic farming among other information

CONCLUSION

Agriculture is the backbone of human civilization. The history of agriculture begins with the rise of human civilization but with the arrival of mass media it has got pace. It is helping in the sustainable development of the agricultural sector and to visualise its importance across the globe. Mass media in various ways highlight the modern agronomy and suggests the farmers how to use developed seeds for plantation, pesticides and modern fertilizers. It has pushed the farmer towards 'Intensive farming' which deals with cultivation for maximum profit. Mass media serves the agricultural sector by giving a lot of information to the farmers related to farming whereas on the other it is creating a demand for the agriculture globally. The importance of agricultural sector for economic development is well recognised. Agricultural sector provides a wide contribution in the development of national economy however mass media full filling a wide objective of the sector which ultimately enhancing the economy. It is cleared that agriculture and economy is vital for national economic development whereas the mass media giving a momentum in it.

In India, the contribution of mass media in agricultural sector is enormous and its role in agricultural development cannot be ignored. It serves as a catalyst which support in the growth of agriculture and the socio-economic infrastructure. It favours the agriculture system from numerous ways and proved success. It provides valuable information to the farmer about agriculture and new farming process, it helps to maintain a steady demand of food, livestock and raw material in the markets and support in the agricultural study and research. All these functions ultimately help to flourish the agricultural sector and support in the development of socio economic infrastructure of India.

REFERENCES

Arya, R. L., Arya, S., Arya, R., & Kumar, J. (2020). *Fundamentals of Agriculture* (Vol. 1). Scientific Publishers.

Bhatt, B. P., & Krishnamoorthy, P. V. (1965). Radio broadcasting serves rural development. Radio Rural Forum spread throughout India. Report and Paper on Mass Communication UNESCO (No. 48). UNESCO.

Evenson, R. E., Pray, C. E., & Rosegrant, M. W. (1999). *Agricultural Research and Productivity Growth in India*. International Food Policy Research Institute.

Farmers' Portal. (2021a). *About Us*. Retrieved from https://farmer.gov.in/M_Aboutus.aspx

Farmers' Portal. (2021b). *About Wheat*. Retrieved from https://farmer.gov.in/m_cropstaticswheat.aspx

FOA in India. (2021). *India at a Glance*. Retrieved from https://www.fao.org/india/fao-in-india/india-at-a-glance/en/

IBEF. (2021). *Agriculture And Food Industry And Exports*. Retrieved from https://www.ibef.org/exports/agriculture-and-food-industry-india.aspx

Indian Council of Food And Agriculture. (2017). *Role of Media in Agriculture*. National Round Table Conference. Retrieved from https://www.icfa.org.in/assets/doc/reports/RTC_Media.pdf

Kissan Kerala. (n.d.). *About Kissan*. Retrieved from http://www.kissankerala.net/kissan/kissancontents/about.jsp

Manley, W., Foot, K., & Davis, A. (2019). Oxford Dictionary of Agriculture and Land Management (1st ed.). Academic Press.

Mkisan. (n.d.). *About KCC*. Retrieved from https://mkisan.gov.in/aboutkcc.aspx

NCERT. (n.d.). *Contemporary India II. Agriculture*. Retrieved from https://ncert.nic.in/ncerts/l/jess104.pdf

Neill, A. O. (2020). *Distribution of the workforce across economic sectors in India 2020*. Statista. Retrieved from https://www.statista.com/statistics/271320/distribution-of-the-workforce-across-economic-sectors-in-india/

Parija, P., & Mishra, P. (2011). *India Allows Wheat Exports for the First Time in Four Years*. Bloomberg. Retrieved from https://www.bloomberg.com/news/articles/2011-09-08/india-allows-wheat-exports-for-the-first-time-in-four-years-1-

Prasar Bharati. (2016). *National Programme, Farm and Home Unit*. Retrieved from https://prasarbharati.gov.in/all-india-radio-2/#National-Programs

Prasar Bharati. (2021). *About*. Retrieved from https://prasarbharati.gov.in/dd-kisan-homepage/about-dd-kisan/

Sharma. M. (2021). *The Future of Indian Agriculture*. Down to Earth. Retrieved from https://www.downtoearth.org.in/blog/agriculture/the-future-of-indian-agriculture-75384

Small Farmers' Agribusiness Consortium, Ministry of Agriculture and Farmers' Welfare, Government of India. (n.d.). *e-NAM Overview, Vision*. Retrieved from https://www.enam.gov.in/web/

Stephens, A. (2015). *Dictionary of Agriculture* (2nd ed.). Routledge.

TNAU Agritech Portal. (n.d.). *Agricultural Marketing in India*. Retrieved from https://agritech.tnau.ac.in/agricultural_marketing/agrimark_India.html

Tripathy, S. (2011). *Impact of TV programmes on agriculture and rural development*. www.Publishingindia.com

USDA, Foreign Agricultural Service. (2014). *India's Agricultural Exports Climb to Record High*. International Agricultural Trade Report. Retrieved from https://www.fas.usda.gov/data/india-s-agricultural-exports-climb-record-high

KEY TERMS AND DEFINITIONS

Agriculture: Is an art or technique of cultivating soil, growing crops, and raising livestock.

Agronomy: A branch of agricultural science that deals with soil management and crop production.

Economic Development: Growth in economic infrastructure or the development of economy.

Information: Facts, data, or the message.

Marketing: The activities of a company, industries or of any individual public which promotes the buying and selling process.

Mass Media: The media which helps to disseminate information, data, idea, belief, etc., among the masses.

Chapter 3
Analysis on the Present Situation of China's Agricultural Development and the Sustainable Development of the Agricultural Economy

Poshan Yu
https://orcid.org/0000-0003-1069-3675
Soochow University, China & Krirk University, Thailand

Junyi Lv
Independent Researcher, China

Aashrika Ahuja
Independent Researcher, India

ABSTRACT

This chapter aims to study the trends in sustainable agricultural development in China. The deterioration of the world's climate and environment can be counteracted with the trend of development in internet and technological spheres. How to create value out of scientifically advanced agricultural techniques in China so as to achieve sustainable development in the future is the subject of the author's discussion. This chapter will investigate the development of smart and sustainable agricultural techniques that are being employed in China in integration with the internet and information industry. Further, the sustainable ecological development of agriculture in China in the past two decades will be analyzed. Finally, this chapter will provide decision-makers with analysis and suggestions on the way forward and direction with respect to sustainable agricultural development in China in the future.

DOI: 10.4018/978-1-7998-9557-2.ch003

INTRODUCTION

Sustainable agricultural development is crucial for enhancing productivity as well as availability of food and at the same time reviving soil/land fertility, efficient management of resources, strengthening rural livelihood, ensuring participation of all stakeholders leaving no one behind in this transitional phase towards creating food security for all and mitigating impact of climate change on agriculture. Agriculture plays an important role in growth of any economy and since the onset of pandemic in 2020, China determines to rebound by building prosperous society through development of its agricultural economy. China, with large agricultural area being underdeveloped, can only ensure sustainable development for the future by focusing on, the development of agriculture sector thereby serving as an important springboard for the economic take-off as well as providing an effective solution to help the country successfully end poverty (Wang et al., 2008).

History is a testament to the fact that, since 1978, China has gradually focused and developed market-oriented rural reforms. This practice of breaking through the shackles of the traditional system has been a turning point in the historical development of China's agriculture, and a big leap towards the road to sustainable development of China's agriculture economy (Wang, 2004). The successful implementation of economic reforms implemented in China towards market-oriented economy has enabled China to feed nearly 21% of the world's population with less than 9% of the world's arable land (Chen & Pan, 2005). Reforms focusing on agriculture, rural areas and farmers in China have also led to the comprehensive development of the reform of China's economic system and contributed further to the growth of China's economy. Opening doors to foreign trade and foreign investment have also contributed significantly in prosperity of China as a nation and its people.

However, if China's economy wants to continue to develop steadily, the pace of adopting practices related to sustainable development of agriculture and agricultural products needs to increase rapidly. Climate change continues to impact the environment in significant number of ways and the same has been posing a great challenge to the development of agriculture and agricultural products. Increase in global temperatures and precipitation variability, prolonged drought and water shortages, loss of agricultural biodiversity coupled with energy depletion, is taking a great toll on development of China's agriculture and agricultural products. Since these problems have emerged through human interference, it is all the more crucial that human beings themselves find solutions to these pressing issues. Appropriately utilizing renewable energy resources and fixing gaps between proportion of products and services contributing to China's agricultural imports and exports will be key factor for consideration to overcome this dilemma. (Zhang, 2014).

At this point in time when entire humanity is at a crucial crossroads on account of COVID-19, the path of sustainable development of China's agriculture and agricultural products can pave the way for the whole world to break through this impasse. This path is definitely most effective providing a room for great opportunities but also great challenges at the same time. The second section aims to elaborate the impact of deteriorating ecological environment on China's agricultural production and agricultural economy. The author in this section has attempted to analyse the impact of climate and environment on China's agricultural land economy.

The third section attempts to discuss the sustainable utilization of agricultural resources in China. The author investigates the sustainable use of water and soil resources in China as compared to the use of these ecological resources in past few years. The fourth section analyzes the feasibility and sustainable development of agriculture in China. In this section, the author discusses the present situation of the development of sustainable agricultural economy in China from the perspective of theory and development practice. The author also investigates the development of digital and intelligent agriculture in science and technology in the world and in China. On the basis of this study, the author has attempted to evaluate the possibility of sustainable development of agriculture in China.

The fifth section evaluates the opportunities and challenges plaguing the development of sustainable agricultural economy in China. The author investigates the international and Chinese national policies implemented in the agriculture sector in past few years, thereby comparing the impact of different policies on the development of ecological agriculture in China. In addition, the author also attempts to investigate the extent of development of science and technology in agriculture as well as the transfer of labor force in China, so as to provide a basis for evaluating the development of sustainable and ecological agriculture in China. Recommendations and suggestions are further provided towards the end of the chapter.

IMPACT OF CLIMATE AND ENVIRONMENT ON FOOD SECURITY IN CHINA AND THE PRESENT SCENARIO

Climate

Since China is located in the subtropics, on the east coast of the mainland and on the west coast of the Pacific Ocean, it is significantly affected by the monsoon. The summer monsoon period is the most important reason for occurrence of frequent drought and floods. These abnormal changes in weather and precipitation patterns during summer monsoon period coupled with unreasonable production activities causing exploitation of resources have further aggravated ecological damage. The widespread ecological damage has resulted into frequent outbreaks of extreme weather events in China in recent years. If the environment continues to deteriorate further, the resulting damage to the economy cannot be avoided at any cost.

Frequent flooding in China in past few years has caused a serious damage to crop yields. Since June 6, 2008, there have been many incidents of heavy rainfall in southern China, and severe rainstorms and floods have also occurred in some areas. Crops have been affected in area of land covering 1017 thousand hectares, with 67000 houses collapsing leading to a whopping direct economic loss worth 14.45 billion yuan. In 2009, Taiwan suffered heavy rainfall under the influence of strong Typhoon Morakot, resulting into the worst flooding experienced in the south for 50 years. In 2010, there was a rare heavy precipitation in Hainan, which broke all records since 1961. In 2015, the daily rainfall in many places in the south of China created havoc leading to widespread floods. From June to August, there were 18 rainstorms in southern China, which recorded heavy rainfalls. As a result, a total of 24 stations exceeded daily precipitation levels ever recorded in the history (China Meteorological Administration, 2016). In July 2021, a flood that occurred in Henan, China, affected an area of 1021.4 thousand hectares of crops, including a total harvest of 179.8 thousand hectares (International fruit and vegetable mesh, 2021).

In addition to extreme precipitation, the weather in some parts of China has been extremely dry. Some major events relating to droughts in China in past few years have been listed in this section. In 2009, for more than 3 months in a row, there was no effective precipitation in 15 provinces and cities in North China, Huanghuai, Northwest, Jianghuai and other places. This emergency situation on account of draught significantly impacted winter wheat grain yield, including livestock thereby pushing farmers at the brink to shoulder many losses. In 2011, the average annual precipitation in China reached the lowest level recorded for 60 years, creating a record of being the lowest in history since 1951. The spring precipitation was reported lowest since 1951, and the summer precipitation being lowest since 1993. From June 20 to July 20, 2015, the average precipitation in Liaoning and Jilin provinces was 49.5 mm, 65.6% less than the same period in previous year and the lowest since 1961 (China Meteorological Administration, 2016). This frequency of extreme weather and climate change events have led to a sharp decline in China's agricultural production. As a result, China's agricultural economy has suffered huge economic losses. Many studies and assessments over the years, have confirmed that climate change has impacted China's agricultural production to a great extent. Severe climate change caused a sharp decline in China's total agricultural output in the 1990s. The direct economic loss was equivalent to 1% of GDP, and the resulting economic cascading effect was equivalent to 17.8% of GDP (Yuan et al., 2020).

If the carbon fertilization effects of climate change are not taken into account, the resulting economic damages will be devastating. The intensity of the impact however might differ in different regions with underdeveloped areas that are more dependent on agriculture suffering greater economic losses than the developed areas (Yuan et al., 2020).

Agricultural production accounts for a large proportion of China's national economy, and its serious impact cannot be ignored as mentioned above. According to the data of the National Bureau of Statistics, from 2011 to 2019, crops were affected by climate and other natural disaster factors as shown in Table 1:

Table 1.

Agricultural Crops affected by natural disasters (2011-2019)									
Indicators	2011	2012	2013	2014	2015	2016	2017	2018	2019
Affected area of crops (1000 Ha)	32470. 5	24962	31349. 8	24890. 7	21769. 8	26220. 7	18478. 1	20814. 3	19256. 9
Crop failure area (1000 Ha)	2891. 7	1826. 3	3844. 4	3090. 3	2232. 7	2902. 2	1826. 7	2585	2802
Area affected by drought (1000 Ha)	16304. 2	9339. 8	14100. 4	12271. 7	10609. 7	9872. 7	9874. 8	7711. 8	7838
Drought failure area (1000 Ha)	1505. 4	374	1416. 1	1484. 7	1046. 1	1018. 3	752. 4	922. 4	1113. 6
Area affected by floods, geological disasters and typhoons (1000 Ha)	8409. 9	11220. 4	11426. 9	7222	7341. 3	10554. 9	5808. 8	7283. 1	8604. 8
Flood, geological disaster and typhoon failure area (1000 Ha)	872. 8	1095. 3	1828. 9	976. 9	841	1442. 4	766. 1	1009. 9	1480. 8

Environmental Quality

China's total population has exceeded 1.4 billion in the year 2020, and this huge population base has a huge demand for crops such as grain and vegetables. In the past, one of the major ways adopted by Chinese farmers to increase crop yield was to use a large number of chemical fertilizers and pesticides. This has made, China, as the world's leading producer and consumer of pesticides. The use of pesticides reached 1.8077 million tons at the end of 2013, and the use of chemical fertilizers reached 52.51 million tons by the end of 2020 (China's National Bureau of Statistics, 2020). The use of pesticides and chemical fertilizers in China is shown in Figure 1 below, and the application rate has not been improved until recent years.

Figure 1.

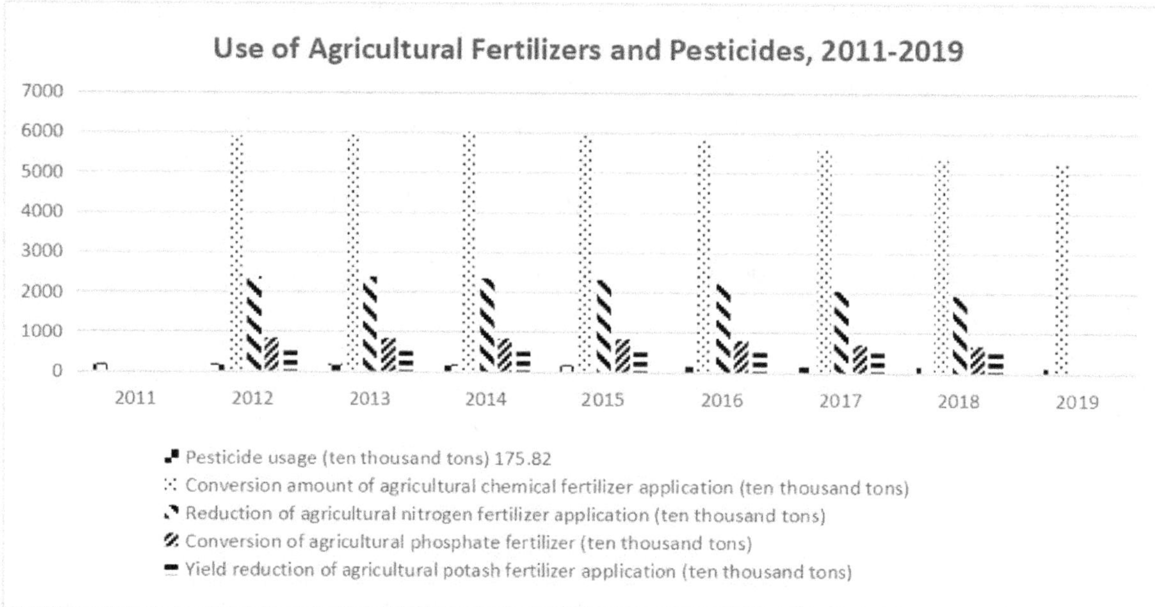

The use of pesticides and fertilizers can effectively increase agricultural productivity and control damage caused by insect pests, for example, the use of pesticides in China avoided the loss of 70.5 million tons of grain, 2.53 million tons of oil seeds and 20.1 million tons of cotton in 2007 (Li et al., 2014). But when we look at long-term use of chemical fertilizers and pesticides, we can conclude that improving agricultural production and at the same time reducing the negative impact of pesticides and chemical fertilizers on the environment is an arduous task and the damage done in this process is irreversible over a period of time.

70% to 80% of the pesticides that are used go directly into the environment (Yan & Peng, 2009). The frequent use of pesticides leads to the gradual increase of drug resistance in various pests, which eventually leads to the continuous increase in concentration of pesticides, resulting in serious pollution and harm to the ecological environment (Chen, 1994). The use of chemical fertilizer also causes pollution of rivers, lakes and underground water sources. According to investigations, overuse of nitrogen fertilizer has been leading to increasing air, soil and water pollution at a large scale. Ammonia volatilization and the release of NOx, CH4 and CO2 will cause not only greenhouse effect, but also the destruction of the ozone layer. Pollution of rivers caused by rampant increase in use of harmful substances such as ammonia nitrogen and nitrates produced by the use of chemical fertilizers is resulting, into eutrophication in large areas of rivers and lakes, further polluting groundwater (Huang et al., 2004).

ANALYSIS OF USE OF AGRICULTURAL RESOURCES IN CHINA

Agricultural production is both a resource provider and a consumer, because it consumes a lot of water and land resources while providing production and essential requisites to sustain life (Rosa et al., 2018). Water resources are an important form of natural resource for economic development, and their distribution across different industries impact the economic development of a region in many ways (Mielke et al., 2010). Land resources on the other hand provide a basis for farmers' income and are a means for realizing social security as well as overall prosperity of the country (Yuan & Qi, 2019). China's agricultural production is facing strong resistance in terms of becoming a source of development from the challenges including insufficiency in availability of water and land resources coupled with poor environmental conditions.

Water

Availability of water resources in China reached at its peak in 2016 and then began to decline. As shown in Figure 2 below, China's total water resources dropped from 3.24664 trillion cubic meters in 2016 to 2.74625 trillion cubic meters in 2018, and per capita water resources also dropped from 2354.92 cubic meters per person in 2016 to 1971.85 cubic meters per person in 2018. In 2019, the total amount of water resources in China reported was 2.9041 trillion cubic meters, and the per capita water resources was 2062.9 cubic meters per person (National Bureau of Statistics of China, 2019).

Figure 2.

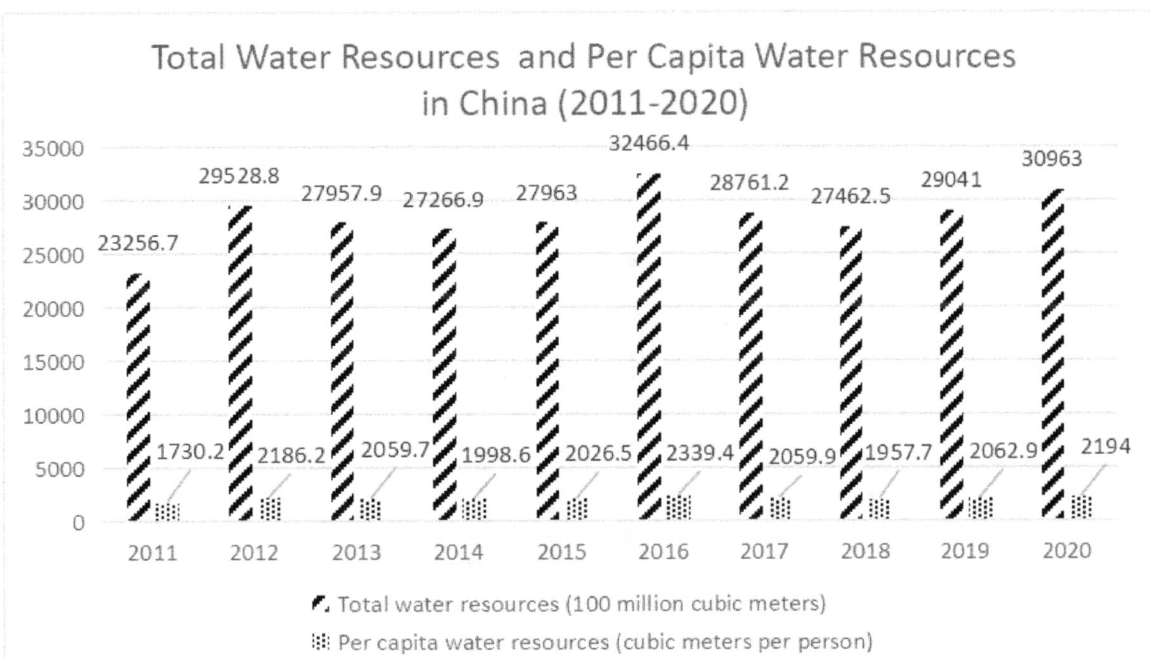

As shown in figure 3 below, agricultural water consumption increased annually from 2010 to 2014. As a result of the movement to protect water resources thereafter including judicious use of water resources, technology of irrigation gradually improved, and agricultural water consumption declined. Agricultural water consumption reached 392.15 billion cubic meters in 2013, compared with 368.23 billion cubic meters in 2019, thus coming down by 6 per cent from the 2013 which was its peak. Per capita water consumption also continued to decline, with a per capita water consumption of 429 cubic meters per person in 2019. However, according to the survey conducted by China National data Network, the proportion of water used in agricultural sector has always been the highest of all industries. Agricultural water consumption in China accounted for 61% of the national water consumption in 2019. The total amount of agricultural water used in 2019 was 367.5 billion cubic meters (National Bureau of Statistics of China, 2019).

Take Hebei province as an example. Hebei province is one of the major producers of grain in China, among which wheat, corn, vegetables, fruits and other high-consumption crops account for a large proportion (Han et al., 2019). Therefore, agriculture is the largest industry in water consumption and consumption in Hebei Province, accounting for more than 70% of the total water consumption in the province (Wang, 2017). However, groundwater resources have been over-exploited in China, and Hebei Province which is located in Heilongjiang area and Piedmont plain area, where groundwater over-exploitation is serious, has been a key area for the implementation of efficient water-saving irrigation techniques. The shortage of water resources not only restricts the agricultural development of Hebei, but also seriously restricts the economy of the whole province (Zhang, 2019).

Figure 3.

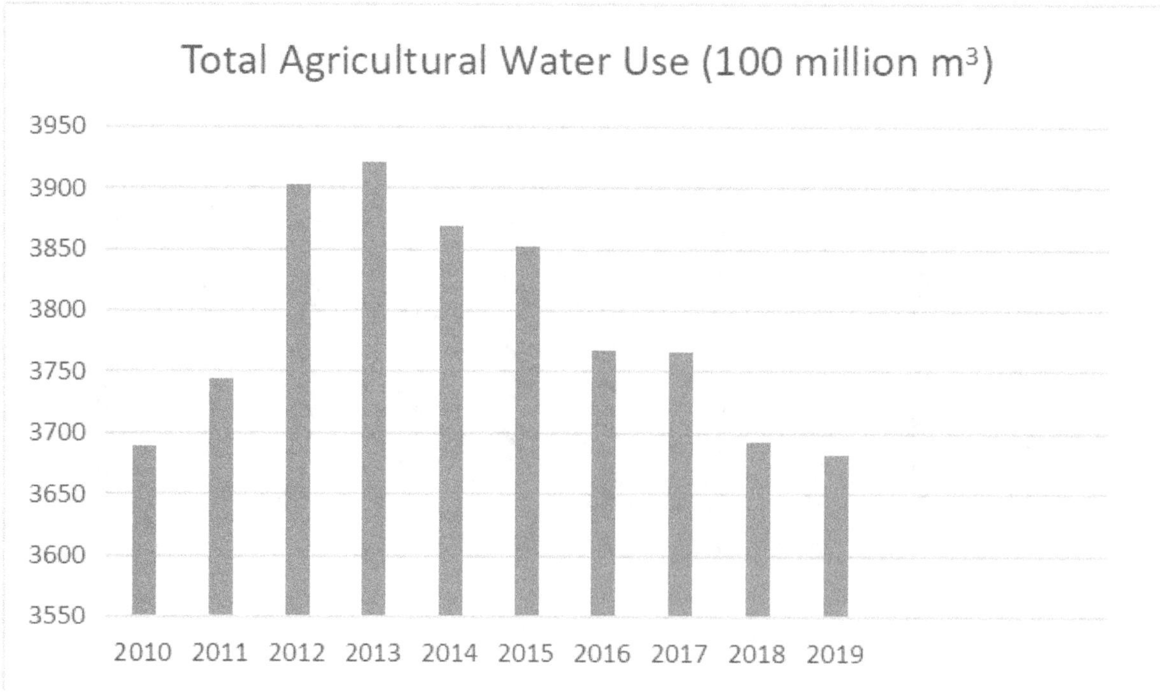

As a result, in China, especially those areas that grow high-consumption crops and take agricultural production as one of the important sources of income, are stepping up efforts to promote efficient water-saving irrigation techniques to improve agricultural water use efficiency (Wang et al., 2013). The development of water-saving agriculture will be an important step to promote the ecological and sustainable development of agriculture. China's policy also puts forward solutions to use water resources in agriculture judiciously. The policy program requires that we focus on the areas that require high-consumption of water for instance crops such as wheat, corn, vegetables and fruits, and strive to optimize the agricultural planting structure to form a modern agricultural production system that matches the carrying capacity of water resources (Zhao, 2018).

Cultivated Land Resources

Land is the most important resource in agricultural production. Unlike other industries, agricultural production, especially crop production, cannot be carried out without land. Since the land is physically fixed, it cannot be moved between regions (Zhong et al., 2007). Therefore, the land productivity of a crop is also affected by specific climate, topography, soil and water conditions.

In addition, as a productive input, land cannot also be moved between seasons. If farmers want to maximize returns, they must practice crop rotation thereby growing different crops in different seasons of the year, or rotate different crops sequentially. In this sense, if additional land resources are not available, farmers cannot significantly expand agricultural production even if it has a comparative advantage. Although technological innovation will help to improve land productivity to some extent, production will still be limited by land restrictions (Funing, et al., 2007).

China is the most populous country in the world but agricultural and land resources are not sufficient to meet the needs of growing population. The area of cultivated land in China accounts for only 14.05% of the total land area, and the per capita cultivated land area is less than 51.27% of the world average. Moreover, China has few reserved resources for cultivated land. The reserve resources of cultivated land in the country is only about 5.3333 million hectares, and most of them are distributed in ecologically fragile areas, which makes it expensive and difficult to supplement cultivated land. In addition, the amount of high-quality arable land in China is also small. Statistics from the 2016 China Land and Resources Bulletin show that the average quality of cultivated land in China was 9.96 in 2015, with superior land accounting for only 2.9% of the total cultivated land, while low-grade land accounted for 17.7% of the total (Liu et al., 2019).

Therefore, increasing the output per unit area of land is not only of great economic value, but also the basis of social stability. In addition, in the cultivated land in China, the proportion of medium-and low-yield land is more than half, and the lack of land nutrients is an important reason why the yield cannot be guaranteed. As the direct value and indirect value of land resources have not reached the optimal state, rural land resources restrict the realization of rural revitalization. Therefore, to further stimulate the vitality of rural land resources and tap the value of rural land is an important way to improve farmers' income and realize agricultural modernization (Zhao, 2019).

Agricultural soil is a precious natural resource, which is non-renewable (Jenny, 1980). Due to pollution, the degradation of agricultural land in China is widespread, of which 19.4% of agricultural land is polluted. As a result, a large part of agricultural land has been abandoned. The salinization of land caused by adverse effects of climate change has also polluted the cultivated land. Salinization has not only reduced soil quality and crop productivity, but also has gradually contributed to reduction in the area of cultivated land (Bhaduri et al., 2016; Cao et al., 2018). At present, more than 1/3 of the arable land in the world has suffered salinization, resulting in a sharp decline in agricultural production, and this problem concerning land degradation is extremely serious (Fan et al., 2012; Gao et al., 2018; Li et al., 2014; Tilman et al., 2011). Salinized area of land is increasing at an annual rate of 10% due to high surface evaporation, low rainfall, inadequate irrigation and other unreasonable human activities (Himabindu et al., 2016; Liu et al., 2017; Mose et al., 2018).

The Chinese government has issued an action plan for soil pollution prevention and remediation, which aims to restore millions of hectares of contaminated agricultural land in the next few years. Among the challenges that are being faced in implementation of the action plan, it is necessary to assess the sustainability of these actions, particularly in arid and semi-arid areas where the rehabilitation of degraded lands is particularly challenging (Hou, et al., 2018).

In order to sustainably use agricultural land resources, Chinese government at all levels has adopted a variety of soil restoration programs and soil control policies in the past decade. China's State Council issued the soil pollution Prevention and Control Action Plan (the Action Plan) in May 2016 to curb soil pollution and clean up existing pollution (Wang, 2016). The action plan requires national and local governments to conduct a more detailed agricultural soil quality survey by 2018 to clean up about 700000 hectares of heavily polluted farmland by 2020, allowing 95 per cent of the contaminated land to be used in a safe way by 2030. In November 2018, China's Ministry of Agriculture issued the Action Plan for Agricultural and Rural pollution Control, which calls for the control of contaminated agricultural land and the implementation of agricultural ecological development.

FEASIBILITY OF PROMOTING ECOLOGICAL AGRICULTURE AND SUSTAINABLE AGRICULTURAL DEVELOPMENT IN CHINA

Sustainable Development Scale and Current Situation of China's Agricultural Economy

Theoretical Research

The following data range from China Knowledge Infrastructure Project (CNKI) database includes literature retrieval, knowledge meta-retrieval and citation retrieval in the shape of a "journal". From the classification of retrieval methods, this study is the scope of literature retrieval. Under the condition that certain reference value is guaranteed for the particular literature, the retrieval subject is "agricultural sustainability", and the results are shown in Figure 4:

Figure 4.

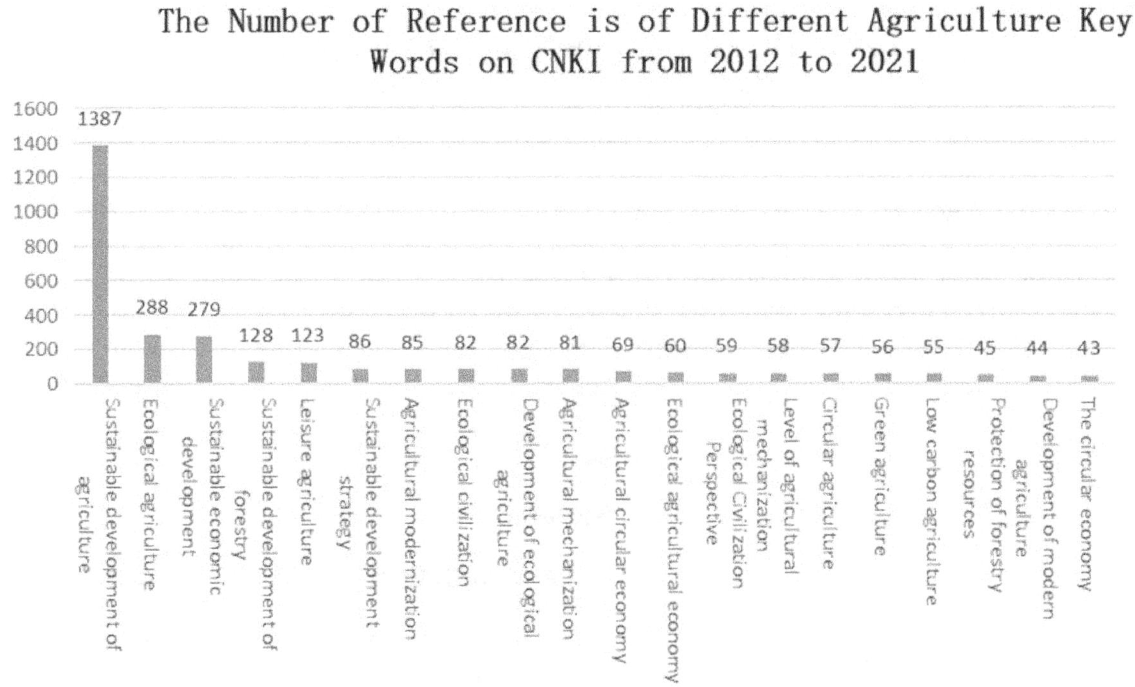

After the analysis according to the retrieval requirements, it is found that the research field of "sustainable development of agriculture" has received a lot of attention by the academic circles in China since 1998. Taking April 2021 as the cut-off point, there are a total of 1387 journal articles with "Agricultural Sustainable Development" as the key words. According to the order of the number of articles published from more to less, it has been found out that there is a total of 20 research contents. Among them, "sustainable agricultural development", "ecological agriculture" and "economic sustainable development" have been ranked in the top three, with 1387 and 279 articles published respectively (China National Knowledge Infrastructure, 2021). Thus, it is clearly visible that the sustainable development of agriculture has been widely considered by the industry with significant time being spent on the academic research on the sustainable development of agriculture, and ecological agriculture. At the same time, the policies promulgated by the government have also played a significant role in promoting the sustainable development of agriculture.

However, there is still a great room for progress in the research on the sustainability of agriculture. As shown in Figure 5, when we classify literature according to the level of research, it can be divided into eight categories: basic research (social sciences), basic and applied basic research (natural sciences), policy research (social sciences), industry guidance (social sciences), engineering technology (natural sciences), industry technical guidance (natural sciences), policy research (natural sciences), advanced science popularization (social sciences) and so on. As shown in figure 6 below, the number of articles published at the advanced popular science (social sciences) level is only 1, and the decimal removal is about 0%. This shows that there is still not enough in-depth research on the sustainable development of agriculture. Once there is a lack of in-depth academic research results in this area, the in-depth development of agricultural sustainability will also be limited by the lack of theoretical and technical support.

Figure 5.

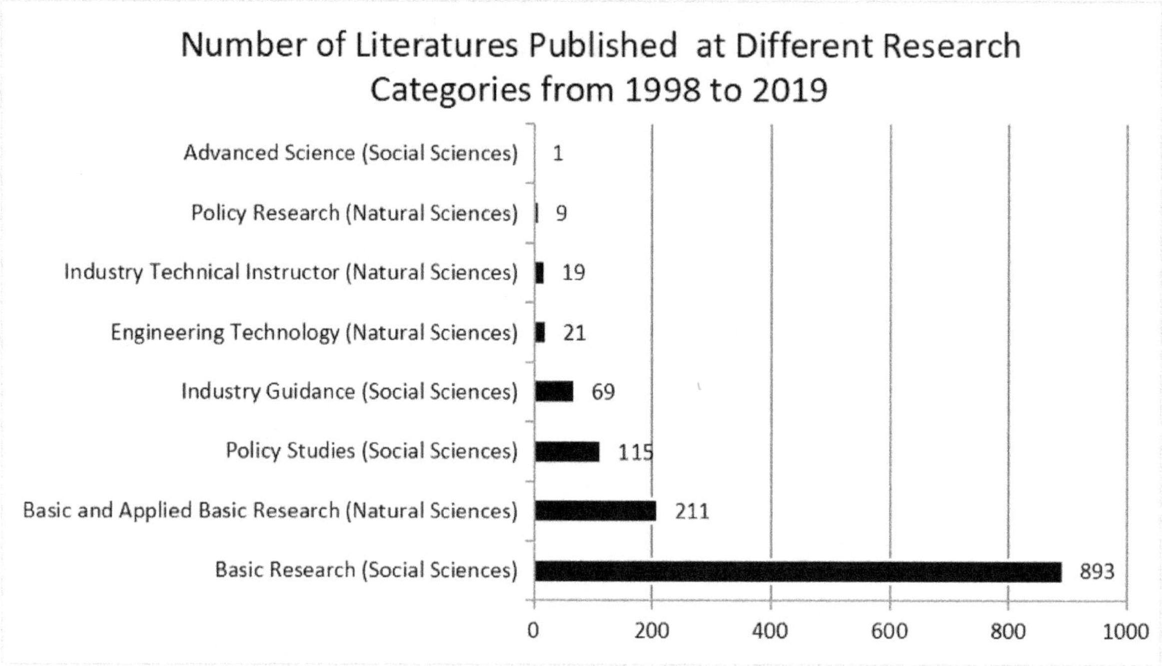

Import and Export of Agricultural Products in China

From the perspective of import and export, the proportion of China's share in the world agricultural trade is increasing year by year. Figure 6 below reflects the quantitative changes in the import and export of China's agricultural products in recent years.

The component of value addition of China's export that includes agricultural products is relatively low. China's export of agricultural products is mainly realized through general trade, which shows that China's agricultural products exports are mostly limited to direct sales, ignoring the processing of products (Meng, 2009). At the same time, it also shows that due to the lack of advanced agricultural products processing technology, China's agricultural products are less competitive compared with those exported by some developed countries (Lu, 2005).

From the perspective of import and export, the import volume of China's agricultural products has always been greater than the export volume. In 2019, China's agricultural exports totaled US $78.57 billion, decreasing 0.9 per cent from the same period last year. Imports of agricultural products on the other hand, totaled 149.85 billion US dollars, with an increase of 9.6 percent over the same period (Zhiyan consultation, 2020). Thus, it can be seen that the trade deficit of agricultural products in China is constantly expanding, and the degree of external dependence of agriculture is increasing. From 2015 to 2019, the trade deficit of agricultural products in China showed an overall upward trend. It reached US $71.28 billion in 2019, an increase of 24.2% over the same period last year (Forward-looking industry research institute, 2020).

At the same time, it is important to note consistency in China's agricultural policy reform as well. In 2019, the guidance on promoting the Revitalization of Rural Industry proposed to guide the financial

Figure 6.

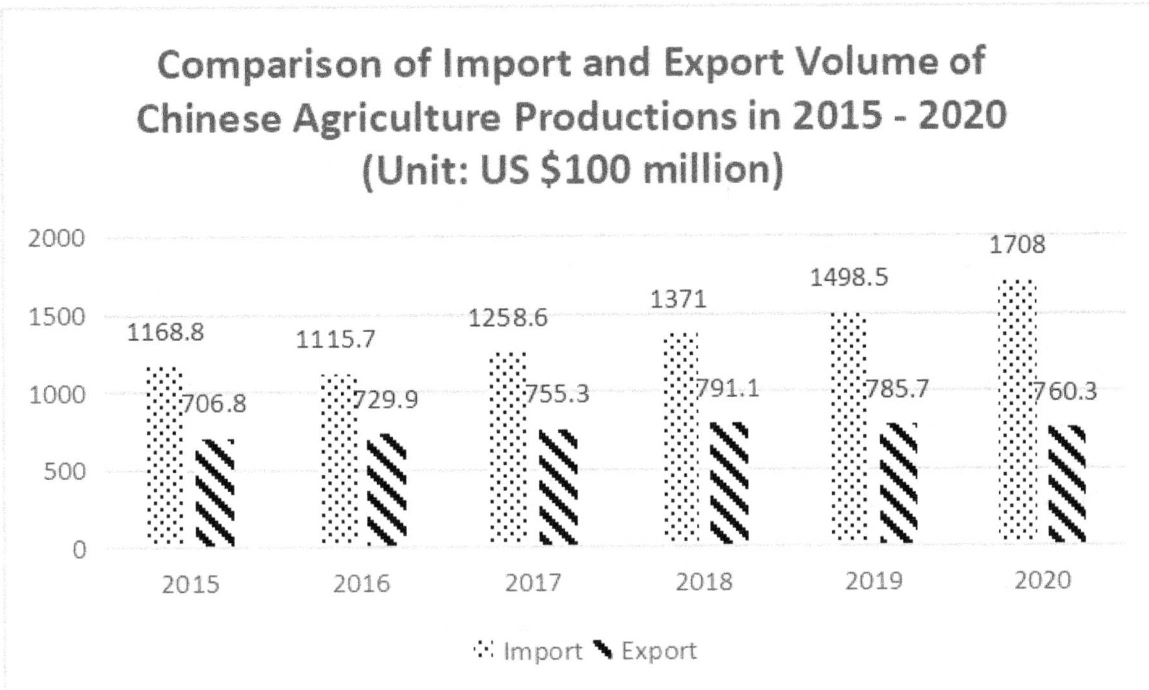

institutions to use deposits mainly for local industries. The preferential financing policies for small and micro enterprises have been applicable to rural industries, innovation and entrepreneurship, and support the listing and financing of qualified agricultural enterprises. All these policy reforms have led to "the biggest increase in economic welfare over a 15-year period in history" (Li, 2019).

Since past few years, the volume of export of agricultural products in China has undergone a significant shift. In 2019, vegetable exports were 15.5 billion US dollars, with a significant upward trend depicting a 1.7 percent rise from the same period last year; fruit exports were 7.45 billion US dollars, with an upward trend with a rise up to 4.1 percent from the same period last year; but the export of aquatic products was 20.66 billion US dollars, depicting a decline of 8.0 percent from the same period last year. Fruit began to show signs of a trade deficit in 2018, reaching US $2.91 billion in 2019, an increase of 1.3 times over the previous year. Although aquatic products maintained a trade surplus of US $1.96 billion, they decreased significantly by 74.1% from the previous year. Table 2 below shows the changes in China's export volume of various agricultural products from 2015 to 2019, indicating that China's export income of agricultural products did not maintain stable growth in recent years. Affected by trade barriers, since 2017, China's exports of aquatic products and vegetable as well as agricultural a product have fluctuated and decreased by a considerable range.

After being heavily impacted by Sino-US economic and trade frictions, imports of agricultural products from the United States decreased from US $16.23 billion in 2018 to US $14.16 billion, recording a drop of 12.8%, and exports to the United States too decreased by 22.1%. In sharp contrast, agricultural cooperation between China and countries along the "Belt and Road Initiative" route has achieved great results. In 2019, the total trade volume of agricultural products between China and the countries along the "Belt and Road Initiative" route reached 60.89 billion US dollars, an increase of 16.5 percent over

Table 2.

Export Value of Goods

Indicators	2015	2016	2017	2018	2019
Export value of aquatic products (million of US dollars)	19567. 63	19996. 27	20407. 4	22001. 36	20326. 57
Export value of vegetables (million of US dollars)	10708. 29	12294. 67	13152. 37	12615. 04	12566. 72
Export value of fresh or frozen vegetables (million of US dollars)	4443. 27	5406. 98	5308. 5	4637. 16	5543. 52

Note: Import and export data are from General Administration of Customs.

the same period last year which proved to be 10.3 percentage points higher than the overall trade growth rate of China's agricultural products. Volume of imports rose up to US $60.88 billion, representing an upward trend of 23.3 percent, and exports further increased up to US $27.01 billion, signifying an upward trend of 9.1 percent. The growth rate of imports and exports was 13.1 and 9.8 percentage points higher than the overall growth rate, respectively (Sun et al., 2020).

According to the 2019 China Agricultural Economic Development report and Prospect of China Economic Network, the trade volume of China's agricultural products is expected to undergo a rapid change in the future (Han et al., 2020). As the growth of imports will dominate the growth of agricultural trade, the net imported products will continue to rise (Xu, 2020). As far as vegetable production is concerned, the same would stabilize, the consumption patterns will optimize and the growth rate of fruit production and consumption would slow down. The large-scale breeding of pork is expected to accelerate, and net imports are expected to decrease. Poultry production and consumption will provide much needed stability to imports; with beef and mutton consumption driving the rapid growth of production leading to beef imports continuously increasing. The growth rate of poultry and egg production is expected to slow down, and the costs are expected to lead to inflation in prices as a whole. While the milk production will enter a period of steady growth, but imports will slow down. The growth rate of production of aquatic products will slow down, and the consumption patterns will lead to an increase in imports. All these factors will push the market to become vulnerable and weak (Zeng, 2020).

Feasibility Analysis

Development of Smart Agriculture and Digital Agriculture

Amidst these challenges, the development of intelligent agriculture and digital agriculture is a great hope that will significantly promote the ecological and sustainable development of agriculture. Digital agriculture that assumes widespread significance in wake of transition to digital economy, takes agricultural big data as the core production factor and uses digital information technology to manage agricultural objects and the whole process of environment, production, sales and management. For instance, it includes usage of robotics in dairy, poultry and beef farming where technology helps in autonomous feeding and milking, egg collection and sorting, autonomous cleaning and early detection of diseases in animals; use of new packaging material, use of biotechnology for data measurement, weather monitoring etc. Basically, it is a new model for rural development to promote rural revitalization, and serves as an important measure to promote the full-cycle transformation of the digital industry chain in agricultural and rural areas (Chen & Sun, 2020). Digital agriculture has expanded the "production + sale" service system of traditional agriculture to promote the commercialization and market-oriented transformation of agriculture, covering data acquisition and transmission (UAV, monitoring, agricultural Internet of things), industrial chain and supply chain analysis, precision and intelligent management, asset optimization and risk management, ecological environmental protection and rural governance (Xu et al., 2021).

Sustainable agricultural practices "SAP" (including crop rotation, improved varieties and residue retention), the safe, smart and sustainable agriculture project, and Good Agriculture Practices (GAP) have made a major breakthrough as well. Studies that have been conducted to map the impact of SAP, have reached a consensus that adoption of SAP can improve agricultural economy and environmental performance significantly (e.g. Adolwa et al., 2019; Kassie et al., 2013; Ma and Abdulai, 2019; Midingoyi et al., 2019; Ndiritu et al., 2014; Rodriguez et al., 2009; Tambo and Mockshell 2018; Teklewold et al., 2013; Wossen et al., 2015; Zeweld et al., 2017).

The digital transformation of agriculture can better solve the efficiency (Liu, 2020), output, quality, product traceability and so on, especially as a result of artificial intelligence (AI) and the Internet of things to better meet the needs of production (Xu, 2015). Digital agriculture has potential to contribute significantly leading to economic, social and environmental benefits, and if promoted in an inclusive manner, it can contribute to the achievement of sustainable development goals and the maintenance as well as improvement of agricultural production.

Smart agriculture has also opened up a new track for digital agriculture in China. It is reported that in May 2021, SAP[1] announced a strategic cooperation with Beijing Zhongdi Industrial Holdings Co., Ltd. With the help of SAP S/4HANA digital system, the two sides will help Zhongdi Xingye transform and upgrade in the fields of intelligent agriculture, smart pastures and intelligent factories, empower product channels, create new brands, and realize the integrated development of the whole industry chain.

Development of "Internet Plus" Agriculture in China

Transition to Internet plus agriculture in Rural economy has provided a great thrust to the movement of Sustainable Development Agriculture. Application of mobile internet, Internet of Things, cloud computing and big data has resulted into great improvement in agricultural production, processing, trade, services, consumption and tourism in rural areas (Yang, 2018). The integration of traditional agriculture and emerging technology indeed has enabled traditional agriculture to enter a new phase. Intelligent agriculture breaks the existing agricultural production mode and combines Internet of things, technology, 3s (remote sensing, positioning, geographic information) technology, wireless communication technology and high-end agricultural science and technology to achieve a series of precise and intelligent control processes, such as remote control of agricultural production, remote monitoring of agricultural growth environment, and early warning of disasters (Xi, 2019).

In addition, "Internet of things + agriculture" is another market-oriented feature of sustainable smart agriculture (Liu, 2017). With the basic support of the "Internet of things + Agriculture", agriculture and agricultural products realize the informationization of agricultural management by using modern information technology to realize the transparency and integration of production, logistics and sales, and break the geographical restrictions on the circulation of agricultural products. It has greatly reduced the gap between the production and the consumer end, greatly improved the circulation of agricultural products, and has resulted into a multi-sales model with B2B, B2C and O2O as the main representatives (Ye, 2014).

Business big data monitoring shows that online retail sales in rural area reached 1.79 trillion yuan in 2020, an increase of 8.9 percent over the same period last year. E-commerce has accelerated the industrialization and digital development of agriculture, as a result of which, there has been an effective rural revitalization and poverty alleviation. In the first quarter of 2021, China's online retail market maintained steady growth, being strongly driven by consumption. Not only this, but rural e-commerce has also opened up new opportunities for rural businesses to thrive and innovate thus promoting both local and global sales of agricultural products (Wu, 2021).

The Ministry of Commerce will continue to subsidize the certification of "three products and one standard", build the brand of agricultural products, cultivate rural e-commerce talents, and promote the development of rural e-commerce industrialization (Liu, 2020). According to data from the Ministry of Commerce, as shown in Figure 7, rural online retail sales reached 180 billion yuan in 2014, broke through the trillion mark in 2017, and reached 1.79 trillion yuan in 2020, an increase of 8.9 percent over the same period last year. The e-commerce of agricultural products has flourished only on the basis of the innovative practices in the domain of Internet and the Internet of things.

OPPORTUNITIES AND CHALLENGES IN PROMOTING SUSTAINABLE AND ECOLOGICAL AGRICULTURAL DEVELOPMENT IN CHINA

Opportunity

Policy Factors

In terms of policy, on May 27, 2015, China officially released the National Plan for Sustainable Agricultural Development (2015-2030). This is a programmatic document issued by the government of China, laying down a detailed plan for the sustainable development of agriculture towards 2030. The "Plan" comprehensively considers the resource carrying capacity, environmental capacity, ecology and development of agriculture in various parts of the country, and divides the country into three major areas, namely, optimized development areas, moderate development areas, and protection and development areas. Any measures decided to be implemented, will only succeed if they are in tandem with local conditions, climate and other factors (National Agricultural Sustainable Development Plan (2015-2030), 2015). Under this program, the sustainable development of agriculture has got a clear policy direction.

In 2016, intelligent agriculture was introduced and soon it became a part of important national documents, indicating that intelligent agriculture indicating the wide attention it is getting from the perspective of China's strategic development (Tian et al., 2018). In 2017, with the strong support of the government, the number of rural Internet users in China exceeded 200 million (*China Internet Development Report 2019*). In 2018, a pilot project on the application of big data technology to the production of agricultural products was carried out in 21 provinces and cities across the country, realizing real-time monitoring of Internet technology and guiding the production and marketing of agricultural products. 14 provinces also carried out pilot projects on rural e-commerce, achieving remarkable results in poverty alleviation (Zeng, 2020).

In October 2020, the proposal of the CPC Central Committee on the formulation of the 14th five-year Plan for National Economic and Social Development and the long-term goal of 2035 clearly pointed out that we should vigorously build intelligent agriculture, strengthen rural reform, and develop a new type of rural economy to promote all round rural revitalization. With the strong support of national policies, China's intelligent agriculture has the potential to usher in excellent opportunities for development. The development of rural economy can inject new life into the economy.

Promotion of Internet Big Data

"Internet + Agriculture" can play a great role in promoting the sustainable development of rural economy and can contribute significantly in ridding the farmers of poverty. After China entered the information age, e-commerce has gradually become the main consumption channel, which provides a great strength to the development of rural e-commerce industry (Lin, 2018). At present, the scale of rural development as a result of e-commerce is expanding and "Internet + Agriculture" and "Internet of things + Agriculture" have both proved to be strong links between agricultural resources and the market.

Big data can help China's agriculture to build an intelligent Internet of things. At present, China advocates the combination of fine production of modern agriculture and big data technology in agriculture, so as to expand the actual demand in the market. Under the influence of big data in agriculture, in the process of production and monitoring of agricultural products, we have been able to grasp the weather change, market supply and demand and crop growth and other information, which has greatly reduced the economic burden. And under the construction of big data's intelligent platform, the relevant agro-technical experts can also monitor crop growth timely and accurately, and grasp the yield and quality of agricultural products (Wang et al., 2020). In terms of the sales and operation of crops, big data helped China's agriculture effectively create an omni-directional and corner-free management system, improved the supervision of every detail of China's agriculture in economic management, improved the detection of fluctuations in market sales, enhanced the sales rate of crops, and even reduced the unsalable rate of crops (Lan, 2021).

Challenges

Global Policy Challenges

This entire chain proves that how when we work towards one sustainable development goal, we can achieve the other one as a byproduct and a process. China's innovative practices to adopt sustainable way of agricultural development will directly contribute to realizing Sustainable Development Goal 2 of removing poverty and hunger as a natural byproduct. But at the same time, conflicts may also arise between different policies, thus hindering the sustainable development of agriculture. For example, there may be a tradeoff between promoting economic growth to reduce poverty and achieving the goal of "protecting, restoring and promoting the sustainable use of terrestrial ecosystems" (Sachs et al., 2016). International trade policies and domestic policies for the development of other industries are likely to collide with policies to promote the development of ecological agriculture. With regard to the goal of ecological agriculture, some solutions can create more friendly synergies, while others may increase tensions. This requires policy makers to weigh the impact of their decisions on sustainable agricultural development as well as people who are a part of this movement again and again when issuing guidelines and laws. This will require the integration of a large and complex trade network involving economic, social, environmental, cultural and security issues (Nilsson et al., 2016; Waage et al., 2015).

But when the focus and the heart of all policies becomes the development and growth of people; not leaving any precious life behind, then definitely the most appropriate solutions and strategies in policy making will be adopted. Since people are the ones who are going to be impacted by outcome of any plan and since people are the one who are implementing the plans, it can be said that the more policies are oriented towards enabling each individual today and in future, live respectfully in a dignified manner, more they will tend to succeed.

Inadequate resources and capacity to implement and coordinate sustainable development goals may also hamper the development of sustainable agriculture (Xue et al., 2018). The implementation of sustainable agricultural development is a goal and tests the management capacity of many countries. While the market and social sectors have an important role to play in the implementation of sustainable development goals, the governments in collaboration with various stakeholders needs to play a leading role in raising awareness, mobilizing resources and implementing strategies that are oriented towards overall prosperity of people. The lack of adequate management capacity is in itself a common symptom of underdevelopment in many countries but if the management itself is made to understand that the resulting good or bad outcome of any policy or strategy will impact their life as it would impact any other citizen's life; the way management capacity expands will be phenomenal.

The Loss of Labour Force

The transfer of agricultural labor force in China has experienced two stages: "leaving the land without leaving the hometown" and "leaving the land and leaving the hometown again" (Xie & Xu, 1988). In the 1980s, with the reform of the rural household contract responsibility system, agricultural productivity greatly improved, and a large number of agricultural surplus labor was released. However, limited by the household registration policy at that time, farmers could only stay in rural areas for development (Zhang & Zhang, 2003). At that point in time, the sudden rise of township and business enterprises absorbed the vast majority of surplus labor. In the 1990s, with the growth of market-oriented reform, a large number of farmers began to leave their hometown to work in developed coastal areas, as a result of which a huge wave of "migrant workers" appeared.

In the 21st century, with the industrial transformation and upgradation of the coastal areas, a large number of labor-intensive industries were transferred to the inland areas of the central and western regions, and farmers began to work in the central cities of the province (Yu & Zhan, 2006). With the rapid growth of China's economy, hundreds of millions of farmers were transferred from the agricultural sector (Liu, 2015).

For some time, the characteristics of China's rural areas and farmers engaged in agricultural cultivation and management could be described as follows: rich population, small scale farming and low rural household income. The limited funding opportunities for farmers and low household income restricted the development and mechanization of China's rural agriculture (Zhu & Zhao, 2011). However, with the rapid economic growth in the 1990s, the demand for rural labor in the non-agricultural sector increased. Working in cities was considered to be beneficial as it brought along more economic benefits than staying in the countryside and continuing to engage in agriculture. At present, the vast majority of young and middle-aged workers in China's rural areas have migrated outside for work, and those who remain in the rural areas are middle-aged and elderly farmers and some preschool children who are too young with a low level of education. The massive transfer of rural labor force has led to the rapid growth of rural wages and the opportunity cost of agricultural labor. These changes have led farmers to revamp the scenario by introducing a certain degree of mechanization to replace the labour force (e.g., Huang & Rozelle 1996; McNamara & Weiss 2005). Coupled with the development of science and technology and intelligent agriculture in recent years, the degree of agricultural mechanization has improved.

With the improvement in the agricultural mechanization, the national policy has also put forward higher requirements for the ecological and sustainable development of agriculture. However, in order to achieve the above expectations, a large number of young and qualified labor force is required to participate in agricultural production, but this conflicts with a large number of agricultural labor force moving to cities and other industries.

At the present stage, one of the main sticking points of the high production cost of agricultural products in China is the labor cost. Young and educated labour force tends to work more frequently in the non-farm labour market, and non-farm employment reduces the possibility and intensity of agricultural production. Production is related to the low level of non-agricultural employment (Du, 2017). In the current situation marred by inflationary prices of all kinds of agricultural means of production, the labor cost in agricultural production is also rising. This makes the cost of agricultural production unaffordable, and the profits of agricultural products are also affected likewise (Li & Wang, 2020). Moreover, the high-tech equipment used in the development of ecological agriculture needs to be managed and controlled by qualified experts. Nowadays, most of the farmers engaged in agricultural production have not received professional study and training, which seriously impacts the way technology is being put to use thereby impacting the overall quantity and quality of agricultural products.

Generally speaking, the education level of farmers in China is limited, and there are only a handful of farmers who have expertise to use technology for agricultural production. There is also a serious challenge of ageing of labour force that impacts the productivity of agriculture in China (Yang, 2014). Some scholars have found that the ageing of agricultural labor force will not only reduce the efficiency of cultivated land use, but also hinder the flow of land, resulting in the inability to make full use of agricultural resources. This is disadvantageous to scale operation, which leads to extensive abandonment

(Zhou & Lu, 2014). Farmers' lack of professional knowledge in agricultural production has set a direct bottleneck for the improvement of both the quantity and quality of agricultural products.

CONCLUSION AND RECOMMENDATIONS

Sustainable Development of Agriculture plays an important role in promoting the sustainable development of any economy. On the whole, under the regulation and control of a series of policies of the Chinese central government and the strong support for the sustainable development of agriculture, the trend of stable growth of agricultural production along with stable supply and demand of agricultural products will continue to prosper at the same pace. Coupled with the promotion of intelligent agriculture by the development of the Internet and the transformation of agriculture to digitalization, we can be more optimistic about sustainable development of agriculture in the future.

However, if we need to make a breakthrough in the future, we still need to vigorously develop and intensify the movement of protecting and strengthening intellectual property rights that will stimulate innovative methods and techniques in this sector. Innovative ways of substituting harmful pesticides and insecticides with biopesticides that are environment friendly, innovation in technology to monitor weather patterns as well as taking care of livestock, innovation in plant variety and seed breeding, usage of local and traditional knowledge as a means to add greater value and educating the labour and work force simultaneously. While preventing the crisis of climate and environmental quality, we should pay attention to the shortage of soil and water resources and improve the efficiency of resources which again would require skilled workforce and development of appropriate technology. We need to pay close attention and create full optimum value out of the major opportunities brought about by sustainable development of agriculture through spread of digital information, and science.

At the same time, we also need to think about such questions in order to prepare ahead of time: for example, how should we weigh the pros and cons and adjust policies and guidelines when global and domestic policies in the time of conflict with the ecological development of agriculture? With the increase in non-farm employment leading to a continuous decline in the agricultural labor force, can China continue to make use of its demographic dividend to realize its comparative advantage in labor-intensive agriculture? What is the possible impact of the decline in non-farm employment on technology adoption? The answers to these questions are important because they will have a significant impact not only on China's domestic production and the welfare of the rural population, but also on other sectors, such as industry and services (which will require labour in the rural economy) and international markets (a signal to China that China should produce and export large amounts of labour-intensive fruits and vegetables) (Huanga et al., 2008).

REFERENCES

Adolwa, I. S., Schwarze, S., & Buerkert, A. (2019). Impacts of integrated soil fertility management on yield and household income: The case of Tamale (Ghana) and Kakamega (Kenya). *Ecological Economics*, *161*, 186–192. doi:10.1016/j.ecolecon.2019.03.023

Cao, D., Li, Y. Y., Liu, B. H., Kong, F. J., & Tran, L. S. P. (2018). Adaptive mechanisms of soybean grown on salt-affected soils. *Land Degradation & Development*, *29*(4), 1054–1064. doi:10.1002/ldr.2754

Chen, S. Y., & Sun, J. K. (2020). Creating something out of nothing and changing lanes to be driven by innovation— Enlightenment from the investigation and study of the development of digital economy in Guizhou Province. *Yunnan Science and Technology Management*, *033*(002), 34–36.

Chen, Z. m. (1994). Discussion on rational use of pesticides to avoid resistance of pests. *Agricultural Science and Technology*, 4.

Cheng, Y. F., & Pan, Y. J. (2005). The area of cultivated land may lead to the problem of food and clothing. *Well-off*, (2), 56–56.

China Meterological Administration. (2016). *The latest related information of drought and flood disasters.* http://www.cma.gov.cn/

China National Knowledge Infrastructure. (2021). *Sustainable development of agriculture*. https://www.cnki.net/gycnki/gycnki.htm

China's National Bureau of Statistics. (2019). *Agricultural water consumption data.* https://data.stats.gov.cn/index.htm

China's National Bureau of Statistics. (2020). *Data on the use of agricultural chemical fertilizers and pesticides.* https://data.stats.gov.cn/index.htm

Hou, D. Y., Ding, Z. Y., Guang, H., & Long, H. (2018). A sustainability assessment framework for agricultural land remediation in China. *Land Degradation & Development*, *29*(4), 1005–1018. doi:10.1002/ldr.2748

D'Odorico, P., Davis, K. F., Rosa, L., Carr, J. A., Chiarelli, D., Dell'Angelo, J., Gephart, J., MacDonald, G. K., Seekell, D. A., Suweis, S., & Rulli, M. C. (2018). The global food-energy-water nexus. *Reviews of Geophysics*, *56*(3), 456–531. doi:10.1029/2017RG000591

Fan, X., Pedroli, B., Liu, G., Liu, Q., Liu, H., & Shu, L. (2012). Soil salinity development in the Yellow River Delta in relation to groundwater dynamics. *Land Degradation & Development*, *23*(2), 175–189. doi:10.1002/ldr.1071

Forward-looking industry research institute. (2020). *In 2020 the development of Chinese agricultural products import and export industry market size and trend analysis*. https://www.sohu.com/a/413530053_473133

Zhong, F. N., Zhu, J., & Xie, Z. Q. (2007). How Mobile Are Resources in Chinese Agriculture? — Implications for China's Agricultural Trade Policy. China & World Economy, 15(1), 43 - 61.

Gao, P., Wang, Y. M., Li, P. F., Zhao, G. J., Sun, W. Y., & Mu, X. M. (2018). Land degradation changes in the Yellow River Delta and its response to the streamflow-sediment fluxes since 1976. *Land Degradation & Development*, *29*(9), 3212–3220. doi:10.1002/ldr.3049

Han, Y. P., Qu, C., & Jia, D. D. (2019). Analysis of water footprint and water consumption structure of main crops in Hebei province. *Guangai Paishui Xuebao*, 10.

Himabindu, Y., Chakradhar, T., Reddy, M. C., Kanygin, A., Redding, K. E., & Chandrasekhar, T. (2016). Salt-tolerant genes from halophytes are potential key players of salt tolerance in glycophytes. *Environmental and Experimental Botany*, *124*, 39–63. doi:10.1016/j.envexpbot.2015.11.010

Huang, G. Q., Wang, X. X., Qian, H. Y., Zhang, T. L., & Zhao, Q. G. (2004). Negative effects of chemical fertilizer application on agricultural ecological environment and countermeasures. *Ecology & Environment*.

Huang, J., Wu, Y., & Rozelle, S. (2010). Moving off the farm and intensifying agricultural production in shandong: A case study of rural labor market linkages in china. *Agricultural Economics*, *40*(2), 203–218. doi:10.1111/j.1574-0862.2009.00370.x

Huang, J., & Rozelle, S. (1996). Chinese rice production potential, consumption and trade. *Journal of Chinese Rural Economy*, (4), 21–27.

International fruit and vegetable mesh. (2021). *Henan crops the affected area of over 15 million mu, rejection of 2.697 million mu*. https://www.sohu.com/a/480242165_617283

Jenny, H. (1980). *The soil resource: Origin and behavior*. Springer Science & Business Media.

Bhaduri, D., Saha, A., Desai, D., & Meena, H. N. (2016). Restoration of carbon and microbial activity in salt-induced soil by application of peanut shell biochar during short-term incubation study. *Chemosphere*, *148*, 86–98.

Zhan, J. T., Xu, T., Zhang, Y. Y., Yang, X. L., Qu, Z. Q., & Tao, T. (2017). The Effects of Agricultural R&D on Chinese Agricultural Productivity Growth: New Evidence of Convergence and Implications for Agricultural R&D Policy. *Canadian Journal of Agricultural Economics*, *65*(3), 453–475. doi:10.1111/cjag.12137

Kassie, M., Jaleta, M., Shiferaw, B., Mmbando, F., & Mekuria, M. (2013). Adoption of interrelated sustainable agricultural practices in smallholder systems: Evidence from rural Tanzania. *Technological Forecasting and Social Change*, *80*(3), 525–540. doi:10.1016/j.techfore.2012.08.007

Lan, H. Y. (2018). Research on the important impact of big data on agricultural economic management in the process of building smart agriculture. *Modern Business*, *2021*(05), 40–42.

Li, G. X. (2019). Several issues to be considered in promoting the Revitalization of Rural Industry interpret the guidance on promoting the Revitalization of Rural Industry. *China Rural Science and Technology*, (8), 8–11.

Li,, X., & Jiang,, W., & Duan, D. (2020). Spatio-temporal analysis of irrigation water use coefficients in China. *Journal of Environmental Management*, 262. PMID:32090879

Li, X. H., & Wang, Y. S. (2020). Problems and countermeasures of sustainable development of Agricultural economy in China. *Journal of Business Culture*, *471*(18), 24-25.

Li, H., Zeng, E. Y., & Jing, Y. (2014). Mitigating pesticide pollution in china requires law enforcement, farmer training, and technological innovation. *Environmental Toxicology and Chemistry*, *33*(5), 963–971. doi:10.1002/etc.2549 PMID:24753037

Li, J., Pu, L., Zhu, M., Zhang, J., Li, P., Dai, X., Xu, Y., & Liu, L. (2014). Evolution of soil properties following reclamation in coastal areas: A review. *Geoderma, 226*, 130–139. doi:10.1016/j.geoderma.2014.02.003

Liang, C., Li, D., Yuan, Z., Liao, Y., Nie, X., Huang, B., Wu, X., & Xie, Z. (2019). Assessing urban flood and drought risks under climate change, China. *Hydrological Processes, 33*(9), 1349–1361. doi:10.1002/hyp.13405

Lin, M. (2018). Exploration and construction of rural e-commerce development model -- based on the investigation of E-commerce development in Wugong County. *Enterprise Reform and Management, 328*(11), 71-73.

Liu, D., Gong, Q. W., & Yang, W. J. (2019). The evolution and optimization path of China's cultivated land protection policy in the past 40 years of reform and opening up. *Zhejiang Land and Resources, 189*(01), 21–22.

Liu, S. J. (2015). Key areas to improve productivity under the new normal economy. *The Trends of China's Industrial Economy,* (7), 7-9.

Liu, W. Y. (2020). Use Internet thinking to promote the transformation and upgrading of walnut industry. *Horticulture in the North,* (16).

Liu, Y. D. (2017). *Internet of Things and smart agriculture*. Hohai University.

Liu, Y. S. (2020). Efficiency analysis and countermeasures of agricultural digital transformation. *Economic Review (Kansas City, Mo.).*

Liu, G. M., Zhang, X. C., Wang, X. P., Shao, H. B., Yang, J. S., & Wang, X. P. (2017). Soil enzymes as indicators of saline soil fertility under various soil amendments. *Agriculture, Ecosystems & Environment, 237*, 274–279. doi:10.1016/j.agee.2017.01.004

Liu, Y., Hu, W., S Jetté-Nantel, & Tian, Z. (2014). The influence of labor price change on agricultural machinery usage in chinese agriculture. *Canadian Journal of Agricultural Economics/Revue canadienne d\"agroeconomie, 62*(2), 219-243.

Ma, W., & Wang, X. (2020). Internet use, sustainable agricultural practices and rural incomes: Evidence from china. *The Australian Journal of Agricultural and Resource Economics, 64*(4), 1087–1112. doi:10.1111/1467-8489.12390

Ma, W., & Abdulai, A. (2019). IPM adoption, cooperative membership and farm economic performance: Insight from apple farmers in China. *China Agricultural Economic Review, 11*(2), 218–236. doi:10.1108/CAER-12-2017-0251

McNamara, K. T., & Weiss, C. (2005). Farm household income and on and off-farm diversification. *Journal of Agricultural and Applied Economics, 37*(1), 37–48. doi:10.1017/S1074070800007082

Meng, L. (2009). *Study on the influence of RMB appreciation on the export trade of agricultural products in China* [Doctoral dissertation]. Hunan Agricultural University.

Midingoyi, S. G., Kassie, M., Muriithi, B., Diiro, G., & Ekesi, S. (2019). Do farmers and the environment benefit from adopting integrated pest management practices? Evidence from Kenya. *Journal of Agricultural Economics*, *70*(2), 452–470. doi:10.1111/1477-9552.12306

Mielke, E., Diaz Anadon, L., & Narayanamurti, V. (2010). *Water consumption of energy resource extraction, processing, and conversion. Discussion paper for Energy Technology Innovation and Policy*. Harvard Kennedy School.

Mose, A. A., Mostaf, A., & David, J. B., & Lam-Son, P. T. (2018). Salt stress tolerance mechanisms and potential applications of legumes for sustainable reclamation of salt-degraded soils. *Land Degradation & Development*, *29*, 3783–3794.

Ndiritu, S. W., Kassie, M., & Shiferaw, B. (2014). Are there systematic gender differences in the adoption of sustainable agricultural intensification practices? Evidence from Kenya. *Food Policy*, *49*, 117–127. doi:10.1016/j.foodpol.2014.06.010

Nilsson, M., Griggs, D., & Visbeck, M. (2016). Policy: Map the interactions between Sustainable Development Goals. *Nature*, *534*(7607), 320–322. doi:10.1038/534320a PMID:27306173

Qiang, L., Huang, J., Luo, R., & Liu, C. (2013). China's labor transition and the future of china's rural wages and employment. *China & World Economy*, (03), 4–24.

Robinson, G. M. (2009). Towards sustainable agriculture: Current debates. *Geography Compass*, *3*(5), 1757–1773. doi:10.1111/j.1749-8198.2009.00268.x

Rodriguez, J. M., Molnar, J. J., Fazio, R. A., Sydnor, E., & Lowe, M. J. (2009). Barriers to adoption of sustainable agriculture practices: Change agent perspectives. *Renewable Agriculture and Food Systems*, *24*(1), 60–71. doi:10.1017/S1742170508002421

Silver, R. (2005). The gap and policy between China's agricultural products processing technology and developed countries. *Scientific Management Research*, *023*(006), 93–96.

Sun, S. K., Yin, Y. L., Wu, P. T., Wang, Y. B., Luan, X. B., & Li, C. (2019). Geographical Evolution of Agricultural Production in China and Its Effects on Water Stress, Economy and the Environment: The Virtual Water Perspective. *Water Resources Research*, *55*(5), 4014–4029. doi:10.1029/2018WR023379

Sachs, J., Schmidt-Traub, G., Kroll, C., Durand-Delacre, D., & Teksoz, K. (2016). *SDG index and dashboard-global report*. Bertelsmann Stiftung and Sustainable Development Solutions Network.

Shao, H., Chu, L., Lu, H., Qi, W., Chen, X., Liu, J., Kuang, S., Tang, B., & Wong, V. (2019). Towards sustainable agriculture for the salt-affected soil. *Land Degradation & Development*, *30*(5), 574–579. doi:10.1002/ldr.3218

Song, L. C. (1991). Analysis on the variation and influence of drought and flood during the active period of summer monsoon in China. *Disaster Science*, (03), 23–28.

Sun, S. F., Xin, X., Liu, M., Lu, Z. W., Qin, Y., Cai, H. L., & Ma, L. (2019). *China agricultural Economic Development Report and outlook*. China Economic Daily.

Tambo, J. A., & Mockshell, J. (2018). Differential impacts of conservation agriculture technology options on household income in Sub-Saharan Africa. *Ecological Economics*, *151*, 95–105. doi:10.1016/j.ecolecon.2018.05.005

Teklewold, H., Kassie, M., & Shiferaw, B. (2013). Adoption of multiple sustainable agricultural practices in rural Ethiopia. *Journal of Agricultural Economics*, *64*(3), 597–623. doi:10.1111/1477-9552.12011

Ye, T., Ye, L., Wang, S., Feng, Z. M., Dong, L. M., & Yin, Y. L. (2018). Building agricultural products traceability system to promote the development of smart agriculture. *High Technology and Industrialization*, *000*(005), 42–45.

Wang, T. M., Wu, J. G., Kou, X. J., Oliver, C., Mou, P., & Ge, J. P. (2010). Ecologically asynchronous agricultural practice erodes sustainability of the loess plateau of china. *Ecological Applications*, *20*(4), 1126–1135. doi:10.1890/09-0229.1 PMID:20597295

Tilman, D., Balzer, C., Hill, J., & Befort, B. L. (2011). Global food demand and the sustainable intensification of agriculture. *Proceedings of the National Academy of Sciences of the United States of America*, *108*(50), 20260–20264. doi:10.1073/pnas.1116437108 PMID:22106295

Waage, J., Yap, C., Bell, S., Levy, C., Mace, G., Pegram, T., Unterhalter, E., Dasandi, N., Hudson, D., Kock, R., Mayhew, S., Marx, C., & Poole, N. (2015). Governing the UN Sustainable Development Goals: Interactions, infrastructures, and institutions. *The Lancet. Global Health*, *3*(5), e251–ee252. doi:10.1016/S2214-109X(15)70112-9 PMID:25833663

Wang, F., Chen, W. K., & Chen, J. A. (2008). The Construction of New Countryside and the Development of Modern Agriculture in underdeveloped areas-- A Summary of the National Annual meeting of Young and Middle-aged Agricultural economists in 2007. *Agricultural Economic problems*, (1), 107-109.

Wang, G. R. (2017). *Study on benefit Evaluation and Popularization of High efficiency Water use Technology Model for main crops in Hebei Province* (Doctoral dissertation). Northeast Agricultural University.

Wang, W. H. (2016). Soil control is in action. *China Rural Science and Technology*, (10), 52–55.

Wang, X., Yu, J. F., Zhao, X. L., Huo, W. B., Qi, P., & Li, Y. H. (2020). *The invention relates to a management system based on agricultural data and production and an agricultural data monitoring terminal.* CN111352370A.

Wang, Z. G. (2004). Measurement of rural marketization degree in China in 2002 and its influencing factors. *Annual conference of Chinese Society of Agricultural Economics.*

Wang, J., Mendelsohn, R., Dinar, A., Huang, J., & Zhang, L. (2010). The impact of climate change on china's agriculture. *Agricultural Economics*, *40*(3), 323–337. doi:10.1111/j.1574-0862.2009.00379.x

Wang, X., Yamauchi, F., & Huang, J. (2016). Rising wages, mechanization, and the substitution between capital and labor: Evidence from small scale farm system in china. *Agricultural Economics*, *47*(3), 309–317. doi:10.1111/agec.12231

Wossen, T., Berger, T., & Di Falco, S. (2015). Social capital, risk preference and adoption of improved farm land management practices in Ethiopia. *Agricultural Economics*, *46*(1), 81–97. doi:10.1111/agec.12142

Wu, Y. N. (2021). *2020 rural online retail sales reached 1.79 trillion yuan. China's e-commerce market continues to lead the world.* https://www.chinairn.com/news/20210124/121901569.shtml

Xi, X. J. (2019). Research on the application of Internet of things production mode in intelligent agriculture. Scientific and technological. *Innovation*, (33), 77–78.

Xie, Y. K., & Xu, Y. M. (1988). Agricultural labor transfer and urbanization in China. *Journal of Renmin University of China*, (01), 80–86.

Xu, F. (2020). Analysis on the import trade structure of China's agricultural products-- ternary decomposition based on trade growth. *China's Circulation economy*, *34*(6), 96-104.

Xu, K., Yang, H. J., & Lu, Z. Y. (2021). The Prospect of the Development of Digital Agriculture during the 14th five-year Plan. *China Engineering Consulting*, (4), 4.

Xu, Z. Q. (2015). Development trend of China's manufacturing informatization based on the Internet of Things era. *Economic Management (Abstracts)*, (8), 00283–00283.

Xue, L., Weng, L., & Yu, H. (2018). Addressing policy challenges in implementing Sustainable Development Goals through an adaptive governance approach: A view from transitional China. *Sustainable Development*, *26*(2), 150–158. doi:10.1002d.1726

Wanowaa, Furaha, & Yan. (2020). Effects of Planting Field on Groundwater and Surface Water Pollution in China. *CLEAN - Soil Air Water.*

Yan, H. H., & Peng, S. Q. (2009). CREB signaling pathway mediates changes in cognitive function induced by organophosphorus pesticides in rats and its possible mechanism. *The Fifth National Conference of Chinese Society of Toxicology*.

Yang, G., Ding, L., Huo, L., & Hu, Y. (2021). Coordinated development of agricultural water resources and the socio-economy in Shanxi province considering uncertainty. *Irrigation and Drainage*, *2021*(4), 1–10. doi:10.1002/ird.2581

Yang, H. P. (2018). Research on the Promotion of Agricultural products under the Mode of Internet + Agriculture-- taking native products as an example. *China Management Informatization*, *376*(10), 147-148.

Yang, J. (2014). *The impact of "aging" agricultural Labor force on Agricultural Production* (Dissertation). Northwest Normal University.

Ye, G. X. (2014). High-fidelity Agricultural Products E-commerce development analysis and research in Sichuan Province. *Sichuan Agriculture and Agricultural Machinery*, (4), 12–12.

Ye, Y., Jiang, S., Wen, T., & Tang, P. (2020). *The method and device of big data analysis in agriculture.* CN111582324A.

Yu, S. Y., & Zhan, Y. P. (2006). The phenomenon of shortage of migrant workers and the regional transfer of labor-intensive industries in China. *Business Research*, (24), 105–107.

Yu, Y., Lian, X. Y., & Jiang, Y. H. (2017). Risk-based prioritization method for the classification of groundwater pesticide pollution from agricultural regions. *Integrated Environmental Assessment and Management*, *13*(6), 137. PMID:28577331

Yu, C., Huang, X., Chen, H., Huang, G., Ni, S., Wright, J. S., Hall, J., Ciais, P., Zhang, J., Xiao, Y., Sun, Z., Wang, X., & Yu, L. (2018). Assessing the impacts of extreme agricultural droughts in China under climate and socioeconomic changes. *Earth's Future*.

Liu, Y., Li, N., Zhang, Z. T., Huang, C. F., Chen, X., & Wang, F. (2020). Climate change effects on agricultural production: The regional and sectoral economic consequences in China. *Earth's Future, 6*(5), 689–703. doi:10.1002/2017EF000768

Yuan, S., & Qi, W. (2019). The predicament and way out of promoting the value of rural land resources from the perspective of rural revitalization. *The 16th Shenyang Science Annual Meeting*.

Yue, D., Jiang, W., He, B., Zheng, C., & Kai, J. (2018). Change in intensity and frequency of extreme precipitation and its possible teleconnection with large-scale climate index over the China from 1960 to 2015. *Journal of Geophysical Research, D, Atmospheres, 123*(4), 2068–2081. doi:10.1002/2017JD027078

Zeng, S. Q. (2020). Information leads the Future-- an Analysis of the China Agriculture Outlook report (2020-2029). *Agricultural Products Market, 939*(9), 8-10.

Zeng, Y. S. (2020). The long-term goal and strategic plan of agricultural and rural modernization. *Social Governance, 55*(11), 22-24.

Zeweld, W., Van Huylenbroeck, G., Tesfay, G., & Speelman, S. (2017). Smallholder farmers' behavioural intentions towards sustainable agricultural practices. *Journal of Environmental Management, 187*, 71–81. doi:10.1016/j.jenvman.2016.11.014 PMID:27883941

Zhang, H. K. (2014). Imbalance of China's agricultural products international trade and development countermeasures. *Shopping Mall Modernization, 000*(033), 2–2.

Zhang, L. S., & Zhang, P. L. (2003). stimulating rural economy; Increasing rural income. *Construction of Small Towns, 000*(004), 6–8.

Zhang, Y. Q. (2019). Environmental impact analysis of high-efficiency water-saving irrigation in groundwater over-exploitation area. *Henan Water Conservancy and South-to-North Water Transfer, 338*(8), 24-35.

Zhao, G. L. (2019). Study on the mechanism of enhancing the value of rural land resources. *Journal of Inner Mongolia University of Finance and Economics, 17*(1), 8–12.

Zhao, H. J. (2018). How farmers can better participate in land consolidation. *Chinese and Foreign Entrepreneurs, 596*(06), 213.

Zhao, H. M. (2018). Activating endogenous power to promote rural revitalization in Hebei Province. *Hainan Today, 242*(09), 26.

Zhao, L., Shao, N., Dou, X. B., & Tao, W. Y. (2019). Analysis of China's agricultural products market in 2018 and prospects for 2019. *Agricultural Outlook, 15*(1), 6-9: 17.

Zheng, H., & Ma, W. (2021). The role of resource reallocation in promoting total factor productivity growth: Insights from China's agricultural sector. *Review of Development Economics, 00*(4), 1–22. doi:10.1111/rode.12826

Zhiyan Consultation. (2020). *Analysis on the development trend of China's agricultural product output and agricultural product import and export trade in 2019*. https://www.chyxx.com/industry/202007/883979.html

Zhou, D., & Lu, Y. (2014). Impact of aging of agricultural labor force on land use efficiency. *Tech Rich Guide*, (14), 52–52.

Zhu X. C. & J. Zhao. (2011). Subsidy policy of the purchase of machinery has accelerated progression of Chinese mechanization during the last seven years. *Chinese Agricultural News of Farmers' Daily*.

ENDNOTE

1 SAP refers to a business. The full name of the company is System Applications and Products. The Chinese name is 思爱普. This is a German independent software company. There is a lot of cooperation with Chinese agricultural companies in the development of smart agriculture.

Chapter 4
Contributions of Sustainable Biomass and Bioenergy in Agriculture Transitions Towards a Circular Economy

Oloiva Maria Tavira
Universidade de Trás os Montes e Alto Douro, Portugal

José Tadeu Marques Aranha
CITAB, Inov4Agro, University of Trás-os-Montes and Alto Douro, Portugal

Maria Raquel Lucas
Universidade de Évora, Portugal

ABSTRACT

The production of bioenergy and biofertilizers based on animal and plant biomass is a crucial pillar in circular economy (CE). CE conceptual model and main aims are closely related to the 3 "R" (reduce, reuse, and recycle) rule, which is to improve the use of resources, minimize waste, and assure sustainability. Although bioenergy offers many opportunities and could be an alternative to fossil fuels use, the path for a broader implementation of this type of activity is still long. This study marks the starting point or direction of research to be taken, ensuring the existence of benefits from plant and animal biomass for the production of bioenergy and biofertilizer, as well as the contributions of this type of production to the circular economy and the mitigation of the climate change impacts.

INTRODUCTION

Bioenergy and bio fertilizers production from animal and plant biomass is an issue in the Circular Economy (CE). This article aimed to provide information focusing on CE in the use of biomass for the production

DOI: 10.4018/978-1-7998-9557-2.ch004

of bioenergy and bio fertilizers. Several studies address the reuse of agro-waste as raw material for the production of bioenergy and bio fertilizers.

The bibliography consulted for this paper frequently mentions economic aspects associated with these resources, having as source the animal and vegetal biomass, in which, for the bioenergy conversion process, several types of technologies based on Anaerobic Digestion (AD) are applied. gasifiers, pyrolysis, combustion, liquefaction and fermentation.

The technology to be used to produce bioenergy depends on the type of biomass. The DA process can be of two types: i) digestion and ii) fermentation. In fermentation, ethanol is produced, while in digestion there is a combination of carbon dioxide and methane, resulting in the production of biogas (Balaman et al., 2018). Waste from this process can be used to create biofertilizers for soil improvement.

The viability of the activity of producing bioenergy, biofertilizers, biogas in a given region depends on the technology to be applied, the profitability (profitability), the amount of biomass to be used, the business viability, regulatory policies, legislation, the financial resources of investors, effective strategies related to waste management, the type of equipment to be invested, the organization of the logistic and supply chain (Buchmann-Duck & Beazley, 2020; Sherwood, 2020; Vaskalis et al., 2019).

44 articles were selected for a general literature review which was derived on 4 stages: (1) general aspects of bioenergy production based on plant and animal biomass, (2) economic aspect of bioenergy production, (3) technological process for bioenergy production, (4) methodology applied to bioenergy production.

Figure 1. Themes associated with Circular Economy based on the articles included in this study
Source:(Own elaboration based on the purpose of the study investigation)

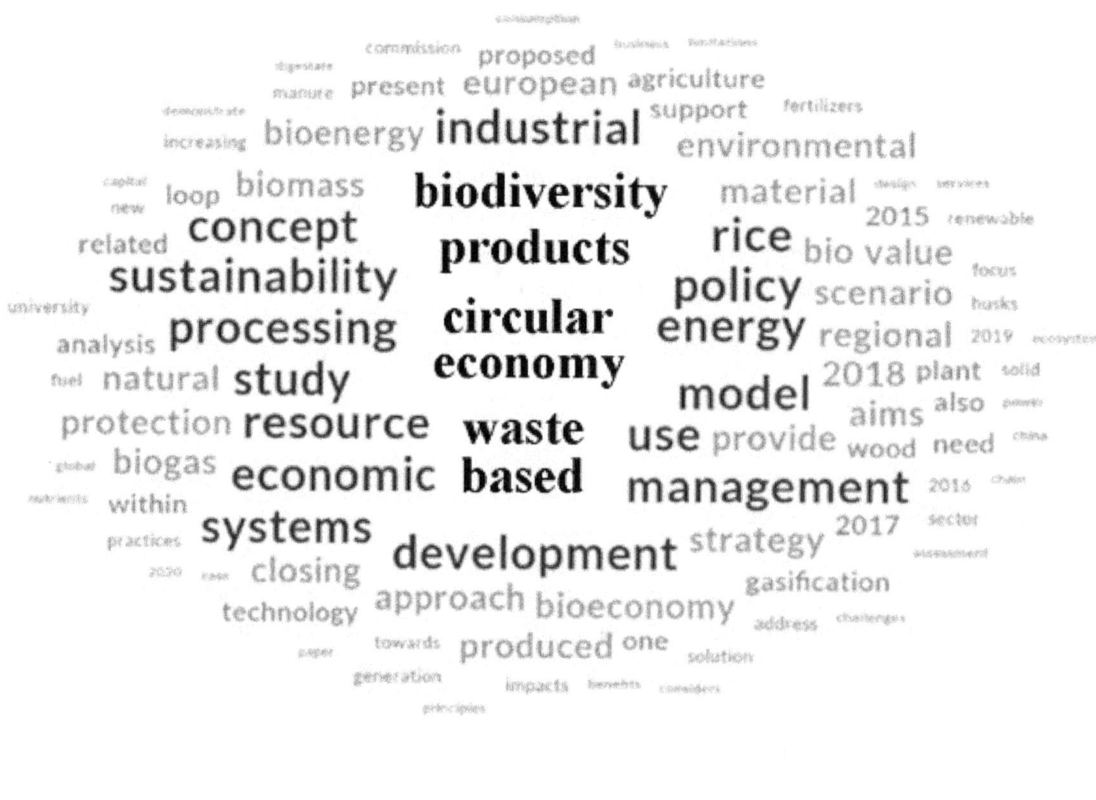

Circular economy (CE) is defined as the economic and conceptual model focused on practical actions of the 3 "R" rule: reduce, reuse and recycle, whose purpose is to improve the use of resources and minimize waste in various activities, guaranteed environmental and social sustainability (Kirchherr, Reike, & Hekkert, 2017), as schematic presented in Figure 1.

The papers analyzed correspond to the period between 2015 and 2020 and focus on the CE approach for bioenergy production, on recommendations to governments for the creation of regulatory policies and incentives for bioenergy production and chemical processes and technological innovation for the conversion of biomass in bioenergy through laboratory experiments (Jarre, Petit-Boix, Priefer, Meyer, & Leipold, 2020; Molina-Moreno, Leyva-Diaz, Llorens-Montes, & Cortes-Garcia, 2017; Sherwood, 2020; Vanhamaki et al., 2019; Vaskalis et al., 2019; Yazan et al., 2018). Animal and vegetal biomass has a potential for bioenergy production using several technological processes: thermochemical pyrolysis, gasification, DA, combustion, liquefaction and fermentation that consequently promote CE (Antoniou et al., 2019; Balaman, Wright, Scott, & Matopoulos, 2018; Barčić, Kuzman, Hárova, & Oblak, 2019; Milanovic et al., 2020; Vaskalis, Skoulou, Stavropoulos, & Zabaniotou, 2019; Yazan et al., 2018). General aspects are sought: economic, technological and methodological processes applied for bioenergy production.

Circular economy scenarios based on computer simulation and modelling, complemented by qualitative approaches in European countries, support economic feasibility studies for bioenergy production without neglecting biodiversity and other environmental aspects (Hoo et al, 2020., Balaman et al., 2018).

This chapter will provide information focusing CE on the use of biomass for the production of bioenergy and biofertilizers and, to explore findings and trends addressing the reuse of agro-waste as raw material for the production of bioenergy and biofertilizers

METHODOLOGY

This study presents a review of papers published in the Scopus and Web of Science (WOS) databases. The search focus on previously selected keywords and the Boolean commands AND and OR. The inclusion criteria were: studies aimed at the agricultural, livestock, and forestry sector in all abstract, keywords, full-text, and/or in the abstract the words EC, plant and/or animal biomass, bioenergy, and/or bio fertilizers.

In this sense, both the Scopus database and the Web Of Science were defined as a query string using the words: "circular economy" AND "bioenergy" OR "bio fertilizer", as presented in Table 1.

Table 1. Search criteria

Data base	Query string	Result
Scopus	(SRCTITLE ("circular economy") OR TITLE ("circular economy") AND TITLE-ABS-KEY (bioenergy) OR TITLE-ABS-KEY (biofertilizer)) AND PUBYEAR > 2014 AND PUBYEAR < 2021	34
WOS	TI="circular economy" AND ((AB=bioenergy OR TI=bioenergy OR KP=bioenergy OR AK=bioenergy)) OR ((AB=bioefertilizer OR TI= bioefertilizer OR KP= bioefertilizer OR AK= bioefertilizer)).	29
	Total Result	63

Source:(Own elaboration based on the purpose of the study investigation)

Consultations were limited to the period between 2015 and 2020 and, only scientific publications in English were considered. Thematic areas were: business, management and accounting, agricultural and biological sciences, social sciences, energy, environmental science, chemical engineering, engineering, biochemistry, genetics and molecular biology, earth, and planetary sciences, immunology, and microbiology.

The publications were then managed and processed using the EndNote software.

In the EndNote software, articles were categorized with a rating from 1 to 5 to identify the most relevant articles, taking into account the keywords included in the abstracts.

Thus, a score of 5 corresponds to articles that included all the inclusion criteria, e.g., in the abstract, contain the keywords circular economy, plant biomass, animal biomass, bioenergy, and bio fertilizers. Score 4 corresponds to articles that contained only circular economy, plant biomass, animal biomass, bioenergy. Score 3 corresponds to articles that contained only three keywords: circular economy, plant biomass, animal biomass of the inclusion criteria. A score of 2 corresponds to articles mentioning only two of the five keywords circular economy, plant biomass of the inclusion criteria. A score of 1 corresponds to articles containing only the circular economy keyword.

All papers that did not meet the criteria previously referenced as inclusion were excluded.

After identifying the papers that met the defined criteria, they were exported to the NVivo qualitative data analysis software to analyse the full texts in greater detail and rigour.

In NVivo software, it was possible to structure the information from the literature included in thematic categories. Through the papers' full reading, it was possible to identify the main themes and cross the information with the literature included in the study. This qualitative analysis enabled the identification of gaps in the literature, contradictions, similarities, differences, problems, and solutions.

In a second thematic analysis stage, using a NVivo software routine, they were accounted the most frequent words in all reviewed papers.

RESULTS

In a total of 63 publications, 34 results were obtained from the Scopus database, 29 from the Web Of Sciences database were downloaded from internet. From those, 19 publications were discarded because they were duplicated, leaving 28 publications from Scopus and 16 publications from WOS. Figure 2 depicts the flowchart.

After running the flowchart shown in Figure 2, we removed the duplicated jobs. This selection resulted in 44 papers. In a second stage, we submitted the selected papers to inclusion/exclusion criteria to score them according to the number of keywords related to this research. This stage led to eliminating six of them. Achieved results from criteria database management and processing are presented in Figure 3.

Analyzing the number of papers published per year between 2015 and 2020, it was found that the 2020 was the year in which many publications related to the topic under study were produced in opposite to 2016 was the year in which less literature related to the topic under study was produced. The comparison between these two years showed a 93.33% increase in publications (Figure 4), which can be considered satisfactory.

In terms of the type of consulted publication, articles were the most common, followed by literature reviews, book chapters, conference papers and books, as shown in Figure 5.

The analysed publications addressed many words that are related to the circular economy, such as production, waste, energy, biomass, process, supply, circular, sustainable, economy, cost, gas, wood, technology, fuel, plant, policy, biogas, manure, electricity, industry, renewable, model and emissions. These words are associated with the production of bioenergy and bio fertilizers from agroforestry residues. In a total, they were accounted 23 most frequent word, as shown in Table 2.

Figure 2. Literature review protocol flowchart
Source:(Own elaboration based on the purpose of the study investigation)

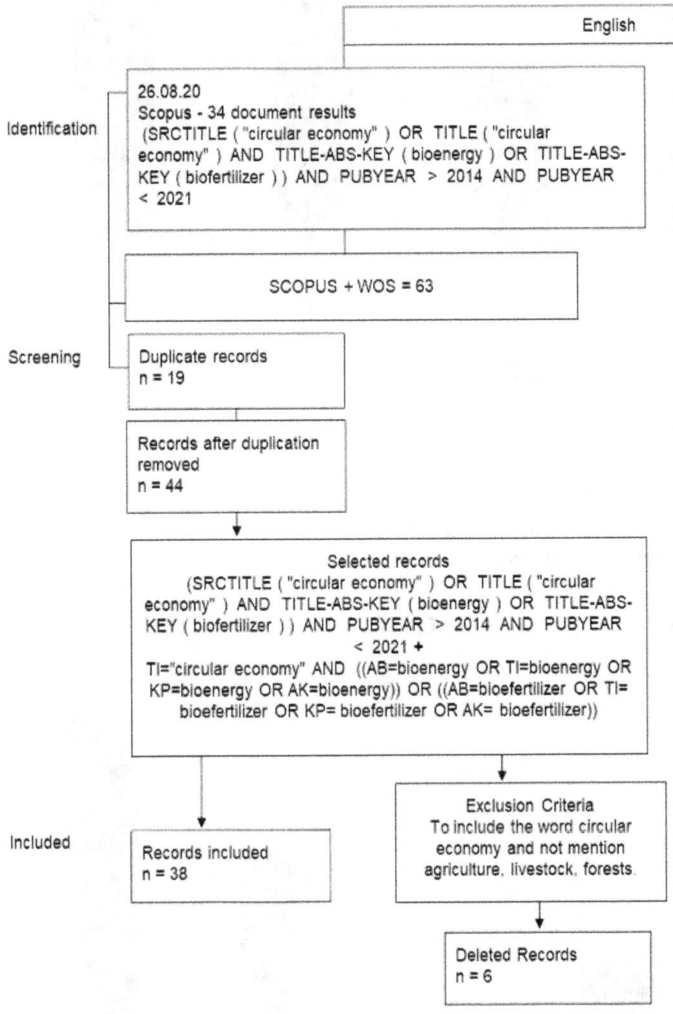

According to table 2, the keywords: production, waste, energy, biomass processes, supply, circular, sustainability, economy, cost, gas, wood, technology, fuel, planting, policy, biogas, manure, electricity, industry, renewable, model, and emissions are the main keywords in the literature addressed from 2015 to 2020.

Figure 3. Number of paper per classification score
Source:(Own elaboration based on the purpose of the study investigation)

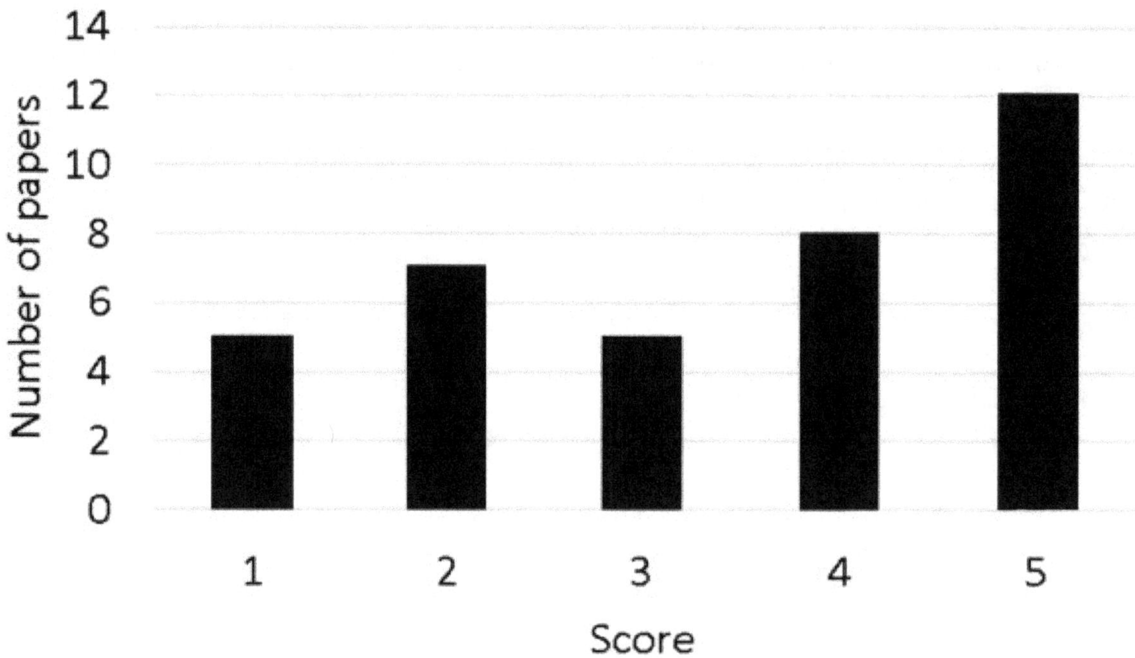

Figure 4. Number of papers per year and according to selected keywords
Source:(Own elaboration based on the purpose of the study investigation)

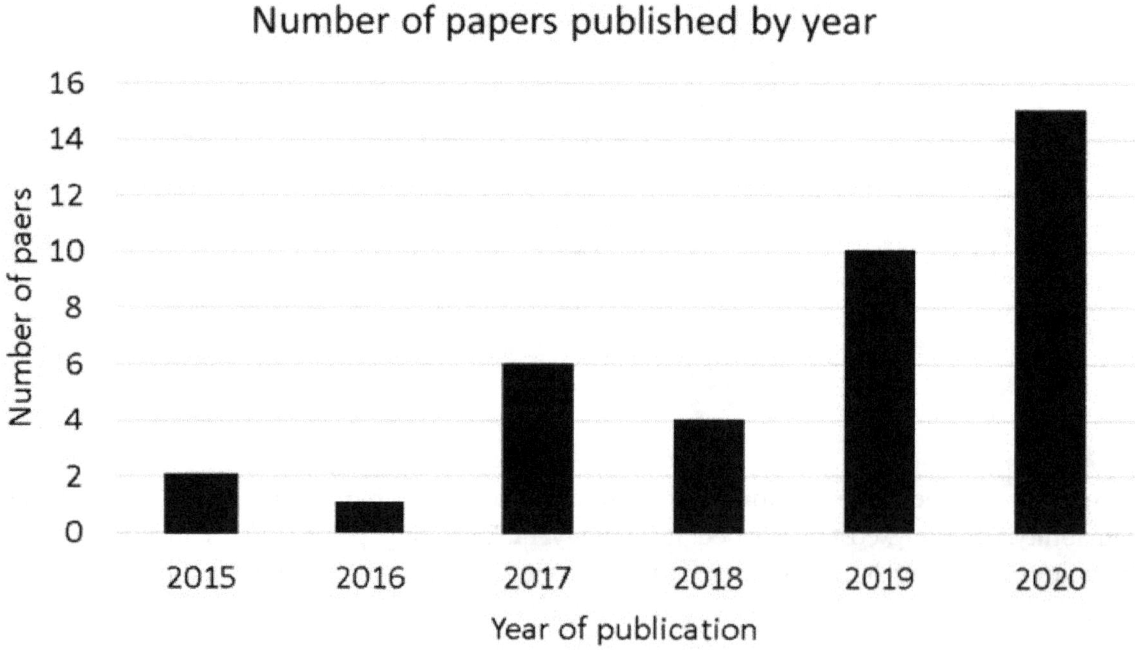

Figure 5. Number of paper per type of publication
Source:(Own elaboration based on the purpose of the study investigation)

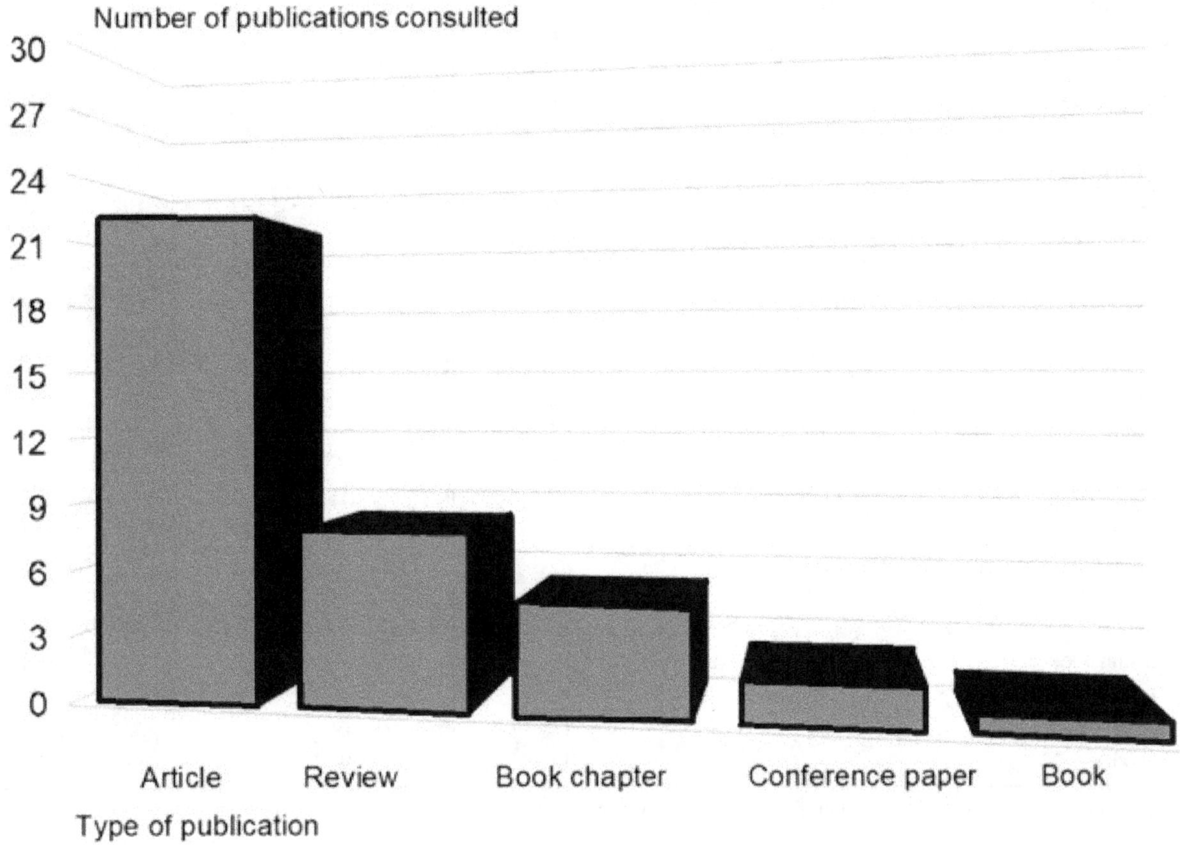

2020, compared to all other years, was the year in which more keywords were discussed, with emphasis on the wood, followed by the: fuel, policies, circular, gas, waste, biomass, and sustainability. In 2016, very few papers addressed the topics under study.

Table 2. Most frequent words in the analysed articles per publication year

Most discussed words in the analysed publications	2015	2016	2017	2018	2019	2020
Biogas	1	0	10	46	4	39
Biomass	1	0	3	33	13	51
Circular	1	1	5	12	28	53
Cost	10	0	3	55	11	22
Economy	1	1	5	11	35	47
Electricity	16	1	7	41	10	26
Emissions	13	0	6	45	6	29
Energy	17	1	7	19	20	36
Fuel	16	2	7	3	13	60
Gas	2	1	9	16	18	53
Industry	12	0	11	15	21	42
Manure	1	0	18	51	2	28
Model	15	2	2	6	21	24
Plant	14	0	5	25	17	40
Policy	16	0	1	12	14	57
Process	12	2	13	10	23	41
Production	9	1	7	27	16	40
Renewable	21	0	12	19	10	38
Supply	9	0	1	51	7	32
Sustainable	3	1	7	16	23	51
Technology	6	0	7	36	13	39
Waste	11	0	6	10	22	51
Wood	13	1	0	5	17	64

Source:(Own elaboration based on the purpose of the study investigation)

DISCUSSION

General Aspects of Bioenergy Production Based on Plant and Animal Biomass

The reviewed papers do not address aspects related to the health hazard associated with technology implementation for bioenergy production. They soft mention environmental issues and do not address social related issues such as how many jobs could be created in these areas and where these focused technologies are due in EC.

The dynamic of themes under analysis was higher in 2020, followed by 2019 and 2015. The least relevant years were 2016 and 2017, as previous presented in Table 2.

The results achieved for 2020 and 2019 were expected, as they were the most prolific years. The result for 2015, compared to 2016 and 2017, was not expected, as it was the year with the fewest publications. These results show that study trends are somewhat fluctuating. According to table 2, the keywords economy, sustainable and circular were the least frequent in 2015 and the most frequent in 2020.

Of the publications analyzed, several fully highlighted essential points related to bioenergy and bio fertilizers production from waste, which gave rise to new by-products. In turn, plant and/or animal-based waste, as well as those resulting from the industrial and urban activity, became the raw material (Chojnacka, Moustakas, & Witek-Krowiak, 2020; Merkulova, Kononova, & Titomir, 2017; Owonubi,

Linganiso, Motaung, & Songca, 2019; Pan et al., 2015; Sherwood, 2020; Šupín, Loučanová, & Olšiaková, 2019; Vanhamaki et al., 2019; Velvizhi et al., 2020; Werle, Sobek, & Kaczor, 2019).

The authors point to the application of technology from Anaerobic Digestion (AD), combustion, gasifiers as appropriate means for the transformation of organic matter residues into bioenergy, heat, biogas (Antoniou et al., 2019; Chojnacka et al., 2020; Hoo, Hashim, & Ho, 2020; Pan et al., 2015; Stiles, Styles, Chapman, Esteves, Bywater, Melville, Silkina, Lupatsch, Fuentes Grünewald, et al., 2018; Vaskalis et al., 2019; Wall, McDonagh, & Murphy, 2017) as shown in Figure 5.

Figure 6. Bioenergy production based on plant and animal waste (Adapted from (Hoo et al., 2020; Kapoor et al., 2020; Milanovic et al., 2020; Pan et al., 2015; Soares et al., 2020; Xu et al., 2018; Zabaniotou, Rovas, Libutti, & Monteleone, 2015)
Source:(Own elaboration based on the purpose of the study investigation)

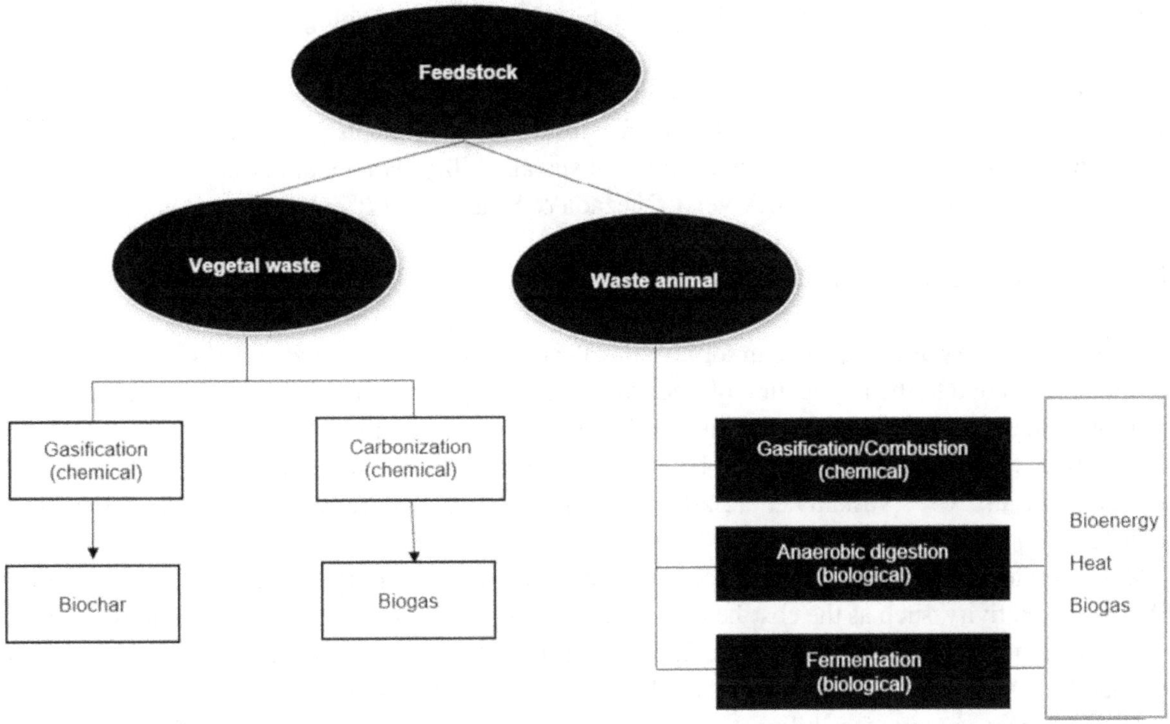

The AD process takes place with the decomposition of biological matter into biogas, from which bioenergy is obtained. This technique extracts an organic matter useful for the improvement of soils composed of nutrients based on potassium, nitrogen and phosphorus (Balaman et al., 2018; Chojnacka et al., 2020; Hoo et al., 2020; Pan et al., 2015; Stiles, Styles, Chapm).

According to Pan et al. (2015) and (Milanovic, 2020) manure-based biomass is not economically viable to be used due to the reduced production of biogas derived from activities on farms. Therefore, the authors suggest the combined use of different bio-based biomass of animal and plant origin, industrial with urban, to produce more biogas to make projects of this nature profitable. Along AD process, the carbon to nitrogen ratio called C/N, to be viable, needs to vary from 20 to 30. It is also essential that there is a combination of animal biomass with urban biomass, ie, urban sewage with chicken and swine waste to produce 400 dm3 per Kg VS. The same author suggests the co-digestion technology as an alternative to AD, as it can allow more profitability to be obtained in the production of biogas, use of bio-waste and mitigation of environmental problems. In this process, nitrogen, potassium and phosphorus are obtained. These are useful substances to improve agricultural soils fertility, leading to the reduction of expenses related to waste management and climate problems mitigation. With this technology, the production of pernicious substances and gases such as ammonium and lipids do not occur, which constitutes a benefit for the environment. Other no less important issues related to economic aspects, regulatory policies, legal forum issues, supply chain and environmental issues were also highlighted in this paper, which led to considering it a complete article having as reference the protocol applied in this study. Thus, this paper could be further used as a basis for future research related to the technological application for the production of bioenergy, biogas, bio fertilizers based on anaerobic digestion, fermentation, gasifiers between others.

According to Zabaniotou et al. (2015), the combination of the technological system of pyrolysis and biochar based on olive groves (pruning, bagasse) for the extraction of bioenergy and bio fertilizers for soil improvement, mainly in Mediterranean countries and for both small and large farms. The use of this type of technological innovation increases the feasibility of investments in bioenergy production, also contributing to increased social and environmental sustainability (Hoo et al., 2020; Molina-Moreno et al., 2017; Vanhamaki et al., 2019; CA Vega-Quezada & Blanco, 2017; Zabaniotou et al., 2015).

The Economic Issue of Bioenergy Production

The economic issue was highlighted in some of the 38 articles that were included in this literature review. The authors pointed to the application of adequate technology for the production of biogas, bioenergy, and biofertilizers as the solution to accelerate the circular economy model (Priyadarshini & Abhilash, 2020b; Stiles, Styles, Chapman, Esteves, Bywater, Melville, Silkina, Lupatsch, Grunewald, et al., 2018; Vanhamaki et al., 2019; Vaskalis et al., 2019; Velvizhi et al., 2020; Yazan et al., 2018; Zabaniotou et al., 2015).

Many studies have applied various economic methods to measure the efficiency and performance of bioenergy activity, such as the cost-benefit analysis method (Barros, Salvador, de Francisco, & Piekarski, 2020; Buchmann-Duck & Beazley, 2020; Cristhian, María, & Hugo, 2017; Marshall et al., 2020 Priyadarshini & Abhilash, 2020a; Soares et al., 2020; C. Vega -Quezada, Blanco, & Romero, 2017(b); Xu et al., 2018), economic feasibility study of the biogas logistics chain (Hoo et al., 2020; Stiles, Styles, Chapman, Esteves, Bywater, Melville, Silkina, Lupatsch, Grunewald, et al., 2018(b); Vaskalis et al., 2019; Yazan et al., 2018), economic evaluation of the supply chain of crops for the production of bioenergy and biofertilizers (Balaman et al., 2018; Marques, Cunha, De Meyer, & Navare, 2020; Palmieri, Suardi, Alfano, & Pari, 2020), SWOT analysis of pyrolysis-biochar (Zabaniotou et al., 2015) and integrated supply chains (Balaman et al., 2018).

Currently, the price of renewable energy is still high due to the existence of few producers and lack of cooperation between them (Yazan, 2018), but the scenario is changing as many subsidies that were previously allocated to fossil energy were cut and taxes paid by activities related to this type of energy increased, decreasing the negative impact proven by this energy source (Pan et al., 2015).

The profitability and competitiveness of bioenergy from organic biomass depend on the application of economic incentives and price subsidies, which also happens in activities related to fossil energy (Palmieri, 2020).

The allocation of credit lines, development of programs to encourage increased investment in these activities, market regulation, the reduction of obstacles related to the production of bioenergy are fundamental actions to encourage the creation of projects that promote the activity of bioenergy (Kapoor, 2020; 2020; Marques, 2020; Hoo, 2020; Nibbi, 2019). As an example, the ESCO project is one of the actions implemented by the governments of several countries that have been successful (Pan et al., 2015).

The reduction of subsidies to fossil energy prices has caused a decrease in GHG pollution (Palmieri, 2020), and it is suggested that these subsidies be directed towards renewable energies to boost bioenergy production activities (Pan et al., 2015).

The challenges presented by bioenergy production are surmountable if institutions, financial programs, and regulatory policies focus on these goals (Hoo, 2020; Pan, 2015). This activity is economically and financially viable if governments intervene in structuring the supply chain, in carrying out public-private partnerships and environmental literacy, also including internationalization (Nibbi, Chiaramonti, & Palchetti, 2019; Pan et al., 2015).

Kapoor et al. (2020) to answer the research problem, addressed the Circular Economy for the production of biogas from agro-waste and point out the following factors as important: i) elimination of barriers in the supply chain., ii) deepening of technical knowledge for foster the application of the circular economy model., iii) implementation of corrective policies to promote activities of the circular economy model., iv) regulation and structuring of activity based on the circular economy model focused on the production of bioenergy., v) use of promising technologies for the recovery of agro-waste.

The biomass and bioenergy waste supply chain in the circular economy scenario works when there is a B2B relationship, resulting in the balance of the bioenergy supply chain (Barčić, 2019).

Similarities, Contradictions and Weaknesses

When analysing the papers' content and structure, we noted some contradictions regarding the use of biologically-based resources and their residues for bioenergy production.

For some authors (Jagger, 2016, Molina-Moreno et al., 2017, Vanhamaki et al., 2019) local bioenergy production is of great importance as it reduces dependence on fossil fuels, enabling food production, thus contributing to the economic, social and environmental farms' sustainability, under a CE conception.

In the opposite direction, some authors (Hoo et al., 2020) state that biogas production based on anaerobic digestion cogeneration is a risk for people living close to bio refineries. This theory makes few invest in this activity, which drives away the financiers and makes this activity presents weak economic sustainability.

In the literature consulted, few case studies (Hoo et al., 2020; Palmieri et al., 2020; Sherwood, 2020; Soares et al., 2020; Xu et al., 2018; Zabaniotou et al., 2015) are reported. The presentation or formulation of theoretical rather than practical aspects regarding the production of bioenergy based on animal and plant biomass in a CE approach was common. This is because there is still insufficient literature related to this topic studied (Yazan et al., 2018, Buchmann-Duck and Beazley, 2020).

FUTURE RESEARCH DIRECTIONS

Few research on the circular economy, bioenergy and biofertilizers production is presented for African countries. Climate change and world population increase, point out these countries like the ones that most need new production models as well as applying the 3 R concept. The future work to be developed will be based on morphological, edaphic-climatic and anthropogenic characteristics of Angola, in order to present models of circular economy and sustainable agroforestry production with a strong component in bioenergy and biofertilizers.

CONCLUSION

This paper demonstrates that the circular economy model, for practical applicability, still has a long way to go, despite the important steps that have already been taken. It was identified which areas are produced more articles. Worldwide, Europe produces the most literature on the EC theme, namely Italy.

It was noticed that the studies were mostly carried out in Europe and Asia. The studies portrayed aspects related to the process of converting plant and animal biomass for the production of bioenergy and bio fertilizers. Bioenergy was highlighted in the articles, which did not happen much with bio fertilizers. Waste management was also highly considered in the papers. Most of the noted issues are related to economics associated with an investment, cost-benefit analysis, technological, chemical, and environmental issues.

Although bioenergy offers many opportunities and is seen as an alternative to fossil fuels, what was found is that the model and operating method for the broader implementation of this type of activity is still rising. The bibliographical review now presented marks the starting point or direction of research for those who want to be updated based on this theme. There are benefits from conducting additional research about the circular economy related to plant and animal biomass use in the production of bio-energy and bio fertilizers.

Some gaps were found, such as the lack of investigation that includes the presentation of statistical data in bioenergy production activities, which offers an opportunity for future research.

Therefore, the following is concluded: with this literature review, the scientific knowledge related to the study of the circular economy for bioenergy production has been enriched. We hope that this study will serve as a basis for future investigations in the area and will also serve as a guide and sequence for future work.

It was clear through the literature review that the circular economy model is very studied and little applied. From the 38 analyzed papers, just 4 of them reported concrete cases of circular economy and not a simulation.

Conflicts of Interest: The authors declare no conflict of interest.

ACKNOWLEDGMENT

Funding: This work is supported by National Funds by FCT - Portuguese Foundation for Science and Technology, under the project UIDB/04033/2020.

REFERENCES

Antoniou, N., Monlau, F., Sambusiti, C., Ficara, E., Barakat, A., & Zabaniotou, A. (2019). Contribution to Circular Economy options of mixed agricultural wastes management: Coupling anaerobic digestion with gasification for enhanced energy and material recovery. *Journal of Cleaner Production, 209*, 505–514. doi:10.1016/j.jclepro.2018.10.055

Balaman, S. Y., Wright, D. G., Scott, J., & Matopoulos, A. (2018). Network design and technology management for waste to energy production: An integrated optimization framework under the principles of circular economy. *Energy, 143*, 911–933. doi:10.1016/j.energy.2017.11.058

Barčić, A. P., Kuzman, M. K., Haviarova, E., & Oblak, L. (2019). *Circular economy & Sharing collaborative economy principles: A case study conducted in wood-based sector*. Woodema, Ia-Int Assoc Econ & Manag Wood Processing & Furn Manuf.

Barros, M. V., Salvador, R., de Francisco, A. C., & Piekarski, C. M. (2020). Mapping of research lines on circular economy practices in agriculture: From waste to energy. *Renewable & Sustainable Energy Reviews, 131*, 109958. Advance online publication. doi:10.1016/j.rser.2020.109958

Buchmann-Duck, J., & Beazley, K. F. (2020). An urgent call for circular economy advocates to acknowledge its limitations in conserving biodiversity. *The Science of the Total Environment, 727*, 9. doi:10.1016/j.scitotenv.2020.138602 PMID:32325313

Chojnacka, K., Moustakas, K., & Witek-Krowiak, A. (2020). Bio-based fertilizers: A practical approach towards circular economy. *Bioresource Technology, 295*, 122223. Advance online publication. doi:10.1016/j.biortech.2019.122223 PMID:31623921

Cristhian, V. Q., María, B., & Hugo, R. (2017). Production of Bioenergy in the Framework of Circular Economy: A Sustainable Circular System in Ecuador. In Advances in Biofeedstocks and Biofuels (Vol. 1, pp. 1-31). Wiley. doi:10.1002/9781119117322.ch1

Hoo, P. Y., Hashim, H., & Ho, W. S. (2020). Towards circular economy: Economic feasibility of waste to biomethane injection through proposed feed-in tariff. *Journal of Cleaner Production, 270*, 122160. Advance online publication. doi:10.1016/j.jclepro.2020.122160

Jarre, M., Petit-Boix, A., Priefer, C., Meyer, R., & Leipold, S. (2020). Transforming the bio-based sector towards a circular economy - What can we learn from wood cascading? *Forest Policy and Economics, 110*, 12. doi:10.1016/j.forpol.2019.01.017

Kapoor, R., Ghosh, P., Kumar, M., Sengupta, S., Gupta, A., Kumar, S. S., Vijay, V., Kumar, V., Kumar Vijay, V., & Pant, D. (2020). Valorization of agricultural waste for biogas based circular economy in India: A research outlook. *Bioresource Technology, 304*, 11. doi:10.1016/j.biortech.2020.123036 PMID:32107150

Kirchherr, J., Reike, D., & Hekkert, M. (2017). Conceptualizing the circular economy: An analysis of 114 definitions. *Resources, Conservation and Recycling, 127*, 221–232. doi:10.1016/j.resconrec.2017.09.005

Marques, A., Cunha, J., De Meyer, A., & Navare, K. (2020). Contribution Towards a Comprehensive Methodology for Wood-Based Biomass Material Flow Analysis in a Circular Economy Setting. *Forests*, *11*(1), 21. doi:10.3390/f11010106

Marshall, R., Lag-Brotons, A. J., Inam, E. J., Herbert, B. M. J., Hurst, L., & Semple, K. T. (2020). From Bioenergy By-products to Alternative Fertilisers: Pursuing a Circular Economy. In RSC Green Chemistry (pp. 289-314). Royal Society of Chemistry.

Merkulova, T., Kononova, K., & Titomir, O. (2017). Development trends of circular economy: Case study of Ukraine. *Rivista di Studi sulla Sostenibilità*, (2), 51–66. doi:10.3280/RISS2017-002005

Milanovic, T., Popovic, V., Vuckovic, S., Rakascan, N., Popovic, S., & Petkovic, Z. (2020). Analysis of soybean production and biogas yield to improve eco-marketing and circular economy. *Ekonomika Poljoprivreda-Economics of Agriculture*, *67*(1), 141–156. doi:10.5937/ekoPolj2001141M

Molina-Moreno, V., Leyva-Diaz, J. C., Llorens-Montes, F. J., & Cortes-Garcia, F. J. (2017). Design of Indicators of Circular Economy as Instruments for the Evaluation of Sustainability and Efficiency in Wastewater from Pig Farming Industry. *Water (Switzerland)*, *9*(9), 13. doi:10.3390/w9090653

Nibbi, L., Chiaramonti, D., & Palchetti, E. (2019). Project BBChina: a new Master Program in three Chinese Universities on Bio-Based Circular Economy; from Fields to Bioenergy, Biofuel and Bioproducts. In J. Yan, H. X. Yang, H. Li, & X. Chen (Eds.), *Innovative Solutions for Energy Transitions* (Vol. 158, pp. 1261–1266). Elsevier Science Bv. doi:10.1016/j.egypro.2019.01.416

Owonubi, S. J., Linganiso, L. Z., Motaung, T. E., & Songca, S. P. (2019). Wood and its byproducts: Cascading utilization for biomass (re)generation. In "Waste-to-Profit" (W-t-P): Circular Economy in the Construction Industry for a Sustainable Future (Vol. 2, pp. 293-311). Nova Science Publishers, Inc.

Palmieri, N., Suardi, A., Alfano, V., & Pari, L. (2020). Circular Economy Model: Insights from a Case Study in South Italy. *Sustainability (Switzerland)*, *12*(8), 11. doi:10.3390u12083466

Pan, S. Y., Du, M. A., Huang, I. T., Liu, I. H., Chang, E. E., & Chiang, P. C. (2015). Strategies on implementation of waste-to-energy (WTE) supply chain for circular economy system: A review. *Journal of Cleaner Production*, *108*, 409–421. doi:10.1016/j.jclepro.2015.06.124

Priyadarshini, P., & Abhilash, P. C. (2020a). Circular economy practices within energy and waste management sectors of India: A meta-analysis. *Bioresource Technology*, *304*, 11. doi:10.1016/j.biortech.2020.123018 PMID:32087547

Priyadarshini, P., & Abhilash, P. C. (2020b). Fostering sustainable land restoration through circular economy-governed transitions. *Restoration Ecology*, *28*(4), 719–723. doi:10.1111/rec.13181

Sherwood, J. (2020). The significance of biomass in a circular economy. *Bioresource Technology*, *300*, 122755. Advance online publication. doi:10.1016/j.biortech.2020.122755 PMID:31956060

Soares, C. A., Viancelli, A., Michelon, W., Sbardelloto, M., Camargo, A. F., Vargas, G. D. L. P., Fongaro, G., & Treichel, H. (2020). Biogas yield prospection from swine manure and placenta in real-scale systems on circular economy approach. *Biocatalysis and Agricultural Biotechnology*, *25*, 101598. Advance online publication. doi:10.1016/j.bcab.2020.101598

Stiles, W. A. V., Styles, D., Chapman, S. P., Esteves, S., Bywater, A., Melville, L., Silkina, A., Lupatsch, I., Fuentes Grünewald, C., Lovitt, R., Chaloner, T., Bull, A., Morris, C., & Llewellyn, C. A. (2018). Using microalgae in the circular economy to valorise anaerobic digestate: Challenges and opportunities. *Bioresource Technology*, *267*, 732–742. doi:10.1016/j.biortech.2018.07.100 PMID:30076074

Šupín, M., Loučanová, E., & Olšiaková, M. (2019). *Sustainable bioenergy policy for the period after 2020.* Paper presented at the 12th WoodEMA Annual International Scientific Conference - "Digitalisation and Circular Economy: Forestry and Forestry Based Industry Implications".

Vanhamaki, S., Medkova, K., Malamakis, A., Kontogianni, S., Marisova, E., Dellago, D. H., & Moussiopoulos, N. (2019). Bio-based circular economy in European national and regional strategies. *International Journal of Sustainable Development and Planning*, *14*(1), 31–43. doi:10.2495/SDP-V14-N1-31-43

Vaskalis, I., Skoulou, V., Stavropoulos, G., & Zabaniotou, A. (2019). Towards circular economy solutions for the management of rice processing residues to bioenergy via gasification. *Sustainability (Switzerland)*, *11*(22), 6433. Advance online publication. doi:10.3390u11226433

Vega-Quezada, C., Blanco, M., & Romero, H. (2017). Synergies between agriculture and bioenergy in Latin American countries: A circular economy strategy for bioenergy production in Ecuador. *New Biotechnology*, *39*, 81–89. doi:10.1016/j.nbt.2016.06.730 PMID:27287049

Vega-Quezada, C. A., & Blanco, M. (2017). Photobioreactors within the framework of circular economy and poverty eradication: Bioenergy production in the Ecuador-Peru border integration zone. In Photobioreactors: Advancements, Applications and Research (pp. 275-298). Nova Science Publishers, Inc.

Velvizhi, G., Shanthakumar, S., Das, B., Pugazhendhi, A., Priya, T. S., Ashok, B., Nanthagopal, K., Vignesh, R., & Karthick, C. (2020). Biodegradable and non-biodegradable fraction of municipal solid waste for multifaceted applications through a closed loop integrated refinery platform: Paving a path towards circular economy. *The Science of the Total Environment*, *731*, 138049. Advance online publication. doi:10.1016/j.scitotenv.2020.138049 PMID:32408201

Wall, D. M., McDonagh, S., & Murphy, J. D. (2017). Cascading biomethane energy systems for sustainable green gas production in a circular economy. *Bioresource Technology*, *243*, 1207–1215. doi:10.1016/j.biortech.2017.07.115 PMID:28803063

Werle, S., Sobek, S., & Kaczor, Z. (2019). *The role of the thermal processes of waste biomass in the circular economy perspective.* Paper presented at the 32nd International Conference on Efficiency, Cost, Optimization, Simulation and Environmental Impact of Energy Systems, ECOS 2019.

Xu, X. B., Ma, Z., Chen, Y. Q., Gu, X. M., Liu, Q. Y., Wang, Y. T., Sun, M., & Chang, D. H. (2018). Circular economy pattern of livestock manure management in Longyou, China. *Journal of Material Cycles and Waste Management*, *20*(2), 1050–1062. doi:10.100710163-017-0667-4

Yazan, D. M., Cafagna, D., Fraccascia, L., Mes, M., Pontrandolfo, P., & Zijm, H. (2018). Economic sustainability of biogas production from animal manure: A regional circular economy model. *Management Research Review*, *41*(5), 605–624. doi:10.1108/MRR-02-2018-0053

Zabaniotou, A., Rovas, D., Libutti, A., & Monteleone, M. (2015). Boosting circular economy and closing the loop in agriculture: Case study of a small-scale pyrolysis-biochar based system integrated in an olive farm in symbiosis with an olive mill. *Environmental Development*, *14*, 22–36. doi:10.1016/j.envdev.2014.12.002

ADDITIONAL READING

Jain, A., Sarsaiya, S., Kumar Awasthi, M., Singh, R., Rajput, R., Mishra, U. C., Chen, J., & Shi, J. (2022). Bioenergy and bio-products from bio-waste and its associated modern circular economy: Current research trends, challenges, and future outlooks. *J Fuel.*, *307*, 121859. doi:10.1016/j.fuel.2021.121859

Kirchherr, J., Reike, D., & Hekkert, M. (2017). Conceptualizing the circular economy: An analysis OF 114 definitions. *Resources, Conservation and Recycling*, *127*, 221–232. doi:10.1016/j.resconrec.2017.09.005

Kostas, E. T., Adams, J. M. M., Ruiz, H. A., Durán-Jiménez, G., & Lye, G. J. (2021). Macroalgal biorefinery concepts for the circular bioeconomy: A review on biotechnological developments and future perspectives. J. *Renewable & Sustainable Energy Reviews*, *151*, 111553. doi:10.1016/j.rser.2021.111553

Mativenga, P. T., Sultan, A. A. M., Agwa-Ejon, J., & Mbohwa, C. (2017). - Composites in a Circular Economy: A Study of United Kingdom and South Africa. *Procedia CIRP*, *61*, 691–696. doi:10.1016/j.procir.2016.11.270

Zailan, R., Lim, J. S., Manan, Z. A., Alwi, S. R. W., Mohammadi-ivatloo, B., & Jamaluddin, K. (2021). Malaysia scenario of biomass supply chain-cogeneration system and optimization modeling development: A review. J. *Renewable & Sustainable Energy Reviews*, *148*, 111289. doi:10.1016/j.rser.2021.111289

Chapter 5
Crises in Agriculture: Concepts, Responses, and Recovery Strategies

Anastasiia Sliusarenko
ISCAP, Polytechnic Institute of Porto, Portugal

Oscar Bernardes
ISCAP, Polytechnic Institute of Porto, Portugal

ABSTRACT

COVID-19 is a pandemic of the 21st century, a disease that shook the world and altered the lives of entire communities. Due to the enormous negative influence on the economy, it permanently alters the way an organization operates, leading businesses to develop crisis management techniques and implement new innovative practices. Agriculture is no exception. Given the sector's constant growth, which is not only due to population growth but also to continuous lifestyle changes, it is critical to implement recovery plans at the organizational and government levels. Thus, this chapter provides an overview of crisis management, including its key characteristics and framework; analyzes the importance of innovation in the agricultural sector; provides an overview of the agricultural sector; examines the impact of the pandemic on this sector and some recovery strategies; and examines the attitude of agricultural professionals toward the COVID-19 crisis.

INTRODUCTION

The SARS-CoV-2, also called COVID-19, gained this designation in order to not create unnecessary panic among the population of different countries, such as Asia that suffered great impact of SARS in 2003[1] (WHO, 2020a). Despite of its outbreak in Wuhan province in China in late 2019, it was reported to World Health Organization´s Chinese office only on 31st of December the same year. The virus spread that started with 282 confirmed cases, mostly in China and some in Japan, Republic of Korea and Thailand, turned out to be the most devastating pandemic of the XXI century counting around 173.005.553 cases and 3.727.605 death worldwide (WHO, 2020b). The most harmful impact on health of populations was

DOI: 10.4018/978-1-7998-9557-2.ch005

registered in Americas and Europe (43% and 31% cumulative cases respectively) (WHO, 2020b), and as a consequence world's economy suffered irrecoverable damage.

As a preventive measure, massive lockdowns and household isolations, gradually adopted across and within countries, had a major influence on collapse of economic activity worldwide (FAO, 2020). All companies (directly or not) experienced COVID-19 impact on their activity, through factories' shut-downs, labor shortages due to cash flow stress, and disruptions in the supply chain (Reid et al., 2020 as cite in Rowan & Galanakis, 2020).

While the most of the governments were concentrated on reducing the transmission and further spread of this proven to be deadly virus, the whole industries had to rethink their business models and transition to sustainable supply chains in the aftermath of COVID-19 that includes rethinking vulnerabilities created by over-reliance on 'just-in-time' or 'business-as-usual' practices (Sarkis et al., 2020 in Rowan & Galanakis, 2020).

There is no doubt that economic recovery is not expected to occur shortly (FAO, 2020a) but there are some industries, such as those belonging to agrifood sector, that are crucial to survival of populations. According to FAO report, the most destructive impact of current pandemic is being observed in the group of low-income food-deficit countries, which are not self-sufficient enough to produce the food they consume. Eventually, the number of people without regular access to basic food supplies will increase (FAO, 2020a).

In 2018, the agro-sector in European Union reached € 1.098 billion turnovers and employed 4.24 million people (Saguy et al., 2018 as cite in Rowan & Galanakis, 2020). The importance of sector is undeniable, that is why most of the countries granted economic stimulus (European Commission provided €750 billion to help mitigate the shock from COVID-19 pandemic). As for other priorities established by EU (European Union) there are climate actions, health program and digital strategies, created in order to ensure continuity of supply chain for food, medical products and services across European Union.

Beside existing stimulus, each company had to make urgent decision in order to survive first shock of COVID-19 and to prevent possible pandemics impacts in future. In such situations, crisis management and recovery strategies play imperative role.

The main purpose of present research paper is to study managers' attitude towards crisis provoked by COVID-19 virus spread and strategies adapted by organizations in agro-sector (agriculture and agri-food) in order to minimize pandemic impact. In order to achieve the main goal, theoretical overview was structured in two parts (Figure 1): Crisis (concept and management) and Response (possible strategies for better recovery).

Figure 1. Theoretical background structure

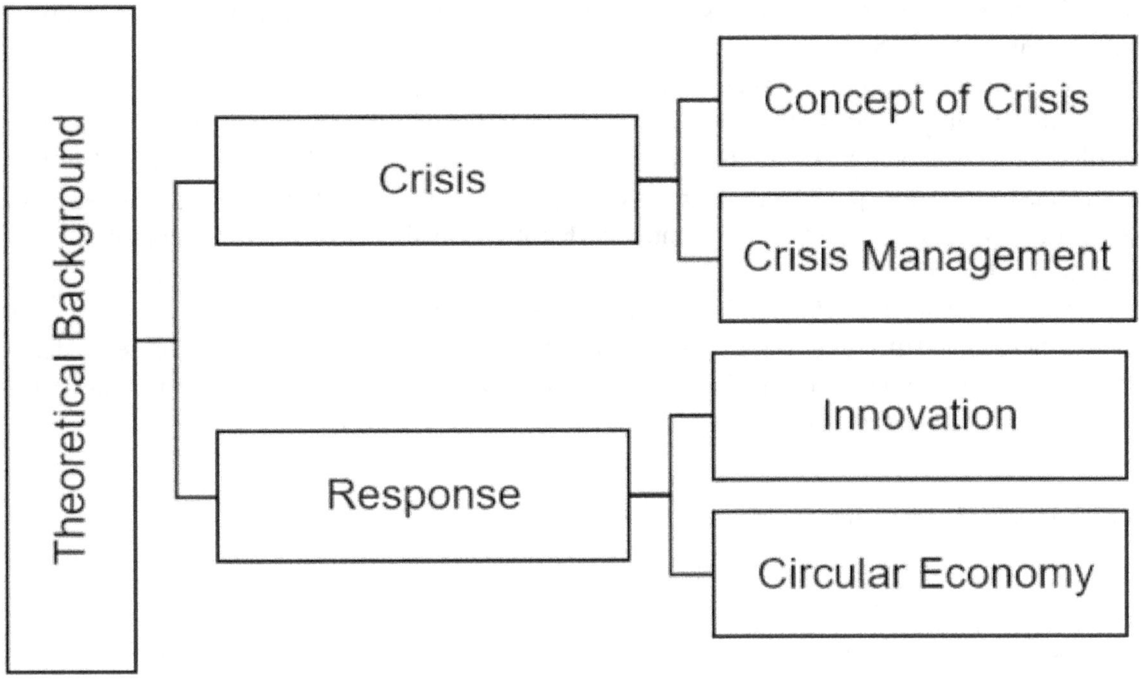

While crisis management provides framework for detecting crisis nature, building strategies and, as a consequence, the capability for an effective response that safeguards the interests of its key stakeholders, the response strategies are important for crisis aftermath, creating value auditioning mechanisms and evoke improvements to "new normality" of organizations.

CONCEPTUAL/ THEORETICAL BACKGROUND

Company Crisis Concept, Types and Stages

The most frequent use of term "crisis" is related to natural disasters such as earthquakes (Romero-Meza & Blanco-Vidal, 2011) or political crisis, demanding fast reaction on governmental level (Győrffy, 2013). It can be a threat of nuclear war, an embargo on the export of oil or wheat to hostile countries. The danger to the physical integrity of citizens can also be considered a crisis, for example, in situations that involve hijacking or kidnaping, or it can also originate from an employment and economic crisis (Rosenthal & Kouzmin, 1997). However, for the purpose of present research, the concept of organizational crisis is being analyzed.

There are numerous definitions for crisis management mentioned in literature through the past years. The term "crisis" has it origin in Greek word "krisis", which means judgement, choice or decision (Paraskevas, 2006). Boin et al. (2008) and Dekker & Hansen (2004) described crisis as an unpredictable event, characterized by complex situations of chaos and stress, usually with a lack of reliable information (Broekema et al., 2017). Due to crisis' uncertain nature, according to Jindal et al. (2015), decision-making process tends to be complicated, especially when the origin of crisis is unknown. Authors define three main crisis characteristics, such as uncertainty creation, threat to important goals and unexpected nature (Jindal et al., 2015).

As for organizational crisis, it can be defined as: "I…I a low-probability, high-impact event that threatens the viability of the organization and is characterized by ambiguity of cause, effect, and means of resolution, as well as by a belief that decisions must be made swiftly". (Pearson and Clair 1998 as cited in Paraskevas, 2006, p. 893). To the main characteristics of a crisis, Krystek (1987) added a limited duration of unwanted and unexpected event, and that it can lead to the company's complete downfall and failure (Vallaster, 2017).

For instance Bénaben (2016), besides defining crisis as a disruption within the state of a system which reveals instability and discontinuity, considered that it requires a specific treatment in order to deal with consequences and obtain new acceptable state of considered system. In order to make accurate decision, crisis has to be characterized (Figure 2).

Figure 2. Crisis characteristics and types

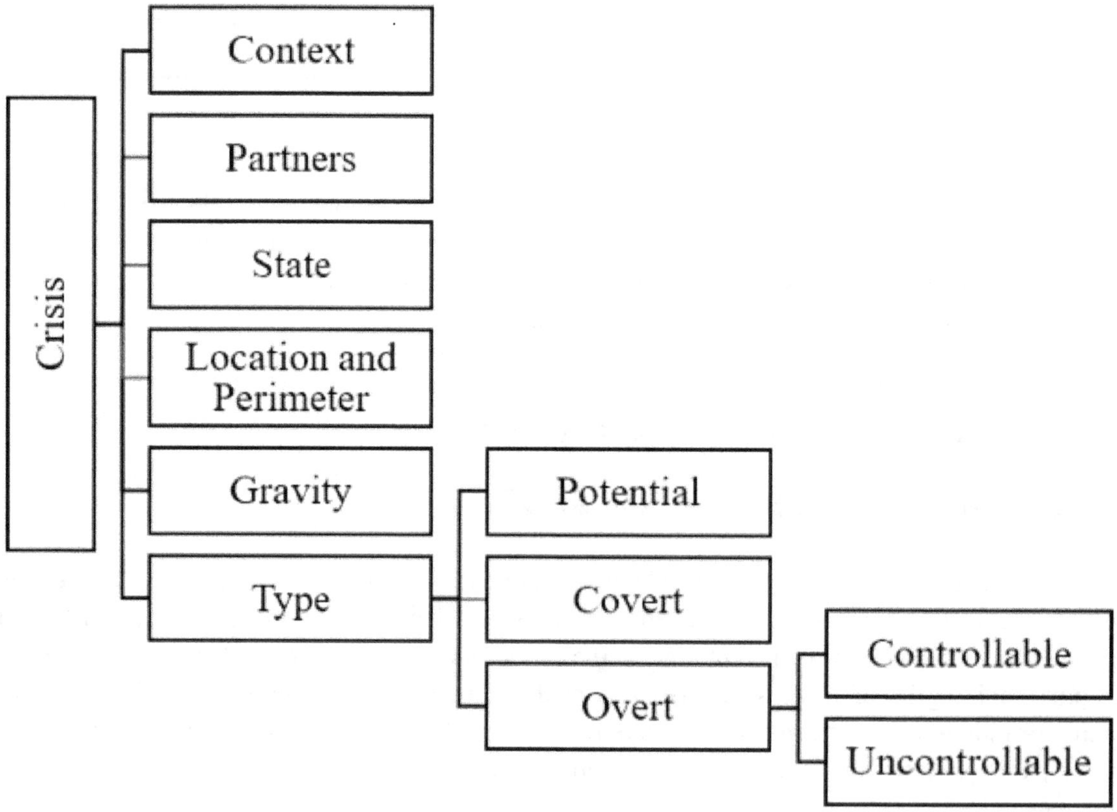

Bénaben (2016) proposed six criterions for primary crisis characterization:

- The context of a crisis – the description on intrinsic risks and impacted elements (population, goods etc.)

- Partners – the potential role that all possible stakeholder can play in crisis resolution and assist in crisis management.
- State – the actual effects and potential risks of crisis event.
- Location and perimeter – geographical area.
- Gravity - evaluation of crisis, some authors suggest money, victims or even minutes on TV news as a criterion.
- Type - the nature of crisis.

In other words, crisis in general terms can be analyzed through location, gravity and type, and for crisis management, such characteristics as context, partners and objectives should be considered (Bénaben, 2016).
As for crisis types, there are three of them that can be found in chosen literature:

- Potential crisis – not actual crisis, but possible. No reliable indications of crisis are evident, but performance is endangered by the occurrence of potential threat. There is still time before actual crisis occurs.
- Covert (latent crisis) – occurs if no action was taken throughout the potential crisis, possibility of temporary difficulties in achievement of organizational goals, organizational potential is not fully utilized to overcome crisis. In this case, all actions should be aimed to identify and minimize the latent crisis.
- Overt (evident crisis) – occurs when all preventive measures for covert crisis are not efficient. This type of crisis can be controllable (immediate, urgent and accurate crisis management decisions should be taken, organization is still in the position to overcome crisis) and uncontrollable (final stage of the crisis process, when organization has no potential to overcome crisis and it should be resolved) (Kuzmanova, 2016).

When analyzing stages of the crisis, authors defend four main phases:

- Preconditions (prodromal crisis) – warning stage, where the set of several small events that take place before actual crisis occurs. This chain of events usually leads to a significant incidence, also called "trigger event".
- Trigger event (acute crisis stage) – event that sets crisis in motion and turns it evident to the key stakeholders of the organization.
- Crisis (chronic crisis stage) – where crisis produces the greatest damage to the organization and its stakeholders, such as employees, management, stockholders, customers, suppliers, local community or even government regulators.
- Postcrisis (crisis resolution state) – when the acute state is over and organization achieves a new "normality". At this stage, management should analyze the crisis event and determine what strategies can be established in order to prevent future crisis events (Crandall et al., 2014), (Giannacouroua et al., 2015).

Even after analyzing and characterizing crisis, crisis management cannot base its decisions only on internal factors. There are external factors that can be decisive for crisis management. Not only organization's performance can play a crucial role, but also such factors as markets, competitors or government policies. Smircich and Stubbard (1985) affirm that the environment (internal or external) can be represented as a set of events and relationships between them, which are perceived and understood through certain cognitive strategies affecting business strategy and decisions. In some cases, managers seem to form an image of the environment and respond to that image, rather than to facts. (Weick, 1969 as cited in Giannacouroua et al., 2015). Since the impact of a crisis or disaster [can be] on internal (business units, staff, managers, shareholders) and external (other agencies and organizations, general public, media, tourists) level (Ritchie, 2004), crisis management strategies and approaches might become a very complex procedure.

Crisis Management

It is highly important to detect crisis on its early stages, and when potential threats are detected, the efficiency of company's management is crucial for faster and less harmful recovery. As Pearson and Clair (1998) defined it, *"organizational crisis management is a systematic attempt by organizational members with external stakeholders to avert crises or to effectively manage those that occur"* (Crandall & Mensah, 2008, p. 18). Other authors present crisis management as *"the process by which an organization deals with any major unpredictable event threatening to harm the organization, its stakeholders or the general public"* (Vargo & Seville, 2011, as cite in Vallaster, 2017, p. 510) In any case, the prime purpose of crisis management is to mitigate the negative impacts of a crisis, keeping positive relationships with stakeholders, community and focus on more sustainable development practices.

The one of common characteristics of crisis in most of the sources is its unpredictability. The unexpected situations (such as a case of a COVID-19 pandemic), especially in the unprepared companies can generate panic state amid employees, and even for its management. That is why it is essential, as a first step in controlling crises, create a crisis management team. In accordance with company's size and nature, crisis management team can consist of a team leader, security director, legal counselor, media spokesperson, internal communication specialist, human resources director, employee representative and security specialist. If needed, an external adviser or field consultant may also be hired (Robinson, 2005 as cite in Vardarlıer, 2016).

In several papers and books Crandall et al. (2008, 2010 p. 11, 2014 p. 15) approach crisis management as a four stages framework with two perspectives: Internal and External Landscape (Table 1).

The first stage framework, or the **Landscape Survey**, is the primary step for management personnel where the evaluation of internal and external environments takes place. While the internal environment is focused on the level of preparation of the company to crisis events, organizational culture and other company's policies, the external landscape is analyzed in order to understand the situation in industry, political factors and other implications that cannot be control from within the organization.

Only after analyzing all available information, management can start with development of a **Strategic Plan**. On this particular stage, crisis management team is formed, worst-case scenarios are discussed and crisis management plan is established. At the same time, on external level, stakeholders and governmental agencies adapt their activities to manage existing crisis and prevent its further development.

The third stage, designated as a **Crisis Management**, occurs on the most critical phase of a crisis event. At that point, primary (owners, employees, customers, suppliers and local community) and secondary (other communities and organizations interesting in organization) stakeholders are managed in order to align their efforts to contain crisis and resume normal operations. The external landscape is responsible for stakeholders' reactions and include negative media coverage, web-based criticism and obstructing government regularizations. According to authors, this stage is the only one that is reactive, it can only commence when actual crisis occurs, when other three stages are of a proactive nature.

Table 1. Crisis management framework

	Landscape survey	Strategic Planning	Crisis Management	Organizational Learning
The internal landscape	◆ Propensity for crisis management ◆ Organizational culture ◆ Ethical environment Company safety polices	◆ Form crisis management team ◆ Develop worst-case scenarios ◆ Formulate crisis management plan ◆ Conduct mock disasters and training	◆ Primary stakeholders management ◆ Secondary stakeholders management	◆ Organizational learning ◆ Organizational renewal ◆ Evaluate success or failure of crisis management planning
The external landscape	◆ Industry vulnerability ◆ Degree of political stability ◆ Globalization implications ◆ Technological implications	◆ Existing government regulations ◆ Industry standards	◆ Reaction of stakeholders ◆ Negative media coverage ◆ Public and Web criticism ◆ Impeding government regulations	◆ Industry renewal ◆ New government regulations ◆ New stakeholder outlook

At the end, the final phase is the **Organizational Learning**. At this stage, on internal level, organization must evaluate how crisis was handled and if any adjustments to a crisis management plan are needed. It is crucial for organization to learn from the crisis in order to avoid the same mistakes in future. On the external level, the whole industries can reevaluate and renew their procedures, external stakeholders and may change their attitude towards company and the way it overcame the crisis event (Crandall & Mensah, 2008).

Acquier et al. (2008) defend that for a better recovery from crisis it is crucial, yet difficult, to correctly diagnose the nature of crisis and be able to react quickly and accordingly. When organization adopts the same positioning and strategies for every crisis, it can induce itself in a harmful situation (Acquier ey al., 2008). The crisis response itself can be considered as a complex system, where different people, from different departments have to work together as a team (crisis management team), focusing on several response tasks (as detection, prevention, damage limitation etc.). This system is characterized by its ability to learn from the environment and adapt (change) all organizational structure and behavior on individual level (Paraskevas, 2006).

Small and medium sized enterprises (SMEs) can be more vulnerable to crises, comparing to larger organizations (OECD, 2009). Considering that crisis is seen as an unexpected event, smaller companies can experience lack of human and financial resources or even management capacities to deal with crisis event. Knowing that SMEs account for over 99% of all enterprise in all European Union (Eurostat, 2021), they employ more than a half of the labor force in private center, it is highly important for governments to support them in times of crisis (Giannacouroua et al., 2015).

The recent study mentioned by Rowan et al. (2020), shows that COVID-19 pandemic is likely to change organizational behaviors and strategies. Some of the strategies that might be adopted by companies are: forced mergers and acquisition transactions for weaker companies (with lower valuation), consolidation of similar sized companies to withstand market uncertainties and secure customers, adjustments and improvements in supply-chain structures, searching for value adding solutions (for example better food security to satisfy customer needs). In the end, all companies will share the similar characteristic (inherent to competitive markets) – need for innovation, as a survival requirement (Rowan & Galanakis, 2020).

Response to Crisis

Innovation as an Advantage in Crisis Aftermath

It is a well-known fact that innovation, although a critical measure, has a significant impact on economic growth, motivating companies to increase performance and giving them an advantage in uncertain and threatening environment (Shakina & Barajas, 2020). According to the "first-mover-advantage" theory (Lieberman and Montgomery, 1988), innovative companies have a competitive advantage and leading position, when presenting new products in the industry. As a result, innovative organizations can obtain strategic benefit, establish the market's power, and take the potential bounce (Nguyen et al., 2020). The strong relationship between environmental uncertainty (crisis) and innovation has already been proven through research, due to its capacity to reduce environmental impermanence and to provide sense of stability. According to Miller (1987), companies in more unstable environment have a higher need for innovation, in order to remain competitive and survive (Giannacouroua et al., 2015).

The COVID-19 pandemic, indeed, can be considered turbulent, uncertain and unstable environment. At the same time, this pandemic can be a trigger factor that gives opportunity for companies to overcome limitations and obstructions that they used to experience in their normal work environments (Bjorklund et al., 2020). It is quite possible that the COVID-19 crisis will be a driver to the next wave of innovators with economic and social impact, at the same time encouraging cooperation between academic world and industry to identify the next disruptive technology (Rowan & Galanakis, 2020).

The innovation per se can be divided in two modes: Science, Technology and Innovation (based on the production and use of technical knowledge) and Doing, Using and Interacting (based on informal processes of learning and experimentation) (Gibson & Naquin, 2011). The most common areas of innovation, affordable to almost any company are value offering (most frequently through new products and services), bundling products and rebranding (implementation of ideas that were considered too risky and discarded in pre-crisis time), exploration of new market segments, new business models, diversified portfolios (Gibson & Naquin, 2011). Innovative measures can help reduce the fixed costs or even change company's positioning towards its customer through, for example, prosocial activities (food donations, free deliveries to elderly people (Bjorklund et al., 2020). Open innovation can be considered an interesting method, since it brings new ideas from out of organizational boundaries, giving an opportunity to take advantage of knowledge of others (outside in), and share your knowledge with other companies (inside out) (Chesbrough, 2020).

Undoubtedly, digital technology in our modern world, offers innumerous possibilities in research area for data acquisition and analysis. It helps producing safe and healthy food, and offers improved sustainability (Bakalis et al., 2020). Artificial Intelligence, on the other hand, is being used to develop new foods and flavors (Rowan & Galanakis, 2020).

Considering that even before pandemic the customers' purchasing behavior was changing (in 2015 in United Kingdom, average time spent online was 21h per week, half of the time via smartphone) (Bakalis et al., 2020), since the COVID-19 virus spread, most of the population worldwide remained in lockdown, in order to prevent infection, depending on online retail and new innovative delivery systems (Rowan & Galanakis, 2020).

The agro-sector is not an exception. Since the 20th century, Agriculture 3.0 got a massive development through new computer program and robotic technics, giving an opportunity to operate more efficiently, to reduce use of chemicals, to improve irrigation precision and much more. Nowadays, current technologies like the Internet of Things, Artificial Intelligence, Cloud Computing and all other products of Digital Era allowed to agrifood sector to step into Agriculture 4.0 (Zhai et al., 2020). The innovative computational tools created for agrifood sector aim to agile management decision-making process through data analysis (energy consumption, general productivity etc.) and strategic information, at the same time, gives possibility to produce in more ecological way without any extra costs. Even from customer perspective, "green" positioning presents more advantage for companies comparing to other players in the industry (Zocca et al., 2019).

The customer perspective is highly important for any industry. In order to be competitive, companies have to offer more than just a product - the entire purchasing system should be considered (from transparency of information to the post-sale services). The importance of communication and interaction with customer is growing. In search of other opinions, customers often form groups or communities with other costumers, and these communities turned into an important tool for companies in order to not only learn customers' behavior, but also gain new ideas for future innovations based on customers' preferences.

This special kind of e-commerce, denominated social commerce, allows interaction between company and customer in a social environment such as Facebook, Twitter, etc. Besides the opportunity to interact with customers, it provides better visibility on social media and create closer relationships with customers (Sturialea & Scuderi, 2013).

However, the innovation process can be challenging, especially in agro-sector due to a large number of actors throughout the supply chain. The information flow often suffers disruptions due to their heterogeneity and dynamically changing business relations. In this case, common platforms and software could be very useful. Systems, that could allowed access to necessary information for every stakeholders, gain efficiency in production and distribution and as a result to reduce waste and costs of the final product (Kaloxylos et al., 2013).

Circular Economy in Agro Sector

The problem of a food waste in Europe is not a recent issue. According to data collected in 2017, Europe was generating about 1.3 billion tons of waste annually, with 700 million tons generated in agriculture (Toop et al., 2017). For instance, in 2013 Italy was responsible for 244 thousand tons of waste, due to failure in selling, on average, between 1% and 1.2% of grocery stores and supermarkets turnover. Around the same time, British food markets produced 366 thousand tons of annual waste throughout retail and distribution stages. The most common sources of increasing amounts of waste are associated with logistics (location problems, the waste allocations or even vehicle routing problems (Fancello et al., 2017), the lack of innovation in production system and out of date technology (Toop et al., 2017).

One possible solution for reducing the food waste is integrating circular economy in the food production system, with adaption of food waste management practices, design of industrial routs, and transformation of food waste into feedstock, bio-chemicals or even energy. Besides the obvious advantage for reducing waste, circular economy can add value to agrifood production (Borrello et al., 2020).

Circular economy can be defined as an industrial system that is restorative or regenerative by intention and design (EMF, 2015), enhancing the continuous flow of technical and biological materials in the value circle while keeping products, components and materials at their highest utility and value at all times and reducing waste to a minimum (EMF, 2013). The transformation of a linear economy into a circular one, however, can be complex. It requires innovative perspective in fields like logistics (shifting to reverse mode), new customer-supplier relationship dynamics, new forms of organizational and marketing strategies connected to various value chains (Donner et al., 2020).

The new circular economy models, according to Ghisellini (2016, as cited in Borrelo et al., 2020, p. 1), could provide and effective way to defy societal challenges and to "|…| contribute to sustainable development goals mostly by promoting a break through approach to actualize the sustainability of industrial systems and rethink the organization of supply chains" (Borrello et al., 2020). There are various sustainability strategies like creating of take-back systems for customers who wishes to reuse or recycle the product, choosing functionality over the ownership (produce-as-a-service approach), where customer is offered a functionality without owning the product. (Geissdoerfer et al. 2018, as cited in Borrello et al. 2020). Circular economy models, generally, have six major patterns: repair and maintenance, reuse and redistribution, refurbishment and remanufacturing, recycling, cascading and repurposing, and organic feedstock (Donner et al., 2020).

The implementation of new circular economy models can be challenging. Due to low market prices of products, low quantities and seasonality or high transportation cost and water content, the production from agro-waste into value-added products tend to be highly expensive. The heterogeneity of resources, instability in volume and quality over time and regions can be considered challenging factors. However, the major challenge, according to author, is that the implementation of circular economy requires a change on a system level, including involvement of all the actors through the value chain (Donner et al., 2020).

THE AGRO-SECTOR OVERVIEW

The most of agriculture industries are located in Africa, North and South Americas and Asia, and worldwide they occupy more than 48 million km^2 (37.2% of the total land area) (Lyasnikov, 2019). Even though, more than 800 million of people worldwide do not have food access. The agriculture sector is heterogenic and its scale, generally, depends on geographical position. For instance, while most farms (74%) in Europe and Americas have over than 10 ha, in Asia 85% are smaller (Lyasnikov, 2019).

Recent studies (Silva, 2018) show that demand for food worldwide will increase by 70%, while share of agriculture in world GDP (Gross domestic product) is expected to decrease to 3%. There are four prime affecting factors – demographics, scarcity of natural resources, climate change and food waste (Anishchenko, 2019).

The importance of this sector is undeniable, not only as a key to survival of humanity, but also from economical point of view. At the moment, in the European Union (still including United Kingdom), the agriculture is the largest manufacturing sector that employs around 4,72 million people and represents € 1.2 trillion of annual turnover according to Food & Drink Europe (European Commission, 2019). The sector accounts of more than 285 thousand SMEs and micro SMEs in the European Union, where 0,9% of the companies contribute to almost 50% of the activity (Bakalis et al., 2020).

As European Commission (2019) has established in the 2019 report, European food market is highly influenced by social demand, whether related to health, environment, climate change, animal welfare and lifestyle. Massive market shifts were expected and while some activities like dairy or meat industries were expected to decline, others like olive oil and wine production and their export volume could increase. Despite of possible decrease in income by 2025 (due to the fall of pig meat, wheat, maize and soya beans prices), by the 2030 an average EU farm income could increase, the labor force outflow from agricultural sector was expected to decline and technological progress in machinery and equipment (together with decision-support tools) prognosed to be implemented (European Commission, DG Agriculture and Rural Development, 2019).

Even though around 79% of food produced worldwide is consumed in the cities (Galli et al., 2020) with the highest concentration of population, small farms play very important role in quality food production, at the same time providing diversification of rural economy, management of natural resources, supporting employment and family income. In spite of their small family household nature, small farms may be empowered through the connections to the bigger markets, value chains and innovation (Tisenkopfs et al., 2020).

The new modernization provided by Agriculture 4.0, such as new sensing, automation and digital technologies, can easily increase efficiency and flexibility of the agrifood sector. Innovation and stock minimization contribute for transformation of agricultural industry from traditional to "on-demand" system, increasing dependence on more effective transportation, that was highly affected during the COVID-19 pandemic, inhibiting or complicating the food access for large number of people. This proved that one of the greatest challenges in crisis planning is a development of a resilient food system, where one failure in normal functioning of any of the actors, can provide the failure of the entire food supply chain (Bakalis et al., 2020).

Covid-19 Pandemic in Origin of a Food Crisis

The COVID-19 pandemic, besides direct threat to human health and well-being, evoked colossal food crisis worldwide. Lockdowns and social distancing measures caused supply chain disruption and, therefore, increase food losses especially in high value and nutrient-rich products like dairy, meats, fruits and vegetables (Cattaneoa et al., 2020). Empty shelfs in supermarkets were reported in many countries. If before the pandemic 50% of food was purchased in supermarkets and the other half in food services, after virus spread almost 100% of customers shift their preferences to supermarkets, that were not able to cope with demand. The same shift of purchasing habits, alongside to confinement measures, were responsible for cease of a big part of food services (Bakalis et al., 2020), resulted in induced unemployment that contributed to food insecurity, especially in low-income households.

Table 2.

Short-term impact	Medium-term impact	Long-term impact
•Higher demand for food at the start of the pandemic	•Reduction of budget support for agriculture and investments in the sector due to general economic crisis	•Strengthening of "green trends", increase of more biological agriculture and animal farming, use of resource and energy-saving technologies, less dependent on external supply of resources and materials
•Decrease in demand for food from the hotel business, tourism and catering	•Decrease in demand for food due to a decrease in real incomes of the population	•Increasing the role of regional and local food supply systems operating outside global markets and in this regard, the growth of the importance of small and medium-sized businesses, agricultural cooperation, small-format trade, direct links between food producers and consumers
•Disruptions in the supply of resources and services for agriculture, as well as in the "food chain" due to restrictions in the operation of transport, complication in logistics, closing of retail markets	•Growth of production costs due to the rise in prices for imported capital goods and materials	•Deurbanization, a more even distribution of the population throughout the country to reduce risks of epidemics and other emergencies
•Closure of a number of processing plants with a high concentration of workers		
•Shortage of seasonal workers due to reduced (restricted) migration		
•Disruption of the functioning of global agri-food markets due to export restrictions		

The catastrophic impact of COVID-19 crisis on agrifood sector stroked mostly small and medium-sized agribusiness enterprises, and, as it was mentioned in previous section, these companies are critical to the functioning of entire food system in Europe. Because of short-term cash flow shortages and decreasing of operational capacities, agrifood sector suffered employment and income crisis. Food and Agriculture Organization of United Nations expects even further negative impact, especially on most vulnerable population groups and countries, small family farms, small-scale food producers, seasonal and migrant workers, elderly population and others (FAO, 2020b).

This negative impact of COVID-19 virus spread, however, cannot be seen as a singular event, but an extended crisis that gradually affected different parts of world, sectors, populations and lifestyle patterns, and can be divided in three stages (Table 2): short-term, medium-term and long-term impact (Petrikov, 2020)

The first, short-term impact was experienced worldwide at the very beginning of pandemic, between March and June depending on the start point of first wave in every country (ex. mid-April in Portugal and in June in Russian Federation). The medium-term impact (decreasing income and purchasing power together with governments struggle to keep companies from bankruptcy) is something that characterizes current daily life of populations in different countries. As for long-term impacts, these are possible outcome of this crisis, that should be analyzed by all actors of agrifood sector.

Besides the transition to a "new normality" and necessary adjustments that should be considered for agrifood sector, there are actions that should be taken on organizational and governmental levels. Measures like support for innovation (machinery, microbiological industry, plan protection products, etc.) could reduce dependence on external suppliers, creation of new digital platforms could be an alternative to traditional retail chains and facilitate market access for small local businesses (Petrikov, 2020). In addition, the financial support for agricultural sector is essential.

Recovery Strategies

The sudden drop in demand and decrease in clients purchase power created a high pressure on governments and financial institutions (FAO, 2020b), requiring creation of different recovery strategies designed on organizational and governmental level (Table 3)

All of these and other measures were conceived in order to allow close to normal functioning and faster recovery of small and medium-sized businesses. Through better coordination of all stakeholders, including industries and governments, could be achieved global trade flow with minimum of logistics disruptions. Automation should provide efficient production even with a reduced number of workforce and agile distribution and market access even for smallholders. New regulations have to aim to maximize health and safety to both customers and business holders, and to determine new food standards where information can be used in order to provide results. Innovation tools such as food apps need to provide more extensive information regarding nutrition and ingredients and highly regulated to protect both producers and clients and serve for more than merely marketing purposes (FAO, 2020a).

Table 3. Recovery strategies

Organizational level	Governmental level
New digital channels Enhance productivity and market visibility **Short food supply chains** Direct sales, minimized logistics disruptions, new distribution technologies, development of locally sourced foods, allocate efficiently resources, suppliers, retailers, home delivery via small distribution centers in cities **Innovation** Enhance market transparency through IT, e-commerce app, apps for ordering and delivering, interconnection within agrifood sector, waste reduction/processing/reusing **New layout** Of workplaces, manufacturing facilities, retails, ensure safety of workers and customers.	**Urban agriculture** Enhance better resilience, better food access, population decentralization **Regulations** Public sector involvement, supervision, infrastructure development, institutional support **Financial Support Stimulus** New loan schemes interest free for six month (United Kingdom) or five to ten years (Japan), dedicated credit lines for small family farmers below market rate (Brazil), credit package about 3.3% of GDP for small farm household (Thailand), monthly grants of €9 to €15 thousand for SMEs (Germany), renegotiation of existing loans and new schemes for lending (Italy), simplified procedures and access for support (USA), more flexibility and six-month moratorium on bankruptcy applications (Russian Federation) **Digital transactions** Internet banking, mobile money systems, reducing transaction fees, regulation for expansion of the use of technology

Source: Personal adaptation from (FAO, 2020) & (Bakalis et al., 2020)

CONCLUSION

The COVID-19 virus spread caused negative economic impact on almost every industry around the world affecting mostly small and medium-sized enterprises. And since the COVID-19 pandemic is not a singular event but an extend crisis, its influence tends to spread gradually affecting different parts of the world, their economy and populations. The changes in demand, customers' lifestyles and purchase behavior were some of the reasons for agro-sector to rethink and readapt to this new reality. The implementation of new technologies, development of recovery strategies on organizational and governmental levels are tend to mitigate the pandemic impact outcome.

This chapter motivates various lines for further research, that could be conducted in different fields like economy, sociology or technology among others. There are questions that surfaced during the data analysis and could be included in future papers. For example, what is the basis of differences in managers' awareness towards crisis like COVID-19? Or what factors influence decision making process regarding crisis management? What prevents managers from reaching out to external experts? Is it costs of their services or risks related to confidential and critical internal information? What kind of incentives have to be created to motivate more innovation? How can be reduced the costs of Circular Economy and its implementation?

REFERENCES

Acquier, A., Gand, S., & Szpirglas, M. (2008). From Stakeholder to Stakeholder Management in Crisis Episodes: A Case Study in a Public Transportation Company. *Journal of Contingencies and Crisis Management*, *16*(2), 101–114. doi:10.1111/j.1468-5973.2008.00538.x

Anishchenko, A. (2019). Agriculture 4.0 as a promising model for scientific and technological development of agrarian sector in modern Russia. *Food Policy and Security*, 129-140.

Bakalis, S., Valdramidis, V., Argyropoulos, D., Ahrne, L., Chen, J., Cullen, P., . . . LeBail, A. (2020). Perspectives from CO+RE: How COVID-19 changed our food systems and food security paradigms. *Current Research in Food Science*, 166–172.

Bénaben, F. (2016). A Formal Framework for Crisis Management Describing Information Flows and Functional Structure. *Procedia Engineering*, *159*, 353–356. doi:10.1016/j.proeng.2016.08.208

Bjorklund, T., Mikkonen, M., Mattila, P. & Marel, F. (2020). Expanding entrepreneurial solution spaces in times of crisis: Business model experimentation amongst packaged food and beverage ventures. *Journal of Business Venturing Insights*, 1-14.

Borrello, M., Pascucci, S., Caracciolo, F., Lombardi, A., & Cembalo, L. (2020). Consumers are willing to participate in circular business models: A practice theory perspective to food provisioning. *Journal of Cleaner Production*, *259*, 1–14. doi:10.1016/j.jclepro.2020.121013

Broekema, W., Kleef, D., & Steen, T. (2017). What Factors Drive Organizational Learning From Crisis? Insights From the Dutch Food Safety Services' Response to Four Veterinary Crises. *Journal of Contingencies and Crisis Management Volume*, 326-340.

Cattaneoa, A., Sáncheza, M. V., Toreroa, M., & Vosb, R. (2020). Reducing food loss and waste: Five challenges for policy and research. *Food Policy*, 1–9. PMID:33012956

Chesbrough, H. (2020). To recover faster from Covid-19, open up: Managerial implications from an open innovation perspective. *Industrial Marketing Management*, *88*, 410–413. doi:10.1016/j.indmarman.2020.04.010

Crandall, W. R., & Mensah, E. C. (2008). Crisis management and sustainable development: a framework and proposed research agenda. *International Journal of Sustainable Strategic Management*, 16-34.

Crandall, W. R., Parnell, J. A., & Spillan, J. E. (2014). *Crisis Management: Leading in the New Strategy Landscape*. SAGE.

Donner, M., Gohier, R., & Vries, H. (2020). A new circular business model typology for creating value from agrowaste. *The Science of the Total Environment*, *716*, 137065. doi:10.1016/j.scitotenv.2020.137065 PMID:32044489

European Commission, DG Agriculture and Rural Development. (2019). *EU agricultural outlook for markets and income, 2019-2030.* Brussels: EU.

European Commission, DG Agriculture and Rural Development. (2020). Financial needs in the agriculture and agrifood sectors in Portugal. Brussels: European Commission - Directorate General, Agriculture and Rural Development.

Eurostat. (2021). *Entrepreneurship and Small and medium-sized enterprises (SMEs).* Eurostat.

Fancello, G., Mola, F., Frigau, L., Serra, P., Mancini, S., & Fadda, P. (2017). A new management scheme to support reverse logistics processes in the agrifood distribution sector. *Transportation Research Procedia, 25*, 695–715. doi:10.1016/j.trpro.2017.05.452

FAO. (2020a). *COVID-19 crisis and support for agrifood: Public sector responses through the financial sector.* Rome: UN.

FAO. (2020b). *COVID-19 global economic recession: Avoiding hunger must be at the center of the economic stimulus*. United Nations.

FAO. (2020c). *Review of agri-food trade policy in post-Soviet countries 2017-2018*. FAO - The Food and Agriculture Organization.

Galli, A., Pires, S. M., Iha, K., Alves, A. A., Lin, D., Mancini, M. S., & Teles, F. (2020). Sustainable food transition in Portugal: Assessing the Footprint of dietary choices and gaps in national and local food policies. *The Science of the Total Environment*, *749*, 1–15. doi:10.1016/j.scitotenv.2020.141307 PMID:32846345

Gibson, D. V., & Naquin, H. (2011). Investing in innovation to enable global competitiveness: The case of Portugal. *Technological Forecasting and Social Change*, *78*(8), 1299–1309. doi:10.1016/j.techfore.2011.04.004

Győrffy, D. (2013). Crisis Management in the EU, Prospects for the de-politicization of Economic Policy. *Focus – Crisis Management in the EU*, 119-132.

Jindal, S., Laveena, & Aggarwal, A. (2015). A comparitive study of crisis management- toyota v/s General motors. *Scholedge International Journal of Management & Development*, 1-12.

Kaloxylos, A., Wolfert, J., Verwaart, T., Terol, C. M., Brewsterd, C., Robbemond, R., & Sundmaker, H. (2013). The Use of Future Internet Technologies in the Agriculture and Food Sectors: Integrating the Supply Chain. *Procedia Technology*, 51 – 60.

Kuzmanova, M. (2016). Contemporary problems related to crisis management of Organizations. *Trakia Journal of Sciences*, *14*(3), 256–261. doi:10.15547/tjs.2016.03.010

Lyasnikov, N. (2019). Global challenges and threats of agrarian sector development in Russia. *Food Policy and Security*, 85 – 96.

Nguyen, S. K., Vo, X. V., & Vo, T. M. (2020). Innovative strategies and corporate profitability: The positive resources dependence from political network. *Heliyon*, *6*(4), 1–10. doi:10.1016/j.heliyon.2020.e03788 PMID:32373728

OECD. (2009). *The Impact of the Global Crisis on SME and Entrepreneurship Financing and Policy Responses*. OECD Publishing.

Paraskevas, A. (2006). Crisis management or crisis response system? A complexity science approach to organizational crises. *Management Decision*, *44*(7), 892–907. doi:10.1108/00251740610680587

Petrikov, A. V. (2020). Adaptation of the agri-food sector to post-pandemic reality. Free Economic Society of Russia, 99-105.

Polit, D., & Hungler, B. (1995). *Fundamentação de Pesquisa de Enfermagem*. Artes Médicas.

Ritchie, B. W. (2004). Chaos, crises and disasters: A strategic approach to crisis management in the tourism industry. *Tourism Management*, *25*(6), 669–683. doi:10.1016/j.tourman.2003.09.004

Romero-Meza, R., & Blanco-Vidal, C. (2011). Management of Crises: Lessons from Earthquake and Mining Crises in Chile. *Vikalpa*, 97–99.

Rosenthal, U., & Kouzmin, A. (1997). Crises and Crisis Management: Toward Comprehensive Government Decision Making. *Journal of Public Administration: Research and Theory*, *7*(2), 277–304. doi:10.1093/oxfordjournals.jpart.a024349

Rowan, N. J., & Galanakis, C. M. (2020). Unlocking challenges and opportunities presented by COVID-19 pandemic for cross-cutting disruption in agri-food and green deal innovations: Quo Vadis? *The Science of the Total Environment*, *748*, 1–15. doi:10.1016/j.scitotenv.2020.141362 PMID:32823223

Shakina, E., & Barajas, A. (2020). "Innovate or Perish?": Companies under crisis. *European Research on Management and Business Economics*, 1–10.

Silva, G. (2018). *Feeding the world in 2050 and beyond – Part 1: Productivity challenges*. Michigan State University.

Sturialea, L., & Scuderi, A. (2013). Evaluation of Social Media Actions for the Agrifood System. *Procedia Technology*, 200 – 208.

Tisenkopfs, T., Adamsone-Fiskovica, A., Kilis, E. S., Grivins, M., Pinto-Correia, T., & Bjørkhau, H. (2020). Territorial fitting of small farms in Europe. *Global Food Security*, *26*, 1–9. doi:10.1016/j.gfs.2020.100425

Toop, T., Ward, S., Oldfield, T., Hull, M., Kirby, M., & Theodorou, M. K. (2017). AgroCycle – developing a circular economy in agriculture. *Energy Procedia*, *123*, 76–80. doi:10.1016/j.egypro.2017.07.269

Vallaster, C. (2017). Managing a Company Crisis through Strategic Corporate Social Responsibility: A Practice-Based Analysis. *Corporate Social Responsibility and Environmental Management, 24*(6), 509–523. doi:10.1002/csr.1424

Vardarlıer, P. (2016). Strategic approach to human resources management during crisis. *Procedia: Social and Behavioral Sciences, 235*, 463–472. doi:10.1016/j.sbspro.2016.11.057

WHO. (2020a). *Naming the coronavirus disease (COVID-19) and the virus that causes it.* Retrieved from World Health Organization: https://www.who.int/emergencies/diseases/novel-coronavirus-2019/technical-guidance/naming-the-coronavirus-disease-(covid-2019)-and-the-virus-that-causes-it

WHO. (2020b). *Novel Coronavirus (2019-nCoV) Situation Report- 1.* Retrieved from World Health Organization: www.who.int/docs/default-source/coronaviruse/situation-reports/20200121-sitrep-1-2019-ncov.pdf?sfvrsn=20a99c10_4

Zhai, Z., Martínez, J. F., Beltran, V., & Martínez, N. L. (2020). Decision support systems for agriculture 4.0: Survey and challenges. *Computers and Electronics in Agriculture, 170*, 1–16. doi:10.1016/j.compag.2020.105256

Zocca, R., Gaspar, P. P., Silva, P. D., Santos, F. D., Andrade, L. P., & Nunes, J. (2019). Decision-making computationally aided in the management of energy sources used in agrifood industries. *Energy Procedia, 161*, 100–107. doi:10.1016/j.egypro.2019.02.063

KEY TERMS AND DEFINITIONS

Chronic Crisis Stage: Where the greatest harm is done to the firm and its stakeholders, including employees, management, investors, customers, suppliers, the local community, and even government regulators.

Covert: Occurs when no action is taken during a potential crisis, when temporary obstacles in achieving organizational goals are encountered, and when organizational capacity is not fully leveraged to overcome the crisis.

Crisis: An unforeseen incident marked by complicated settings of turmoil and tension, frequently accompanied by a dearth of accurate information.

Overt: Occurs when all precautionary measures available to deal with concealed crisis are ineffective.

Postcrisis: When the acute stage has passed and the organization has reverted to "normalcy".

Preconditions: Warning stage, where the set of several small events that take place before actual crisis occurs.

Trigger Event: The event that initiates the crisis and makes it visible to the organization's main stakeholders.

ENDNOTE

1 WHO, Naming the coronavirus disease (COVID-19) and the virus that causes it, 2020.

Chapter 6
Effect of Climate Change on Agricultural Production

Helena Esteves Correia
https://orcid.org/0000-0003-1274-5141
Agrarian School of Viseu, Polytechnic Institute of Viseu, Portugal

Daniela de Vasconcelos Teixeira Aguiar da Costa
Agrarian School of Viseu, Polytechnic Institute of Viseu, Portugal & CERNAS, Portugal

ABSTRACT

Agricultural production is influenced by environmental factors such as temperature, air humidity, soil water, light intensity, and CO_2 concentration. However, climate change has influenced the values of average temperature, precipitation, global atmospheric CO_2 concentration, or ozone level. Thus, climate change could lead to different situations on plants and consequently influence agricultural production. With this chapter, the authors intend to research how climate change influences some plant metabolisms (such as photosynthesis, photorespiration, transpiration, among others) and therefore agricultural production.

INTRODUCTION

Climate change they are associated with variations in precipitation patterns, temperature increase and CO_2 concentration (Jiang et al., 2020). However, these climate changes are regionally different, for example in Central Europe the increase in temperature will lead to increases in the frequency and duration of the summer drought periods, while in Central and Northern Europe the increase in rainfall during the winter and spring increases the risk of flooding (Kreuzwieser & Gessler, 2010).

As a result of the lack of water, oxygen, or a rise in temperature, the physiological activity of plants will be affected (Kreuzwieser & Gessler, 2010). Thus, agricultural productivity is influenced by these climate changes, particularly by the lack of water, increased temperature, increased CO_2 concentration (Jiang et al., 2020; Neto et al., 2021). While increasing CO_2 concentration can benefit or lessen the effects of drought, increasing temperature could delete that benefit (Neto et al., 2021). Ainsworth et al. (2019) relate that the ability for terrestrial ecosystems to buffer anthropogenic emissions will diminish

DOI: 10.4018/978-1-7998-9557-2.ch006

with rising temperature and drought stress and, consequently, that terrestrial plants absorb each time fewer of CO_2 (Tkemaladze & Makhashvili, 2016).

Physiological changes in plants that occur with climate change are: in photosynthesis, respiration, synthesis of primary metabolites (proteins, carbohydrates and fats) and secondary metabolites (volatile oils, polyphenols, steroids, alkaloids, etc.), a degradation of nitrogenous compounds such as amino acids, proteins, nucleic acids, the consequent accumulation of ammonia in tissues, cells, its intoxication and decomposition of subcellular structures and plant death (Tkemaladze & Makhashvili, 2016). The impacts of climate change on plant nutrient balance occur through interactions between the effects of climate change on nutrient uptake and allocation, which is important for food quality and productivity under global climate change (Wang et al., 2019).

The effects of climate change on plants can be mitigated by simple adaptation strategies that involve cultivation and reproduction techniques. Some of these techniques involve sowing earlier spring crops (Tubiello et al., 2000), or using drought-resistant plants (Tkemaladze & Makhashvili, 2016), or with the exogenous application of growth regulators (Fahad et al., 2016; Jahan et al., 2019).

The main objective of this work is to characterize some effects resulting from climate change on plant metabolism and consequently on production. With this characterization, it was possible to identify some techniques to be used to mitigate the effects of climate change on agricultural production. For the realization of this chapter was based on literature review, our review is based on studies carried out in all around the world published in the period 2000- 2021. The keywords "high temperature", "drought", "high CO_2" and "mitigation", were used in combination with either "climate changes" or "agricultural production".

The chapter is organized into five sections. The introduction (Section 1) describes the effects of climate change on distinct factors such as carbon dioxide, temperature and water. Section 2 contains a review of the effect of carbon dioxide on agricultural production. In Section 3 the effect of temperature change on plant metabolisms and mitigation measures to try to reduce these effects are studied. In Section 4 the effect of drought on plant production and some mitigation measures to try to reduce these effects are studied. Section 5 presents the main conclusions as well as areas for further research.

EFFECT OF DIOXIDE CARBON ON AGRICULTURAL PRODUCTION

The concentration of CO_2 is increasing and is predicted to reach 550 ppm by 2050 (Bahrami et al., 2017). Este rising CO_2 contribute to the greening of the planet, while O_3 pollution has likely reduced terrestrial net primary productivity (Ainsworth et al., 2019). Photosynthesis is the principal process related with the primary production in the Biosphere. It fixes about 100bn tons for CO_2 each year of the 7000bn tons of CO_2 present in the atmosphere. The CO_2 assimilated by the photosynthesis is the support of crop production (Baslam et al., 2020). The intensity of photosynthesis increases with the increment of CO_2 in the atmosphere and attain maximum when your concentration is 0.8-1.0%. As the concentration of CO_2 in the atmosphere is $\approx$ 0.038%, plants have more potential for assimilation, i.e. it can effectively resist to the increase of CO_2 in the atmosphere (Tkemaladze & Makhashvili, 2016). There by a high atmospheric CO_2 concentration stimulates plant growth (AbdElgawad et al., 2016).

Thus, the role photosynthesis plays in the environment, by assimilate atmospheric CO_2 may be important to reduce the climate changing effects of atmospheric greenhouse gasses. When pairing the reestablishment of plant communities with other cultural techniques necessary to reduce soil loss, photosynthesis can promote soil health by providing an increase in soil organic matter and, thus sequestering atmospheric greenhouse gasses and favouring food production through soil regeneration (Craggs, 2016).

While increment dioxide carbon concentration over the past 150 years has been accompanied by higher CO_2 assimilation and storage in terrestrial ecosystems, exists evidence that temperatures change and drought stress can constraint the ability of future ecosystems to barrier against atmospheric emissions (Ainsworth et al., 2019). Such increase in carbon source, will affect plant metabolism, growth, and development (fertilizing effect), especially under favourable water and nutrient conditions. This effect may be brief, and differ among plant groups, namely between C3 and C4 type of plants photosynthesis (AbdElgawad et al., 2016).

Forests are one of the world's greatest sinks of atmospheric pollution. Forests represent about 50% of plant productivity and they cover 30% of Earth's terrestrial surface. However, the devastation of old forests for other purposes is continuing worldwide. In some parts of the world the loss are still growing. This plant loss result in less photosynthesis and the deforestation discharges the stored carbon to the atmosphere (Craggs, 2016).

As a result of a high concentration of CO_2 are changes in stomatal opening, could result in leaf stomata remaining open for less time, thus decreasing plant water uptake and reducing water vapor loss from the leaves (Jiang et al., 2020), elimination of photorespiration and increase in carbohydrates through CO_2 fixation (AbdElgawad et al., 2016). Influence flowering time regulation, involving a key flowering gene, MOTHER OF FT AND TFL1 (MFT) (Jagadish et al., 2016). Impacts postharvest quality causing malformation, occurrence of common scab, and changes in sugars substances on potatoes (Moretti et al., 2010). Promote seed yield, root biomass, aboveground biomass, leaf area and leaf biomass. The stimulation of CO_2 concentration in root biomass were greater in deeper soil layers (Uddin et al., 2018b). Which allow better access to sub-soil water during grain filling period, when additional water is converted into additional yield with high efficiency in Mediterranean-type dryland agro-ecosystems (Uddin et al., 2018a). Thus, that stimulation of belowground biomass may help to mitigate the impact of surface drought on biomass and grain yield if sufficient water is available in the sub-soil (Uddin et al., 2018b). Rodrigues et al. (2021) found that the great concentration of CO_2 mitigates the water insufficiency by antioxidant and osmoregulation activity, as well as the gathering of sugar-alcohols in the green prop roots, which can be responsible for the growth in biomass together with the cell proliferation.

High CO_2 concentration significantly affected plant nutrient ratios, which varied with plant characteristics and functional organs. So, for example great CO_2 concentration in rice and wheat enlarged C:N ratio in entire plant, and increased C:P ratio for rice. The changes in ratio were consequence to an increase in C level and a reduction in N level for both crops, and a reduction in P level by for rice (Wang et al., 2019). Despite a high concentration of CO_2 can promote aboveground biomass and grain yield in *Triticum aestivum* L. 'TRISO'. However, it can negatively affect the quality characteristics of the whole grain, such as a decrease in the total grain protein concentration and in the composition and concentration of proteins and amino acids, with greater reductions in non-essential amino acids than in essential ones. Decrease in minerals such as manganese, iron, cadmium and silicon while potassium, molybdenum and lead increased. The level of fructose and fructane, as well as the quantities per area of total and individual non-structural carbohydrates, except starch, enlarged expressively in the grain. The same happens with the amount of lipids. Relatively to the flour mixing and rheological properties, enlarged in gluten resistance under high CO_2 (Högy et al., 2009).

Bahrami et al. (2017) found that a high concentration of CO_2 in wheat caused a greater percentage of N remained as unassimilated [NO_3^-] in the tissue and this was significantly correlated with decreased grain protein. Thus, this high CO_2 concentration due affects the nitrate absorption capacity with direct negative action for grain quality.

In grapes although the increase in atmospheric CO_2 concentration advance grape maturity, decreased the elapsed time between fruit set and mid-veraison and between mid-veraison and maturity, respectively but delay amino acidic maturity, and did not affect anthocyanin concentration (Arrizabalaga-Arriazu et al., 2020).

However, an increase in CO_2 brings with it climate change, which includes global warming and extreme phenomena. These stress factors disturb the growth and metabolic pathways of plants. To neutralize the impacts of abiotic and biotic stresses, plants used diverse protection approaches, counting volatile organic compounds emission. Plants are producing diverse compounds such as terpenes, aromatic compounds, alcohols, and aldehydes (Copolovici et al., 2021). Gleadow et al. (2009) found an increase in CO_2 concentration in *Trifolium repens* enhance the ratio of total cyanogenic glycosides to total protein ratio was nearly two times higher in leaves. What could make *Trifolium repens* rich pastures bad for animals if the CO_2 concentration continues to rise.

EFFECT OF TEMPERATURE ON AGRICULTURAL PRODUCTION

Higher temperatures expected with global warming and the possibilities for more extreme events will have consequences on plant productivity (Hatfield & Prueger, 2015). The expected global warming comprises the rise in the global average air temperature. At the end of this century, the increases in air temperature will possibly be around 1.4–5.8 °C, comparative to the temperatures of 1980–1999, (Krishnan et al., 2011).

With current temperatures and the regularity of heat events projected to increase worldwide with climate change, produce reductions due to greater temperatures during the grain-filling stage could considerably destabilize future global food security (Asseng et al., 2011).

Temperature is one of the main factors that affects the rate of plant development (Hatfield & Prueger, 2015). Great temperature causes physiological, biochemical and molecular modifications that unfavourably affect tree evolution and productivity by reducing photosynthesis (Song et al., 2014a).

High temperature influences some metabolisms and consequently the production of some substances and resulting in lower crop yields (Jiang et al., 2020). So high temperatures affects photosynthesis (Moretti et al., 2010), accelerated crop growth rate, shortening the growing period for both corn and soybean (Jiang et al., 2020), affect the pollination (Hatfield & Prueger, 2015), hastened berry ripening, sugar accumulation, and malic acid breakdown (Arrizabalaga-Arriazu et al., 2020), flavonoids contents, firmness and antioxidant activity (Moretti et al., 2010).

Heat stress leads to changes in lipid-based signalling cascades and alterations in calcium transport and availability (Horváth et al., 2012).

Effect of Temperature on Phenological States

When analysing the effect of high temperature on phenological states, it is verified that pollination is one of the most sensitive phenological stages to temperature extremes across all species and during this developmental stage temperature extremes would greatly affect production (Hatfield & Prueger, 2015).

Growing temperature and raised levels of CO_2 are key global warming factors that could disturb plant fitness and flowering associated events, advancing flowering (Jagadish et al., 2016), which may occasionally lead to complete sterility. The impact of great temperatures at night is more disturbing than day-time or mean daily temperatures. Humidity also has a determinant role in growing the spikelet sterility at augmented temperature (Shah et al., 2011). In rice high night temperature was more detrimental for grain formation and yield while he high day temperature posed more negative effects on physiological attributes (Fahad et al., 2016).

In rice the number of panicles will be smaller by the earlier the maximum tillering stage. The dry-matter production is reduced by the elevation of leaf senescence during the ripening period. The 1000-grain weight declines due to the earlier termination of grain thickening growth. The percentage of ripened grains declines with the increase in the percentage of sterile spikelets (Oh-e et al., 2007).

The action of temperature perturbation was more evident on the tip of the cob maize compared to the base and older organs are well developed in vascular elements that sustain turgor pressure for wall extensibility in the growing cell wall and while younger reproductive structures become sensitive to stress on account of less vascularisation (Suwa et al., 2010).

As a strategy to resist high temperatures, it may go through to select for plants which shed pollen during the cooler periods of the day or are indeterminate so flowering occurs over a longer period of the growing season (Hatfield & Prueger, 2015).

Effect of Temperature on Physiology

Heat stress meaningly decrease seed germination and seedling growing, cell turgidity, and plant water-use efficiency (Akter & Islam, 2017). Influence plant productivity correlated with photosynthetic rates. Decrease photosynthetic rates in sugar maple (*Acer saccharum* Marsh.) (Gunderson et al., 2000). In rice reduced leaf area, above, and below-ground biomass, photosynthesis (Fahad et al., 2016). Djanaguiraman et al. (2018) report that the decrease in photosynthetic rate in wheat plants 'Chinese Spring' under high temperature stress is an interplay between thylakoid membrane damage, thylakoid membrane lipid composition, oxidative damage of cell organelle, and stomatal and non-stomatal restrictions. Under elevated temperature stress, cumulation of TAGs with no modification in unsaturation index as plastoglobules show lipid remodelling under stress.

The changes in lipid species indicates increases in activities of desaturating, oxidizing, glycosylating and acylating enzymes under high stress. Cumulative effect of high temperature stress is generation of reactive oxygen species, cell organelle damage, plasma and cell organelle membrane damage and reduced antioxidant enzyme activity indicating the imbalance among reactive oxygen species and antioxidant defence structure.

Alayafi (2020) also shown that the oxidative damage to cell components was established by greater levels of hydrogen peroxide, lipid peroxidation, electrolyte leakage, total oxidant status, and oxidative stress index. Furthermore, acute heat stress expressively reduced the photosynthetic pigment substances, and nutrient substances in tomato seedling leaves.

However, respiration showed little dependence on prevailing temperature during the growing season (Gunderson et al., 2000). Heskel et al. (2016) when analysing measurements in a comprehensive database for 231 species spanning 7 biomes, they demonstrate that temperature-dependent increases in leaf respiration do not follow a commonly used exponential function. Instead, they obtained a slowing function as leaves warm, reproducing a decreasing sensitivity to greater temperatures that is extraordinarily constant across all biomes and plant types.

The photosynthesis was inhibited in poplar with high temperature treatment (12 hours), that photosystems could be inhibited, together with the limitation of stomatal factors altering net photosynthetic rate. While electron transport rate was not changed, the characteristics of four genes (PETA, PETB, PETM and ATPA) for the redox chain was up-regulated at six hours heat stress, suggesting that cyclic electron transport might be encouraged by great temperature. In the Calvin cycle, three genes related with carboxylation were expressively inhibited, signifying that control of carboxylation processes is the probable reason for the failure in Rubisco activity, which causes photosynthesis to fall at higher temperatures (Song et al., 2014a). Decreased Rubisco activation state also was occurred with at high temperature in rice, wheat and maize. In turn, reduced Rubisco stimulation was not associated to changed quantities of Rubisco activase, but correlated with modifications in the rate of electron transport. Thus, Rubisco is as an objective important for increasing the photosynthetic performance of C3 (wheat and rice) and C4 (maize) cereal crops under progressively variable and climate changes (Perdomo et al., 2017). In 'Cabernet Sauvignon' leaves high temperatures promote the overclosure of PSII reaction centers results in the photoinhibition of PSII, while the stimulation of cyclic electron flow under heat stress contributes to the generation of a proton gradient (Sun et al., 2017). Heat stress (more 12 hours) caused reduced electron transport, damaged photosystems, activated the glycolate pathway and caused H_2O_2 production; as a result, photosynthetic capacity did not recover completely (Song et al., 2014a).

A transcription factor HYR (HIGHER YIELD RICE) is a master regulator, directly activating photosynthesis genes, cascades of transcription factors and other downstream genes involved in photosynthetic carbon metabolism and yield stability under environmental stress conditions (Ambavaram et al., 2014). Song, et al. (2014b) identified genes which included HSFA3, DREB2C, NAC1, MYB 60 and DOF1 that play important roles in regulating stomatal dynamics and heat stress responses in poplar under high temperature treatment.

The high temperature influencing photosynthesis will lead to a decrease in yield, for example Hatfield & Prueger (2015) verified grain yield in maize was significantly reduced by as much as 80–90%, whereas in Andean lupin that heat stress can have a significant effect on the quantity and quality of seed yield (Zou, 2009), can caused reductions in grain wheat production of up to 50% that can be attributed to increased leaf senescence (Asseng et al., 2011). Wheat production is projected to decrease by 6% for each °C of additional temperature rise and become more mutable over space and time (Asseng et al., 2015).

High temperature hastened berry ripening, sugar accumulation, and malic acid breakdown (Arrizabalaga-Arriazu et al., 2020).

Mitigation of Temperature Effects

The capacity of photosynthetic organisms to adjust to rises in temperatures is becoming more relevant with global warming. Temperature stress is recognized to encourage heat-shock proteins (HSPs) several of which act as chaperones. Usually, it has been supposed that protein denaturation is a trigger for HSP stimulation (Horváth et al., 2012). Protection of structural proteins, enzymes and membranes and expression of heat shock proteins (HSPs) are some of the biochemical processes that can impart thermotolerance (Shah et al., 2011).

Substitution of heat-sensitive cultivars by heat-tolerant ones, modification of sowing time, selection of varieties with a growth duration permitting prevention of peak stress periods, and exogenous application of plant hormones are some of the adaptive procedures that will help in the mitigation of estimate yield decrease in consequence of climate change (Shah et al., 2011).

Plant hormones also play fundamental roles in the capacity of plants to adapt to climate change by mediating growth, development and nutrient transport. Hormones circulate with specific pathways to controlling sites where they answer to stress at awfully low levels. All biological activities are impacted by both phytohormones (Fahad et al., 2015).

So, the exogenous application of plant growth regulators he can helpful in alleviating the adverse effects of high temperature. Among plant growth regulators combinations, the ascorbic acid + alpha-tocopherol + methyljasmonates + brassinosteroids was effective treatment for rice cultivars under high temperature stress. Therefore the greatest grain production by ascorbic acid + alpha-tocopherol + methyljasmonates + brassinosteroids treated plants was consequence of enhanced photosynthesis, spikelet fertility and grain filling, which recompensed the difficulties of high temperature levels (Fahad et al., 2016).

A pre-foliar supplementation of salicylic acid alleviated the heat stress-induced oxidative damage through the enhancement of photosynthetic function together with accelerated antioxidant enzymes activity (SOD, POD, CAT, and APX), and therefore, triggering more scavenging reactive oxygen species which protects the photosynthesis apparatus to help plants to survive in stressful conditions (Jahan et al., 2019).

The use of exogenous ascorbic acid (0.5 mM) could be suggested to encourage heat stress tolerance in tomato seedlings. Ascorbic acid assumed a priming influence on tomato roots and, significantly, improved heat stress effects on seedlings through decreasing the oxidative damage and growing the substances of ascorbic acid, proline, photosynthetic pigments, and upregulation of heat shock proteins in leaves (Alayafi, 2020).

EFFECT OF DROUGHT AND ITS EFFECT ON AGRICULTURAL PRODUCTION

Modifications in the water cycle in answer to the climate change over the twenty-first century will not be identical. The difference in precipitation (propensity to rise the rainfall intensity, but a reduction in its frequency (Chen & Dai, 2018)) among wet and dry areas and between wet and dry periods will possibly rise, while there may be regional exclusions. Global warming is contributing to the climate and on land much of that heat goes into drying. A natural drought must hence set in faster, become more intense, and longer. Droughts may be wider as a result (Trenberth et al., 2014). Increased temperature under climate change resulted in higher evapotranspiration; therefore, less water was lost to drainage or runoff (Jiang et al., 2020). Water and soil salinity probable to be augmented in parallel with the temperatures rise as a result of growing up the evapotranspiration, therefore irrigation demand will be augmented radically (Ashour & Al-Najar, 2012).

Drought and predicted changes in climate, such as high temperature, may affect the growth and productivity of crop plants and generate varying responses, including morphological, physiological (Figure 1), biochemical and molecular changes. Physiologically water deficit negatively affects transpiration in wheat (Shi et al., 2014), photosynthesis (Neto et al., 2021), decreased leaf Rubisco activity (Awasthi et al., 2014) in *Medicago sativa* (Aranjuelo et al., 2011), and consequently influence the yield and the biochemical seed-filling mechanisms (Awasthi et al., 2014), reducing seed size and number, eventually affecting the commercial trait '100 seed weight' and seed quality. Seed filling is impacted by numerous metabolic developments occurring in the leaves, particularly production and translocation of photoassimilates, introducing precursors for biosynthesis of seed reserves, minerals and other functional elements.

These developments are extremely sensitive to drought, in consequence of contribution of array of various enzymes and transporters, situated in the leaves and seeds (Sehgal et al., 2018).

Figure 1. Plant responses to drought (-) decrease, (+) increase

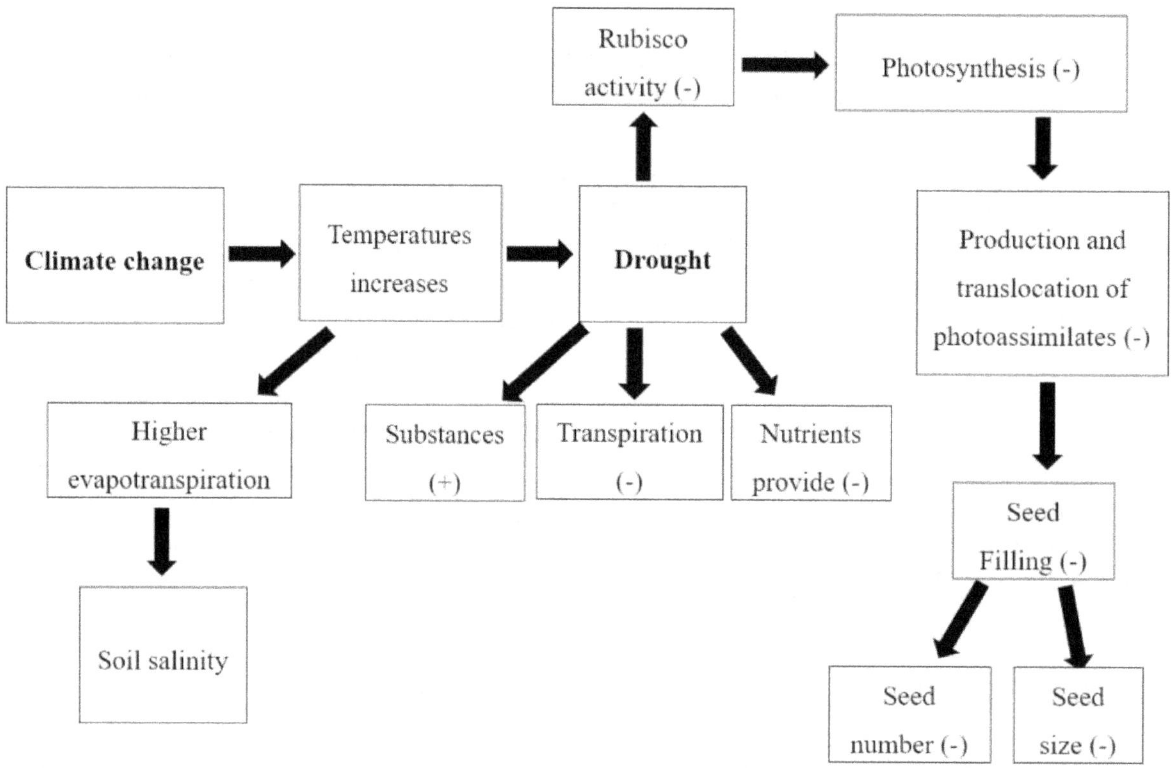

The drought can affect, plant height; rooting density, root fresh and dry weights, shoot fresh and dry weights, cell membrane thermostability, leaf temperature, osmotic and turgor potentials, peroxidase activity and the leaf area in maize (Ali et al., 2011). This reduction in leaf area in cotton can be by approximately 30% (Karademir et al., 2012). It can also affect the chlorophyll a content for example wheat cultivars with high transpiration efficiency also have a high chlorophyll content (Fotovat et al., 2007).

Drought can increase some substances production as ethylene, (Danish et al., 2020a), malate in the roots in barley (Sicher et al., 2012), proline in the nodule in alfafa (a major compound involved in osmotic regulation) and soluble sugar (D-pinitol) levels to contribute towards the decrease in osmotic potential (Ψs) (Aranjuelo et al., 2011).

Drought fight of plants is influenced by their physiological and anatomic-morphological particularities what is replicated in reduced request for water, capacity of cell compression and water use from the soil. Among different morphological peculiarities (Figure 2), scarcity of leaf surface and the ability of the root system to deepen into the soil are most noteworthy. Structurally drought resilient plants are characterized by small cells and a solid network of vascular bundles. They have a well-built cuticle, few small-sized stomata which precludes extra water disappearance and defends plants from warmness (Tkemaladze & Makhashvili, 2016).

Figure 2. Some anatomical effects that allow plants to resist drought
(Adapted Tkemaladze & Makhashvili, 2016)

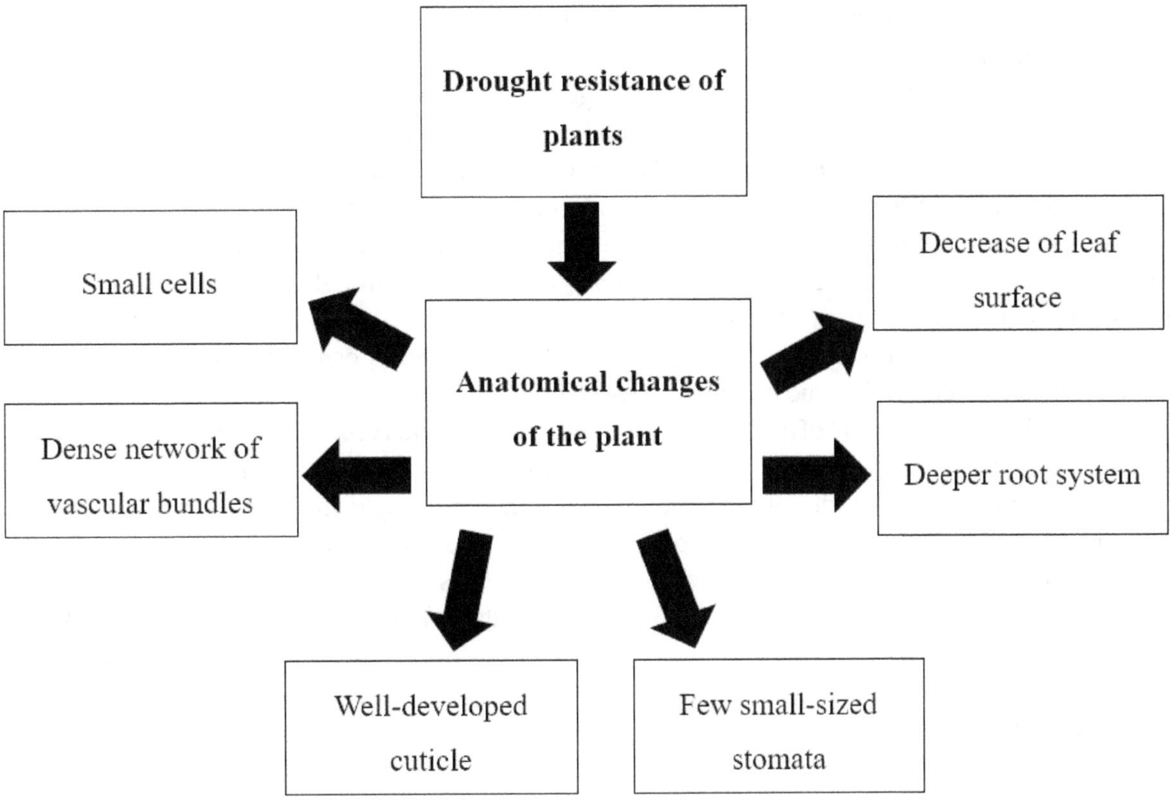

Mitigation of Drought Effects

To mitigate the effects of drought on plants, some treatments can be carried out, such as the inoculation of rhizobacteria into the rhizosphere of some plants. Thus, the inoculation in the rhizosphere of wheat and maize with plant growth-promoting rhizobacteria *Bacillus* sp. and *Enterobacter* sp. led to a delay in the onset of drought symptoms in plants, leading to increased root system growth. These plant growth-promoting rhizobacteria are produced and excreted indole-3-acetic acid and salicylic acid (Jochum et al., 2019). Plant growth promoting rhizobacteria *Achromobacter xylosoxidans* with higher rate of biochar in maize led to an increase in photosynthetic rate, stomatal conductivity and grain yield per plant under water stress conditions (Danish et al., 2020b). Danish et al. (2020a) also reported *Leclercia adecarboxylata* is drought tolerant ACC deaminase containing PGPR that might have the capacity to alleviate drought stress by increasing root elongation, nitrogen, phosphorus and potassium assimilation and possibility decreasing ethylene in plants.

The application of growth regulators as benzyl amino purine and salicylic acid they also decrease the effects of water stress on plants. For example, the application of benzyl amino purine (50 mg L^{-1}) in maize led to decreased effects of water stress, leading to an increase in the root fresh and dry weights, cell membrane thermostability, and chlorophyll a and b contents (Ali et al., 2011). Bulegon et al. (2019) also found the applications in soybean of foliar spray with plant regulator (containing auxin, gibberellin and cytokinin) and *Azospirillum brasilense* mitigate the effects of drought stress on photosynthesis and culminate in lower yield losses.

On the other hand, exogenous application of salicylic acid and potassium in *Vigna radiata* L. mitigated the adverse effects of drought stress significantly (Majeed et al., 2016). For example, this application in spinach under drought, increased shoot and root lengths, dry weights of plants, chlorophyll a, chlorophyll b and total chlorophyll. This superior growth, under drought conditions, was related with better root elongation and nutrients (N, P and K) assimilation (Gilani et al., 2020).

The application of other substances such as ϒ aminobutyric acid, trehalose, chitosan and yeast extract can also mitigate the effects of drought on plants. Thus Krishnan et al. (2013) was found that gamma aminobutyric acid application (50 mm) in *Lolium perenne* mitigated drought stress damage by maintaining higher relative water content and membrane stability. In *Phaseolus vulgaris* L. this approach carried by better performances in growth, water status, membrane integrity, osmotic adjustment, antioxidant protection and nutrient achievement (Abd El-Gawad et al., 2021).

Whereas the foliar treatment of trehalose in *Trigonella foenum graecum* plants led to increases in growth parameters, photosynthetic pigments, yield components, carbohydrate, protein, total phenolic, flavonoids contents, and antioxidant activity of the yielded seeds either in drought stressed plants (Sadak, 2016).

Although the mixture of yeast extract plus chitosan in stressed garlic plants directed to rise plant height, ascorbic acid concentration, and relative water level, as well as the chlorophyll a and b levels, controlled the proline content and concentration of antioxidant enzymes (Abdelaal et al., 2021).

CONCLUSION

Climate change is a current problem that influences abiotic characteristics such as the concentration of CO_2 in the atmosphere, temperature and drought. One of the most important metabolisms in the plant is photosynthesis. Plants through this metabolism assimilate CO_2, which may be important in reducing the effects of climate change from atmospheric greenhouse gases. However, rising temperatures and increasing drought can limit the ability of plants to fix CO_2. The high concentration of CO_2 stimulates production but lead to changes in the composition of grains and consequently in their nutritional value. CO_2 concentration may also play an important role in drought tolerance. High temperatures and drought will influence agricultural production through different metabolisms such as photosynthesis, the photosynthetic pigment contents, the leaf area, the above, and below-ground biomass and consequently yield. It can also lead to the production of secondary metabolites that could decrease the nutritional value of crops. But, respiration presented little dependency on predominant temperature during the developing season.

With this work we have seen some ways of mitigating the effects of climate change on plants, such as the use of thermosensitive cultivars for heat tolerant, adjustment of the sowing season, choosing varieties with growth duration that allows avoiding peak periods of stress and application exogenous plant regulators such as salicylic acid. With this work we have seen some ways of mitigating the effects of climate change on plants, such as the use of thermosensitive cultivars for heat tolerant, adjustment of the sowing season, choosing varieties with growth duration that allows avoiding peak periods of stress and application exogenous plant regulators such as salicylic acid.

However, in the future it is important to focus research on as many of these techniques as possible to mitigate the effects of climate change on plant production. For example, to study the best bacteria to inoculate into the rhizosphere of some plants. With this inoculation we will be able to reduce the adverse effects of drought.

REFERENCES

Abd El-Gawad, H. G., Mukherjee, S., Farag, R., Abd Elbar, O. H., Hikal, M., Abou El-Yazied, A., Abd Elhady, S. A., Helal, N., ElKelish, A., El Nahhas, N., Azab, E., Ismail, I. A., Mbarki, S., & Ibrahim, M. F. M. (2021). Exogenous γ-aminobutyric acid (GABA)-induced signaling events and field performance associated with mitigation of drought stress in *Phaseolus vulgaris* L. *Plant Signaling & Behavior*, *16*(2), 1853384. doi:10.1080/15592324.2020.1853384 PMID:33356834

Abdelaal, K., Attia, K. A., Niedbała, G., Wojciechowski, T., Hafez, Y., Alamery, S., Alateeq, T. K., & Arafa, S. A. (2021). Mitigation of Drought Damages by Exogenous Chitosan and Yeast Extract with Modulating the Photosynthetic Pigments, Antioxidant Defense System and Improving the Productivity of Garlic Plants. *Horticulturae*, *7*(11), 510. doi:10.3390/horticulturae7110510

AbdElgawad, H., Zinta, G., Beemster, G. T. S., Janssens, I. A., & Asard, H. (2016). Future Climate CO_2 Levels Mitigate Stress Impact on Plants: Increased Defense or Decreased Challenge? *Frontiers in Plant Science*, *7*, 556. doi:10.3389/fpls.2016.00556 PMID:27200030

Ainsworth, E. A., Lemonnier, P., & Wedow, J. M. (2019). The influence of rising tropospheric carbon dioxide and ozone on plant productivity. *Plant Biology*, *22*(S1, Suppl. 1), 5–11. doi:10.1111/plb.12973 PMID:30734441

Akter, N., & Rafiqul Islam, M. (2017). Heat stress effects and management in wheat. A review. *Agronomy for Sustainable Development*, *37*(5), 37. doi:10.100713593-017-0443-9

Alayafi, A. A. M. (2020). Exogenous ascorbic acid induces systemic heat stress tolerance in tomato seedlings: Transcriptional regulation mechanism. *Environmental Science and Pollution Research International*, *27*(16), 19186–19199. doi:10.100711356-019-06195-7 PMID:31448379

Ali, Z., Basra, S. M. A., Munir, H., Mahmood, A., & Yousaf, S. (2011). Mitigation of Drought Stress in Maize by Natural and Synthetic Growth Promoters. *Journal of Agriculture & Social Sciences*, *7*(2), 8.

Ambavaram, M. M. R., Basu, S., Krishnan, A., Ramegowda, V., Batlang, U., Rahman, L., Baisakh, N., & Pereira, A. (2014). Coordinated regulation of photosynthesis in rice increases yield and tolerance to environmental stress. *Nature Communications*, *5*(1), 5302. doi:10.1038/ncomms6302 PMID:25358745

Aranjuelo, I., Molero, G., Erice, G., Avice, J. C., & Nogués, S. (2011). Plant physiology and proteomics reveals the leaf response to drought in alfalfa (*Medicago sativa* L.). *Journal of Experimental Botany*, *62*(1), 111–123. doi:10.1093/jxb/erq249 PMID:20797998

Arrizabalaga-Arriazu, M., Gomès, E., Morales, F., Irigoyen, J. J., Pascual, I., & Hilbert, G. (2020). High Temperature and Elevated Carbon Dioxide Modify Berry Composition of Different Clones of Grapevine (*Vitis vinifera* L.) cv. Tempranillo. *Frontiers in Plant Science*, *11*, 1888. doi:10.3389/fpls.2020.603687 PMID:33335536

Ashour, E. K., & Al-Najar, H. (2012). The Impact of Climate Change and Soil Salinity in Irrigation Water Demand in the Gaza Strip. *Journal of Earth Science & Climatic Change, 03*(02). Advance online publication. doi:10.4172/2157-7617.1000120

Asseng, S., Ewert, F., Martre, P., Rötter, R. P., Lobell, D. B., Cammarano, D., Kimball, B. A., Ottman, M. J., Wall, G. W., White, J. W., Reynolds, M. P., Alderman, P. D., Prasad, P. V. V., Aggarwal, P. K., Anothai, J., Basso, B., Biernath, C., Challinor, A. J., De Sanctis, G., ... Zhu, Y. (2015). Rising temperatures reduce global wheat production. *Nature Climate Change, 5*(2), 143–147. doi:10.1038/nclimate2470

Asseng, S., Foster, I., & Turner, N. C. (2011). The impact of temperature variability on wheat yields. *Global Change Biology, 17*(2), 997–1012. doi:10.1111/j.1365-2486.2010.02262.x

Awasthi, R., Kaushal, N., Vadez, V., Turner, N. C., Berger, J., Siddique, K. H. M., & Nayyar, H. (2014). Individual and combined effects of transient drought and heat stress on carbon assimilation and seed filling in chickpea. *Functional Plant Biology: FPB, 41*(11), 1148–1167. doi:10.1071/FP13340 PMID:32481065

Bahrami, H., De Kok, L. J., Armstrong, R., Fitzgerald, G. J., Bourgault, M., Henty, S., Tausz, M., & Tausz-Posch, S. (2017). The proportion of nitrate in leaf nitrogen, but not changes in root growth, are associated with decreased grain protein in wheat under elevated [CO_2]. *Journal of Plant Physiology, 216*, 44–51. doi:10.1016/j.jplph.2017.05.011 PMID:28575746

Baslam, M., Mitsui, T., Hodges, M., Priesack, E., Herritt, M. T., Aranjuelo, I., & Sanz-Sáez, Á. (2020). Photosynthesis in a Changing Global Climate: Scaling Up and Scaling Down in Crops. *Frontiers in Plant Science, 11*, 882. doi:10.3389/fpls.2020.00882 PMID:32733499

Bulegon, L. G., Guimarães, V. F., Battistus, A. G., Inagaki, A. M., & da Costa, N. V. (2019). Mitigation of drought stress effects on soybean gas exchanges induced by *Azospirillum brasilense* and plant regulators. *Pesquisa Agropecuária Tropical, 49*. Advance online publication. doi:10.1590/1983-40632019v4952807

Chen, D., & Dai, A. (2018). Dependence of estimated precipitation frequency and intensity on data resolution. *Climate Dynamics, 50*(9), 3625–3647. doi:10.100700382-017-3830-7

Copolovici, L., Popitanu, A., & Copolovici, D. (2021). Volatile organic compound emission and residual substances from plants in light of the globally increasing CO_2 level. *Current Opinion in Environmental Science & Health, 19*, 100216. doi:10.1016/j.coesh.2020.10.004

Craggs, G. (2016). Photosynthesis and its Role in Climate Change and Soil Regeneration. *Future Directions International*, 1–6.

Danish, S., Zafar-Ul-Hye, M., Hussain, S., Riaz, M., & Qayyum, M. F. (2020). Mitigation of drought stress in maize through inoculation with drought tolerant ACC deaminase containing PGPR under axenic conditions. *Pakistan Journal of Botany, 52*(1). Advance online publication. doi:10.30848/PJB2020-1(7)

Danish, S., Zafar-ul-Hye, M., Mohsin, F., & Hussain, M. (2020). ACC-deaminase producing plant growth promoting rhizobacteria and biochar mitigate adverse effects of drought stress on maize growth. *PLoS One, 15*(4), e0230615. doi:10.1371/journal.pone.0230615 PMID:32251430

Djanaguiraman, M., Boyle, D. L., Welti, R., Jagadish, S. V. K., & Prasad, P. V. V. (2018). Decreased photosynthetic rate under high temperature in wheat is due to lipid desaturation, oxidation, acylation, and damage of organelles. *BMC Plant Biology*, *18*(1), 55. doi:10.118612870-018-1263-z PMID:29621997

Fahad, S., Hussain, S., Saud, S., Hassan, S., Ihsan, Z., Shah, A. N., Wu, C., Yousaf, M., Nasim, W., Alharby, H., Alghabari, F., & Huang, J. (2016). Exogenously Applied Plant Growth Regulators Enhance the Morpho-Physiological Growth and Yield of Rice under High Temperature. *Frontiers in Plant Science*, *7*, 1250. doi:10.3389/fpls.2016.01250 PMID:27625658

Fahad, S., Nie, L., Chen, Y., Wu, C., Xiong, D., Saud, S., Hongyan, L., Cui, K., & Huang, J. (2015). Crop Plant Hormones and Environmental Stress. In E. Lichtfouse (Ed.), *Sustainable Agriculture Reviews* (Vol. 15, pp. 371–400). Springer International Publishing., doi:10.1007/978-3-319-09132-7_10

Fotovat, R., Valizadeh, M., & Toorchi, M. (2007). Association between water-use efficiency components and total chlorophyll content (SPAD) in wheat (*Triticum aestivum* L.) under well-watered and drought stress conditions. *Journal of Food Agriculture and Environment*, *225–227*. Advance online publication. doi:10.1234/4.2007.1166

Gilani, M., Danish, S., Ahmed, N., Rahi, A. A., Akrem, A., Younis, U., Irshad, I., & Iqbal, R. K. (2020). Mitigation of drought stress in spinach using individual and combined applications of salicylic acid and potassium. *Pakistan Journal of Botany*, *52*(5). Advance online publication. doi:10.30848/PJB2020-5(18)

Gleadow, R. M., Edwards, E. J., & Evans, J. R. (2009). Changes in Nutritional Value of Cyanogenic *Trifolium repens* Grown at Elevated Atmospheric CO_2. *Journal of Chemical Ecology*, *35*(4), 476–478. doi:10.100710886-009-9617-5 PMID:19352773

Gunderson, C. A., Norby, R. J., & Wullschleger, S. D. (2000). Acclimation of photosynthesis and respiration to simulated climatic warming in northern and southern populations of *Acer saccharum*: Laboratory and field evidence. *Tree Physiology*, *20*(2), 87–96. doi:10.1093/treephys/20.2.87 PMID:12651476

Hatfield, J. L., & Prueger, J. H. (2015). Temperature extremes: Effect on plant growth and development. *Weather and Climate Extremes*, *10*, 4–10. doi:10.1016/j.wace.2015.08.001

Heskel, M. A., O'Sullivan, O. S., Reich, P. B., Tjoelker, M. G., Weerasinghe, L. K., Penillard, A., Egerton, J. J. G., Creek, D., Bloomfield, K. J., Xiang, J., Sinca, F., Stangl, Z. R., Martinez-de la Torre, A., Griffin, K. L., Huntingford, C., Hurry, V., Meir, P., Turnbull, M. H., & Atkin, O. K. (2016). Convergence in the temperature response of leaf respiration across biomes and plant functional types. *Proceedings of the National Academy of Sciences of the United States of America*, *113*(14), 3832–3837. doi:10.1073/pnas.1520282113 PMID:27001849

Högy, P., Wieser, H., Köhler, P., Schwadorf, K., Breuer, J., Franzaring, J., Muntifering, R., & Fangmeier, A. (2009). Effects of elevated CO_2 on grain yield and quality of wheat: Results from a 3-year free-air CO_2 enrichment experiment. *Plant Biology*, *11*(s1), 60–69. doi:10.1111/j.1438-8677.2009.00230.x PMID:19778369

Horváth, I., Glatz, A., Nakamoto, H., Mishkind, M. L., Munnik, T., Saidi, Y., Goloubinoff, P., Harwood, J. L., & Vigh, L. (2012). Heat shock response in photosynthetic organisms: Membrane and lipid connections. *Progress in Lipid Research*, *51*(3), 208–220. doi:10.1016/j.plipres.2012.02.002 PMID:22484828

Jagadish, S. V. K., Bahuguna, R. N., Djanaguiraman, M., Gamuyao, R., Prasad, P. V. V., & Craufurd, P. Q. (2016). Implications of High Temperature and Elevated CO_2 on Flowering Time in Plants. *Frontiers in Plant Science*, *7*, 913. doi:10.3389/fpls.2016.00913 PMID:27446143

Jahan, M. S., Wang, Y., Shu, S., Zhong, M., Chen, Z., Wu, J., Sun, J., & Guo, S. (2019). Exogenous salicylic acid increases the heat tolerance in Tomato (*Solanum lycopersicum* L) by enhancing photosynthesis efficiency and improving antioxidant defense system through scavenging of reactive oxygen species. *Scientia Horticulturae*, *247*, 421–429. doi:10.1016/j.scienta.2018.12.047

Jiang, Q., Qi, Z., Xue, L., Bukovsky, M., Madramootoo, C. A., & Smith, W. (2020). Assessing climate change impacts on greenhouse gas emissions, N losses in drainage and crop production in a subsurface drained field. *The Science of the Total Environment*, *705*, 135969. doi:10.1016/j.scitotenv.2019.135969 PMID:31838422

Jochum, M. D., McWilliams, K. L., Borrego, E. J., Kolomiets, M. V., Niu, G., Pierson, E. A., & Jo, Y.-K. (2019). Bioprospecting Plant Growth-Promoting Rhizobacteria That Mitigate Drought Stress in Grasses. *Frontiers in Microbiology*, *10*, 2106. doi:10.3389/fmicb.2019.02106 PMID:31552009

Karademir, C., Karademir, E., Çopur, O., & Gencer, O. (2012). Effect of Drought Stress on Leaf Area in Cotton (*Gossypium hirsutum* L.). *11th Meeting of Inter-Regional Cooperative Research Network on Cotton for the Mediterranean and Middle East Regions*.

Kreuzwieser, J., & Gessler, A. (2010). Global climate change and tree nutrition: Influence of water availability. *Tree Physiology*, *30*(9), 1221–1234. doi:10.1093/treephys/tpq055 PMID:20581013

Krishnan, P., Ramakrishnan, B., Reddy, K. R., & Reddy, V. R. (2011). Chapter three—High-Temperature Effects on Rice Growth, Yield, and Grain Quality. In D. L. Sparks (Ed.), *Advances in Agronomy* (Vol. 111, pp. 87–206). Academic Press., doi:10.1016/B978-0-12-387689-8.00004-7

Krishnan, S., Laskowski, K., Shukla, V., & Merewitz, E. B. (2013). Mitigation of Drought Stress Damage by Exogenous Application of a Non-Protein Amino Acid γ– Aminobutyric Acid on Perennial Ryegrass. *Journal of the American Society for Horticultural Science*, *138*(5), 358–366. doi:10.21273/JASHS.138.5.358

Majeed, S., Akram, M., Latif, M., Ijaz, M., & Hussain, M. (2016). Mitigation of drought stress by foliar application of salicylic acid and potassium in mungbean (*Vigna radiata* L.). *Legume Research -. International Journal (Toronto, Ont.)*, *39*(2). Advance online publication. doi:10.18805/lr.v39i2.9532

Moretti, C. L., Mattos, L. M., Calbo, A. G., & Sargent, S. A. (2010). Climate changes and potential impacts on postharvest quality of fruit and vegetable crops: A review. *Food Research International*, *43*(7), 1824–1832. doi:10.1016/j.foodres.2009.10.013

Neto, J. B., Silva, J. R. I., Bezerra, C. W. F., Amorim, T. L., Hermínio, P. J., Morato, R. P., Albuquerque-Silva, M. M., & Simões, V. J. L. P. (2021). Impactos da seca e das mudanças climáticas no metabolismo das plantas: Uma revisão. *Research. Social Development*, *10*(8), e16710817060–e16710817060. doi:10.33448/rsd-v10i8.17060

Oh-e, I., Saitoh, K., & Kuroda, T. (2007). Effects of High Temperature on Growth, Yield and Dry-Matter Production of Rice Grown in the Paddy Field. *Plant Production Science*, *10*(4), 412–422. doi:10.1626/pps.10.412

Perdomo, J. A., Capó-Bauçà, S., Carmo-Silva, E., & Galmés, J. (2017). Rubisco and Rubisco Activase Play an Important Role in the Biochemical Limitations of Photosynthesis in Rice, Wheat, and Maize under High Temperature and Water Deficit. *Frontiers in Plant Science*, *8*, 490. doi:10.3389/fpls.2017.00490 PMID:28450871

Rodrigues, T. de S., Arge, L. W. P., Guedes, F. A. de F., Travassos-Lins, J., de Souza, A. P., Cocuron, J.-C., Buckeridge, M. S., Grossi-de-Sá, M. F., & Alves-Ferreira, M. (2021). *Integrative Analysis of Sorghum Bicolor Green Prop Roots Under Elevated CO_2 and Water Deficit Conditions* (p. 2021.03.12.435150). doi:10.1101/2021.03.12.435150

Sadak, M. S. S. (2016). Mitigation of drought stress on Fenugreek plant by foliar application of trehalose. *International Journal of Chemtech Research*, *9*(2), 147–155.

Sehgal, A., Sita, K., Siddique, K. H. M., Kumar, R., Bhogireddy, S., Varshney, R. K., HanumanthaRao, B., Nair, R. M., Prasad, P. V. V., & Nayyar, H. (1705). HanumanthaRao, B., Nair, R. M., Prasad, P. V. V., & Nayyar, H. (2018). Drought or/and Heat-Stress Effects on Seed Filling in Food Crops: Impacts on Functional Biochemistry, Seed Yields, and Nutritional Quality. *Frontiers in Plant Science*, *9*, 1705. Advance online publication. doi:10.3389/fpls.2018.01705 PMID:30542357

Shah, F., Huang, J., Cui, K., Nie, L., Shah, T., Chen, C., & Wang, K. (2011). Impact of high-temperature stress on rice plant and its traits related to tolerance. *The Journal of Agricultural Science*, *149*(5), 545–556. doi:10.1017/S0021859611000360

Shi, J., Yasuor, H., Yermiyahu, U., Zuo, Q., & Ben-Gal, A. (2014). Dynamic responses of wheat to drought and nitrogen stresses during re-watering cycles. *Agricultural Water Management*, *146*, 163–172. doi:10.1016/j.agwat.2014.08.006

Sicher, R., Timlin, D., & Bailey, B. (2012). Responses of growth and primary metabolism of water-stressed barley roots to rehydration. *Journal of Plant Physiology*, *169*(7), 686–695. doi:10.1016/j.jplph.2012.01.002 PMID:22285575

Song, Y., Chen, Q., Ci, D., Shao, X., & Zhang, D. (2014). Effects of high temperature on photosynthesis and related gene expression in poplar. *BMC Plant Biology*, *14*(1), 111. doi:10.1186/1471-2229-14-111 PMID:24774695

Song, Y., Ci, D., Tian, M., & Zhang, D. (2014). Comparison of the physiological effects and transcriptome responses of *Populus simonii* under different abiotic stresses. *Plant Molecular Biology*, *86*(1–2), 139–156. doi:10.100711103-014-0218-5 PMID:25002226

Sun, Y., Geng, Q., Du, Y., Yang, X., & Zhai, H. (2017). Induction of cyclic electron flow around photosystem I during heat stress in grape leaves. *Plant Science*, *256*, 65–71. doi:10.1016/j.plantsci.2016.12.004 PMID:28167040

Suwa, R., Hakata, H., Hara, H., El-Shemy, H. A., Adu-Gyamfi, J. J., Nguyen, N. T., Kanai, S., Lightfoot, D. A., Mohapatra, P. K., & Fujita, K. (2010). High temperature effects on photosynthate partitioning and sugar metabolism during ear expansion in maize (*Zea mays* L.) genotypes. *Plant Physiology and Biochemistry*, *48*(2), 124–130. doi:10.1016/j.plaphy.2009.12.010 PMID:20106675

Tkemaladze, G. S., & Makhashvili, K. A. (2016). Climate changes and photosynthesis. *Annals of Agrarian Science*, *14*(2), 119–126. doi:10.1016/j.aasci.2016.05.012

Trenberth, K. E., Dai, A., van der Schrier, G., Jones, P. D., Barichivich, J., Briffa, K. R., & Sheffield, J. (2014). Global warming and changes in drought. *Nature Climate Change*, *4*(1), 17–22. doi:10.1038/nclimate2067

Tubiello, F. N., Donatelli, M., Rosenzweig, C., & Stockle, C. O. (2000). Effects of climate change and elevated CO_2 on cropping systems: Model predictions at two Italian locations. *European Journal of Agronomy*, *13*(2), 179–189. doi:10.1016/S1161-0301(00)00073-3

Uddin, S., Löw, M., Parvin, S., Fitzgerald, G. J., Tausz-Posch, S., Armstrong, R., O'Leary, G., & Tausz, M. (2018). Elevated [CO_2] mitigates the effect of surface drought by stimulating root growth to access sub-soil water. *PLoS One*, *13*(6), e0198928. doi:10.1371/journal.pone.0198928 PMID:29902235

Uddin, S., Löw, M., Parvin, S., Fitzgerald, G. J., Tausz-Posch, S., Armstrong, R., & Tausz, M. (2018). Yield of canola (*Brassica napus* L.) benefits more from elevated CO_2 when access to deeper soil water is improved. *Environmental and Experimental Botany*, *155*, 518–528. doi:10.1016/j.envexpbot.2018.07.017

Wang, J., Liu, X., Zhang, X., Li, L., Lam, S. K., & Pan, G. (2019). Changes in plant C, N and P ratios under elevated [CO_2] and canopy warming in a rice-winter wheat rotation system. *Scientific Reports*, *9*(1), 5424. doi:10.103841598-019-41944-1 PMID:30931987

Zou, L. (2009). *Effects of gradual and sudden heat stress on seed quality of Andean lupin, Lupinus mutabilis* [Master Degree]. University of Helsinki - Faculty of Agriculture and Forestry.

Chapter 7
Effect of Global Warming on Food Security:
An Indian Perspective

Arti Yadav
Indian Council of Social Science Research, India

Badar Alam Iqbal
University of South Africa, South Africa

ABSTRACT

This chapter will aim to explicate the challenges posed by global warming or the climate change conditions on food security especially from the point of view of India. The negative impact of global warming has been seen, especially in developing economies, on the agricultural yields leading towards food insecurity. The four pillars of food security (i.e., availability, accessibility, utilization, and stability) are having an impact on climate change. The present study will begin by highlighting the concept of global warming. It will further provide an overview of the Indian food security system followed by the impact of global warming on the food security level in India. The study will also highlight the global warming and food security scenario in the present situation of the ongoing COVID-19 pandemic in India.

INTRODUCTION

Food is the basic necessity for the survival of humans on the earth and the prevailing system of food consisting of Production, Processing, Packaging, Storage, Transport, Retail, Consumption, Waste and Loss like stages, feeds, and provides livelihood to over one billion population around the world. There has been an increase of more than 30 percent of the food supply per capita since 1961 (Stuart, 2015). However, in recent years, global warming has posed out various issues cocerning to world agriculture and the Food security system. The food system gets affected by both the climate-related factors (increasing temperature, water level, land degradation, etc.) as well as non-climate factors (demand, income growth, population level, etc.). The climate and non-climate factors are also impacting the food security system in

DOI: 10.4018/978-1-7998-9557-2.ch007

terms of its availability, accessibility, utilization, and stability (Raza et al., 2019). The impact of climate change has been seen both ways positively as well as negatively; in higher latitude regions the climate change has affected positively in recent decades for the yield of some crops like wheat, sugar beets, and maize, while it has affected negatively in lower latitude regions for the yield of crops like wheat and maize (Ali et al., 2017). Drylands mostly those in Africa, South America, and high mountain regions of Asia have seen the significant impact of climate change on their food security system (IPCC, 2019). The effects of global warming have been largely seen on the agriculture economies than on the industrial countries. Food alone is responsible for 26 percent of global greenhouse gas emissions (Ritchie, 2019). The developing economies are mostly agriculture-based economies and are based mostly in the warmer parts of the Earth. So, any increase in the temperature beyond the limit will rather than increasing reduce the output of agriculture (European Environment Agency, 2015). Further, food production also contributes the global warming as before reaching the plates, it goes through the process of production, packaging, transportation, preparation, and serving (FAO, 2017). All these activities lead to directly or indirectly in the generation of greenhouse gases in the atmosphere. For instance, in the case of the European Union, agriculture accounted for 10% of the total greenhouse gas emissions in 2012 (Olivier and Peters, 2020).

Not only ecological systems but also human life got affected by the acceleration of global warming, therefore, making global warming a significant issue both at the national and international level (World Meteorological Organization, 2021). In addition to that rising and growing population leads to having an impact on the livelihoods and environment as the competition is rising between energy, water, agriculture, forestry, fisheries, transport mining, and other sectors (FAO, 2017). In the case of India, the majority of the regions are considered to have tropical climates. The different regions in India show a wide diversity of temperatures. However, the monsoon season dominates the climate of India (Pradhan, Singh, and Singh, 2019). The effect of global warming on the climate of India has lead to climate disasters in the form that according to a study, out of 35 states, 27 are being disaster-prone and the most frequent disaster is on the foods (Ministry of Home Affairs, Government of India, 2011). Moreover, the present situation of pandemic covid-19 has only added to the plight of climate change over the food security system worldwide and India is not an exception. The pandemic has hit India at the beginning of 2020 and since then to curb its spread various steps have been taken by the government including lockdown to protect human lives. However, ordinary people and the local food security system have a radical impact of the adopted measures in India (Pothan, Taguchi and Santini, 2020). In light of the above considerations, the present study aims to examine the impact of climate change on food security in India. For this, the impact of carbon emission will be analyzed on the agriculture value added in India. Additionally, it will scrutinize the effect of climate change on food security during COVID-19 and try to explore the remedial measures and opportunities in the post-pandemic situation.

LITERATURE REVIEW

Global warming through affecting the climate has given rise to various issues and challenges in terms of the achievement of food and nutritional security. Due to climate change, there has been the emergence of issues like degradation of resources like soil, biodiversity, and water; the rise of sea level; melting of glaciers; change in the pattern of temperature and rainfall, and submergence of coastal areas (UNFCCC, 2007; Nunez, 2019). In several agro-climatic regions, the impact of climate change can easily be seen on the productivity of agriculture (Singh, 2016). Climate change is also one of the reasons which are affecting food security globally; other factors are food and water scarcity and consistently rising population (Ruwoldt, 2013). Ericksen, Ingram and Liverman (2009) exhibited that the food security issue will not be solved by taking into consideration only the technical aspect, there is a need for a food system that should not be just based on agricultural practices but should be more integrated. Further, food security issues also emerged due to variability and volatility of the price, therefore, there is a need to have more focus on the food system governance (Kalkuhl, Braun and Torero, 2016; Termeer et.al, 2018). Cassia et.al. (2018) found that the selective pressures building a new Earth's landscape are due to the accelerating rate of climate change along with habitat fragmentation caused by human activity. In addition to that, the authors found that climate change is multidimensional in terms of variation in frequency, duration, and intensity of parameters like precipitation and temperature, leading to changes in the seasons and Earth's life. Zhao et.al. (2017) assessed the impact of global temperature increase on yields of the four crops through compiling four analytical methods like global grid-based and local point-based models, field-warming experiments, and statistical regressions. The study found that the multi-method analyses improved the confidence in assessments of future climate impacts on global major crops, with important implications for developing crop and region-specific adaptation strategies to ensure future food supply of an increasing world population and independent methods consistently estimated negative temperature impacts on yields of four major crops at the global scale, generally underpinned by similar impacts at country and site scales.

Various studies have also found that in the case of developing countries, the impact of climate change will have an adverse impact on the food security system as it will increase their dependence on imports (Ludwig et.al., 2007; Ahmed and Suphachalasai, 2014). The impact will be in a disproportional form which will highly impact the vulnerable section of the society leading to an increase in inequality. The food instability issue, in the long run, will be one of the factors defining the socio-economic development path of a nation along with climate-related matters (Schmidhuber and Tubiello, 2007). Nelson et.al. (2009) studied the impact of climate change on agriculture and costs of adaptation through detailed modeling of crop growth and the global agriculture model under climate change. The study found that climate change affects negatively both human well-being and agriculture as production gets affected along with a decline in the yields of the crop, an increase in the prices of meat and crop, and a fall in the cereals consumption. This will have a huge impact through increased child malnutrition and reduction in calorie intake. Ranuzzi and Srivastava (2012) examined the multiple effects on food production and food security of global warming and climate change through applying the Pressure-Impact-Response Framework. To lower the emission from the agriculture sector, the study proposes suitable mitigation strategies for the integration of the overall agenda and approach. It also assessed the inter linkages between food production and ecosystem services in terms of the contribution towards food security. The study suggests that to address the future uncertainties of climate change, Strategies used by farming communities to cope with climate crises in the past will be instrumental. Bandara and Cai (2014) through applying the global dynamic computable general equilibrium model studied the impact of climate change on crop productivity, food prices, and food security in South Asian economies especially Bangladesh, India, Nepal, Pakistan, and Sri Lanka. The study found for all South Asian economies climate change has a negative and significant impact on food production and already rising prices. Moreover, those South Asian economies which are closer to the equator have severe effects compared to the higher altitude countries (Sarkar, 2018). Chakrabarty (2016) studied the impact of climate change on the food security of India in terms of its availability, accessibility, and absorption capacity. The author found that the issue of climate change will put forth challenges for the food security system in India and suggests for the espousal of sustainable agriculture practices in terms of better efforts towards public health and urban food security system, measures in the long run for respite during natural disaster and efforts towards the security of livelihood.

Rao, Prasad and Mohapatra (2019) studied the impact of climate change on India in terms of its impact through cyclones, draughts, floods, hailstorms, coastal salinity, and heatwaves. The study suggests that to combat uncommon incidents due to climate change, a comprehensive assessment and understanding of the adaptation options is highly significant to maintain the sustainability of life. Further, the responses of climate adaptation strategies have shown positive impacts in the Indian scenario. Kaur and Kaur (2017) analyzed the impact of climate change on Indian agriculture and food security. The result shows that for addressing food insecurity in India, the policy initiatives should be based on agriculture productivity enhancement through strategies based on improved irrigation facilities, for addressing food insecurity adaption of direct interventionist measures and developed infrastructure facilities. Sukhwani, Deshkar and Shaw (2020) studied the impact of COVID-19 on the urban-rural partnership concerning the food system in Nagpur city of India. Through using the survey method at two different periods the study found that the concerns related to food and grocery started to rise during the pandemic and people found government sources such as their apps and websites as most reliable. Henry (2020) suggested that to address the long-term issues related to food security such as population growth, climate change, and degradation of the environment, plant sciences will play a significant role. In addition to that, there is a need for policies both at the public and private levels which should support the technology and infrastructure at the regional level in the post-pandemic scenario.

OVERVIEW OF THE CONCEPT OF GLOBAL WARMING

Global warming occurs when there is a significant change in the atmosphere of the Earth due to an increase in the average temperature affecting the global climate in a sustained manner (Stevens, 2011). The increase in the global temperature is mainly due to the enhanced greenhouse effect in the atmosphere causing entrapment of solar radiations at the higher level. The imbalance in the Earth's temperature is due to the increase of carbon dioxide (around two-third) in the climate structure (Zillman and Sherwood, 2017). The greenhouse gases mainly consist of synthetic fluorinated gases, water vapor, nitrous oxide, methane, and carbon dioxide that are heat-trapping pollutants and their impact is known as the greenhouse effect (US EPA, 2021).

Human activities impact the climate by generating greenhouse gases, affecting the atmosphere through the emission of pollutants, and changing the nature of the land surfaces like for farming, clearing of forests, which alters Earth's energy balance (The Royal Society, 2020). Compared to the past 8,00,000 years, the average carbon dioxide in the atmosphere was around 409.8 parts per million at the global level in 2019 (Lindsey, 2020). The global annual temperature has increased in total by a little more than 1 degree Celsius since the industrial revolution and for the last forty years, a rise by 0.18 degree Celsius per decade has been noticed (Lindsey and Dahlman, 2021). In addition to that since 1880, there have been nine out of ten warmest years, between 2005 to 2015, there have been five warmest years (MacMillan and Turrentine, 2021).

The rise in the temperature might seem small, however, it takes an enormous amount of heat energy to raise Earth's average yearly surface temperature even a little amount given the size and tremendous heat capacity of the global oceans (Figure 1) (Dahlman and Lindsey, 2020). Reduction in the snow cover, extremes in the seasonal and regional snow cover, heavy rainfall intensification, sea ice, changes in the habitat ranges for the plants and animals are some of the effects of global warming (Buis, 2019). Moreover, the spread is not evenly in terms of the impact of climate change such as an increase in temperature at different levels as land areas are generally warmer than the oceans, therefore, risks associated with climate are higher for the disadvantaged communities, people, and at lower latitudes (McMichael et.al, 2003).

Figure 1. Global surface temperature (1880-2020)
Source: Lindsey and Dahlman, 2021

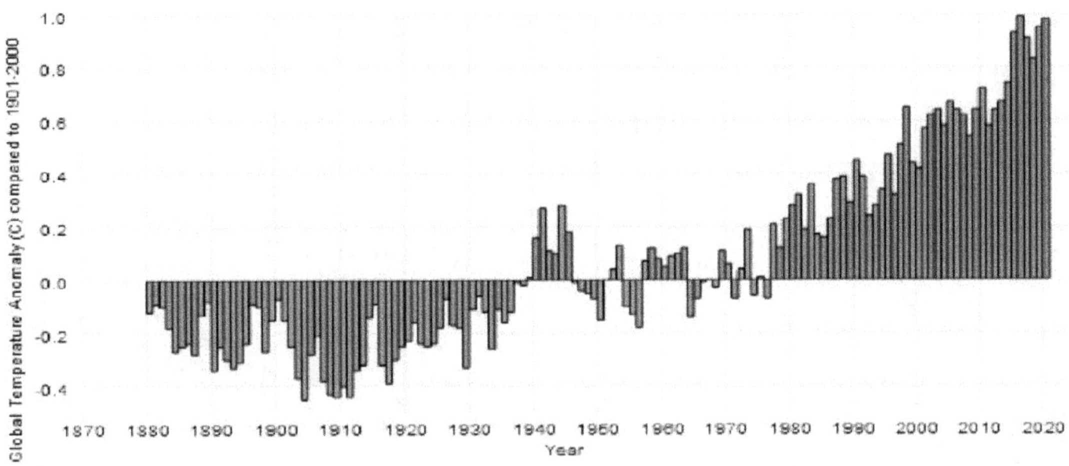

Climate change also poses risks to economic growth at the global level. As per a study, for each degree Celsius increase in global warming the United States could lose 2.3 percent of its Gross Domestic Product (International Monetary Fund, 2019). The economic impacts mostly consist of coastal flooding and erosion due to the rising sea level; coastal groundwater supplies salinity changes leading to stress in freshwater, affecting the marine ecosystem, wildfires, droughts, and an increase in the intensity of tropical cyclones (Swiss Re, 2021). The impact of loss to economic growth affects all people in one way or another, however, mostly to the underprivileged section for which climate change is often a key driver of poverty, displacement, hunger, and social unrest (UNHCR, 2021). Further, with warmer temperatures and longer growing seasons, some areas might experience favorable changes in the cropping patterns leading to increases in crop yields, however, on agriculture production, the average impact of climate change is negative i.e., it surpasses the positive aspects (Gornall, 2010). In large developing economies like India and China, the impact of climate change is based on the investment measures and the policy initiatives leading to significant differences in production across regions. On smallholders, the impact of climate change can be severe due to limited opportunities for adaptation and limited resources (IPCC, 2018).

HOW GLOBAL WARMING AFFECTS FOOD SECURITY: INDIAN SCENARIO

The impact of climate change can be seen on the changing pattern of global food production in terms of an increase in food insecurity level, whose direct impact can be seen on an increase in the food prices and indirect impact will be on poverty level amplification (World Water Development Report, 2020). All the dimensions of food security (availability, accessibility, utilization, and systems stability) get affected by climate change (Figure 2). The impact of climate change is both short-term (changes in weather conditions in the extreme and intense form) and long-term (changing pattern of temperature and precipitation) (FAO, 2016). According to a report by FAO (2017), global agriculture production will have to increase by 60 percent by 2050 to meet the growing demand for agriculture and allied products. Moreover, growth in the demand for dairy products, meat, fish and feed, etc. is also expected to skew food demand due to rising incomes and economic development. The need for increased global food production of about 90 percent is expected mainly in the developing economies and whose own production is expected to increase by 55 percent in 2050 (International Monetary Fund, 2020).

Table 1. Impact of climate change on crop production in South Asia (Estimated 2050s)

Crops	Crop production (2000 Year)	Crop as percentage of total Production	Projected yield improvement No climate change (% per annum)	Crop production 2050s, No Climate Change	Crop Production 2050s with Climate Change and No Co2 Fertilization Effect	Average Annual yield change with Climate Change
Rice(mmt)	120	48	0.9	169	145	-0.2
Wheat(mmt)	97	38	1.6	191	103	-1.3
Maize(mmt)	16	6	0.6	19	16	0.1
Millet (mmt)	11	4	1.5	12	11	0.0
Sorghum (mmt)	8	3	1.2	10	8	1.4
Total (mmt)	252			401	282	

Source: The World Bank Report (2013)

Table 1 shows the impact of climate change on crop production in the South Asian region. The instant risk of climate change for agriculture is on the livelihood of the associated section of the society due to the generation of different kinds of pests and diseases, growth in crop failure cases, livestock loss, and non-availability of requisite and planting material (Pretty and Bharucha, 2014). Climate change is going to increase the vulnerability of the agriculture sector for developed as well as developing economies. According to World Food Summit (1996) "Food security exists when all people, at all times, have physical and economic access to sufficient, safe and nutritious food that meets their dietary needs and food preferences for an active and healthy life". Hence, there are four key dimensions of food security: Availability, Accessibility, Utilization, and Stability. Despite a decade of global commitment hunger and food insecurity still persists at astounding rates with nearly 750 million people experienced severe food insecurity in 2019 and the number of undernourished or food-insecure people is rising, with climate shocks a major contributor (Hobert and Negra, 2020). The development with respect to food security till now across the globe has been threatened by climate change. According to IPCC AR5 (2014), the key risks having a direct impact on the security of food are rural livelihood and income loss, marine and coastal ecosystem loss, inland water ecosystem and terrestrial loss, breakdown of food systems, and food insecurity.

Figure 2. Impact of climate change on food security
Source: Food and Agriculture Organization (FAO) of the United Nations (2016)

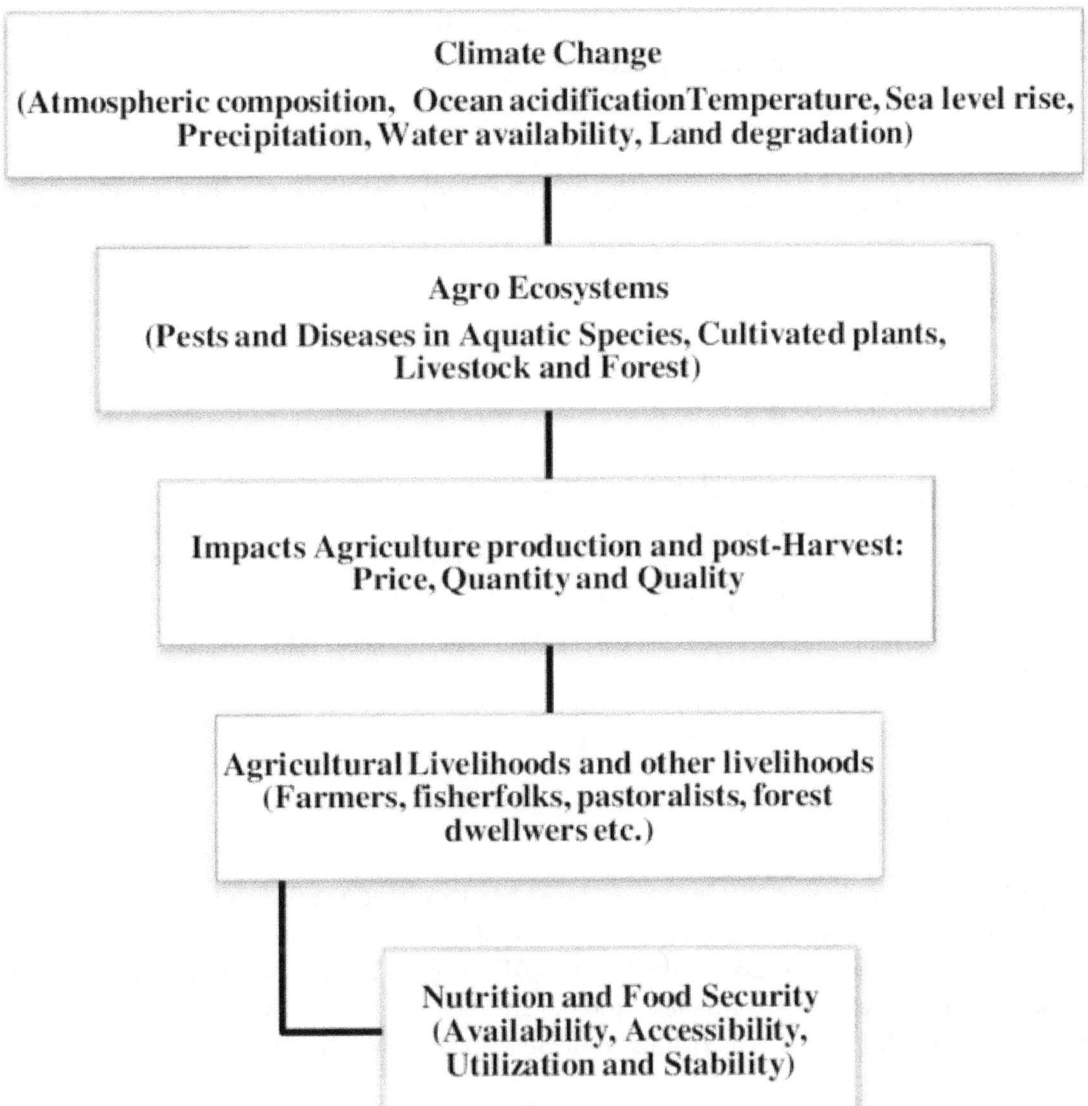

Moreover, the significance of climate change for food security is based on different regions like in the southern part of Africa food insecurity is influenced by the climate in terms of short-lived shock, underlying and ongoing issues. While in the Indo-Gangetic Plain of India, the direct effects of climate change as factors influencing food security are lower than drivers like the availability and quality of groundwater for irrigation and labour issues (Gregory, Ingram and Brklacich, 2005; Food and Agriculture

Organization of the United States, 2019). In the case of India providing access to sufficient nutritious food to the people is still a concern for the authorities even when in terms of purchasing power parity it is the third-largest economy in the world (Iqbal et. al, 2019). The sectoral contribution towards GDP is the highest in the service sector (49%) followed by Industry (27%) and Agriculture (14%). Since 1960, the share of agriculture has declined from 41 percent to 15 percent in 2019 (Figure 3). The vulnerability to climate change is not uniform as multiple socio-economic and bio-physical factors affect food systems and therefore food security (Chikaire, Nnadi and Ajaero, 2010). The biodiversity, ecosystems, and the systems of humans on earth have already been affected by the acceleration in global warming and its associated changes in precipitation (Sintayehu, 2018).

Figure 3. India's share of agriculture and allied activities as a percentage of GDP (1960-2019)
Source: World Bank Database (2020)

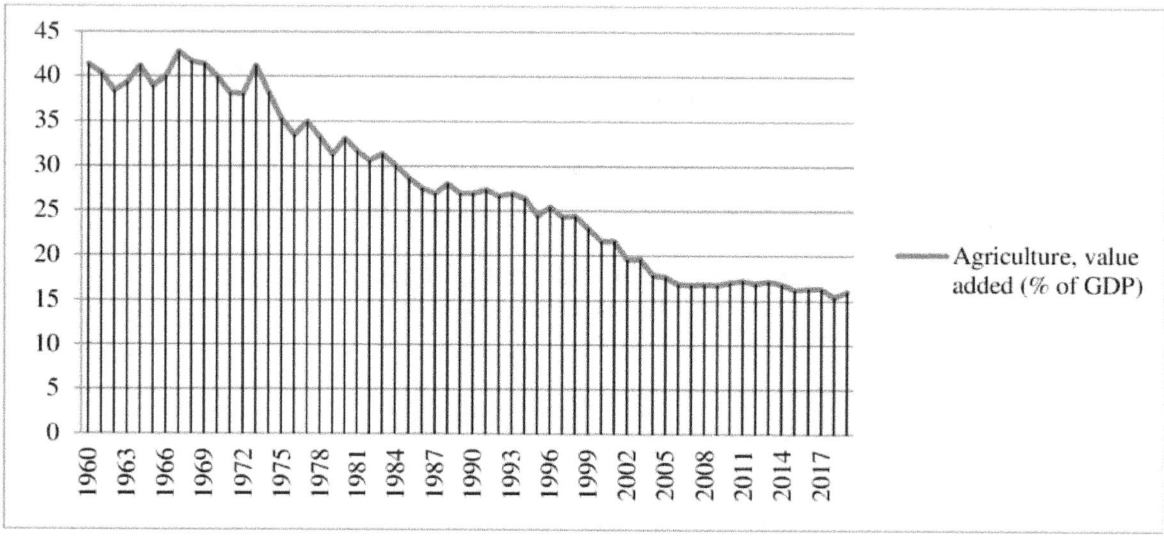

Since 1960, the overall food production in India has increased significantly because of the green revolution. However, the net availability of per capita food per day has decreased to 487 gm in 2018 from 503 gm in 1997 (Department of Economic Affairs, 2018; Kapuria and Saha, 2020). Food and nutrition security is still a matter of concern for the country even when it is working positively towards achieving the Millennium Development Goals (Ministry of Statistics and Programme Implementation and The World Food Programme, 2019).

Climate change is also one of the factors responsible for increasing the level of food insecurity in India. The change in climate conditions is mostly because of the increase in greenhouse gases concentration especially carbon emission, nitrous oxide, and methane. India is the world's third-largest emitter of greenhouse gases (Pathak, 2015). The direct effect of greenhouse gases can be seen in the changing climatic condition in terms of temperature, precipitation, sea level, land degradation, etc. According to the World Bank Group Database (Figure 4), there has been an increase of around 0.59 degree Celsius in the mean annual temperature of India since 1971. Moreover, a change in monsoon precipitation has been observed over the years with a slight decline from the original.

Figure 4. Trend and pattern of temperature and rainfall (1901-2016)
Source: World Bank Group, Climate Change Knowledge Portal (2020)

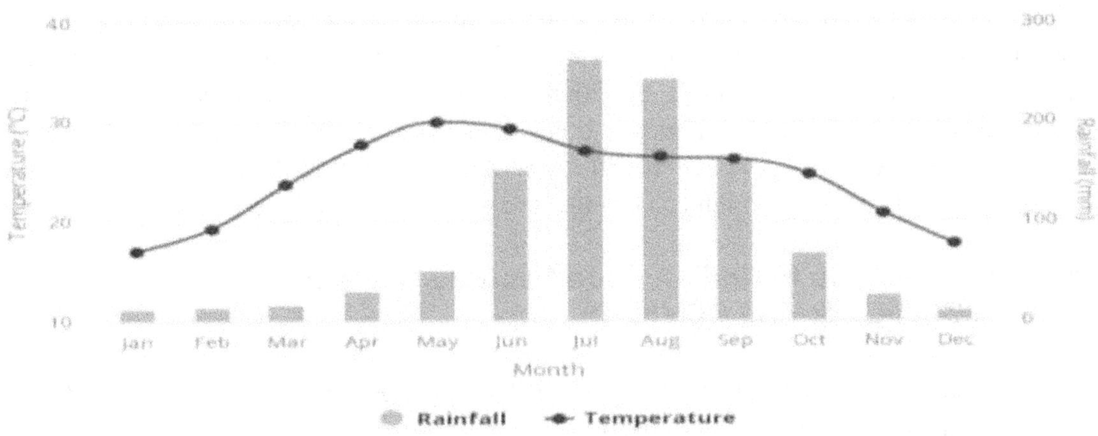

On the Global Climate Risk Index 2019, India is at 14th place and due to extreme climate change (storms, floods, droughts, and heatwaves) it suffered both economic (about 13.8 billion) and human (2726 deaths) loss (Eckstein, Hutfils and Winges, 2018). Within the last 15 years, the year 2018 was one of the warmest years in India and it is expected that the degree of warming is going to be more severe in the future (Indian Meteorological Department, 2018). The impact can be easily seen on agriculture production as according to an estimate the yield of major crops is expected to decline by 25 percent by 2030 (Cline, 2008).

Table 2. Expected impact on the major crops of climate change in India

Year	Season	Temperature Increase indegree Celsius		Rainfall Change inpercentage)	
		Lowest	Highest	Lowest	Highest
2020s	Rabi	1.08	1.54	-1.95	4.36
	Kharif	0.87	1.12	1.81	5.10
2050s	Rabi	2.54	3.18	-9.22	3.82
	Kharif	1.81	2.37	7.18	10.52
2080s	Rabi	4.14	6.31	-24.83	-4.50
	Kharif	2.91	4.62	10.10	15.18

Source: Aggarwal, IARI (2003)

According to a study, climate change will affect significantly the production of major crops such as wheat, barley, mustard, chickpea, and rice in India (Table 2) due to an increase in temperature and change in the level of rainfall. The extent of the impact, however, will depend on their location for instance, in case of a rise in production due to a temperature rise of 2-4 degrees celsius; the eastern region will be highly affected leading to fewer grains and shorter durations for grain filling. Further, Indian agriculture which is highly dependent on rainfall will be affected by the shortage of water availability as it is expected that water availability will decline to 1140 m^3 per year in 2050 from 1820 m^3 per year in 2001 (Chattopadhyay, 2008). A larger portion of the population in India will have an impact as there will be a shift in crop patterns, productivity, livelihood activities, trade policies, water conservation issues, and food security. A loss ranging between 9% to 25% in farm-level net revenue is expected for a 2-3.5°C rise in the level of temperature (FAO, 2008). According to a report of the World Bank (2013), India's summer monsoon will become highly unpredictable in the next decades with an expected rise of 20 degree Celsius in the world's average. It will further lead to a shift in the pattern of rain and due to extreme heat, a significant reduction in crop yields is expected for India by the 2040s.

Issues and Challenges for Indian Food Security System and Climate Change during COVID-19

Even though countries all around the world are already struggling to manage the challenges of climate change, there is an emergence of a major pandemic in the form of Covid-19 which is taking a toll on human lives. According to the Global Risks Report, both climate change and COVID-19 are in the top 10 risks which will have a long-term impact (Whiting, 2020). Economies around the world get affected not only health-wise but economically as well due to the global supply chain and trade restrictions during the pandemic. Every sector saw a negative impact of the pandemic, be it primary, secondary, or tertiary (FAO, 2015). The impact of Covid-19 added a new unfavorable dimension to the climate, weather, and water-related crisis affecting wellbeing and health of humans. Various kinds of restrictions lead to disruptions and downturns in every sector including the generation of food insecurity levels (The World Bank, 2021). Through the cascading chain of interrelated events, climate change has also created uncertainty for the achievement of the Sustainable Development Goals (International Labour Organization, 2021).

On the other hand, the global lockdowns during covid-19 have a positive effect on climate change as compared to 2019 there has been a fall of around 17 percent in greenhouse gases emission in April 2020, however, this change is only for a very short time as the economies started resuming work, the level of emission rose in June by 5 percent compared to the same period in the previous year (Borunda, 2020). For India, the level of emission dropped by 26 percent during the same time (Le Quéré et.al., 2020). However, the drop in the emission level was just temporary as the activities started resuming. Along with the covid-19 crisis, extreme weather events were also happened such as in the western part of India cyclone Nisarga came in June and after a few weeks cyclone Amphan lead to loss of lives and property (Yasir and Schultz, 2020). Further, attack by the Locust swarms, along with the increased temperature level, destroyed the crops which became a big threat to livelihood and food security in India (The World Bank, 2020).

Figure 5. Impact of COVID-19 on agriculture and climate in India
Source: Prepared by authors

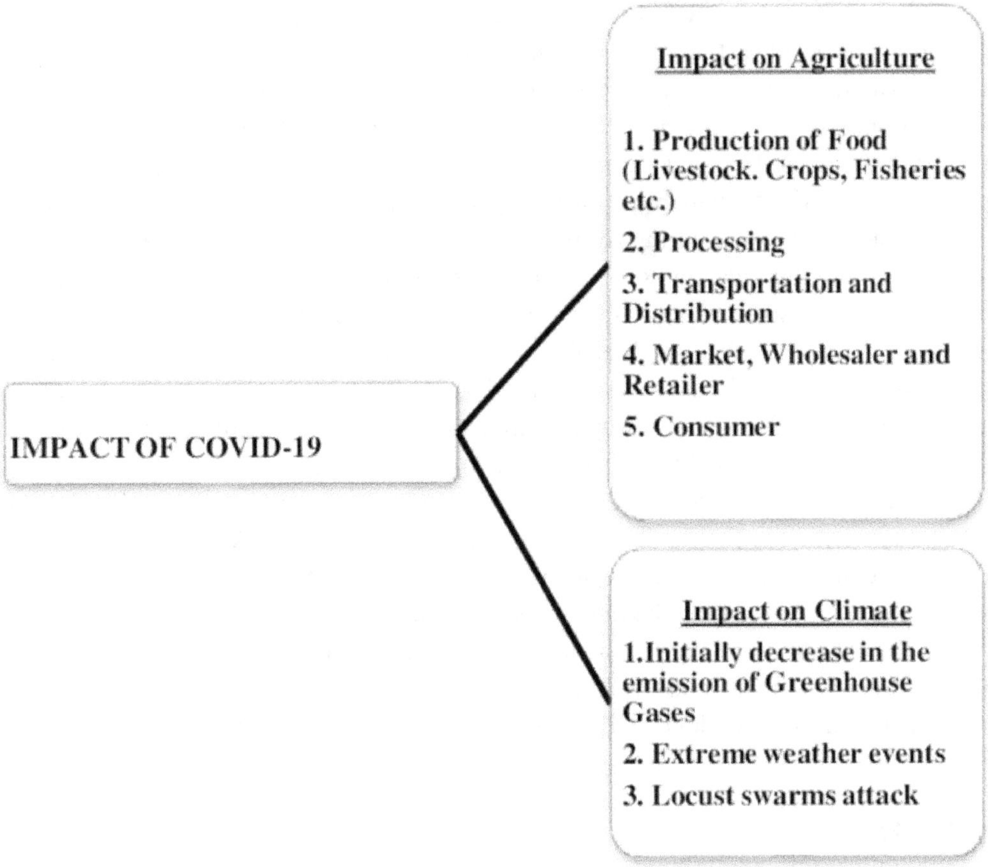

Food Security is also affected by the pandemic in the form of various challenges such as disrupted supply chains, loss of income, remittances, and an increase in food price inflation. According to the U.N. World Food Programme, there were around 135 million people before the pandemic who were facing food insecurity and the pandemic is expected to increase this burden by adding around 130 million people who will be facing the acute level of food insecurity (The World Bank, 2020). In India, after the outburst of the pandemic, the immediate step was the imposition of lockdown and restrictions on the movement of transport vehicles leading to disrupting the supply chain (Figure 5) (EY, 2020). Initially, the availability of food grains didn't get affected as there are sufficient buffers of food gains (58.4 million tons) and pulses (3 million tons), however, disruption of the supply chains has severely affected 78 percent of the food consumption which have a short shelf life like milk, vegetables, fruits, eggs, sugar and meat (Narayanan, 2020). The loss has been for every person involved such as consumers, farmers, and traders. The pandemic has shown the world that the environmental crisis is mostly because of the overconsumption and economic system which overemphasizes the growth aspect as during the lockdown temporary decline in the pollution is observed. Recently various climate disasters and the ongoing pandemic has made it clear that planetary boundaries should be respected as the survival of humans solely depends upon the health of the planet earth (Gardiner, 2020).

History always backed the claim that crisis leads to the generation of some kind of innovation to handle it. So, the same can be expected from COVID-19. It is and will eventually generate various ideas to handle the crisis and various events are already happening in this regard. To ensure food security around the world concerned parties such as businesses, government, communities targeting production, markets, and distribution system are adopting new ways to reach the people. For instance, in the case of India, Kisan Rath mobile application was launched by the National Informatics Centre to help traders and farmers reach the markets to sell their vegetables and fruits (Hawkes, 2020).

Further, to maintain the sustainability of food security during climate change and post-pandemic, there is a need to tackle issues (resulting from changing temperature and precipitation; and new types of diseases and pests) faced in local and global supplies of food. Priority should be given to maintain the sustainability of natural resources such as soil quality, water availability, livestock management, and effective use of solar energy. An increased level of research and development investment in the agriculture sector is the need of time to maintain food security during climate changing and the pandemic situation.

The most vulnerable section from climate change or Covid-19 pandemic is those who were already suffering from undernutrition and hunger. So, the inequities raise by these situations must be restrained by providing direct benefit to the concerned section through eliminating or limiting the level of transferring hierarchy. Further, to restrict the level of emission as suggested by Amjath et.al. (2020), there is a need for a climate ethics code for every sector i.e., primary, secondary, and tertiary.

Last but not the least, other than the authoritative efforts, every human living on this planet should think and work to not harm its surrounding environment and maintain its sustainability. Also, this is the responsibility of the individuals to follow the guidelines or requests made by the authorities, as in most cases; it is the lack of adoption which leads to the failure of a good policy.

CONCLUSION

This study presents an assessment of global warming/climate change on the food security system in India. Global warming and food security have a global link with the feature of inseparability as both have a significant impact on each other. Food and nutritional security are highly affected due to the events generated by the situation of global warming. In addition to that, it tries to find out the impact of the pandemic outbreak concerning the food security system in India and overviewed the measures taken by the authorities. Therefore, it is endowed with original input as very few studies have taken into consideration this issue from the perspective of the pandemic. The study observed that the level of climate change is affecting the food security level in India significantly, as an increase in the frequency of extreme weather events is seen recently. Further, the outbreak of pandemic has only doubled the risk of food security, however, during the initial period of the pandemic decline in the level of greenhouse gases has been seen. But as soon as the things are getting back on track, there has been an increase observed in the emission level of the greenhouse gases and that too with an increased level of intensity. The study suggests that although the authorities are trying to implement various policies to curb the effect of both climate change and Covid-19 pandemic, there is a need for efforts from an individual point of view as well. Further research can be done to understand the effectiveness of various measures taken by authorities to curtail the effect of climate change and to maintain food security during COVID-19 at the state level in India. In addition to that various empirical techniques can be applied to study the intensity of global warming across various regions of India.

REFERENCES

Aggarwal, P. K. (2003). Impact of climate change on Indian agriculture. *Journal of Plant Biology*, *30*(2), 189–198.

Ahmed, M., & Suphachalasai, S. (2014). *Assessing the Costs of Climate Change and Adaptation in South Asia.* Asian Development Bank and UKaid. Retrieved from https://www.preventionweb.net/files/38999_assessingcostsclimatechangeandadapt.pdf

Ali, S., Liu, Y., Ishaq, M., Shah, T., Ilyas, A., & Din, I. U. (2017). Climate Change and Its Impact on the Yield of Major Food Crops: Evidence from Pakistan. *Foods*, *6*(6), 39. DOI: doi:10.3390/foods6060039

Amjath-Babu, T. S., Krupnik, T. J., Thilsted, S. H., & McDonald, A. J. (2020). Key indicators for monitoring food system disruptions caused by the COVID-19 pandemic: Insights from Bangladesh towards effective response. *Food Security*, 1–8. doi:10.1007/s12571-020-01083-2

Bandara, J. S. & Cai, Y. (2014). The impact of climate change on food crop productivity, food prices and food security in South Asia. Economic Analysis and Policy, 44(4). doi:10.1016/j.eap.2014.09.005

Borunda, A. (2020). Plunge in carbon emissions from lockdowns will not slow climate change. *National Geographic*. Retrieved from https://www.nationalgeographic.com/ science/ article/plunge-in-carbon-emissions-lockdowns-will-not-slow-climate-change

Buis, A. (2019). *A Degree of Concern: Why Global Temperatures Matter*. NASA's Global Climate Change. Retrieved from https://climate.nasa.gov/news/2865/a-degree-of-concern-why-global-temperatures-matter/

Cassia, R., Nocioni, M., Aragunda, N. C., & Lamattina, L. (2018). Climate Change and the Impact of Greenhouse Gasses: CO_2 and NO, Friends and Foes of Plant Oxidative Stress. *Frontiers in Plant Science*. Retrieved from https://www.frontiersin.org/articles/ 10.3389/ fpls.2018.00273/full

Chaktrabarty, M. (2016). Climate Change and Food Security in India. *ORF Issue Brief*, 157. Retrieved from https://www.orfonline.org/wp content/uploads/2016/09/ ORF_ IssueBrief _1571.pdf

Chattopadhyay, N. (2008). Climate Change and Food Security in India. *International Symposium on Climate Change and Food Security in South Asia*. Retrieved from http://wamis.org/agm/meetings/rsama08/S402-Chattopadhyay-Climate-change_Food-Security.pdf

Chikaire, J., Nnadi, F. N., & Ajaero, C. (2010). Impacts of Climate Change on Food Security and Poverty Reduction. *International Journal of Natural and Applied Sciences*, *6*(4).

Cline, W. R. (2008). Global Warming and Agriculture. Finance and Development. *International Monetary Fund, 45*(1). Retrieved from http:/www.imf.org/external/pubs/ft/fandd/2008/03/cline.htm

Dahlman, L., & Lindsey, R. (2020). *Climate Change: Ocean Heat Content*. Retrieved from https://www.climate.gov/news-features/understanding-climate/climate-change-ocean-heat-content

Department of Economic Affairs. (2018). *Economic survey 2017-18 (Statistical Appendix)*. Ministry of Finance, Government of India, Table 1.19. Retrieved from http://mofapp.nic.in:8080 /economicsurvey/appendixPDF/tab1.19.pdf

Eckstein, D., Hutfils, M. L. & Winges, M. (2018). *Global Climate Risk Index 2019*. Germanwatch e.V. Retrieved from https://www.germanwatch.org/sites/germanwatch.org /files/Global%20Climate%20 Risk%20Index%202019_2.pdf

Ericksen, P. J., Ingram, J. S. I., & Liverman, D. M. (2009). Food security and global environmental change: Emerging challenges. *Environmental Science & Policy*, *12*(4), 373–377. Advance online publication. doi:10.1016/j.envsci.2009.04.007

European Environment Agency. (2015). *Agriculture and Climate Change*. Retrieved from https://www.eea.europa.eu/signals/signals-2015/articles/agriculture-and-climate-change

EY. (2020). *Managing the impact of COVID-19 on India's supply chains- Now, Next and Beyond*. Retrieved from https://assets.ey.com/content/dam/ey-sites/ey-com/en_in/topics/ government-and-public-sector/2020/09/managing-the-impact-of-covid-19-on-india-supply-chains.pdf

FAO. (2015). *Climate change and food security: Risks and responses*. Retrieved from https://www.fao.org/3/i5188e/i5188e.pdf

FAO. (2017). *The Future of food and agriculture: Trends and Agriculture*. Retrieved from https://www.fao.org/3/i6583e/i6583e.pdf

FAO. (2017). *Water for Sustainable Food and Agriculture: A report produced for the G20 Presidency of Germany*. Retrieved from https://www.fao.org/3/i7959e/i7959e.pdf

FAO (Food and Agriculture Organization of the United States). (2008). *Climate Change and Food Security: A Framework Document*. Retrieved from http://www.fao.org/ forestry/ 15538-079b31d45081f-e9c3dbc6ff34de4807e4.pdf

Food and Agriculture Organization of the United Nations. (2016). *Climate change and food security: Risks and responses*. Retrieved from https://www.fao.org/3/a-i5188e.pdf

Food and Agriculture Organization of the United States. (2019). *The State of Food Security and Nutrition in the World: Safeguarding Against Economic Slowdowns and Downturns*. Retrieved from https://www.fao.org/3/ca5162en/ca5162en.pdf

Gardiner, B. (2020). Pollution made COVID-19 worse. Now, lockdowns are clearing the air. *National Geographic*. Retrieved from https://www.nationalgeographic.com/science/ 2020/04/pollution-made-the-pandemic-worse-but-lockdowns-clean-the-sky/

Gornall, J., Betts, R., Burke, E., Clark, R., Camp, J., Willett, K., & Wiltshire, A. (2010). Implications of climate change for agriculture productivity in the early twenty-first century. *Philosophical Transactions of the Royal Society of London. Series B, Biological Sciences*, *365*(1554), 2973–2989. doi:10.1098/rstb.2010.0158 PMID:20713397

Gregory, P. J., Ingram, J. S., & Brklacich, M. (2005). Climate Change and Food Security. *Philosophical Transactions of the Royal Society of London. Series B, Biological Sciences*, *360*(1463), 2139–2148. doi:10.1098/rstb.2005.1745 PMID:16433099

Hawkes, C. (2020). *COVID-19 and the promise of food system and innovation. Chapter 29 in Book COVID-19 and Global Food Security by John Swinnen and McDermott*. International Food Policy Research Institute. Retrieved from https://ebrary.ifpri.org/ utils/getfile/collection /p15738coll2/id/133762/ filename/133971.pdf

Henry, R. (2020). Innovations in Agriculture and Food Supply in Response to the COVID-19 Pandemic. *Molecular Plant*, *13*(8), 1095–1097. doi:10.1016/j.molp.2020.07.011 PMID:32711125

Hobert, R., & Negra, C. (2020). *Climate Change and the Future of Food*. United Nations Foundation. Retrieved from https://unfoundation.org/blog/post/climate-change-and-the-future-of-food/

Indian Meteorological Department. (2018). *Press Release*. Retrieved from https://mausam.imd .gov. inpages/press_release_view.php?ff=20160921_pr_56

International Labour Organization. (2021). *Covid-19 and the world of work: Sectoral Impact, responses and recommendations*. Retrieved from https://www.ilo.org/ global/topics/ coronavirus/sectoral/lang--en/ index.htm

International Monetary Fund. (2019). *The Economics of Climate. Finance and Development*. Retrieved from https://www.imf.org/external/pubs/ft/fandd/2019/12/pdf/fd1219.pdf

International Monetary Fund. (2020). *World Economic Outlook: A Long and Difficult Ascent*. Retrieved from file:///C:/Users/dell/AppData/Local/Temp/text.pdf

IPCC. (2014). *Climate change 2014: synthesis report. Contribution of Working Groups I, II and III to the Fifth Assessment Report of the Intergovernmental Panel on Climate Change. Core Writing Team* (R. K. Pachauri & L. A. Meyer, Eds., Vol. 151). IPCC.

IPCC. (2018). Summary for Policymakers. In *Global Warming of 1.5°C. An IPCC Special Report on the impacts of global warming of 1.5°C above pre-industrial levels and related global greenhouse gas emission pathways, in the context of strengthening the global response to the threat of climate change, sustainable development, and efforts to eradicate poverty*. World Meteorological Organization.

IPCC (Intergovernmental Panel on Climate Change). (2019). *Climate Change and Land. IPCC Special Report on Climate Change, Desertification, Land Degradation, Sustainable Land Management, Food Security, and Greenhouse gas fluxes in Terrestrial Ecosystems*. Retrieved from https://www.ipcc.ch/site/ assets/uploads/2019/08/4.-SPM_Approved _ Microsite_FINAL.pdf

Iqbal, B. A., Hassan, M., Yadav, A., & Shriaan, A. A. (2019). Agriculture Sector and Economic Growth of India: An Empirical Association. *Asian Profile*, *47*(1).

Kalkuhl, M., von Braun, J., & Torero, M. (2016). Volatile and Extreme Food Prices, Food Security, and Policy: An Overview. In M. Kalkuhl, J. von Braun, & M. Torero (Eds.), *Food Price Volatility and Its Implications for Food Security and Policy*. Springer., doi:10.1007/978-3-319-28201-5_1

Kapuria, P., & Saha, R. (n.d.). *Resource use efficiency and productivity: An analysis of India's food grain sector*. ORF. Retrieved from https://www.orfonline .org/research/ resource-use-efficiency-and-productivity-an-analysis-of-indias-food-grain-sector-59894/

Kaur, H. & Kaur, S. (2017). Climate Change Impact on Agriculture and Food Security in India. *Journal of Business Thought, 7*. Retrieved from file:///C:/Users/dell/AppData/Local/Temp/21329-27060-1-SM-1.pdf

Le Quéré, C., Jackson, R. B., Jones, M. W., Smith, A. J. P., Abernethy, S., Andrew, R. M., De-Gol, A. J., Willis, D. R., Shan, Y., Canadell, J. G., Friedlingstein, P., Creutzig, F., & Peters, G. P. (2020). Temporary reduction in daily global CO_2 emissions during the COVID-19 forced confinement. *Nature Climate Change, 10*(7), 647–653. doi:10.103841558-020-0797-x

Lindsey, R. (2020). *Climate Change: Atmospheric Carbon Dioxide*. Retrieved from https://www.climate.gov/news-features/understanding-climate/climate-change-atmospheric-carbon-dioxide

Lindsey, R., & Dahlman, L. (2021). *Climate Change: Global Temperature.* Retrieved from https://www.climate.gov/news-features/understanding-climate/climate-change-global-temperature

Ludwig, F., Scheltinga, C. T. V., Verhagen, J., Kruijt, B., Ierland, E. V., Dellink, R., Bruin, K. D., Bruin, K. D., & Kabat, P. (2007). *Climate change impacts on Developing Countries - EU Accountability. Policy Department Economic and Scientific Policy*. European Parliament. Retrieved from https://www.europarl.europa.eu/RegData/etudes/ etudes/join/ 2007/393511/IPOL-ENVI_ET(2007)393511_EN.pdf

MacMillan, A., & Turrentine, J. (2021). *Global Warming 101.* NRDC. Retrieved from https://www.nrdc.org/stories/global-warming-101

McMichael, A. J., Lendrum, D. H. C., Corvalan, C. F., Ebi, K. L., Githeko, A. K., Scheraga, J. D., & Woodward, A. (2003). *Climate change and human health.* World Health Organization. Retrieved from https://www.who.int/globalchange /publications/ climch ange .pdf?ua=1

Ministry of Home Affairs, Government of India. (2011). *Disaster Management in India.* Retrieved from file:///C:/Users/dell/AppData/Local/ Temp/disaster_ management _in_india.pdf

Ministry of Statistics and Programme Implementation and The World Food Programme. (2019). *Food and Nutrition Security Analysis, India, 2019.* Retrieved from http://mospi.nic.in/ sites/default/files/publication_reports/document%281% 29.pdf

Narayanan, S. (2020). *How India's agrifood supply chains fared during the COVID-19 lockdown, from farm to fork.* International Food Policy Research Institute. Retrieved from https://www.ifpri.org/blog/how-indias-agrifood-supply-chains-fared-during-covid-19-lockdown-farm-fork

Nelson, G. C., Rosegrant, M. W., Koo, J., Robertson, R., Sulser, T., Zhu, T., Ringler, C., Msangi, S., Palazzo, A., Batka, M., Magalhaes, M., Santos, R. V., Ewing, M., & Lee, D. (2009). *Climate Change: Impact on agriculture and Costs of Adaptation.* Food Policy Report. International Food Policy Research Institute. Retrieved from http://ebrary.ifpri.org/utils/getfile/collection/p15738coll2/id/130648/filename/130821.pdf

Nunez, C. (2019). Sea level rise, explained. Reference. *National Geographic*. Retrieved from https://www.nationalgeographic.com/environment/global-warming/sea-level-rise/

Olivier, J. G. J., & Peters, J. A. H. W. (2020). *Trends in Global CO_2 and Total Greenhouse Gas Emissions 2019 Report*. PBL Netherlands Environmental Assessment Agency. Retrieved from https://www.pbl.nl/sites/default/files/downloads/pbl-2020-trends-in-global-co2-and-total-greenhouse-gas-emissions-2019-report_4068.pdf

Pathak, H. (2015). Greenhouse Gas Emission from Indian Agriculture: Trends, Drivers and Mitigation Strategies. *Proceedings of the Indian National Science Academy*, *81*(5), 1133–1149. doi:10.16943/ptinsa/2015/v81i5/48333

Pradhan, R., Singh, N., & Singh, R. P. (2019). Onset of summer monsoon in Northeast India is preceded by enhanced transpiration. *Scientific Reports*, *9*(1), 18646. doi:10.103841598-019-55186-8 PMID:31819130

Pretty, J., & Bharucha, Z. P. (2014). Sustainable intensification in agricultural systems. *Annals of Botany*, *114*(8), 1571–1596. doi:10.1093/aob/mcu205 PMID:25351192

Ranuzzi, A., & Srivastava, R. (2012). *Impact of Climate Change on Agriculture and Food Security*. ICRIER Policy Series, 16. Retrieved from http://www.icrier.org/pdf/ Policy_Series_No_16.pdf

Raza, A., Razzaq, A., Mehmood, S. S., Zou, X., Zhang, X., Lv, Y., & Xu, J. (2019). Impact of Climate Change on Crops Adaptation and Strategies to Tackle Its Outcome: A Review. *Plants (Basel, Switzerland)*, *8*(2), 34. doi:10.3390/plants8020034 PMID:30704089

Ritchie, H. (2019). *Food production is responsible for one-quarter of the world's greenhouse gases emissions*. Our World in Data. Retrieved from https://ourworldindata.org/food-ghg-emissions

Ruwoldt, G. (2013). Food insecurity and climate change. *American Journal of Public Health*, *103*(1), e1. doi:10.2105/AJPH.2012.301082 PMID:23153137

Sarkar, S. (2018). *South and Southeast Asia most at risk by climate change*. Retrieved from https://www.thethirdpole.net/2018/12/05/south-asia-climate-change/

Schmidhuber, J., & Tubiello, F. (2007). Global Food Security under Climate Change. *Proceedings of the National Academy of Sciences of the United States of America, 104*(50), 19703-19708. Retrieved from https://www.jstor.org/stable/25450779

Singh, G. (2016). Climate Change and Food Security in India: Challenges and Opportunities. *Irrigation and Drainage*, *65*, 5–10. doi:10.1002/ird.2038

Sintayehu, D. W. (2018). Impact of climate change on biodiversity and associated key ecosystem services in Africa: A systematic review. *Ecosystem Health and Sustainability*, *4*(9), 225–239. Advance online publication. doi:10.1080/20964129.2018.1530054

Srinivasa Rao, Ch., Prasad, R. S., & Mohapatra, T. (2019). *Climate Change and Indian Agriculture: Impacts, Coping Strategies, Programmes and Policy. Technical Bulletin/Policy Document 2019. Indian Council of Agricultural Research, Ministry of Agriculture and Farmers' Welfare and Ministry of Environment, Forestry and Climate Change*. Government of India.

Stevens, A. N. P. (2011). Factors Affecting Global Climate. *Nature Education Knowledge, 3*(10), 18. Retrieved from https://www.nature.com/scitable/knowledge/library/factors-affecting-global-climate-17079163/

Stuart, T. (2015). *Food: How much does the world need?* World Economic Forum. Retrieved from https://www.weforum.org/agenda/2015/05/food-how-much-does-the-world-need/

Sukhwani, V., Deshkar, S., & Shaw, R. (2020). COVID-19 Lockdown, Food Systems and Urban–Rural Partnership: Case of Nagpur, India. *International Journal of Environmental Research and Public Health, 17*(16), 5710. doi:10.3390/ijerph17165710 PMID:32784648

Swiss Re. (2020). *World economy set to lose up to 18% GDP from climate change if no action taken, reveals Swiss Re Institute's stress-test analysis.* Retrieved from https://www. swissre.com/media/news-releases/nr-20210422-economics-of-climate-change-risks.html

Termeer, C. J. A. M., Drimie, S., Ingram, J., Pereira, L., & Whittingham, M. J. (2018). A diagnostic framework for food system governance arrangements: The case of South Africa. *NJAS Wageningen Journal of Life Sciences, 84*(1), 85–93. Advance online publication. doi:10.1016/j.njas.2017.08.001

The Royal Society. (2020). *The Basics of Climate Change.* Retrieved from https://royalsociety.org/topics-policy/projects/climate-change-evidence-causes/basics-of-climate-change/

The World Bank. (2020a). *Food Security and COVID-19.* Retrieved from https://www.worldbank. org/en/topic/agriculture/brief/food-security-and-covid-19

The World Bank. (2020b). *World Bank Announces $500 Million to Fight Locusts, Preserve Food Security and Protect Livelihoods.* Press Release. Retrieved from https://www.worldbank. org/ en/news/press-release/2020/05/21/world-bank-announces-500-million-to-fight-locusts-preserve-food-security-and-protect-livelihoods

The World Bank. (2021). *Climate Smart Agriculture.* Retrieved from https://www.worldbank. org/en/topic/climate-smart-agriculture

The World Bank Report. (2013). *Turn Down the Heat: Climate Extremes, Regional Impacts, and the Case for Resilience.* Retrieved from https://www.worldbank.org/content/ dam/ Worldbank/document/Full_Report_Vol_2_Turn_Down_The_Heat_%20Climate_Extremes_Regional_Impacts_Case_for_Resilience_Print%20version_FINAL.pdf

UNFCCC (United Nations Framework Convention on Climate Change). (2007). *Climate change: Impacts, vulnerability and adaptation in developing countries.* Retrieved from https://unfccc.int/resource/docs/publications/impacts.pdf

UNHCR. (2021). *Climate change and disaster displacement.* Retrieved from https://www. unhcr.org/climate-change-and-disasters.html

US EPA. (2021). *Overview of Greenhouse Gases.* Retrieved from https://www. epa. gov/ ghgemissions/overview-greenhouse-gases

Whiting, K. (2020). *Coronavirus isn't an outlier, it's part of our interconnected viral age.* World Economic Forum. Retrieved from https://www.weforum.org/agenda/2020/03/coronav irus-global-epidemics-health-pandemic-covid-19/

World Bank Database. (2020). *Agriculture, forestry, and fishing, value added (% of GDP).* Retrieved from https://data.worldbank.org/indicator/NV.AGR.TOTL.ZS

World Bank Group. (2020). *Climate Change Knowledge Portal Country India*. Retrieved from https://climateknowledgeportal.worldbank.org/country/india

World Food Summit. (1996). *Rome Declaration on World Food Security*. FAO.

World Meteorological Organization. (2021). *Climate change indicators and impacts worsened in 2020.* Retrieved from https://public.wmo.int/en/media/press-release/climate-change-indicators-and-impacts-worsened-2020

World Water Development Report. (2020). *The United Nations world water development report 2020: water and climate change.* UNESCO World Water Assessment Programme. Retrieved from https://unesdoc.unesco.org/ark:/48223/pf0000372985.locale=en

Yasir, S., & Schultz, K. (2020). Powerful Cyclone Lashes Mumbai as City Grapples With Coronavirus. *The New York Times*. Retrieved from https://www.nytimes.com/ 2020/06/03/world/asia/cyclone-nisarga-india-mumbai.html

Zhao, C., Liu, B., Piao, S., Wang, X., Lobell, D. B., Huang, Y., Huang, M., Yao, Y., Bassu, S., Ciais, P., Durand, J. L., Elliott, J., Ewert, F., Janssens, I. A., Li, T., Lin, E., Liu, Q., Marte, P., Muller, C., ... Asseng, S. (2017). Temperature increase reduces global yields of major crops in four independent estimates. *Proceedings of the National Academy of Sciences of the United States of America*, *114*(35), 9326–9331. doi:10.1073/pnas.1701762114 PMID:28811375

Zillman, J., & Sherwood, S. (2017). *The enhanced greenhouse effect*. Australian Academy of Science. Retrieved from https://www.science.org.au/curious/earth-environment/enhanced-greenhouse-effect

Chapter 8
Labour Drivers in the Agricultural Sector of the European Union:
The Social Role of Farms

Vítor João Pereira Domingues Martinho
Polytechnic Institute of Viseu, Portugal

ABSTRACT

The social role of the farms is, especially, relevant in the rural areas where the socioeconomic problems are, often, more visible. In this perspective, this study aims to investigate the interrelationships of the labour input with other variables inside the farms and assess how the sector may create more employment in a sustainable way. For that, the labour input was, first, correlated with other farm variables and after analysed through factor analysis approaches and cross-section econometric methodologies, considering as basis the Cobb-Douglas and Verdoorn-Kaldor models. The main findings highlight relevant insights to improve the social dimension of the European Union farms. The labour input growth rate is positively influenced by the total output growth rates and negatively impacted by the total productivity growth. The effects from the investment and from the subsidies are residual or not significant.

INTRODUCTION

The agricultural sector has several dimensions, including those associated with the social contributions for a more sustainable development. However, the social dimension is often compromised by the economic objectives, more in a sector where the costs are always a concern (Martinho, 2020b). The social and environmental contributions from the farms are particularly important in regions with less economic dynamics and where the agriculture plays a relevant role in the creation of employment (Martinho, 2020a).

The agricultural labour in the European Union is impacted by different variables, nonetheless the diverse Common Agricultural Policy (CAP) reforms have been one of the main determinants of the number of working hours used by the farms (Dupraz & Latruffe, 2015).

DOI: 10.4018/978-1-7998-9557-2.ch008

The labour is, indeed, an important factor of production for the agriculture, jointly with the capital and land resources (Salvioni et al., 2020), and is correlated with other farm variables (Strelecek et al., 2007) and indicators (Todorovic et al., 2020), such as the productivity (González & Fernández, 2017) and the efficiency (Martinho, 2017). Often, it is the paid labour that more positively impact the efficiency (Trnkova & Kroupova, 2020).

Frequently, the concerns with the agricultural labour are about its relationship with the efficiency, where the impacts seem to be different from paid and family working hours (Kostov et al., 2019). In these assessments, the concept of eco-efficiency (Martinsson & Hansson, 2021), and the concerns with the sustainability (Thomassen et al., 2009) have its relevance.

This preliminary survey of the literature shows that there is, still, field to be explored about the main interrelationships between the labour and other farm variables, in the European Union context and through models from the economic theory.

In this perspective, the main objective of this research is to analyse the relationships between the labour and other farm variables, considering data at farm level from the European Union Farm Accountancy Data Network (FADN, 2021), for the period 2017-2019. This statistical information was assessed through several approaches, namely those related with matrices of correlation, factor analysis, models from the economic theory (Cobb-Douglas and Verdoorn-Kaldor) and regression approaches. The Cobb-Douglas model is the typical approach from the theory of production and the Verdoorn-Kaldor developments capture dynamic effects. The benchmark of the results from the two models will provide interesting insights, for the several stakeholders, about the relationships of the labour with other variables and about their main determinants.

INSIGHTS FROM THE LITERATURE

The agriculture played a central role in the societies of the previous centuries (Horrell et al., 2020). In the last century, the agricultural sector lost its relative importance (comparatively with other sectors) in the regional development (Alonso Villa & Juste Carrion, 2018), nonetheless maintains its relevance in many regions (Melchor-Ferrer, 2020) and on the socio-economic characteristics worldwide (Wilk et al., 2019), specifically for a more sustainable development (Baer-Nawrocka & Poczta, 2018).

The transformations in the societies and in the farming sector has freed labour to other sectors, regions and countries (Greenwood, 2008), solving, for example, problems of skilled labour shortage in the farms of some countries (Kvartiuk, 2015). Some of these transitions are needed to improve the economic performance (Herman, 2016). The agricultural dynamics follow, in general, the trends of the surrounding contexts (Meyfroidt et al., 2016). However, often remain maladjustments that require policy interventions (Poczta & Pawlak, 2011), namely in frameworks where the agriculture still maintains a significant importance for economic and social evolution (Wlodarczyk-Spiewak & Korpysa, 2006). In the transition processes the benchmark with other realities may bring positive insights (Zbarsky et al., 2020).

In the contexts of the farming dynamics the labour appears as a determinant variable, jointly with others (Latruffe et al., 2004), because its social dimension and due the relationships with other indicators, such as the productivity and the efficiency (Beyer & Hinke, 2020). The labour productivity contributes for improvements in the competitiveness (Matkovski et al., 2019) and technical efficiency (Spicka & Smutka, 2014). In the European Union milk sector, the efficiency is positively impacted mainly by the paid labour (Trnkova & Kroupova, 2020).

In fact, if on the one hand the number of working hours represents costs for the farmers (Bijttebier et al., 2018), and in this way, they are interest in reducing the labour input, on the other hand the farm labour input contributes to make more balanced the relationships between the social, environmental and economic dimensions.

The CAP instruments maintain the employment in the farms (Olper et al., 2014) and have their impacts on the European Union agricultural dynamics (González & Fernández, 2017) and, consequently, on the several indicators related with the labour input (Constantin et al., 2021), however not always as expected (Carmen Cuerva, 2012). The farming and rural policies need constant adjustments to the specific realities (Mujcinovic et al., 2021). On the other hand, in the processes of policy design and implementation the involvement of the local stakeholders is crucial (Moyano Pesquera et al., 2017).

The level of education, training and skills held by the workers in the several economic sectors is determinant for an adjusted competitiveness (Doytch & Eren, 2012), including in the agriculture. In any case, the farm labour force is a complex production factor (Fouka & Schlapfer, 2020) that is interrelated with other factors beyond the productivity and the efficiency (Galdeano-Gomez et al., 2006), some of them difficult to manage (Giannakis & Bruggeman, 2018) and to control because depend of external variables (Jambor et al., 2016). This is true for the agricultural sector and the other economic activities (Rodriguez-Pose & Tselios, 2009).

DATA ANALYSIS

With data from the European Union Farm Accountancy Data Network (FADN, 2021), on average for the period 2017-2019 and across the agricultural regions, were obtained the results presented in table 1. This statistical information is published by the FADN per farm (for each agricultural region is published the data for a representative farm obtained by a weighting system) and is relative to the following variables: labour input (hours); unpaid labour input (hours); paid labour input (hours); total utilised agricultural area (ha); total livestock units (LU); total output (euros); total output/total input; total inputs (ratio); total intermediate consumption (euros); subsidies on investments (euros); farm net income (euros); gross investment on fixed assets (euros); total subsidies, excluding on investments (euros). These variables have been selected considering the objectives of this research of assessing the farm variables that affect the level of labour input.

The representative farm of the German region of Brandenburg is between the top European Union agricultural regions with greater results for a relevant part of the variables, exception for unpaid labour, total productivity (total output/total input), subsidies on investments and farm net income (table 1). Similar context happens for the farm of the German region of Thüringen. The German region of Mecklenburg-Vorpommern has, also, a great score for the following farming variables: total utilised agricultural area; total output, total input, total intermediate consumption; gross investment on fixed assets; and total subsidies (excluding on investments). Slovakia has, too, great scores for the total labour, paid labour and area. Slovakia and the top German regions for the variables analysed is where the farms are larger and where is used more labour input, namely the paid labour. Between the farms with more unpaid labour appear those from the Belgium regions of Vlaanderen and Wallonie, jointly with Rheinland-Pfalz from Germany. The farms from Liguria and Trentino (Italy) are among those with higher total productivity. The Netherlands is where the farms have greater farm net income.

Table 1 (a). Dimensions of farm variables, on average over the period 2017-2019 and across the European Union agricultural regions

Country	Region	Labour input	Unpaid labour input	Paid labour Input	Total Utilised Agricultural Area	Total livestock units
Austria	Austria	3476	3188	287	33	30
Belgium	Vlaanderen	5298	3953	1344	39	162
Belgium	Wallonie	4124	3778	346	74	97
Bulgaria	Severen tsentralen	6212	2271	3941	90	24
Bulgaria	Severoiztochen	6010	1691	4319	92	26
Bulgaria	Severozapaden	5595	2422	3173	101	15
Bulgaria	Yugoiztochen	6037	2012	4025	80	27
Bulgaria	Yugozapaden	4563	2782	1781	33	13
Bulgaria	Yuzhen tsentralen	4682	2367	2315	34	19
Croatia	Jadranska Hrvatska	3604	3112	492	12	4
Croatia	Kontinentalna Hrvatska	2670	2332	339	16	10
Cyprus	Cyprus	2893	2101	792	10	13
Czechia	Czechia	10738	2803	7935	193	94
Denmark	Denmark	3551	1592	1959	111	157
Estonia	Estonia	4062	1652	2409	139	37
Finland	Etelä-Suomi	2274	1748	526	67	23
Finland	Pohjanmaa	2536	2046	490	65	34
Finland	Pohjois-Suomi	3166	2494	672	77	35
Finland	Sisä-Suomi	2728	2285	442	57	31
France	Alsace	3141	2088	1053	51	40
France	Aquitaine	3526	2133	1392	53	37
France	Auvergne	2541	2293	248	105	101
France	Basse-Normandie	3042	2432	609	107	130
France	Bourgogne	3453	2251	1202	131	79
France	Bretagne	3400	2388	1013	68	194
France	Centre	2815	2069	747	137	43
France	Champagne-Ardenne	3085	1969	1116	84	28
France	Corse	3963	2023	1939	83	30
France	Franche-Comté	3156	2628	528	127	96
France	Guadeloupe	2186	1672	513	8	9
France	Haute-Normandie	3103	2150	954	123	75
France	La Réunion	2948	1982	966	9	17
France	Languedoc-Roussillon	3328	1985	1343	47	10
France	Limousin	2601	2276	326	105	112
France	Lorraine	2905	2397	508	156	103
France	Martinique	3527	1748	1779	10	5
France	Midi-Pyrénées	2612	2124	489	83	49
France	Nord-Pas-de-Calais	3253	2199	1053	81	71
France	Pays de la Loire	3699	2527	1172	95	159
France	Picardie	2902	2200	702	130	51
France	Poitou-Charentes	2893	2026	866	106	56

Continued on following page

Table 1 (a). Continued

Country	Region	Labour input	Unpaid labour input	Paid labour Input	Total Utilised Agricultural Area	Total livestock units
France	Provence-Alpes-Côte d'Azur	4928	2303	2626	42	11
France	Rhône-Alpes	3365	2208	1157	67	57
France	Île-de-France	3557	2089	1468	152	5
Germany	Baden-Württemberg	4442	3032	1410	54	54
Germany	Bayern	3572	3104	468	51	58
Germany	Brandenburg	14491	2421	12070	481	238
Germany	Hessen	3927	2849	1078	83	61
Germany	Mecklenburg-Vorpommern	10643	2269	8374	467	169
Germany	Niedersachsen	4473	3164	1309	84	139
Germany	Nordrhein-Westfalen	4248	2971	1277	58	114
Germany	Rheinland-Pfalz	5376	3747	1630	57	28
Germany	Saarland	3935	3416	518	118	64
Germany	Sachsen	12652	2556	10096	293	175
Germany	Sachsen-Anhalt	11322	2728	8594	431	171
Germany	Thüringen	16545	2152	14393	432	229
Greece	Ipiros-Peloponissos-Nissi Ioniou	2204	1745	459	7	5
Greece	Makedonia-Thraki	2351	1833	518	12	5
Greece	Sterea Ellas-Nissi Egaeou-Kriti	2744	2270	475	10	7
Greece	Thessalia	2386	1950	436	10	6
Hungary	Alföld	3097	1362	1735	39	17
Hungary	Dunántúl	3643	1582	2061	53	22
Hungary	Észak-Magyarország	3162	1298	1864	55	11
Ireland	Ireland	2355	2176	179	48	65
Italy	Abruzzo	2831	2518	313	15	6
Italy	Alto Adige	3386	2687	699	10	10
Italy	Basilicata	3451	2499	952	33	9
Italy	Calabria	2969	2098	871	11	4
Italy	Campania	2900	2104	796	14	14
Italy	Emilia-Romagna	3289	2507	782	26	24
Italy	Friuli-Venezia Giulia	3563	2904	660	21	23
Italy	Lazio	3199	2339	860	21	12
Italy	Liguria	3246	2599	647	8	4
Italy	Lombardia	3538	3045	493	30	85
Italy	Marche	2634	2488	146	23	7
Italy	Molise	2761	2485	276	22	14
Italy	Piemonte	3493	3093	400	25	25
Italy	Puglia	2544	1642	903	17	2
Italy	Sardegna	2681	2389	292	46	21
Italy	Sicilia	2498	1703	794	19	7
Italy	Toscana	3888	3014	873	25	6
Italy	Trentino	2631	2348	283	7	6
Italy	Umbria	2714	2090	624	26	13

Continued on following page

Table 1 (a). Continued

Country	Region	Labour input	Unpaid labour input	Paid labour Input	Total Utilised Agricultural Area	Total livestock units
Italy	Valle d'Aosta	4201	3595	606	46	25
Italy	Veneto	3242	2735	508	17	58
Latvia	Latvia	3840	2334	1506	67	24
Lithuania	Lithuania	3378	2745	634	48	12
Luxembourg	Luxembourg	3734	2976	758	86	125
Malta	Malta	2752	2453	299	3	14
Netherlands	The Netherlands	6289	3278	3010	39	146
Poland	Malopolska i Pogórze	3338	3067	271	11	6
Poland	Mazowsze i Podlasie	3502	3201	301	16	11
Poland	Pomorze i Mazury	3615	3070	545	38	19
Poland	Wielkopolska and Slask	3514	2930	584	26	16
Portugal	Alentejo e Algarve	3732	1803	1929	62	21
Portugal	Açores e Madeira	2306	1884	422	11	15
Portugal	Norte e Centro	3023	2495	528	15	13
Portugal	Ribatejo e Oeste	3074	2338	737	16	16
Romania	Bucuresti-Ilfov	2088	1126	963	25	2
Romania	Centru	3273	2575	698	14	10
Romania	Nord-Est	2717	2418	299	11	6
Romania	Nord-Vest	3244	2712	532	11	7
Romania	Sud-Est	3145	2360	785	24	7
Romania	Sud-Muntenia	2727	2214	512	18	5
Romania	Sud-Vest-Oltenia	3370	3180	190	11	4
Romania	Vest	3275	2878	397	22	8
Slovakia	Slovakia	19668	1617	18051	442	128
Slovenia	Slovenia	2226	2155	71	10	11
Spain	Andalucía	3142	1549	1593	37	15
Spain	Aragón	2854	1659	1195	76	61
Spain	Asturias	3356	3089	267	29	51
Spain	Canarias	7541	2808	4733	3	11
Spain	Cantabria	3006	2836	170	38	52
Spain	Castilla y León	2740	2185	555	68	45
Spain	Castilla-La Mancha	3702	2287	1415	61	20
Spain	Cataluña	3545	2034	1511	42	93
Spain	Comunidad Valenciana	2432	1252	1179	15	17
Spain	Extremadura	4497	3052	1445	72	42
Spain	Galicia	2331	2156	175	24	41
Spain	Islas Baleares	3444	2169	1275	58	21
Spain	La Rioja	2768	1677	1092	31	15
Spain	Madrid	4400	3391	1010	66	29
Spain	Murcia	5059	2206	2853	35	39
Spain	Navarra	2375	1720	655	65	41
Spain	País Vasco	3138	2576	561	41	33

Continued on following page

Table 1 (a). Continued

Country	Region	Labour input	Unpaid labour input	Paid labour Input	Total Utilised Agricultural Area	Total livestock units
Sweden	Län i norra Sverige	3573	2737	835	92	54
Sweden	Skogs- och mellanbygdslän	3154	2541	613	107	77
Sweden	Slättbyggdslän	3085	2238	847	106	56
United Kingdom	England - East Region	6908	2551	4358	171	114
United Kingdom	England - North Region	5315	2880	2435	156	163
United Kingdom	England - West Region	5632	3048	2584	116	121
United Kingdom	Northern Ireland	3445	3083	362	77	100
United Kingdom	Scotland	4562	3403	1158	329	146
United Kingdom	Wales	3793	3168	625	126	163

Table 1 (b).

Total output	Total output/Total input	Total Inputs	Total intermediate consumption	Subsidies on investments	Farm Net Income	Gross Investment on fixed assets	Total subsidies, excluding on investments
99516	1	88962	59628	672	31316	22901	20600
350513	1	283585	214409	2747	82847	67229	16795
180718	1	165874	118224	2186	44522	27880	30425
98410	1	100931	53854	198	21596	13838	22147
114004	1	113687	58586	117	23104	14085	22616
94397	1	98003	50769	287	20301	18173	22603
79434	1	83308	47602	333	20330	14836	23477
34333	1	31583	18873	188	15058	6202	11931
41281	1	43130	27161	70	12920	7269	13893
24235	1	19736	11797	0	9968	2657	5860
26251	1	22851	16552	0	9870	4025	7180
40326	1	33866	26131	34	10309	2882	4906
319298	1	374856	244493	6312	42307	55575	93484
470080	1	465811	308447	412	40167	50590	39905
132733	1	148678	102070	2872	16660	30302	30086
103203	1	132047	90609	974	18597	25895	47393
135295	1	166834	118021	1730	26506	36100	57175
137301	1	190433	133892	2284	23153	44564	75278
112770	1	144959	105307	1268	20581	23247	52564
173255	1	153893	95087	841	34999	25808	16837
175361	1	172003	113299	2258	23094	29365	19740
121515	1	141779	96214	1595	26971	32019	47229
224456	1	221333	153756	1808	37258	43475	35013
208315	1	202437	120504	1330	47235	41670	42041
303476	1	270797	202290	1746	54146	39870	21647
206968	1	202068	137936	937	39230	33309	35876
235301	1	195176	109775	880	62452	31382	24091

Continued on following page

Table 1 (b). Continued

Total output	Total output/Total input	Total Inputs	Total intermediate consumption	Subsidies on investments	Farm Net Income	Gross Investment on fixed assets	Total subsidies, excluding on investments
141422	1	140093	79472	7557	34608	29513	27588
214261	1	208028	136775	548	48046	53251	43524
46717	1	54779	40953	507	17790	4481	25753
279508	1	266905	177872	1569	42865	39544	32917
96070	1	88052	64773	3936	30767	15372	19915
135165	1	125295	77162	2731	23992	19924	13316
98891	1	120737	83126	1884	25822	25197	46998
205455	1	213153	146681	849	39557	33901	48872
81998	1	99547	59136	1809	30881	11199	47172
112107	1	121313	82912	1548	22444	23018	31879
235665	1	221714	150030	1293	35854	43689	23223
287311	1	268105	194385	2333	49150	38566	30113
276932	1	251283	165273	553	58836	41314	35998
216859	1	190789	131796	1374	55880	32483	31073
198389	1	169242	99910	1778	45874	31606	16828
162779	1	156690	102438	1893	33454	32302	27152
294522	1	268818	171214	1405	64301	55663	40132
179330	1	165759	109463	290	38209	34081	24744
163834	1	148391	102931	274	39776	34826	25050
860292	1	1000334	610051	1340	40578	136723	194393
175336	1	174638	119398	604	33086	35944	32842
781472	1	888890	535001	341	34618	151832	159018
331117	1	301540	216826	40	60881	43295	31222
270153	1	248690	179887	183	45515	40818	24289
193706	1	161574	104294	46	57310	30446	21548
160598	1	163617	115465	128	37063	32874	42662
733352	1	821552	508546	4486	31376	122795	125196
758371	1	864666	509157	461	33820	109159	154951
984147	1	1140791	671197	3521	28058	172886	196035
22132	1	15020	8926	33	10256	2121	3497
23187	1	21237	14387	30	10186	964	8375
20725	1	16457	11066	37	9827	662	5980
23191	1	19579	13981	37	12622	1040	9470
67812	1	59297	43246	118	22417	7722	14686
99757	1	98510	70472	97	20079	13522	19956
60769	1	53497	36342	22	24565	7433	18418
75708	1	67983	55581	682	26796	12002	18573
41342	2	24505	16600	68	21921	1392	5320
81824	2	51717	31647	267	38787	20300	7203
51962	2	33292	21113	20	28768	1947	10306
30919	2	20446	9974	4	19286	1504	9313
57173	2	34103	23236	39	30631	1059	8752

Continued on following page

Table 1 (b). Continued

Total output	Total output/Total input	Total Inputs	Total intermediate consumption	Subsidies on investments	Farm Net Income	Gross Investment on fixed assets	Total subsidies, excluding on investments
114182	2	75193	55729	17	48927	221	10799
110228	2	69945	48308	315	48438	11599	7785
65453	1	44559	27416	113	29116	3914	8954
76514	2	38437	25425	223	40479	-359	3325
174521	2	115827	93881	2	71497	3144	15066
40663	1	28472	19995	4	21081	3083	9395
42201	2	25586	18488	114	26225	1364	9691
95639	2	62388	48656	64	45883	4445	13168
44504	1	30612	18761	4	21552	3944	8317
47488	1	33960	25995	348	26863	2835	12487
43070	2	28280	15984	71	21022	1431	6872
87892	1	62204	38225	145	33733	4065	8603
57803	2	31432	20588	2	30212	2517	4157
52538	1	40131	25820	73	27407	5013	15862
63475	1	52217	34382	976	31881	8321	19515
120051	2	73516	53734	209	56506	12236	10435
70146	1	73234	50337	2547	14981	19812	16622
38893	1	40269	25835	2592	11716	13346	10854
231229	1	246990	156751	17167	58311	70513	53301
38288	1	29790	24915	136	11044	1958	2409
565104	1	486380	337437	1575	93347	82402	17418
18361	1	15262	10429	174	6328	2631	3515
26183	1	21885	15497	228	9258	3847	5257
48826	1	44687	33716	282	15096	5759	11885
42168	1	38176	27856	239	11316	4822	8035
51015	1	34740	19151	87	29555	5118	13483
23142	1	18927	13091	1094	13948	6232	8896
32051	1	24252	17546	687	14115	3081	6177
60835	2	40289	30086	636	26458	4717	5399
34905	2	16152	9286	592	23592	945	4332
19309	1	14844	11071	19	8214	981	4028
15207	1	11022	8315	2	6622	656	2577
15684	1	11446	8674	1	6778	1207	2849
26215	1	20470	14222	4	10736	3849	5665
21817	1	16770	11887	115	8702	921	3726
12515	1	9686	7344	10	4898	-1963	2300
25016	1	19033	12928	1	10578	1310	5126
591473	1	715733	448234	4477	16843	108787	142748
28458	1	28307	18613	357	6396	11443	7151
84623	2	53677	33212	8	44034	1619	12319
88921	1	67627	44274	43	38447	6137	17198
63482	1	57281	49295	5	21278	846	16141

Continued on following page

Table 1 (b). Continued

Total output	Total output/Total input	Total Inputs	Total intermediate consumption	Subsidies on investments	Farm Net Income	Gross Investment on fixed assets	Total subsidies, excluding on investments
117105	1	96944	51896	142	47191	809	24440
61434	1	54166	45093	83	22345	3899	15460
95462	1	77763	61758	3	31943	3152	14401
76168	1	51339	35437	41	35134	5032	10567
112648	1	95074	66269	472	31826	13489	13774
65680	2	44349	31360	3	24180	1236	3644
82311	1	59538	42710	6	36429	1415	12948
59581	1	43801	36628	10	25464	2322	9788
66358	1	53057	33877	442	29391	253	14549
77213	2	49501	30230	75	36133	8347	7241
63533	1	48940	37793	21	22271	1923	7829
117060	2	77851	45661	7	49349	983	8948
102773	1	90561	65846	529	28786	21217	17387
97407	1	76736	56302	428	36772	11003	15613
162646	1	205469	151653	78	24447	33521	67219
195775	1	216278	160513	86	23725	42100	44251
217312	1	233628	162915	21	20120	40093	36598
377376	1	370717	251018	625	52675	63826	46523
303171	1	299663	219267	1214	45883	50614	41516
262638	1	255710	180073	1028	41598	43859	33668
129404	1	126175	99404	710	30225	20554	27029
224139	1	243753	172258	1493	34529	50093	53866
184030	1	182514	143972	1699	34978	32518	31775

In general, the several variables improved the average dimension (over the European Union agricultural regions) form 2017 to 2019 (table 2), exception for the following variables: labour input (decreased slightly in 2018); unpaid labour input (decreased over the three years); paid labour input (decreased very slightly in 2018); farm net income (decreased in 2018); and gross investment on fixed assets (decreased in 2019).

Table 2. Dimensions of farm variables, on average over the European Union agricultural regions and across the period 2017-2019

Year	Labour input	Unpaid labour input	Paid labour Input	Total Utilised Agricultural Area	Total livestock units	Total output	Total output/ Total input	Total Inputs	Total intermediate consumption	Subsidies on investments	Farm Net Income	Gross Investment on fixed assets	Total subsidies, excluding on investments
2017	4019	2431	1588	73	50	147637	1	142928	93651	786	30738	22944	26712
2018	4014	2427	1587	74	50	150254	1	148057	98094	974	29553	24443	27608
2019	4034	2405	1629	74	50	157399	1	153474	102240	1056	31519	23684	27798

On average over the period considered and the agricultural regions, the European Union farms used 4022 hours of labour input, 2421 hours of unpaid labour and 1601 hours of paid labour (table 3). These farms have, on average, 74 hectares, 50 livestock units and a total factor productivity ratio (total output/total input) of around 1. They received 939 euros from subsidies on investments and 27373 euros from total subsidies (excluding on investments). The total subsidies (excluding on investments) represent 90% when compared with the farm net income and 18% when compared with the total output.

Table 3. Summary statistics of farm variables, over the period 2017-2019 and across the European Union agricultural regions

Variable	Observations	Mean	Standard Deviation	Min	Max
Labour input	402	4022	2589	1651	20183
Unpaid labour input	402	2421	559	400	4006
Paid labour Input	402	1601	2592	37	18580
Total Utilised Agricultural Area	402	74	91	3	490
Total livestock units	402	50	53	2	264
Total output	402	151764	170472	7837	1052398
Total output/Total input	402	1	0	1	2
Total Inputs	402	148153	192551	5644	1180252
Total intermediate consumption	402	97995	119001	4140	694049
Subsidies on investments	402	939	1964	0	17957
Farm Net Income	402	30603	17151	-434	105583
Gross Investment on fixed assets	402	23691	30799	-8332	181201
Total subsidies, excluding on investments	402	27373	33892	1484	206141

RESULTS

The results presented in table 4 for the coefficient of correlation (Pearson, 1896; Pearson & Filon, 1898) between the several variables considered were obtained following the procedures proposed by the Stata software (StataCorp, 2017a, 2017b; Stata, 2021).

In this table 4, the strongest and positive correlations between the labour input and the other variables are found for the paid labour, total utilised agricultural area, total inputs and total subsidies (excluding on investments). This means that the European Union farms with more labour input are the same with more area, more total costs and that receive more total subsidies (without those for the investments). The weakest correlations among the labour input and the other variables are verified for the unpaid labour, total livestock units, subsidies on investments and farm net income.

In practice, the total subsidies (excluding on investments) are paid for the less competitive farms, as aimed by the Common Agricultural Policy measures, and they are associated with the use of labour input, what is interesting in a social perspective, nonetheless they are correlated, also, with the use of more inputs, what brings serious concerns in an environmental viewpoint.

It seems that the CAP instruments should be more input-oriented (Martinho, 2015). This means that the total subsidies (excluding on investments) paid to the farmers should be conditioned to an effective

Table 4 (a). Pairwise matrix of correlation for farm variables, over the period 2017-2019 and across the European Union agricultural regions

	Labour input	Unpaid labour input	Paid labour Input	Total Utilised Agricultural Area	Total livestock units	Total output
Labour input	1.000					
Unpaid labour input	0.1036*	1.000				
	(0.038)					
Paid labour Input	0.9768*	-0.1120*	1.000			
	(0.000)	(0.025)				
Total Utilised Agricultural Area	0.8096*	0.007	0.8074*	1.000		
	(0.000)	(0.895)	(0.000)			
Total livestock units	0.5745*	0.2429*	0.5216*	0.7304*	1.000	
	(0.000)	(0.000)	(0.000)	(0.000)		
Total output	0.7976*	0.085	0.7787*	0.8659*	0.8320*	1.000
	(0.000)	(0.090)	(0.000)	(0.000)	(0.000)	
Total output/Total input	-0.3262*	-0.060	-0.3130*	-0.4831*	-0.4421*	-0.4029*
	(0.000)	(0.229)	(0.000)	(0.000)	(0.000)	(0.000)
Total Inputs	0.8295*	0.053	0.8173*	0.9039*	0.8122*	0.9898*
	(0.000)	(0.292)	(0.000)	(0.000)	(0.000)	(0.000)
Total intermediate consumption	0.8060*	0.082	0.7875*	0.8919*	0.8450*	0.9906*
	(0.000)	(0.100)	(0.000)	(0.000)	(0.000)	(0.000)
Subsidies on investments	0.2590*	0.045	0.2490*	0.2562*	0.3212*	0.2915*
	(0.000)	(0.369)	(0.000)	(0.000)	(0.000)	(0.000)
Farm Net Income	0.1374*	0.2460*	0.084	0.1786*	0.4655*	0.4376*
	(0.006)	(0.000)	(0.092)	(0.000)	(0.000)	(0.000)
Gross Investment on fixed assets	0.7612*	0.094	0.7402*	0.8748*	0.8049*	0.9478*
	(0.000)	(0.059)	(0.000)	(0.000)	(0.000)	(0.000)
Total subsidies, excluding on investments	0.8137*	-0.009	0.8147*	0.9260*	0.7009*	0.8722*
	(0.000)	(0.865)	(0.000)	(0.000)	(0.000)	(0.000)

Note: *, statistically significant at 5%.

reduction in the level of inputs use with environmental impacts. In this group of resources whose quantities employed in the farms should be mitigated appear the fertilisers, crop protection products and energy.

Factor Analysis

To better assess the relationships of the several farm variables with the labour input use, it was carried out in this subsection a factor analysis (Kim & Mueller, 1978a, 1978b), following the Stata software and Torres-Reyna (n.d.) procedures.

With the factor analysis were found three factors (table 5) that jointly explain 88% of the total variance. The majority of the variables are correlated with the factor1 (table 6), showing the interrelationships between them, exception for the ratio total output/total input and subsidies on investments that

Table 4 (b).

Total output/Total input	Total Inputs	Total intermediate consumption	Subsidies on investments	Farm Net Income	Gross Investment on fixed assets	Total subsidies, excluding on investments
1.000						
-0.4567*	1.000					
(0.000)						
-0.4728*	0.9956*	1.000				
(0.000)	(0.000)					
-0.3295*	0.3060*	0.3107*	1.000			
(0.000)	(0.000)	(0.000)				
0.065	0.3246*	0.3572*	0.2011*	1.000		
(0.196)	(0.000)	(0.000)	(0.000)			
-0.5173*	0.9581*	0.9594*	0.3972*	0.3387*	1.000	
(0.000)	(0.000)	(0.000)	(0.000)	(0.000)		
-0.5428*	0.9254*	0.9107*	0.3250*	0.1511*	0.8950*	1.000
(0.000)	(0.000)	(0.000)	(0.000)	(0.002)	(0.000)	

Note: *, statistically significant at 5%.

are correlated with the factor2 and for the farm net income that is correlated with the factor3. These findings confirm those obtained before that the variables farm net income, subsidies on investments and the ratio total output/total input are weakly correlated with the others farm variables. Table 7 shows the adequacy of the model.

Table 5. Factor analysis, for farm variables, through principal-component factors and orthogonal varimax (Kaiser off), on average over the period 2017-2019 and across the European Union agricultural regions

Factor	Variance	Difference	Proportion	Cumulative
Factor1	7.318	5.675	0.610	0.610
Factor2	1.643	0.035	0.137	0.747
Factor3	1.608	.	0.134	0.881

Table 6. Rotated factor loadings (pattern matrix) and unique variances, for farm variables, on average over the period 2017-2019 and across the European Union agricultural regions

Variable	Factor1	Factor2	Factor3	Uniqueness
Labour input	0.926	0.041	-0.034	0.140
Paid labour Input	0.926	0.029	-0.096	0.133
Total Utilised Agricultural Area	0.920	0.206	0.066	0.107
Total livestock units	0.678	0.306	0.497	0.200
Total output	0.902	0.161	0.367	0.027
Total output/Total input	-0.379	-0.762	0.165	0.249
Total Inputs	0.935	0.200	0.252	0.022
Total intermediate consumption	0.916	0.221	0.297	0.025
Subsidies on investments	0.124	0.784	0.231	0.317
Farm Net Income	0.139	0.001	0.945	0.088
Gross Investment on fixed assets	0.869	0.332	0.293	0.050
Total subsidies, excluding on investments	0.916	0.292	0.033	0.075

Cross-Section Regressions

Tables 8 and 9 present the results for OLS (Ordinary Least Square) cross-section regressions (Galton, 1886), obtained following the Stata software procedures (StataCorp, 2017a, 2017b; Stata, 2021). In table 8 the variables were considered in growth rates, with the labour input growth rate as dependent variable, considering as basis the Verdoorn-Kaldor laws (Verdoorn, 1949; Kaldor, 1966, 1975) and following, for example, Martinho (2020c). To benchmark, table 9 shows the results for a model based on the Cobb and Douglas (1928) theory of production.

Table 8 reveals that the greatest and positive marginal impact on the labour input growth rate come from the total output growth rate (when the total output growth rates change 1 percentage point the labour

Table 7. Kaiser-Meyer-Olkin measure of sampling adequacy, for farm variables, on average over the period 2017-2019 and across the European Union agricultural regions

Variable	kmo
Labour input	0.859
Paid labour Input	0.835
Total Utilised Agricultural Area	0.958
Total livestock units	0.903
Total output	0.641
Total output/Total input	0.880
Total Inputs	0.648
Total intermediate consumption	0.905
Subsidies on investments	0.268
Farm Net Income	0.207
Gross Investment on fixed assets	0.960
Total subsidies, excluding on investments	0.630
Overall	0.730

Table 8. Cross-section regression results with the labour input as dependent variables, with the variables in growth rates, on average over the period 2017-2019 and across the European Union agricultural regions

Variable	OLS robust model
Constant	-0.014* (-4.450) [0.000]
Total output growth rate	0.345* (4.370) [0.000]
Total output/Total input growth rate	-0.529* (-4.070) [0.000]
Gross Investment on fixed assets growth rate	-0.007* (-3.830) [0.000]
Subsidies on investment growth rate	0.001* (2.400) [0.018]
Total subsidies, excluding on investments growth rate	-0.078 (-1.590) [0.116]
VIF	1.300
Breusch-Pagan / Cook-Weisberg test for heteroskedasticity	17.770* [0.000]
Ramsey RESET test	7.150* [0.000]
R-squared	0.414
Number of observations	113

Note: *, statistically significant at 5%.

Table 9. Cross-section regression results with the total output as dependent variables, with the variables in logarithms, on average over the period 2017-2019 and across the European Union agricultural regions

Variable	OLS model
Constant	0.380 (0.580) [0.563]
Labour input logarithm	0.290* (3.640) [0.000]
Total livestock units logarithm	0.235* (6.500) [0.000]
Total output/Total input logarithm	1.810* (8.840) [0.000]
Gross Investment on fixed assets logarithm	0.237* (7.610) [0.000]
Total subsidies, excluding on investments logarithm	0.563* (8.860) [0.000]
VIF	3.070
Breusch-Pagan / Cook-Weisberg test for heteroskedasticity	2.380 [0.123]
Ramsey RESET test	1.070 [0.366]
Adjusted R-squared	0.899
Number of observations	132

Note: *, statistically significant at 5%.

input growth rate change 0.345 percentage points). The total output/total input ratio growth rates have a marginal effect of -0.529. The subsidies on investments and the gross investment growth rates have an almost null marginal impact on the labour growth rate and the total subsidies (excluding on investments) growth rates have not statistical significance.

With the variables in logarithms and considering the Cobb-Douglas model as basis, table 9 shows that the labour input, livestock units and gross investment have the lowest marginal effects on the total output. For example, when the labour input grows 1% the total output grows 0.290%. In turn, the marginal impact from the total subsidies (excluding on investments) is 0.563.

Benchmarking table 8 and 9, of referring that the total subsidies (without those for the investments) promote more the total output than the labour input. In fact, the current subsidies should be more input-oriented for a more sustainable development, namely mitigating more the environmental implications and promote the social dimensions of the European Union farms.

CONCLUSION

The main objective of this research was to highlight the main relationships of the labour input in the European Union farms with other variables and to assess the main determinants of this agricultural production factor. For that, data from the Farm Accountancy Data Network were considered for the period 2017-2019 and cross the European Union agricultural regions. This statistical information was explored through matrices of correlation, factor analysis and cross-section regressions based on the models from the Cobb-Douglas and Verdoorn-Kaldor developments.

The most dynamic farms (from German agricultural regions, for example) use, in general, more labour inputs, namely paid work force. These findings highlight, as showed by the literature review, the importance of the paid labour for the agricultural indicators performance, namely in terms of technical efficiency. Maybe these results explain the decreased trend, on average, for the unpaid labour over the period 2017-2019. A deeper analysis of the data show that the unpaid labour is the major part of the labour used in the European Union farms. In fact, on average, the farms used 4022 hours of labour input, 2421 hours of unpaid labour and 1601 hours of paid labour. In addition, these farms have, on average, a total factor productivity ratio (total output/total input) of around 1, received 939 euros from subsidies on investments and 27373 euros from total subsidies (excluding on investments). These findings reveal the importance of the subsidies for the sustainability of the farms.

The labour input is positively correlated with the paid labour, area, total inputs and total current subsidies (excluding on investments) and weakly correlated with the unpaid labour. Considering the importance given by the international organizations for the family farming and its relevance for the sustainability, these findings deserve special attention in future studies, namely to identify the main causes and find new solutions to address the respective constraints.

The factor analysis highlighted the correlations between the subsidies on investment and the total factor productivity (ratio total output/total input), showing the importance of the investment and their incentives to promote the competitiveness of the European Union farms.

The results from the regression approaches reveal relevant increasing returns to scale in the European farms, considering the coefficient between the labour growth rates and the total output growth rates close to 0. In fact, this coefficient is expected, from the theory, to range between 0 and 1 and results close to 0 represent higher increasing returns to scale. In addition, the impacts (on growth rates) from the total productivity are strongly negative and the effects from the investment and subsidies are residual or null. In turn, with the variables in logarithm and considering the Cobb-Douglas model as basis, one of the most important impacts on the total output came from the total current subsidies, showing that they more important to maintain the levels of output than to promote improvements in the labour dynamics.

ACKNOWLEDGMENT

This work is funded by National Funds through the FCT - Foundation for Science and Technology, I.P., within the scope of the project Refª UIDB/00681/2020. Furthermore, we would like to thank the CERNAS Research Centre and the Polytechnic Institute of Viseu for their support.

REFERENCES

Alonso Villa, P., & Juste Carrion, J. J. (2018). The automotive sector in Castile and Leon. 50 years of economic growth and productivity: 1961-2011. *Revista De Estudios Regionales*, *113*, 101–136.

Baer-Nawrocka, A., & Poczta, W. (2018). Agricultural transformation in Poland: A long-term analysis. *Berichte über Landwirtschaft*, *96*(1).

Beyer, D., & Hinke, J. (2020). European benchmarking of determinants of profitability for companies with accrual accounting in the agricultural sector. *Agricultural Economics-Zemedelska Ekonomika*, *66*(11), 477–488. doi:10.17221/128/2020-AGRICECON

Bijttebier, J., Ruysschaert, G., Hijbeek, R., Werner, M., Pronk, A. A., Zavattaro, L., Bechini, L., Grignani, C., ten Berge, H., Marchand, F., & Wauters, E. (2018). Adoption of non-inversion tillage across Europe: Use of a behavioural approach in understanding decision making of farmers. *Land Use Policy*, *78*, 460–471. doi:10.1016/j.landusepol.2018.05.044

Carmen Cuerva, M. (2012). Determinants of Labour Productivity Convergence in the European Agricultural Sector. *Agrociencia*, *46*(6), 621–635.

Cobb, C. W., & Douglas, P. H. (1928). A Theory of Production. *The American Economic Review*, *18*(1), 139–165.

Constantin, M., Radulescu, I., Andrei, J., Chivu, L., Erokhin, V., & Gao, T. (2021). A Perspective on Agricultural Labor Productivity and Greenhouse Gas Emissions in Context of the Common Agricultural Policy Exigencies. *Ekonomika Poljoprivreda-Economics of Agriculture*, *68*(1), 53–67. doi:10.5937/ekoPolj2101053C

Doytch, N., & Eren, M. (2012). Institutional Determinants of Sectoral FDI in Eastern European and Central Asian Countries: The Role of Investment Climate and Democracy. *Emerging Markets Finance & Trade*, *48*(sup4), 14–32. doi:10.2753/REE1540-496X4806S402

Dupraz, P., & Latruffe, L. (2015). Trends in family labour, hired labour and contract work on French field crop farms: The role of the Common Agricultural Policy. *Food Policy*, *51*, 104–118. doi:10.1016/j.foodpol.2015.01.003

FADN. (2021). *European Commission | Agri-food data portal | Farm economics*. https://agridata.ec.europa.eu/extensions/FarmEconomyFocus/FADNDatabase.html

Fouka, V., & Schlapfer, A. (2020). Agricultural Returns to Labour and the Origins of Work Ethics. *Economic Journal (London)*, *130*(628), 1081–1113. doi:10.1093/ej/ueaa029

Galdeano-Gomez, E., Cespedes-Lorente, J., & Rodriguez-Rodriguez, M. (2006). Productivity and environmental performance in marketing cooperatives: An analysis of the Spanish horticultural sector. *Journal of Agricultural Economics*, *57*(3), 479–500. doi:10.1111/j.1477-9552.2006.00061.x

Galton, F. (1886). Regression Towards Mediocrity in Hereditary Stature. *Journal of the Anthropological Institute of Great Britain and Ireland*, *15*, 246–263. doi:10.2307/2841583

Giannakis, E., & Bruggeman, A. (2018). Exploring the labour productivity of agricultural systems across European regions: A multilevel approach. *Land Use Policy*, *77*, 94–106. doi:10.1016/j.landusepol.2018.05.037

González, C. H., & Fernández, M. P. R. (2017). Total factor productivity and its components for sheep and goat farms in 37 Southern European regions (2004-2012). *Economía Agraria y Recursos Naturales*, *17*(2), 5–27. doi:10.7201/earn.2017.02.01

Greenwood, M. J. (2008). Family and sex-specific US immigration from Europe, 1870-1910: A panel data study of rates and composition. *Explorations in Economic History*, *45*(4), 356–382. doi:10.1016/j.eeh.2008.02.001

Herman, E. (2016). Productive Employment in Romania: A Major Challenge to the Integration into the European Union. *Amfiteatru Economic*, *18*(42), 335–350.

Horrell, S., Humphries, J., & Weisdorf, J. (2020). Malthus's missing women and children: Demography and wages in historical perspective, England 1280-1850. *European Economic Review*, *129*, 103534. doi:10.1016/j.euroecorev.2020.103534

Jambor, A., Balogh, J., & Kucsera, P. (2016). Country and industry specific determinants of intra-industry agri-food trade in the Baltic Countries. *Agricultural Economics-Zemedelska Ekonomika*, *62*(6), 280–291. doi:10.17221/153/2015-AGRICECON

Kaldor, N. (1966). *Causes of the Slow Rate of Economic Growth of the UK, An Inaugural Lecture*. Cambridge University Press.

Kaldor, N. (1975). Economic Growth and the Verdoorn Law—A Comment on Mr Rowthorn's Article. *Economic Journal (London)*, *85*(340), 891–896. doi:10.2307/2230633

Kim, J.-O., & Mueller, C. W. (1978a). *Factor Analysis: Statistical Methods and Practical Issues*. SAGE Publications, Inc. doi:10.4135/9781412984256

Kim, J.-O., & Mueller, C. W. (1978b). *Introduction to Factor Analysis: What It Is and How To Do It* (1st ed.). SAGE Publications, Inc. doi:10.4135/9781412984652

Kostov, P., Davidova, S., & Bailey, A. (2019). Comparative Efficiency of Family and Corporate Farms: Does Family Labour Matter? *Journal of Agricultural Economics*, *70*(1), 101–115. doi:10.1111/1477-9552.12280

Kvartiuk, V. (2015). Eastern Europe as source for skilled agricultural workers in Germany? *Berichte über Landwirtschaft*, *93*(3).

Latruffe, L., Balcombe, K., Davidova, S., & Zawalinska, K. (2004). Determinants of technical efficiency of crop and livestock farms in Poland. *Applied Economics*, *36*(12), 1255–1263. doi:10.1080/0003684042000176793

Martinho, V. J. P. D. (2015). Output Impacts of the Single Payment Scheme in Portugal: A Regression with Spatial Effects. *Outlook on Agriculture*, *44*(2), 109–118. doi:10.5367/oa.2015.0203

Martinho, V. J. P. D. (2017). Efficiency, total factor productivity and returns to scale in a sustainable perspective: An analysis in the European Union at farm and regional level. *Land Use Policy*, *68*, 232–245. doi:10.1016/j.landusepol.2017.07.040

Martinho, V. J. P. D. (2020a). Agricultural Entrepreneurship in the European Union: Contributions for a Sustainable Development. *Applied Sciences (Basel, Switzerland)*, *10*(6), 2080. doi:10.3390/app10062080

Martinho, V. J. P. D. (2020b). Comparative analysis of energy costs on farms in the European Union: A nonparametric approach. *Energy*, *195*, 116953. doi:10.1016/j.energy.2020.116953

Martinho, V. J. P. D. (2020c). Relationships between agricultural energy and farming indicators. *Renewable & Sustainable Energy Reviews*, *132*, 110096. doi:10.1016/j.rser.2020.110096

Martinsson, E., & Hansson, H. (2021). Adjusting eco-efficiency to greenhouse gas emissions targets at farm level? The case of Swedish dairy farms. *Journal of Environmental Management*, *287*, 112313. doi:10.1016/j.jenvman.2021.112313 PMID:33714041

Matkovski, B., Kalas, B., Zekic, S., & Jeremic, M. (2019). Agri-food competitiveness in South East Europe. *Outlook on Agriculture*, *48*(4), 326–335. doi:10.1177/0030727019854770

Melchor-Ferrer, E. (2020). Determinants of labour productivity growth in Spanish and Portuguese regions: A spatial shift-share approach. *The Annals of Regional Science*, *65*(1), 45–65. doi:10.100700168-019-00973-4

Meyfroidt, P., Schierhorn, F., Prishchepov, A. V., Mueller, D., & Kuemmerle, T. (2016). Drivers, constraints and trade-offs associated with recultivating abandoned cropland in Russia, Ukraine and Kazakhstan. *Global Environmental Change*, *37*, 1–15. doi:10.1016/j.gloenvcha.2016.01.003

Moyano Pesquera, P. B., Miranda Escolar, B., & Gordo Gomez, P. (2017). The involvement of economic and social stakeholders in the revitalization of rural municipalities. *Investigaciones Regionales*, *38*, 91–115.

Mujcinovic, A., Nikolic, A., Tuna, E., Stamenkovska, I. J., Radovic, V., Flynn, P., & McCauley, V. (2021). Is It Possible to Tackle Youth Needs with Agricultural and Rural Development Policies? *Sustainability*, *13*(15), 8410. doi:10.3390u13158410

Olper, A., Raimondi, V., Cavicchioli, D., & Vigani, M. (2014). Do CAP payments reduce farm labour migration? A panel data analysis across EU regions. *European Review of Agriculture Economics*, *41*(5), 843–873. doi:10.1093/erae/jbu002

Pearson, K. (1896). VII. Mathematical contributions to the theory of evolution.—III. Regression, heredity, and panmixia. *Philosophical Transactions of the Royal Society of London. Series A, Containing Papers of a Mathematical or Physical Character*, *187*, 253–318. doi:10.1098/rsta.1896.0007

Pearson, K., & Filon, L. N. G. (1898). VII. Mathematical contributions to the theory of evolution.—IV. On the probable errors of frequency constants and on the influence of random selection on variation and correlation. *Philosophical Transactions of the Royal Society of London. Series A, Containing Papers of a Mathematical or Physical Character*, *191*, 229–311. doi:10.1098/rsta.1898.0007

Poczta, W., & Pawlak, K. (2011). Potential competitiveness and competitive position of the Polish agri-food sector on the Single European Market. *Berichte über Landwirtschaft*, *89*(1), 134–169.

Rodriguez-Pose, A., & Tselios, V. (2009). Education and Income Inequality in the Regions of the European Union. *Journal of Regional Science*, *49*(3), 411–437. doi:10.1111/j.1467-9787.2008.00602.x

Salvioni, C., Henke, R., & Vanni, F. (2020). The Impact of Non-Agricultural Diversification on Financial Performance: Evidence from Family Farms in Italy. *Sustainability*, *12*(2), 486. doi:10.3390u12020486

Spicka, J., & Smutka, L. (2014). The Technical Efficiency of Specialised Milk Farms: A Regional View. *TheScientificWorldJournal*, *985149*, 1–13. Advance online publication. doi:10.1155/2014/985149 PMID:25050408

Stata. (2021). *Stata: Software for Statistics and Data Science*. https://www.stata.com/

StataCorp. (2017a). *Stata 15 Base Reference Manual*. Stata Press.

StataCorp. (2017b). *Stata Statistical Software: Release 15*. StataCorp LLC.

Strelecek, F., Zdenek, R., & Lososova, J. (2007). Influence of farm milk prices in the EU 25 on profitability and production volume indicators. *Agricultural Economics-Zemedelska Ekonomika*, *53*(12), 545–557. doi:10.17221/1224-AGRICECON

Thomassen, M. A., Dolman, M. A., van Calker, K. J., & de Boer, I. J. M. (2009). Relating life cycle assessment indicators to gross value added for Dutch dairy farms. *Ecological Economics*, *68*(8–9), 2278–2284. doi:10.1016/j.ecolecon.2009.02.011

Todorovic, S., Papic, R., Ciaian, P., & Bogdanov, N. (2020). Technical efficiency of arable farms in Serbia: Do subsidies matter? *New Medit*, *19*(4), 81–97. doi:10.30682/nm2004f

Torres-Reyna, O. (n.d.). *Getting Started in Factor Analysis (using Stata)*. http://www.princeton.edu/~otorres/Stata/Factor

Trnkova, G., & Kroupova, Z. Z. (2020). Determinants of Persistent and Transient Technical Efficiency of Milk Production in Eu. *E+M. Ekonomie a Management*, *23*(4), 39–54. doi:10.15240/tul/001/2020-4-003

Verdoorn, P. J. (1949). Fattori che regolano lo sviluppo della produttivita del lavoro. *L'Industria*, *1*, 45–53.

Wilk, E., Krowczyriska, M., & Zagajewski, B. (2019). Modelling the Spatial Distribution of Asbestos-Cement Products in Poland with the Use of the Random Forest Algorithm. *Sustainability*, *11*(16), 4355. doi:10.3390u11164355

Wlodarczyk-Spiewak, K., & Korpysa, J. (2006). Changes of labour market as the main determinant of inequalities in the consumption structure in Polish households. *Transformations in Business & Economics*, *5*(2), 106–121.

Zbarsky, V. K., Trusova, N., Sokil, O. H., Pochernina, N., & Hrytsaienko, M. L. (2020). Social and Economic Determinants for the Development of Resource Potential of Small Forms of Agrarian Production in Ukraine. *Industrial Engineering and Management Systems*, *19*(1), 133–142. doi:10.7232/iems.2020.19.1.133

Chapter 9
Main Factors That Explain the Use of Fertilisers on Farms in the European Union:
Contributions to a More Sustainable Development

Vítor João Pereira Domingues Martinho
Polytechnic Institute of Viseu, Portugal

ABSTRACT

A deeper assessment of the main determinants associated with the use of fertilisers, for example, in the European Union farms may bring relevant insights about the respective frameworks and support to find more sustainable solutions. In this context, the main objective of this study is to identify factors that influence the use of fertilisers in the agricultural sector of the European Union regions. To achieve this objective, statistical information, at farm level, from the European Farm Accountancy Data Network was considered. These data were first analysed through exploratory approaches and after assessed with classification and regression tree methodologies. The results obtained provide interesting insights to promote a more sustainable development in the European farms, namely supporting the policymakers to design more adjusted measures and instruments. In addition, the fertilisers costs on the European Union farms are mainly explained by crop output, costs with inputs, current subsidies, utilised agricultural area, and gross investment.

INTRODUCTION

A rational use of the fertilisers is a concern for the farmers and for the agricultural and environmental organizations, because the associated costs and the adverse impacts on the sustainability. In this perspective, the several contexts of the fertilisers related with the agriculture have called the attention of the different stakeholders, including the scientific community (Martinho, 2021).

DOI: 10.4018/978-1-7998-9557-2.ch009

For the respective assessments diverse approaches and databases have been considered, where classification and regression tree (CART) has its importance as a methodology and the European Union Farm Accountancy Data Network (FADN) as a statistical database (FADN, 2021).

In the frameworks of the fertilisers use in the agricultural sector, the classification and regression tree approach was considered, for example, in the following assessments: implications of conservative agriculture for rice yield in Madagascar (Bruelle et al., 2015); nitrate presence in water beneath land surface of the United States (Burow et al., 2010); implications of tillage on the soil characteristics in Chinese rice fields (Feng et al., 2021); agricultural practices interrelated with the presence of certain diseases in corn productions (Hartman et al., 2020); agricultural sustainability in South-East Asia (Lairez et al., 2020); impacts on rice yield from elevated dioxide carbon concentration (Lv et al., 2020); fertilisers application and risks of water contamination in Canada (Ouellet et al., 2008); factors affecting contaminations by heavy metals in China (Ru et al., 2016); determinants of the rice yield in the Senegal River Valley (Tanaka et al., 2015); impacts on maize productivity in Kenya (Tittonell et al., 2008); emissions from dairy agricultural sector in New Zealand (Vogeler et al., 2012); impacts on soil fertility in Kenya (Willy et al., 2019); factors affecting yield of winter wheat in the Polish context (Wojcik-Gront, 2018).

The consideration of the European or national Farm Accountancy Data Network (statistical information at farm level) for analysis involving the fertilisers was found, for example, in researches related with the following issues: variables impacting fertilisers use in the Lithuanian agricultural sector (Besuspariene & Niskanen, 2020); organic farming practices in the Polish milk production (Borawski et al., 2021); representative farms for the agricultural context in Denmark (Dalgaard et al., 2006); dairy agricultural sector in the Netherlands frameworks (Dolman et al., 2014); efficiency assessment for the Italian farms (Galluzzo, 2017); climate impact analysis (Klein et al., 2012); agricultural energy consumption (Meul et al., 2007); efficiency and total factor productivity (Martinho, 2017); environmental implications from the agriculture (Samson et al., 2012); nitrogen and phosphorus dynamics on farms from Ireland (Thomas et al., 2020).

Nonetheless, few (or maybe none) studies considered the CART methodologies and the approaches here addressed to assess the fertilisers' contexts in the European Union farms, using FADN statistical information. In this way, the main objective of this research is to identify the main farm factors that influence the use of fertilisers in the European Union agricultural sector through CART approaches.

LITERATURE SURVEY

The reduction in the use of fertilizers and crop protection products in the farms of the European Union has been one of the main objectives of agricultural policies to improve the respective sustainability (Spugnoli et al., 2012), but with few desirable impacts in many cases. The Farm to Fork strategy, for instance, aims to reduce 20% in fertiliser applications by 2030 (Montanarella & Panagos, 2021). The agricultural policies are crucial to promote a more sustainable agriculture (Garske et al., 2020), nonetheless they are, often, not enough (Garske & Ekardt, 2021).

Reducing fertilisers and crop protection products has substantial benefits to animal welfare and human health (Sabadas et al., 2018). However, making compatible the several dimensions of the sustainable development is a difficult task in the farming sector (Catarino et al., 2019) around the world (Fischer-Kowalski et al., 2011), because the food security (Tingyu et al., 2020) and the particularities of the sector (Stubenrauch et al., 2021).

Fertilisers and crop protection products are between the main causes of farming environmental effects, namely through the greenhouse gas emissions and biodiversity damage (Detang-Dessendre et al., 2020), including for soil arthropods (Madzaric et al., 2018). Other sources of environmental pollution are associated, for example, with municipal waste or mines (Koprivica et al., 2018). There are also other implications from the fertilisers and pesticides use, such as those related with the Mediterranean Sea eutrophication (Karydis & Kitsiou, 2012) and toxic metals (Koprivica et al., 2018).

The dimension of the consequences of the agronomic practices depends on the training and skills of the farmers (Fanelli, 2020), specifically those required to make management options (Fernando et al., 2010) and implement new strategies (Sawinska et al., 2020). But also on the local/regional/national specific conditions (Firrisa et al., 2014), on land use changes (Krausmann et al., 2003) and on the characteristics of the farming systems (Paladino et al., 2020). For instance, the winter wheat in the European Union accounts for about 25% of the total nitrogen fertiliser applied to crops (Samborski et al., 2016).

In any case, adjusted technologies and approaches may support the several stakeholders to deal with the different challenges created by the negative environmental externalities from the agricultural activities (Karpenstein-Machan, 2001). The wet acid scrubbers, as a cleaning air technology for intensive pig farms, and the respective capture of ammonia are examples of sustainable options (Costantini et al., 2020), as well as, manure management practices in livestock farms (Krizsan et al., 2021), or better farming efficiency (Martinho, 2017), or integrated approaches that consider the legacy soil phosphorous (Rowe et al., 2016), or extensification practices (van Grinsven et al., 2015).

DATA ANALYSIS

Table 1 presents a picture for the following main aggregates of farm variables: fertilisers (euros); labour input (hours); total utilised agricultural area (ha); total livestock units (LU); total output crops & crop production (euros); total crops output (euros/ha); total output livestock & livestock products (euros); total livestock output (euros/LU); total specific costs (euros); total farming overheads (euros); total external factors (euros); subsidies on investments (euros); total assets (euros); gross investment on fixed assets (euros); total subsidies, excluding on investments (euros). These data (on average per farm) were obtained from the European Union Farm Accountancy Data Network (FADN, 2021) for the period 2004-2019 and the European agricultural regions.

In general, the German agricultural regions are those where the farms present higher values, on average over 2004-2019, for a great part of the aggregates of variables considered, namely Mecklenburg-Vorpommern and Thüringen. Other German regions, such as Brandenburg and Sachsen-Anhalt present, also, higher values for some variables. In addition to these agricultural regions, of highlighting the values of Pohjois-Suomi (Finland) for the total livestock output (euros/LU), Denmark for total assets, Corse (France) for subsidies on investments, Martinique (France) for total crops output (euros/ha), Alto Adige (Italy) for total livestock output (euros/LU), Luxembourg for subsidies on investments, The Netherlands for the total assets, Slovakia for labour input, total agricultural utilised area and subsidies on investments and Canarias (Spain) for total crops output (euros/ha) and total livestock output (euros/LU). These frameworks reveal the agricultural dynamics of the German farms, but also their higher costs with fertilisers use and higher amounts received from the total subsidies (excluding on investments). On the other hand, Denmark and The Netherlands farms have higher total assets. Slovakia farms are larger, have more labour input and receive more subsidies on investments.

Table 1 (a). Dimensions of the main aggregates of farm variables, on average over the period 2004-2019 and across the European Union agricultural regions

Country	Region	Fertilisers	Labour input	Total Utilised Agricultural Area	Total livestock units	Total output crops & crop production	Total crops output (€/ha)
Austria	Austria	1934	3390	31	26	20690	677
Belgium	Vlaanderen	7237	5287	36	139	108876	3080
Belgium	Wallonie	9197	4304	69	98	61312	887
Bulgaria	Severen tsentralen	6252	5183	63	17	45900	700
Bulgaria	Severoiztochen	5883	5152	62	18	50046	776
Bulgaria	Severozapaden	6658	5058	68	11	46898	650
Bulgaria	Yugoiztochen	4378	5483	55	16	32606	595
Bulgaria	Yugozapaden	1295	4282	18	9	13396	846
Bulgaria	Yuzhen tsentralen	1912	4488	21	12	16289	849
Croatia	Jadranska Hrvatska	666	3605	12	4	14310	1239
Croatia	Kontinentalna Hrvatska	2344	2835	16	10	15079	959
Cyprus	Cyprus	1532	3016	9	13	15152	1804
Czechia	Czechia	18124	13076	211	103	149624	721
Denmark	Denmark	10867	3308	97	148	124332	1336
Estonia	Estonia	8272	4838	125	36	44578	377
Finland	Etelä-Suomi	6514	2364	59	24	42205	746
Finland	Pohjanmaa	6854	2774	56	34	44077	807
Finland	Pohjois-Suomi	7016	3259	62	33	32081	512
Finland	Sisä-Suomi	4585	2942	49	29	27453	572
France	Alsace	9514	3456	50	31	127158	2647
France	Aquitaine	9028	3857	52	41	118860	2396
France	Auvergne	6900	2485	97	94	19850	204
France	Basse-Normandie	11202	2966	97	116	48171	496
France	Bourgogne	14187	3380	122	75	121296	1009
France	Bretagne	7432	3327	63	187	57731	915
France	Centre	23025	2917	130	39	141429	1139
France	Champagne-Ardenne	18076	3400	90	27	201888	2298
France	Corse	3295	3642	82	36	82028	1021
France	Franche-Comté	11493	2890	117	88	46222	394
France	Guadeloupe	4752	2241	9	8	38423	5424
France	Haute-Normandie	20636	3098	121	73	143852	1213
France	La Réunion	6161	3101	9	14	70719	9250
France	Languedoc-Roussillon	4153	3458	48	12	102479	2319
France	Limousin	6574	2583	99	108	13901	139
France	Lorraine	18543	3052	156	105	79098	508

Continued on following page

Table 1 (a). Continued

Country	Region	Fertilisers	Labour input	Total Utilised Agricultural Area	Total livestock units	Total output crops & crop production	Total crops output (€/ha)
France	Martinique	7542	4318	9	5	88141	10944
France	Midi-Pyrénées	9166	2765	82	52	57097	719
France	Nord-Pas-de-Calais	15051	3220	81	68	120962	1552
France	Pays de la Loire	10092	3602	87	143	75559	868
France	Picardie	25845	3027	131	47	184231	1433
France	Poitou-Charentes	14358	2894	101	54	116783	1199
France	Provence-Alpes-Côte d'Azur	6347	4819	42	11	149145	3745
France	Rhône-Alpes	6791	3445	66	55	73565	1129
France	Île-de-France	27715	3433	141	6	217180	1615
Germany	Baden-Württemberg	6363	4388	50	52	72527	1485
Germany	Bayern	5580	3504	47	55	42058	900
Germany	Brandenburg	41976	15570	430	236	304852	738
Germany	Hamburg	5554	5728	4	0	171016	47619
Germany	Hessen	7995	4041	73	59	68849	963
Germany	Mecklenburg-Vorpommern	78794	13787	486	241	436143	922
Germany	Niedersachsen	11508	4231	76	126	87233	1164
Germany	Nordrhein-Westfalen	7080	4163	55	107	81497	1508
Germany	Rheinland-Pfalz	6759	5345	51	27	117919	2374
Germany	Saarland	11018	3879	111	66	43993	406
Germany	Sachsen	38376	14924	308	177	281953	931
Germany	Sachsen-Anhalt	56519	13102	443	191	399436	923
Germany	Schleswig-Holstein	15808	4236	91	118	82960	921
Germany	Schleswig-Holstein/ Hamburg	16127	4812	103	120	125982	1238
Germany	Thüringen	54124	20217	458	272	433846	965
Greece	Ipiros-Peloponissos-Nissi Ioniou	915	2575	6	5	14660	2504
Greece	Makedonia-Thraki	1881	2616	11	6	17416	1643
Greece	Sterea Ellas-Nissi Egaeou-Kriti	1150	3065	8	7	14567	1955
Greece	Thessalia	1738	2742	10	6	17218	1724
Hungary	Alföld	3442	3185	40	16	37973	978
Hungary	Dunántúl	7802	3860	57	23	55494	991
Hungary	Dél-Alföld	2476	3780	40	16	30053	775
Hungary	Dél-Dunántúl	6774	4422	61	27	49410	813
Hungary	Közép-Dunántúl	6621	5744	84	35	53036	635
Hungary	Közép-Magyarország	3533	4061	55	22	36645	662

Continued on following page

Table 1 (a). Continued

Country	Region	Fertilisers	Labour input	Total Utilised Agricultural Area	Total livestock units	Total output crops & crop production	Total crops output (€/ha)
Hungary	Nyugat-Dunántúl	4992	4940	60	22	47165	776
Hungary	Észak-Alföld	2770	2837	45	15	26523	615
Hungary	Észak-Magyarország	4287	3457	59	14	37494	657
Ireland	Ireland	5158	2428	46	56	9591	205
Italy	Abruzzo	1476	2815	12	5	24524	2057
Italy	Alto Adige	716	3250	8	9	48637	6145
Italy	Basilicata	2186	2925	30	8	28781	1121
Italy	Calabria	1287	2868	9	3	26814	3007
Italy	Campania	1687	2938	10	10	32504	3429
Italy	Emilia-Romagna	3913	3351	23	22	57115	2567
Italy	Friuli-Venezia Giulia	4052	3118	18	21	59704	3426
Italy	Lazio	2019	2929	16	10	36890	2377
Italy	Liguria	2932	3112	5	2	45735	8944
Italy	Lombardia	4002	3576	27	80	57054	2191
Italy	Marche	2465	2614	21	6	29623	1458
Italy	Molise	1762	2546	19	13	21281	1177
Italy	Piemonte	3250	3375	23	23	47057	2104
Italy	Puglia	2099	2243	14	2	29978	2229
Italy	Sardegna	1559	2801	42	21	23205	585
Italy	Sicilia	1900	2421	16	5	30507	2082
Italy	Toscana	2547	3857	23	6	59522	2941
Italy	Trentino	1047	2562	7	5	44190	6918
Italy	Umbria	1695	2638	22	14	28759	1386
Italy	Valle d'Aosta	269	3966	39	24	24714	677
Italy	Veneto	3169	3070	15	34	59972	4144
Latvia	Latvia	4274	4329	66	22	26478	432
Lithuania	Lithuania	3766	3766	44	12	19825	461
Luxembourg	Luxembourg	8688	3781	80	107	42402	534
Malta	Malta	777	3313	3	16	16827	6308
Netherlands	The Netherlands	6609	5889	36	131	198325	5458
Poland	Malopolska i Pogórze	1164	3505	10	7	9850	934
Poland	Mazowsze i Podlasie	1707	3722	15	11	11322	746
Poland	Pomorze i Mazury	4815	3985	37	18	23338	641
Poland	Wielkopolska and Slask	3843	3859	24	17	21461	876
Portugal	Alentejo e Algarve	2313	3429	72	23	22923	405
Portugal	Açores e Madeira	1605	2320	11	15	6500	589

Continued on following page

Table 1 (a). Continued

Country	Region	Fertilisers	Labour input	Total Utilised Agricultural Area	Total livestock units	Total output crops & crop production	Total crops output (€/ha)
Portugal	Entre Douro e Minho/Beira litoral	867	3569	7	11	11243	1612
Portugal	Norte e Centro	845	3095	16	13	14596	1188
Portugal	Ribatejo e Oeste	2805	3461	14	14	36945	3080
Portugal	Tras-os-Montes/Beira interior	494	3172	23	7	9033	535
Romania	Bucuresti-Ilfov	1157	3162	17	4	17702	1030
Romania	Centru	617	3253	11	9	6923	641
Romania	Nord-Est	554	3107	8	5	5806	710
Romania	Nord-Vest	623	3695	9	6	6843	782
Romania	Sud-Est	1206	3451	16	6	11650	701
Romania	Sud-Muntenia	1027	3224	12	6	8950	726
Romania	Sud-Vest-Oltenia	488	3438	8	4	5660	763
Romania	Vest	1200	3140	15	7	11474	737
Slovakia	Slovakia	36066	27232	524	157	279628	558
Slovenia	Slovenia	853	2847	11	11	11081	1048
Spain	Andalucía	3514	3058	32	10	46300	1522
Spain	Aragón	4496	2397	64	54	36730	706
Spain	Asturias	481	3402	24	49	7365	353
Spain	Canarias	4408	5696	3	10	57237	20874
Spain	Cantabria	584	2750	43	54	3470	101
Spain	Castilla y León	5302	2535	65	28	33151	552
Spain	Castilla-La Mancha	3211	3122	54	16	33609	743
Spain	Cataluña	2756	3378	34	81	42437	1449
Spain	Comunidad Valenciana	2016	2205	11	10	33820	3160
Spain	Extremadura	2306	3662	75	33	26590	381
Spain	Galicia	1286	2627	17	39	10272	624
Spain	Islas Baleares	2180	2717	47	20	27923	669
Spain	La Rioja	2685	2544	29	15	47241	1797
Spain	Madrid	2102	3962	70	27	23874	391
Spain	Murcia	3860	3718	28	26	59852	2460
Spain	Navarra	7117	2623	52	39	51847	1165
Spain	País Vasco	2506	3195	33	27	27933	856
Sweden	Län i norra Sverige	3740	3428	86	54	45837	538
Sweden	Skogs- och mellanbygdslän	5865	3184	98	71	47722	498
Sweden	Slättbyggdslän	11529	3080	100	61	87380	900
United Kingdom	England - East Region	21728	6431	160	101	206178	1420

Continued on following page

Table 1 (a). Continued

Country	Region	Fertilisers	Labour input	Total Utilised Agricultural Area	Total livestock units	Total output crops & crop production	Total crops output (€/ha)
United Kingdom	England - North Region	13926	4763	152	149	86096	583
United Kingdom	England - West Region	11969	5488	113	126	88086	812
United Kingdom	Northern Ireland	6778	3659	78	102	16656	214
United Kingdom	Scotland	19189	4777	337	150	76811	232
United Kingdom	Wales	9893	3970	125	167	27210	218

Table 1 (b).

Total output livestock & livestock products	Total livestock output (€/ LU)	Total specific costs	Total farming overheads	Total external factors	Subsidies on investments	Total assets	Gross Investment on fixed assets	Total subsidies, excluding on investments
36358	1391	21968	21902	5682	1224	441796	20483	18937
154351	1104	124585	43220	29166	1713	732897	50896	17382
92916	952	66009	33976	22478	2328	586961	26536	31958
15090	843	23399	12333	20403	255	119923	10243	13311
19767	1017	25239	13043	24730	340	121678	11491	13413
9834	826	21068	12562	17908	325	125385	13596	13345
15438	879	19865	11403	14062	286	97997	9043	12986
7591	867	7837	4676	4535	210	48057	4348	5145
8954	685	10705	6344	6105	165	93519	5223	6880
6161	1354	7328	2698	2324	31	302944	2829	4293
9315	921	12353	3809	2295	65	103749	2901	6180
20081	1615	16871	6588	4666	313	186402	1821	4755
109268	1085	130770	87009	74469	3761	724242	43012	74863
216829	1359	178907	74450	103621	401	2378153	66101	36149
40066	1083	45796	27001	17977	2939	240782	24263	22030
32549	1338	36866	38721	14038	671	429555	24322	44729
51193	1518	45615	45745	15533	1138	473878	32618	52419
61877	1866	53704	46287	14489	1403	404171	36346	65360
51582	1784	41749	39693	12129	946	330890	20689	47405
37861	1236	40736	50858	33943	1010	343364	27115	19086
41139	1016	44148	59603	34693	1664	391862	25519	19979
80537	851	39680	40407	13682	1334	378121	25788	40065
130756	1125	66921	59229	28785	1555	440926	35912	32967
51188	685	51003	54930	39934	1106	471307	33797	40644
191222	1014	104080	68124	27013	1205	400839	32855	22491
34814	884	68038	54065	30396	675	391342	30489	39543
27823	1004	49136	58774	56207	799	730062	34287	30014
29109	780	21575	42671	28959	7135	306617	26409	24733
120694	1356	58068	52778	21360	799	441608	38195	36318
8477	1022	16327	30248	9176	2285	99608	7794	25401
79693	1086	81971	67780	42525	1353	490336	42587	39907
21010	1490	30361	34440	16566	2643	170334	13617	19337
11151	961	23339	43049	26931	2720	310061	17171	13137
72794	671	35835	34445	12375	1438	384891	22406	40272
108440	1029	74348	62729	25858	1621	503893	39165	50747

Continued on following page

Table 1 (b). Continued

Total output livestock & livestock products	Total livestock output (€/ LU)	Total specific costs	Total farming overheads	Total external factors	Subsidies on investments	Total assets	Gross Investment on fixed assets	Total subsidies, excluding on investments
5900	1073	27426	45760	38463	3032	154699	12428	52191
49114	942	40635	40260	16783	1352	314647	21045	30764
81026	1183	76735	56290	32425	1213	478511	37162	27757
148244	1029	89391	66628	30458	1699	423493	34025	30411
51788	1107	87174	65420	43109	498	542360	40183	46249
53003	977	57237	52081	26748	931	428286	27406	32335
8794	570	30921	50220	42200	1857	316692	22444	15150
62178	1131	39929	46687	24364	1504	321916	24895	24018
10225	1540	78567	64086	46810	870	501786	42220	46206
68328	1312	52296	46490	24338	175	621448	26072	21636
78062	1400	45797	39910	14491	394	720772	26898	23351
294377	1231	287389	203656	220857	1843	1616280	113635	157063
162	291	56025	49671	23235	77	300793	17727	1259
73977	1261	57155	47946	24635	472	600438	28959	27534
317242	1351	358375	239017	248719	389	1941664	152795	169227
156458	1231	113080	61223	38445	91	963536	36546	29179
129501	1199	94575	58601	31353	66	788679	31860	23002
34884	1280	45983	45651	25126	109	598442	24747	17342
85380	1292	52723	49038	17359	209	658537	29028	37405
270392	1529	260968	195455	200541	4931	1547212	115175	128233
277864	1459	305509	202720	240155	1554	1813190	128232	163969
152358	1278	101944	67929	40415	0	989534	31689	34824
186785	1552	138988	86443	62840	0	1132576	44425	36851
365704	1347	375017	298750	315826	4708	2209750	157890	201441
5919	1105	5096	2577	1730	28	110485	827	3608
6126	987	8208	5526	3460	110	95152	1005	8531
7031	901	6055	3792	2155	57	105183	775	6057
6331	999	7492	5428	3096	11	102123	479	9084
19048	1138	27153	14265	10330	484	169903	6899	14273
25178	1085	45255	23997	18792	742	239599	13244	20345
13692	843	18604	13008	8824	498	114228	4059	9589
25180	929	36759	24707	19775	929	179529	11733	16066
41138	1158	48667	31761	25306	1455	216231	11128	22534
34318	1509	27613	19143	12404	1277	178563	8950	13725
20399	936	29825	19075	15520	1052	171247	10294	15258
15220	1026	19733	13599	8232	649	109178	4873	10585

Continued on following page

Table 1 (b). Continued

Total output livestock & livestock products	Total livestock output (€/ LU)	Total specific costs	Total farming overheads	Total external factors	Subsidies on investments	Total assets	Gross Investment on fixed assets	Total subsidies, excluding on investments
13585	974	21158	14740	11847	452	160075	7084	16523
44991	789	26987	13343	5041	1005	901894	8471	18971
8139	1255	7960	4852	2982	224	176913	1311	3614
18858	1975	18720	9275	7746	565	549360	18598	4708
8609	1107	9494	6276	5340	44	286727	2832	8843
2967	1081	4726	3362	6190	1	152558	559	6588
12687	1269	11597	5080	5640	24	218017	508	5372
32955	1490	31349	15661	13372	37	527669	3116	8249
18805	954	25802	12528	7525	391	506828	10616	6275
13631	1332	13386	8166	6245	171	284809	4006	5742
3285	1300	10664	4602	3920	98	215198	1543	1630
97106	1205	57715	20864	13200	150	749237	5960	15266
5265	840	9860	7949	3544	97	284091	5123	7614
11246	927	8608	5733	2408	85	229421	2650	7077
26995	1180	25944	11350	5684	161	395495	4947	10550
3447	1712	8442	6824	5977	111	274913	2266	6624
21851	1049	15815	6255	3732	288	324998	5329	9429
5449	1022	8303	4705	5887	120	193704	1639	4481
6804	1058	18715	14219	12909	456	521074	5669	7110
9258	1762	12460	6785	4841	376	590950	14126	3172
10269	774	11399	7809	6265	39	323793	3177	12504
30133	1252	20190	6778	7025	2640	350604	9718	18540
28727	894	29244	12951	8684	307	630240	7330	8624
20482	920	24185	14436	7540	1935	128903	15239	13790
11528	907	13952	6586	3406	2353	102844	10474	8375
119536	1105	73025	48676	23330	14661	1099875	73644	45959
20743	1309	19599	6405	2428	496	194353	2189	4305
202892	1513	170921	99199	79423	737	2208166	71809	17626
6375	904	6330	3406	934	148	87195	2069	2987
10923	992	8950	4183	1248	196	125609	4086	4233
17639	976	19356	8683	3570	244	197336	5861	9621
16121	936	17167	7645	3348	234	183125	5011	6460
10859	454	11305	6959	6787	381	149937	4908	13001
13369	888	8086	2753	2368	1135	77094	4497	8183
12245	1071	11346	3822	1734	505	77922	3661	3773
12246	950	12234	4005	2955	780	97044	3570	5930

Continued on following page

Table 1 (b). Continued

Total output livestock & livestock products	Total livestock output (€/ LU)	Total specific costs	Total farming overheads	Total external factors	Subsidies on investments	Total assets	Gross Investment on fixed assets	Total subsidies, excluding on investments
7904	549	15917	8437	4210	327	84787	5014	4080
4087	621	3376	2134	2427	567	70659	2483	4316
5190	857	4756	4338	3626	555	176263	1582	3734
8765	947	7102	2428	1809	14	46279	946	2792
4184	754	3798	1642	863	23	26410	310	1457
5625	895	4511	2061	1129	12	42167	492	1991
5287	787	6087	2975	2425	49	43956	1582	3181
5861	977	5754	2699	1860	81	42689	1007	2224
3880	912	3131	1713	743	10	29227	-160	1318
5663	790	5953	2746	1742	38	51771	802	2913
149110	980	250400	157387	153147	8747	1103042	87724	137045
9239	803	9986	6095	938	1045	202401	8596	6714
7812	787	13055	9377	10992	33	366817	1790	10383
29153	541	25257	13999	9110	172	285197	5296	14058
51053	1025	32902	9625	2540	43	297989	1736	12138
18205	1865	19628	16322	18616	139	293051	1639	16582
53860	999	31765	9877	3267	615	235323	3797	15219
35232	1269	30414	12986	8352	79	291140	1604	13308
14459	878	13577	7593	6412	22	264623	1803	8493
36826	449	29802	19946	13123	893	368531	11570	11287
9080	924	11260	6104	4872	60	229964	1003	2435
29823	903	21264	7452	6954	13	239476	1040	11408
45754	1180	27295	6354	1469	62	319127	1687	7120
16197	811	15103	8107	6250	599	277145	2544	9116
8047	531	13643	8590	8075	202	360888	4626	5462
33034	1239	21675	6772	4407	99	228651	978	8815
17213	741	20099	14940	14396	34	329133	1570	5890
41782	1084	45956	18651	11707	1389	400844	12968	14780
41004	1496	28316	13534	6314	974	216154	11660	12381
71538	1326	78377	51518	18370	219	462976	32899	58079
92000	1288	84268	47173	22750	57	653162	34613	38161
68499	1129	80422	56938	35664	25	926830	38969	33947
92807	918	135862	79384	66420	490	2167845	48743	46435
137375	916	117773	55306	38051	1095	1320664	41382	41215
131290	1046	108123	54872	40526	1153	1636267	39701	36168
93134	913	62971	24906	9649	1445	1076408	21747	27331

Continued on following page

Table 1 (b). Continued

Total output livestock & livestock products	Total livestock output (€/ LU)	Total specific costs	Total farming overheads	Total external factors	Subsidies on investments	Total assets	Gross Investment on fixed assets	Total subsidies, excluding on investments
103234	691	98574	56306	28891	1514	1501240	43145	56965
120518	726	83079	38087	15582	1132	1078493	29181	39426

The summary statistics showed in table 2 were obtained through the Stata software (StataCorp, 2017a, 2017b; Stata, 2021) and reveal that, on an annual average (2004-2019), the European Union farms consume 8000 euros of fertilisers, use 4200 hours of labour input, use 72 hectares of utilised agricultural area and have 50 livestock units. In turn, these farms have more annual crop output (67780 euros) than livestock output (55516 euros) and higher area productivity than livestock productions per LU. Finally, the European farms received annually 977 euros from subsidies on investments and 25677 euros from total subsidies (excluding on investments.

Table 2. Summary statistics for the main aggregates of farm variables, over the period 2004-2019 and across the European Union agricultural regions

Variable	Observations	Mean	Standard Deviation	Min	Max
Fertilisers	2123	8000	11739	77	113819
Labour input	2123	4200	3402	975	42868
Total Utilised Agricultural Area	2123	72	95	2	615
Total livestock units	2123	50	56	0	355
Total output crops & crop production	2123	67780	81952	1733	669216
Total crops output/ha	2123	1922	4810	37	131506
Total output livestock & livestock products	2123	55516	70597	0	539773
Total livestock output/LU	2121	1076	348	0	2676
Total specific costs	2123	51733	68899	1635	547970
Total farming overheads	2123	35613	47474	1046	397972
Total external factors	2123	26491	49135	273	363442
Subsidies on investments	2123	977	2000	0	19673
Total assets	2123	487171	507108	15821	3118089
Gross Investment on fixed assets	2123	21738	30072	-11668	272353
Total subsidies, excluding on investments	2123	25677	33749	73	228296

RESULTS

Table 3 (a). Pairwise matrix of correlation for the main aggregates of farm variables, over the period 2004-2019 and across the European Union agricultural regions

	Fertilisers	Labour input	Total Utilised Agricultural Area	Total livestock units	Total output crops & crop production	Total crops output/ha
Fertilisers	1.000					
Labour input	0.6849*	1.000				
	(0.000)					
Total Utilised Agricultural Area	0.8838*	0.8262*	1.000			
	(0.000)	(0.000)				
Total livestock units	0.7236*	0.6137*	0.7836*	1.000		
	(0.000)	(0.000)	(0.000)			
Total output crops & crop production	0.9122*	0.7077*	0.8001*	0.6555*	1.000	
	(0.000)	(0.000)	(0.000)	(0.000)		
Total crops output/ha	-0.0564*	0.014	-0.1485*	-0.1444*	0.1038*	1.000
	(0.009)	(0.532)	(0.000)	(0.000)	(0.000)	
Total output livestock & livestock products	0.7678*	0.6194*	0.7637*	0.9446*	0.7358*	-0.1123*
	(0.000)	(0.000)	(0.000)	(0.000)	(0.000)	(0.000)
Total livestock output/LU	0.1849*	0.0826*	0.0861*	0.0818*	0.2109*	0.0897*
	(0.000)	(0.000)	(0.000)	(0.000)	(0.000)	(0.000)
Total specific costs	0.8909*	0.7427*	0.8619*	0.8644*	0.8906*	-0.036
	(0.000)	(0.000)	(0.000)	(0.000)	(0.000)	(0.100)
Total farming overheads	0.9069*	0.7660*	0.8795*	0.8257*	0.9150*	-0.019
	(0.000)	(0.000)	(0.000)	(0.000)	(0.000)	(0.388)
Total external factors	0.8898*	0.7966*	0.8699*	0.7542*	0.9160*	-0.021
	(0.000)	(0.000)	(0.000)	(0.000)	(0.000)	(0.326)
Subsidies on investments	0.2653*	0.3775*	0.3554*	0.3141*	0.2677*	-0.0600*
	(0.000)	(0.000)	(0.000)	(0.000)	(0.000)	(0.006)
Total assets	0.6594*	0.5089*	0.6556*	0.7787*	0.7051*	-0.039
	(0.000)	(0.000)	(0.000)	(0.000)	(0.000)	(0.076)
Gross Investment on fixed assets	0.8769*	0.6698*	0.8238*	0.8060*	0.8651*	-0.0573*
	(0.000)	(0.000)	(0.000)	(0.000)	(0.000)	(0.008)
Total subsidies, excluding on investments	0.8871*	0.7726*	0.9267*	0.7781*	0.8195*	-0.1094*
	(0.000)	(0.000)	(0.000)	(0.000)	(0.000)	(0.000)

Note: *, statistically significant at 5%.

Table 3 (b).

Total output livestock & livestock products	Total livestock output/LU	Total specific costs	Total farming overheads	Total external factors	Subsidies on investments	Total assets	Gross Investment on fixed assets	Total subsidies, excluding on investments
1.000								
0.2742*	1.000							
(0.000)								
0.9350*	0.2599*	1.000						
(0.000)	(0.000)							
0.8858*	0.2376*	0.9487*	1.000					
(0.000)	(0.000)	(0.000)						
0.8376*	0.2101*	0.9375*	0.9530*	1.000				
(0.000)	(0.000)	(0.000)	(0.000)					
0.2971*	0.028	0.3123*	0.3521*	0.3019*	1.000			
(0.000)	(0.197)	(0.000)	(0.000)	(0.000)				
0.8202*	0.2512*	0.8294*	0.7443*	0.7293*	0.2362*	1.000		
(0.000)	(0.000)	(0.000)	(0.000)	(0.000)	(0.000)			
0.8775*	0.2650*	0.9311*	0.9300*	0.8949*	0.4267*	0.7779*	1.000	
(0.000)	(0.000)	(0.000)	(0.000)	(0.000)	(0.000)	(0.000)		
0.8109*	0.2202*	0.8920*	0.9328*	0.9168*	0.3847*	0.6585*	0.8793*	1.000
(0.000)	(0.000)	(0.000)	(0.000)	(0.000)	(0.000)	(0.000)	(0.000)	

Note: *, statistically significant at 5%.

Considering the correlation coefficients (Pearson, 1896; Pearson & Filon, 1898) on table 3, obtained through the Stata software for pairwise matrix, the European Union farms with higher fertilisers' costs are larger, have higher crop output, total specific costs, farming overheads, external factors costs and gross investments and receive higher amounts total subsidies (excluding on investments). These farms with higher fertilisers' costs have negative correlation with the productivities of the area and weaker correlation with the livestock output per LU and the subsidies on investments. These findings mean that the farms that receive relevant amounts of current subsidies are the same that, in general, have higher costs with the fertilisers use, showing that, in fact, the Common Agricultural Policy (CAP) instruments should be more input oriented (Martinho, 2015).

Table 4 shows that there is a positive and strong correlation between the costs with the fertilisers and the quantities of the main vegetal macronutrients used, revealing that the effects from the differences of prices, among regions, on these contexts is residual.

Table 4. Pairwise matrix of correlation for the fertilisers use (in euros and quintal), over the period 2017-2019 and across the European Union agricultural regions

	Fertilisers (€)	Fertiliser N (q)	Fertiliser P2O5 (q)	Fertiliser K2O (q)
Fertilisers (euros)	1.000			
Fertiliser N (quintal)	0.9783*	1.000		
	(0.000)			
Fertiliser P2O5 (quintal)	0.9166*	0.9319*	1.000	
	(0.000)	(0.000)		
Fertiliser K2O (quintal)	0.9031*	0.9156*	0.9117*	1.000
	(0.000)	(0.000)	(0.000)	

Note: *, statistically significant at 5%.

Classifications and Regression Tree Approach

To better understand the impacts from the several farm variables on the costs with the fertilisers use, in this subsection was carried out a deeper assessment through classification and regression tree approaches (Breiman et al., 1984) with the IBM SPSS software (IBM Corp, 2019; IBM SPSS, 2021).

As found before, the most important variables to predict the costs with the fertilisers in the European Union farms, over the period considered, are the following (table 5): crop output, costs with inputs, current subsidies, area and gross investment.

Table 5. Importance of the main aggregates of farm variables in the explanation of the fertilisers use (euros), over the period 2004-2019 and across the European Union agricultural regions

Independent Variable	Normalized Importance
Total output crops & crop production	100%
Total external factors	99%
Total farming overheads	99%
Total subsidies, excluding on investments	98%
Total specific costs	98%
Total Utilised Agricultural Area	97%
Gross Investment on fixed assets	90%
Labour input	76%
Total output livestock & livestock products	72%
Total livestock units	67%
Total assets	7%
Total crops output/ha	2%
Subsidies on investments	2%
Total livestock output/LU	1%

The highest values for the predicted results, of the fertilisers' costs, in the terminal nodes of table 6 are verified for farms with higher area and crop output. In the farms with lower area (less or equal to 81 ha), the highest predicted values are found for farms with larger costs (farming overheads, specific costs and external factors). These findings confirm that the farms with higher area, crop output and input costs spend more money with the fertilizers use.

Looking deeper for the several uses of the utilised agricultural area (table 7), the most important variables to predict the costs with the fertilisers are the cereals, rented area and other field crops. Following node 1, the predicted values for costs with the fertilisers are higher (table 8) in terminal nodes for farms with larger cereals hectares and rented area.

The cereals are important variables to predict the costs with the fertilisers use. Before was found its importance in terms of land use and now, in table 9, in terms of output. Table 9 shows, too, the importance of the oil-seed crops. The highest predicted values are verified in terminal nodes of table 10 for farms with larger cereals output. The lowest predicted result for the dependent variable (1905.6) is found in a terminal node where a random farm has a 38.5% probability of appearing there and is associated with lower wine and grapes and cereals outputs.

Table 6. Classification and regression tree results for the main aggregates of farm variables as independent variables and the fertilisers use (euros) as dependent variable, over the period 2004-2019 and across the European Union agricultural regions

Node	Mean	Standard Deviation	N	Percent	Predicted Mean	Parent Node	Primary Independent Variable	Improvement	Split Values
0	8000	11739	2123	100%	8000				
1	6061	6101	2037	96%	6061	0	Total output crops & crop production	89055378	<= 259353
2	53928	18007	86	4%	53928	0	Total output crops & crop production	89055378	> 259353
3	3591	2751	1555	73%	3591	1	Total Utilised Agricultural Area	18879738	<= 81
4	14028	7054	482	23%	14028	1	Total Utilised Agricultural Area	18879738	> 81
5	1955	1278	949	45%	1955	3	Total farming overheads	3071182	<= 14133
6	6154	2471	606	29%	6154	3	Total farming overheads	3071182	> 14133
7	9260	3675	243	11%	9260	4	Total output crops & crop production	5247141	<= 83376
8	18876	6326	239	11%	18876	4	Total output crops & crop production	5247141	> 83376
9	1381	765	576	27%	1381	5	Total farming overheads	227109	<= 7144
10	2841	1400	373	18%	2841	5	Total farming overheads	227109	> 7144
11	4735	1669	285	13%	4735	6	Total specific costs	510267	<= 38238
12	7413	2384	321	15%	7413	6	Total specific costs	510267	> 38238
13	15772	4352	145	7%	15772	8	Total output crops & crop production	1672317	<= 154962
14	23663	5910	94	4%	23663	8	Total output crops & crop production	1672317	> 154962
15	1066	471	309	15%	1066	9	Total farming overheads	31221	<= 4549
16	1746	872	267	13%	1746	9	Total farming overheads	31221	> 4549
17	1648	1248	73	3%	1648	10	Total output crops & crop production	60819	<= 21299
18	3131	1277	300	14%	3131	10	Total output crops & crop production	60819	> 21299
19	4056	1444	155	7%	4056	11	Total external factors	73769	<= 14255
20	5544	1559	130	6%	5544	11	Total external factors	73769	> 14255
21	6195	1847	151	7%	6195	12	Total external factors	199413	<= 21945
22	8496	2288	170	8%	8496	12	Total external factors	199413	> 21945

Between the specific costs, the crop protection and seeds and plants costs are the most important for the model with the fertilisers' costs as dependent variable (table 11). The highest predicted values in the terminal nodes of table 12 are associated with farms with greater crop protection costs. The lowest result is verified for the terminal node associated with lower crop protection costs, lower other livestock specific costs and lower seeds and plants costs.

Table 7. Importance of the main farm variables related with the area in the explanation of the fertilisers use (euros), over the period 2004-2019 and across the European Union agricultural regions

Independent Variable	Normalized Importance
Cereals	100%
Rented UAA	98%
Other field crops	96%
Forage crops	55%
Permanent grassland	43%
Energy crops	42%
Other permanent crops	32%
Set aside	9%
Total agricultural area out of production	6%
Vegetables and flowers	2%
Orchards	1%
Permanent crops	1%
Vineyards	1%
Permanent crops	1%
Agricultural fallows	0%
Woodland area	0%
Olive groves	0%

The several farming overheads have great importance to predict the costs with the fertilisers use (table 13), however the energy costs appear as the most important. Following node 1, the highest predicted values are found for higher contact work costs, higher other direct inputs and higher energy costs (table 14).

The rent paid and wage paid are the most important variables in the model with the fertilisers costs as dependent variable (table 15). In general, the highest predicted results were obtained for farms with greater rent and wage paid (table 16).

Table 8. Classification and regression tree results for the main farm variables related with the area as independent variables and the fertilisers use (euros) as dependent variable, over the period 2004-2019 and across the European Union agricultural regions

Node	Mean	Standard Deviation	N	Percent	Predicted Mean	Parent Node	Primary Independent Variable	Improvement	Split Values
0	8000	11739	2123	100%	8000				
1	5901	5818	2022	95%	5901	0	Other field crops	88216686	<= 42
2	50025	19158	101	5%	50025	0	Other field crops	88216686	> 42
3	4253	3474	1740	82%	4253	1	Cereals	15954437	<= 36
4	16067	6947	282	13%	16067	1	Cereals	15954437	> 36
5	2771	2108	1197	56%	2771	3	Rented UAA	3967027	<= 31
6	7519	3656	543	26%	7519	3	Rented UAA	3967027	> 31
7	13599	5372	211	10%	13599	4	Cereals	2404026	<= 64
8	23401	5854	71	3%	23401	4	Cereals	2404026	> 64
9	4697	2542	281	13%	4697	5	Vineyards	641456	<= 0.005
10	2180	1530	916	43%	2180	5	Vineyards	641456	> 0.005
11	10753	4097	115	5%	10753	7	Rented UAA	964669	<= 83
12	17009	4700	96	5%	17009	7	Rented UAA	964669	> 83
13	5831	2113	162	8%	5831	9	Woodland area	231727	<= 0.415
14	3153	2253	119	6%	3153	9	Woodland area	231727	> 0.415
15	1385	886	396	19%	1385	10	Vegetables and flowers	207724	<= 0.185
16	2786	1638	520	24%	2786	10	Vegetables and flowers	207724	> 0.185

Table 9. Importance of the main farm variables related with the crop output in the explanation of the fertilisers use (euros), over the period 2004-2019 and across the European Union agricultural regions

Independent Variable	Normalized Importance
Cereals	100%
Oil-seed crops	91%
Forage crops	49%
Sugar beet	44%
Protein crops	44%
Energy crops	23%
Potatoes	10%
Wine and grapes	2%
Other crop output	1%
Industrial crops	1%
Vegetables & flowers	1%
Fruit (excl. Citrus and grapes)	1%
Olives & olive oil	0%
Citrus fruit	0%

Table 10. Classification and regression tree results for the main farm variables related with the crop output as independent variables and the fertilisers use (euros) as dependent variable, over the period 2004-2019 and across the European Union agricultural regions

Node	Mean	Standard Deviation	N	Percent	Predicted Mean	Parent Node	Primary Independent Variable	Improvement	Split Values
0	8000	11739	2123	100%	8000				
1	5976	5941	2030	96%	5976	0	Cereals	89358524	<= 112897
2	52164	18381	93	4%	52164	0	Cereals	89358524	> 112897
3	3741	2831	1639	77%	3741	1	Cereals	20030970	<= 28059
4	15347	6396	391	18%	15347	1	Cereals	20030970	> 28059
5	2853	2158	1293	61%	2853	3	Cereals	2272160	<= 15166
6	7057	2577	346	16%	7057	3	Cereals	2272160	> 15166
7	12574	4122	277	13%	12574	4	Cereals	3441507	<= 55306
8	22086	5920	114	5%	22086	4	Cereals	3441507	> 55306
9	4736	2637	323	15%	4736	5	Wine and grapes	719006	<= 4
10	2227	1524	970	46%	2227	5	Wine and grapes	719006	> 4
11	7531	2432	271	13%	7531	6	Olives & olive oil	132206	<= 0.500
12	5345	2362	75	4%	5345	6	Olives & olive oil	132206	> 0.500
13	10715	2943	139	7%	10715	7	Cereals	454475	<= 37024
14	14447	4296	138	7%	14447	7	Cereals	454475	> 37024
15	3977	2426	215	10%	3977	9	Other crop output	174885	<= 686
16	6249	2385	108	5%	6249	9	Other crop output	174885	> 686
17	1906	1232	817	38%	1906	10	Wine and grapes	251269	<= 12297
18	3940	1777	153	7%	3940	10	Wine and grapes	251269	> 12297
19	6158	1717	100	5%	6158	11	Other crop output	140677	<= 645
20	8334	2433	171	8%	8334	11	Other crop output	140677	> 645

Table 11. Importance of the main farm variables related with the specific costs in the explanation of the fertilisers use (euros), over the period 2004-2019 and across the European Union agricultural regions

Independent Variable	Normalized Importance
Crop protection	100.0%
Seeds and plants	76.7%
Other livestock specific costs	61.7%
Feed for grazing livestock	49.9%
Other crop specific costs	39.2%
Seeds and plants home-grown	34.8%
Feed for pigs & poultry	26.2%
Feed for grazing livestock home-grown	20.7%
Feed for pigs & poultry home-grown	16.4%
Forestry specific costs	1.1%

Table 12. Classification and regression tree results for the main farm variables related with the specific costs as independent variables and the fertilisers use (euros) as dependent variable, over the period 2004-2019 and across the European Union agricultural regions

Node	Mean	Standard Deviation	N	Percent	Predicted Mean	Parent Node	Primary Independent Variable	Improvement	Split Values
0	8000	11739	2123	100%	8000				
1	5721	5443	2007	95%	5721	0	Crop protection	89849182	<= 24693
2	47427	19162	116	5%	47427	0	Crop protection	89849182	> 24693
3	3544	2706	1558	73%	3544	1	Crop protection	15543115	<= 6019
4	13274	5793	449	21%	13274	1	Crop protection	15543115	> 6019
5	2293	1832	1014	48%	2293	3	Seeds and plants	2140725	<= 2648
6	5876	2522	544	26%	5876	3	Seeds and plants	2140725	> 2648
7	10565	3750	316	15%	10565	4	Crop protection	3688640	<= 11924
8	19711	4582	133	6%	19711	4	Crop protection	3688640	> 11924
9	1916	1216	901	42%	1916	5	Other livestock specific costs	542880	<= 3375
10	5304	2861	113	5%	5304	5	Other livestock specific costs	542880	> 3375
11	4481	1941	246	12%	4481	6	Other livestock specific costs	411813	<= 2004
12	7028	2362	298	14%	7028	6	Other livestock specific costs	411813	> 2004
13	8907	2575	201	9%	8907	7	Feed for grazing livestock	715192	<= 25294
14	13463	3729	115	5%	13463	7	Feed for grazing livestock	715192	> 25294
15	1281	651	548	26%	1281	9	Crop protection	265392	<= 1224
16	2901	1232	353	17%	2901	9	Crop protection	265392	> 1224
17	3541	1675	121	6%	3541	11	Crop protection	99054	<= 2860
18	5390	1742	125	6%	5390	11	Crop protection	99054	> 2860
19	6509	2145	221	10%	6509	12	Crop protection	108491	<= 5078
20	8517	2336	77	4%	8517	12	Crop protection	108491	> 5078

Table 13. Importance of the main farm variables related with the farming overheads in the explanation of the fertilisers use (euros), over the period 2004-2019 and across the European Union agricultural regions

Independent Variable	Normalized Importance
Energy	100%
Other direct inputs	97%
Machinery & building current costs	95%
Contract work	90%

Table 14. Classification and regression tree results for the main farm variables related with the farming overheads as independent variables and the fertilisers use (euros) as dependent variable, over the period 2004-2019 and across the European Union agricultural regions

Node	Mean	Standard Deviation	N	Percent	Predicted Mean	Parent Node	Primary Independent Variable	Improvement	Split Values
0	8000	11739	2123	100%	8000				
1	5991	5988	2030	96%	5991	0	Energy	88089813	<= 43778
2	51850	18833	93	4%	51850	0	Energy	88089813	> 43778
3	2902	2246	1325	62%	2902	1	Contract work	17143175	<= 6262
4	11796	6494	705	33%	11796	1	Contract work	17143175	> 6262
5	1609	1051	756	36%	1609	3	Energy	1386440	<= 3740
6	4620	2260	569	27%	4620	3	Energy	1386440	> 3740
7	8478	3283	289	14%	8478	4	Other direct inputs	2538682	<= 14525
8	14100	7146	416	20%	14100	4	Other direct inputs	2538682	> 14525
9	1287	729	516	24%	1287	5	Energy	79564	<= 2741
10	2302	1279	240	11%	2302	5	Energy	79564	> 2741
11	3851	1728	397	19%	3851	6	Energy	366309	<= 6881
12	6396	2347	172	8%	6396	6	Energy	366309	> 6881
13	9857	4405	105	5%	9857	8	Energy	1191397	<= 8525
14	15533	7328	311	15%	15533	8	Energy	1191397	> 8525
15	1134	591	429	20%	1134	9	Contract work	27916	<= 1122
16	2039	869	87	4%	2039	9	Contract work	27916	> 1122
17	2997	1011	175	8%	2997	11	Machinery & building current costs	107361	<= 2292
18	4524	1875	222	10%	4524	11	Machinery & building current costs	107361	> 2292

Table 15. Importance of the main farm variables related with the external factors in the explanation of the fertilisers use (euros), over the period 2004-2019 and across the European Union agricultural regions

Independent Variable	Normalized Importance
Rent paid	100%
Wages paid	95%
Interest paid	60%

Table 16. Classification and regression tree results for the main farm variables related with the external factors as independent variables and the fertilisers use (euros) as dependent variable, over the period 2004-2019 and across the European Union agricultural regions

Node	Mean	Standard Deviation	N	Percent	Predicted Mean	Parent Node	Primary Independent Variable	Improvement	Split Values
0	8000	11739	2123	100%	8000				
1	6031	6057	2034	96%	6031	0	Wages paid	88539264	<= 111869
2	52983	18421	89	4%	52983	0	Wages paid	88539264	> 111869
3	3121	2442	1418	67%	3121	1	Rent paid	18678095	<= 7180
4	12731	6560	616	29%	12731	1	Rent paid	18678095	> 7180
5	2175	1550	1032	49%	2175	3	Rent paid	1599891	<= 2656
6	5652	2589	386	18%	5652	3	Rent paid	1599891	> 2656
7	9954	4125	393	19%	9954	4	Rent paid	3943347	<= 14740
8	17625	7174	223	11%	17625	4	Rent paid	3943347	> 14740
9	1716	1176	753	35%	1716	5	Wages paid	276358	<= 4794
10	3413	1748	279	13%	3413	5	Wages paid	276358	> 4794
11	4781	2153	198	9%	4781	6	Interest paid	145329	<= 1702
12	6569	2695	188	9%	6569	6	Interest paid	145329	> 1702
13	15987	6274	155	7%	15987	8	Rent paid	642015	<= 22972
14	21357	7728	68	3%	21357	8	Rent paid	642015	> 22972
15	1463	874	601	28%	1463	9	Interest paid	89947	<= 385
16	2717	1608	152	7%	2717	9	Interest paid	89947	> 385

CONCLUSION

The main objective of this research was to identify the main factors that influence the costs with the fertilisers in the European Union farms. For that, statistical information from the FADN database was considered, for the period 2004-2019, that was assessed through matrices of correlation and classification and regression tree approaches.

The data analysis highlights that, generally, the German farms present the higher values, on average for the period considered, for great part of the farm variables assessed, showing the agricultural dynamics of these farms, but also the associated higher costs with inputs, such as the fertilisers use. The European Union farms had, on average over the period taken into account, 8000 euros of costs with fertilisers, used 4200 hours of labour, used 72 hectares and had 50 livestock units. These costs with fertilisers represent almost 15% of the total specific costs (also on average).

The results show that crop output, costs with inputs, current subsidies, utilised agricultural area and gross investment are the variables with greater importance to predict the fertilisers' costs in the European Union farms. This means that the larger farms and with greater crop output spend more budgets with fertilisers use, showing that there is some work to do with these agricultural systems to mitigate the impacts from the fertilisers consumption.

In terms of utilised agricultural area, the most important variables, to predict the costs with fertilisers, are those associated with the areas of the cereals, rented area and other field crops, confirming the findings obtained by the literature review, where the winter wheat, for example, consumes a relevant percentage of the fertilisers applied in all crops. The cereals, jointly with the oil-seed crops, appear also with great importance in the group of the crop output.

Between the variables related with the specific costs, the costs associated with the crop protection products and seeds and plants appear as the most important. Among the several farming overheads, the energy costs are the most relevant, however with not a great different relatively to the other variables of this group (other direct inputs, machinery and building current costs and contract work). Finally, the rent paid and wage paid are the most important variables between the external factors.

In terms of practical implications and policy recommendations, the findings highlighted here provide relevant insights to support the design of future policies that intend to improve the agricultural sustainability in European contexts. For future researches, it could be interesting to quantify through regression approaches the marginal impacts of each variable in the fertilisers use.

ACKNOWLEDGMENT

This work is funded by National Funds through the FCT - Foundation for Science and Technology, I.P., within the scope of the project Refª UIDB/00681/2020. Furthermore, we would like to thank the CERNAS Research Centre and the Polytechnic Institute of Viseu for their support.

REFERENCES

Besuspariene, E., & Niskanen, V. A. (2020). The Assessment of Factors Affecting Fertiliser Use on Family Farms in Lithuania. *Studies in Agricultural Economics (Budapest)*, *122*(1), 13–21. doi:10.7896/j.1923

Borawski, P., Borawski, M. B., Parzonko, A., Wicki, L., Rokicki, T., Perkowska, A., & Dunn, J. W. (2021). Development of Organic Milk Production in Poland on the Background of the EU. *Agriculture-Basel, 11*(4), 323. doi:10.3390/agriculture11040323

Breiman, L., Friedman, J., Stone, C. J., & Olshen, R. A. (1984). *Classification and Regression Trees* (1st ed.). Chapman and Hall/CRC.

Bruelle, G., Naudin, K., Scopel, E., Domas, R., Rabeharisoa, L., & Tittonell, P. (2015). Short- to Mid-Term Impact of Conservation Agriculture on Yield Variability of Upland Rice: Evidence from Farmer's Fields in Madagascar. *Experimental Agriculture, 51*(1), 66–84. doi:10.1017/S0014479714000155

Burow, K. R., Nolan, B. T., Rupert, M. G., & Dubrovsky, N. M. (2010). Nitrate in Groundwater of the United States, 1991-2003. *Environmental Science & Technology, 44*(13), 4988–4997. doi:10.1021/es100546y PMID:20540531

Catarino, R., Gaba, S., & Bretagnolle, V. (2019). Experimental and empirical evidence shows that reducing weed control in winter cereal fields is a viable strategy for farmers. *Scientific Reports, 9*(1), 9004. doi:10.103841598-019-45315-8 PMID:31227731

Costantini, M., Bacenetti, J., Coppola, G., Orsi, L., Ganzaroli, A., & Guarino, M. (2020). Improvement of human health and environmental costs in the European Union by air scrubbers in intensive pig farming. *Journal of Cleaner Production, 275*, 124007. doi:10.1016/j.jclepro.2020.124007

Dalgaard, R., Halberg, N., Kristensen, I. S., & Larsen, I. (2006). Modelling representative and coherent Danish farm types based on farm accountancy data for use in environmental assessments. *Agriculture, Ecosystems & Environment, 117*(4), 223–237. doi:10.1016/j.agee.2006.04.002

Detang-Dessendre, C., Guyomard, H., Requillart, V., & Soler, L.-G. (2020). Changing Agricultural Systems and Food Diets to Prevent and Mitigate Global Health Shocks. *Sustainability, 12*(16), 6462. doi:10.3390u12166462

Dolman, M. A., Sonneveld, M. P. W., Mollenhorst, H., & de Boer, I. J. M. (2014). Benchmarking the economic, environmental and societal performance of Dutch dairy farms aiming at internal recycling of nutrients. *Journal of Cleaner Production, 73*, 245–252. doi:10.1016/j.jclepro.2014.02.043

FADN. (2021). *European Commission | Agri-food data portal | Farm economics*. https://agridata.ec.europa.eu/extensions/FarmEconomyFocus/FADNDatabase.html

Fanelli, R. M. (2020). The Spatial and Temporal Variability of the Effects of Agricultural Practices on the Environment. *Environments, 7*(4), 33. doi:10.3390/environments7040033

Feng, J., Yang, T., Li, F., Zhou, X., Xu, C., & Fang, F. (2021). Impact of tillage on the spatial distribution of CH4 and N2O in the soil profile of late rice fields. *Soil & Tillage Research, 211*, 105029. doi:10.1016/j.still.2021.105029

Fernando, A. L., Duarte, M. P., Almeida, J., Boleo, S., & Mendes, B. (2010). Environmental impact assessment of energy crops cultivation in Europe. *Biofuels Bioproducts & Biorefining-Biofpr, 4*(6), 594–604. doi:10.1002/bbb.249

Firrisa, M. T., van Duren, I., & Voinov, A. (2014). Energy efficiency for rapeseed biodiesel production in different farming systems. *Energy Efficiency*, *7*(1), 79–95. doi:10.100712053-013-9201-2

Fischer-Kowalski, M., Singh, S. J., Lauk, C., Remesch, A., Ringhofer, L., & Gruenbuehel, C. M. (2011). Sociometabolic transitions in subsistence communities: Boserup revisited in four comparative case studies. *Human Ecology Review*, *18*(2), 147–158.

Galluzzo, N. (2017). Efficiency Analysis in Different Typologies of Farming in Italian Fadn Dataset. *Ekonomika Poljoprivreda-Economics of Agriculture*, *64*(2), 451–465. doi:10.5937/ekoPolj1702451G

Garske, B., & Ekardt, F. (2021). Economic policy instruments for sustainable phosphorus management: Taking into account climate and biodiversity targets. *Environmental Sciences Europe*, *33*(1), 56. doi:10.118612302-021-00499-7

Garske, B., Stubenrauch, J., & Ekardt, F. (2020). Sustainable phosphorus management in European agricultural and environmental law. *Review of European, Comparative & International Environmental Law*, *29*(1), 107–117. doi:10.1111/reel.12318

Hartman, T., Harbour, J., Tharnish, B., Van Meter, J., & Jackson-Ziems, T. A. (2020). Agronomic Factors Associated with Bacterial Leaf Streak Development Caused by Xanthomonas vasicola pv. Vasculorum in Corn. *Phytopathology*, *110*(6), 1132–1138. doi:10.1094/PHYTO-02-20-0043-R

IBM Corp. (2019). *IBM SPSS Statistics for Windows, Version 26.0*. IBM Corp.

IBM SPSS. (2021, June 10). *SPSS Software*. https://www.ibm.com/analytics/spss-statistics-software

Karpenstein-Machan, M. (2001). Sustainable cultivation concepts for domestic energy production from biomass. *Critical Reviews in Plant Sciences*, *20*(1), 1–14. doi:10.1080/20013591099164

Karydis, M., & Kitsiou, D. (2012). Eutrophication and environmental policy in the Mediterranean Sea: A review. *Environmental Monitoring and Assessment*, *184*(8), 4931–4984. doi:10.100710661-011-2313-2 PMID:21956336

Klein, T., Calanca, P., Holzkamper, A., Lehmann, N., Roesch, A., & Fuhrer, J. (2012). Using farm accountancy data to calibrate a crop model for climate impact studies. *Agricultural Systems*, *111*, 23–33. doi:10.1016/j.agsy.2012.05.001

Koprivica, A., Beljic, C., Vakanjac, B., Vakanjac, V. R., & Ilic, M. C. (2018). The Content of Toxic Metals in Agricultural Produce near a Coal Mine: Case Study KCB in Lazarevac, Serbia. *Minerals (Basel)*, *8*(4), 131. doi:10.3390/min8040131

Krausmann, F., Haberl, H., Schulz, N. B., Erb, K. H., Darge, E., & Gaube, V. (2003). Land-use change and socio-economic metabolism in Austria - Part I: Driving forces of land-use change: 1950-1995. *Land Use Policy*, *20*(1), 1–20. doi:10.1016/S0264-8377(02)00048-0

Krizsan, S. J., Chagas, J. C., Pang, D., & Cabezas-Garcia, E. H. (2021). Sustainability aspects of milk production in Sweden. *Grass and Forage Science*, *76*(2), 205–214. doi:10.1111/gfs.12539

Lairez, J., Lopez-Ridaura, S., Jourdain, D., Falconnier, G. N., Lienhard, P., Striffler, B., Syfongxay, C., & Affholder, F. (2020). Context matters: Agronomic field monitoring and participatory research to identify criteria of farming system sustainability in South-East Asia. *Agricultural Systems*, *182*, 102830. doi:10.1016/j.agsy.2020.102830

Lv, C., Huang, Y., Sun, W., Yu, L., & Zhu, J. (2020). Response of rice yield and yield components to elevated [CO2]: A synthesis of updated data from FACE experiments. *European Journal of Agronomy*, *112*, 125961. doi:10.1016/j.eja.2019.125961

Madzaric, S., Ceglie, F. G., Depalo, L., Al Bitar, L., Mimiola, G., Tittarelli, F., & Burgio, G. (2018). Organic vs. Organic—Soil arthropods as bioindicators of ecological sustainability in greenhouse system experiment under Mediterranean conditions. *Bulletin of Entomological Research*, *108*(5), 625–635. doi:10.1017/S0007485317001158 PMID:29166981

Martinho, V. J. P. D. (2015). Output Impacts of the Single Payment Scheme in Portugal: A Regression with Spatial Effects. *Outlook on Agriculture*, *44*(2), 109–118. doi:10.5367/oa.2015.0203

Martinho, V. J. P. D. (2017). Efficiency, total factor productivity and returns to scale in a sustainable perspective: An analysis in the European Union at farm and regional level. *Land Use Policy*, *68*, 232–245. doi:10.1016/j.landusepol.2017.07.040

Martinho, V. J. P. D. (2021). Direct and indirect energy consumption in farming: Impacts from fertilizer use. *Energy*, *236*, 121504. doi:10.1016/j.energy.2021.121504

Meul, M., Nevens, F., Reheul, D., & Hofman, G. (2007). Energy use efficiency of specialised dairy, arable and pig farms in Flanders. *Agriculture, Ecosystems & Environment*, *119*(1–2), 135–144. doi:10.1016/j.agee.2006.07.002

Montanarella, L., & Panagos, P. (2021). The relevance of sustainable soil management within the European Green Deal. *Land Use Policy*, *100*, 104950. doi:10.1016/j.landusepol.2020.104950

Ouellet, G., Lapen, D. R., Topp, E., Sawada, M., & Edwards, M. (2008). A heuristic model to predict earthworm biomass in agroecosystems based on selected management and soil properties. *Applied Soil Ecology*, *39*(1), 35–45. doi:10.1016/j.apsoil.2007.11.003

Paladino, O., Massabo, M., & Gandoglia, E. (2020). Assessment of Nitrate Hazards in Umbria Region (Italy) Using Field Datasets: Good Agriculture Practices and Farms Sustainability. *Sustainability*, *12*(22), 9497. doi:10.3390u12229497

Pearson, K. (1896). VII. Mathematical contributions to the theory of evolution.—III. Regression, heredity, and panmixia. *Philosophical Transactions of the Royal Society of London. Series A, Containing Papers of a Mathematical or Physical Character*, *187*, 253–318. doi:10.1098/rsta.1896.0007

Pearson, K., & Filon, L. N. G. (1898). VII. Mathematical contributions to the theory of evolution.—IV. On the probable errors of frequency constants and on the influence of random selection on variation and correlation. *Philosophical Transactions of the Royal Society of London. Series A, Containing Papers of a Mathematical or Physical Character*, *191*, 229–311. doi:10.1098/rsta.1898.0007

Rowe, H., Withers, P. J. A., Baas, P., Chan, N. I., Doody, D., Holiman, J., Jacobs, B., Li, H., MacDonald, G. K., McDowell, R., Sharpley, A. N., Shen, J., Taheri, W., Wallenstein, M., & Weintraub, M. N. (2016). Integrating legacy soil phosphorus into sustainable nutrient management strategies for future food, bioenergy and water security. *Nutrient Cycling in Agroecosystems*, *104*(3), 393–412. doi:10.100710705-015-9726-1

Ru, F., Yin, A., Jin, J., Zhang, X., Yang, X., Zhang, M., & Gao, C. (2016). Prediction of cadmium enrichment in reclaimed coastal soils by classification and regression tree. *Estuarine, Coastal and Shelf Science*, *177*, 1–7. doi:10.1016/j.ecss.2016.04.018

Sabadas, M., Dirja, M., & Maxim, M. (2018). Study on Implemented and Future Projects for Sustainable Development in Aghiresu Area, Cluj County. *Scientific Papers-Series E-Land Reclamation Earth Observation & Surveying Environmental Engineering*, *7*, 113–116.

Samborski, S. M., Gozdowski, D., Stepien, M., Walsh, O. S., & Leszczynska, E. (2016). On-farm evaluation of an active optical sensor performance for variable nitrogen application in winter wheat. *European Journal of Agronomy*, *74*, 56–67. doi:10.1016/j.eja.2015.11.020

Samson, E., Van der Werf, H. M. G., Dupraz, P., Ruas, J.-F., & Corson, M. S. (2012). Estimating environmental impacts of agricultural systems with LCA using data from the French Farm Accountancy Data Network (FADN). *Cahiers Agricultures*, *21*(4), 248–257. doi:10.1684/agr.2012.0581

Sawinska, Z., Switek, S., Glowicka-Woloszyn, R., & Kowalczewski, P. L. (2020). Agricultural Practice in Poland Before and after Mandatory IPM Implementation by the European Union. *Sustainability*, *12*(3), 1107. doi:10.3390u12031107

Spugnoli, P., Dainelli, R., D'Avino, L., Mazzoncini, M., & Lazzeri, L. (2012). Sustainability of sunflower cultivation for biodiesel production in Tuscany within the EU Renewable Energy Directive. *Biosystems Engineering*, *112*(1), 49–55. doi:10.1016/j.biosystemseng.2012.02.004

Stata. (2021). *Stata: Software for Statistics and Data Science*. https://www.stata.com/

StataCorp. (2017a). *Stata 15 Base Reference Manual*. Stata Press.

StataCorp. (2017b). *Stata Statistical Software: Release 15*. StataCorp LLC.

Stubenrauch, J., Ekardt, F., Heyl, K., Garske, B., Schott, V. L., & Ober, S. (2021). How to legally overcome the distinction between organic and conventional farming—Governance approaches for sustainable farming on 100% of the land. *Sustainable Production and Consumption*, *28*, 716–725. doi:10.1016/j.spc.2021.06.006

Tanaka, A., Diagne, M., & Saito, K. (2015). Causes of yield stagnation in irrigated lowland rice systems in the Senegal River Valley: Application of dichotomous decision tree analysis. *Field Crops Research*, *176*, 99–107. doi:10.1016/j.fcr.2015.02.020

Thomas, I. A., Buckley, C., Kelly, E., Dillon, E., Lynch, J., Moran, B., Hennessy, T., & Murphy, P. N. C. (2020). Establishing nationally representative benchmarks of farm-gate nitrogen and phosphorus balances and use efficiencies on Irish farms to encourage improvements. *The Science of the Total Environment*, *720*, 137245. doi:10.1016/j.scitotenv.2020.137245 PMID:32325548

Tingyu, L., Zhang, W., Cao, H., Hao, Y., Qingsong, Z., Siyang, R., Zitong, L., Yulong, Y., Wei, Q., Zhenling, C., Xuejun, L., Xiaotang, J., Oenema, O., de Vries, W., & Fusuo, Z. (2020). Region-specific nitrogen management indexes for sustainable cereal production in China. *Environmental Research Communications*, *2*(7), 075002. doi:10.1088/2515-7620/aba12d

Tittonell, P., Shepherd, K. D., Vanlauwe, B., & Giller, K. E. (2008). Unravelling the effects of soil and crop management on maize productivity in smallholder agricultural systems of western Kenya—An application of classification and regression tree analysis. *Agriculture, Ecosystems & Environment*, *123*(1–3), 137–150. doi:10.1016/j.agee.2007.05.005

van Grinsven, H. J. M., Erisman, J. W., de Vries, W., & Westhoek, H. (2015). Potential of extensification of European agriculture for a more sustainable food system, focusing on nitrogen. *Environmental Research Letters*, *10*(2), 025002. doi:10.1088/1748-9326/10/2/025002

Vogeler, I., Beukes, P., Romera, A., & Cichota, R. (2012). Estimating nitrous oxide emissions from a dairy farm using a mechanistic, whole farm model and segregated emission factors for New Zealand. *Soil Research (Collingwood, Vic.)*, *50*(3), 188–194. doi:10.1071/SR12064

Willy, D. K., Muyanga, M., Mbuvi, J., & Jayne, T. (2019). The effect of land use change on soil fertility parameters in densely populated areas of Kenya. *Geoderma*, *343*, 254–262. doi:10.1016/j.geoderma.2019.02.033 PMID:31160827

Wojcik-Gront, E. (2018). Variables influencing yield-scaled Global Warming Potential and yield of winter wheat production. *Field Crops Research*, *227*, 19–29. doi:10.1016/j.fcr.2018.07.015

Chapter 10
Mediterranean Diet as a Healthy, Sustainable, and Secure Food Pattern

Elsa Lamy
Universidade de Évora, Portugal

Cristina Conceição
Universidade de Évora, Portugal

Fernando Capela e Silva
Universidade de Évora, Portugal

Maria Raquel Lucas
Universidade de Évora, Portugal

ABSTRACT

Despite the recognized advantages of Mediterranean Diet (MD), the adherence to it decreased with modern lifestyle, where the time dedicated to acquisition, preparation/confection of food and meals diminished. At the same time, Mediterranean regions face a growth in the levels of non-communicable diseases, such as obesity, diabetes, and hypertension, sometimes together with undernutrition that affects other parts of the population. This chapter make a presentation about MD as a sustainable food system, essential to promote food security, at the same time that the methods of food production and consumption must respect the environment, maintain biodiversity, and economic society valorisation. Also, it shows MD associated with several factors such as gender, marital status, education level, lifestyle, and body weight. Maintaining the traditional MD pattern is crucial for public health, particularly in pandemic contexts such as COVID-19 where it shows the opportunity and relevance of adopt and promote MD as a healthy and sustainable diet.

DOI: 10.4018/978-1-7998-9557-2.ch010

INTRODUCTION

The constant increase in world population forces agricultural production to accompany this growth, to satisfy the corresponding needs in food and animal feed. However, the current production systems, most of them intensive, must be transformed to guarantee the optimization of the use of resources and the preservation of the planet.

The accentuated climate changes, characterized by extreme and unpredictable weather events oblige the adoption of sustainable food production and consumption patterns. The impacts of climate change on agriculture are felt in the decrease in plant growth, in increasing spread of pests and pathogens and in the change in the interactions between the different components of agro-ecosystems (Mall et al., 2017; Smith and Almaraz, 2004). This urgency in implementing sustainable production systems must be associated with changes in consumption behavior, lifestyles and personal choices, which are strongly influenced by cultural values shared by societies (Lucas, 2006). Although there are few studies that explore the differences in individual patterns of food and energy intake between urban and rural areas and between more and less developed countries (Kosaka et al., 2018), there is evidence in the literature that, while in large urban centers, energy intake can be excessive, in rural areas, especially in the case of poorer communities, energy and protein can be in deficit (Mayén et al., 2014). On the other hand, poor people living in urban areas may have a diet based on cheap energy, provided mainly by foods rich in sugar, salt and fat and with a deficient consumption of vegetables, fruits and legumes. Thus, preventing and controlling diseases and reducing hunger and social inequalities in food intake should be considered in the context of a changing planet (Mayén et al., 2014).

In September 2015, the member countries of the United Nations signed the agenda 2030, in which 17 objectives (Sustainable Development Goals – SDGs) were considered to achieve a sustainable development, in the different fields, until 2030 (UN, na; Pradhan et al., 2017; Caiado et al., 2018; Bennich et al., 2020). Among these objectives, 3 are directly linked with food production and consumption: SDG 2 – Zero hunger; SDG 3 - Ensure healthy lives and promote well-being for all at all ages; and SDG 12 - Ensure sustainable consumption and production patterns. Altogether, these goals focus the need of achieving sustainable and responsible food systems, able to feed the world population, while guarantee the adequate input of macro- and micronutrients for all, keeping biodiversity and improving well-being and economic subsistence of societies

It is known that in rich countries, most of the resources and budget allocated to health are used to treat chronic diseases, mainly related with bad eating habits. In total, it is estimated that they contribute to about 71% of all deaths, and that most of them would be avoidable by changing lifestyles. Recent studies indicate that, if the trend in the way food is consumed continues, and if there is no intervention in eating habits and lifestyles, in 2050 food availability will suffer a reduction of around 3.2% per person, with reductions of 4,0% in the consumption of fruits and vegetables, which will result in a decrease of about 28% in the ability to prevent associated deaths and excess weight (Springmann et al., 2016).

Sustainable diets are understood to be those with a low environmental impact, which contribute to food and nutritional security and to the healthy life of present and future generations. Sustainable diets protect and respect biodiversity and ecosystems, optimizing the use of natural and human resources. Moreover, these are diets which are culturally accepted, accessible, economically fair and affordable, nutritionally balanced, safe and healthy (Burlingame and Dernini, 2011).

The Mediterranean Diet (MD) is recognized as a pattern of sustainable production and healthy eating, with beneficial nutritional effects (Estruch et al., 2018). The Mediterranean dietary pattern is in line with all that was referred above, constituting a valuable dietary and lifestyle-pattern for people living in Mediterranean countries. The food and lifestyle pattern associated with MD has more than 5000 years and had its origin in Mediterranean basin, being influenced by the characteristics of the native countries and by the habits and believes brought by the cultures that populated them (Judaism, Christianity and Islam). The way in which food is produced, processed and how it reaches consumers and is prepared and consumed accomplish sustainability principles. Foods's characteristic of this dietary pattern are produced without using intensive approaches, respecting environmental rhythms, seasonality and biodiversity, and consumption is strongly based on local and no- or low-processed products and should therefore be valued and promoted.

Despite the recognized advantages of MD, considerably emphasized in the last years, marked by the UNESCO Intergovernmental Committee recognition as Intangible Cultural Heritage of Humanity (UNESCO, 2010), the adherence to MD decreased with modern lifestyle, where the time dedicated to acquisition and preparation/confection of food, as well as the time dedicated to meals diminished. Even in countries from the Mediterranean region, this decline is evident. At the same time, these regions face a growth in the levels of non-communicable diseases, such as obesity, diabetes and hypertension, sometimes together with undernutrition that affects other part of population. As such, strategies to promote MD lifestyle pattern, in these areas, are, more than ever, necessary.

Due to what was stated above, this chapter will make a presentation about MD as a sustainable food system, essential to promote food security, at the same time that the methods of food production and consumption respects environment, maintaining biodiversity and economic valorization of the societies involved.

After this introductory presentation of MD history and characteristics, different topics will be considered and critically reviewed, including: health benefits of MD pattern, sustainable aspects of production and consumption patterns of MD, geographical variations in the characteristics and adherence to MD, adherence levels, drivers and barriers towards MD, variations in MD adherence according to individuals' demographic characteristics, such as age and socio-economic aspects and, MD and COVID-19 pandemic. All this, with the aim of putting together, in the same chapter, the different aspects why agriculture and food sectors must consider MD in the actual context of climate changes.

HEALTH BENEFITS OF MD PATTERN

Due to all its distinctive characteristics, in 2010, MD was described by UNESCO as "the set of skills, knowledge, rituals, symbols, and traditions, ranging from the landscape to the table, which in the Mediterranean basin concerns the crops, harvesting, picking, fishing, animal husbandry, conservation, processing, cooking, and particularly sharing and consuming of food" (Dernini and Berry, 2015), being, because of that, recognized as humanity intangible heritage.

MD is a plant-based dietary pattern, where foods are mainly from local production, consumed according to their seasonality and with a minimal processing. Because of this, MD is recognized as a sustainable food pattern, with positive health effects (Estruch et al., 2018). Although MD dietary pattern shows some differences among countries, in general it is based in high proportion of vegetables, fruits, non-refined cereals, pulses and nuts. Olives are frequently consumed and olive oil the main fat, added to most of the products, both as sauce and for preserving purposes. The consumption of products from animal origin is limited to 3-4 times per week, being these products mainly fish, eggs and dairy. Red meat is consumed only occasionally and usually associated with festive events. Water is the main drink and wine is the alcoholic drink most consumed, in moderate levels and particularly during meals (Willett et al., 1995). One of the main nutritional characteristics is the high ratio monounsaturated/polyunsaturated fatty acids, with low intake of simple sugars (Bach-Faig et al., 2011).

Because all that was stated above, the health benefits of MD have been evidenced in different studies. The adoption of this pattern was reported as a way to prevent type II diabetes and to improve glycemic control (Schröder, 2007). A large-scale, multicenter, controlled randomized trial – the PREDIMED study - performed in Spain, reported that MD, enriched with extra-virgin olive oil or nuts prevented diabetes, being this better than a low-fat diet (Salas-Salvadó et al., 2014). Another study – the ATTICA study – described that adherence to MD is associated with improvements in glucose homeostasis, insulin levels and insulin resistance (Panagiotakos et al., 2007). The way MD helps in diabetes has been tentatively investigated. There are several common components such as table oils (olive oil) and whole grains, fruit, nuts, vegetables, legumes, protein-rich foods, moderate alcohol consumption, and reduced intake of red/processed meat and simple sugar products that may contribute to this (Esposito et al., 2014). In the particular case of olive oil, in a randomized cross-over trial it was observed that a mean daily consumption of 25mL/day (corresponding to 577 mg/kg polyphenols), for 8 weeks, lowered fasting plasma glucose and glycated hemoglobin (Santangelo et al., 2016).

MD was also reported has having a protective effect against cardiovascular diseases, what appears to be linked to its positive effects in reducing arterial pressure, oxidative stress, improvements in serum lipidic profile and vascular/endothelial function (Salas-Salvadó et al., 2018; Schröder, 2007; Widmer et al., 2015). A study, made on the 80's of the last century, comparing 7 different countries, concluded about lower coronary heart disease mortality rate in the Mediterranean countries, what the authors explained by the low intake of saturated fatty acids by these populations (Keys et al., 1986). Several studies followed this one, most of which were prospective cohort studies and only few randomized controlled trials. Overall, the results demonstrate an inverse association between adherence to MD and the risk of different types of cardiovascular disease, as reviewed by Salas-Salvadó and collaborators (Salas-Salvadó et al., 2018).

Several different studies also suggest beneficial effects of MD in various types of cancer, namely a protective role of MD towards cancer (Verberne et al., 2010). A positive effect of long-term dietary intervention, based on MD supplemented with extra-virgin oil, in the primary prevention of breast cancer was reported (Toledo et al., 2015). Not only for breast cancer, but also for colon (Fasanelli et al., 2019), head and neck (Giraldi et al., 2017), hepatocellular (Turati et al., 2014), and upper digestive tract (Bosetti et al., 2003) there are evidences of benefits from MD adherence.

Another health area where MD has positive effects is in obesity. The PREDIMED study (already referred above) showed results highlighting the benefits of MD in this condition (Álvarez-Álvarez et al., 2019). Some other studies sometimes diverge in finding effects what can be due, not only due to the multifactorial nature of obesity, but also due to some variability in the definition of MD standards and adherence indexes (D'Alessandro and De Pergola, 2018). Except for genetic disorders, or variations at the level of genes involved in metabolism, a considerable part of obesity is due to energy imbalance, i.e., to an excessive intake of energy in relation to its expenditure (Tsakiraki et al., 2010). The benefits in preventing against obesity, attributed to MD, come from the comparisons of obesity prevalence, among world countries, from where higher rates of obesity were observed in United States or United Kingdom, where processed, energy-dense, and high-palatable foods were prevailing. With the spread of this Western type diet, different world regions, including Asian countries, where obesity was traditionally very low, saw a considerable proportion of their population becoming overweight or obese (Chakraborty and Das,

2016). Concerning the countries from Southern Europe, and North of Africa, which were countries with a history of adherence to MD, it is surprising to see, in the last years, considerable high rates of obesity, particularly among children (WHO, 2018). According to experts, this seems to be due to a shift to an increased intake of processed foods, high in saturated fats and refined carbohydrates, and consequently in energy, and simultaneously poor in fiber and nutrients (WHO, 2018).

The high levels of polyphenols in foods characteristic of MD also seems to contribute to the positive effects of MD in obesity [reviewed by (Nani et al., 2021)]. Polyphenols have antioxidant and anti-inflammatory activities, being oxidative stress and inflammation problems associated with obesity. The consumption of these plant secondary compounds, greatly present in products like olive oil, fruits, legumes and nuts, prevents these problems and help in the management of obesity (Nani et al., 2021). Besides the nutritional aspect of MD, its characteristics of *slow-food*, eaten in social context, with time spend during the meal and the physical activity associated, also helps to maintain an energy balance, reducing the amount of energy ingested and increasing the amount expended, preventing adiposity and increased body mass index.

Intimately related with the effects that MD can have on health is the role that this lifestyle/dietary pattern has in guaranteeing food security. The definition of food security considers different dimensions: food availability (availability of sufficient quantities of food with quality); food access (capacity of individuals to acquire the appropriate foods to ensure nutrition); utilization (utilization of food through adequate diet, clean water, sanitation and health care); and stability (access to adequate food, in enough amount, at all times) (FAO, 2006). This means, in a very simplistic way, that food security considers socio-economic, nutritional and food security aspects. Based on that, the importance of MD in a food security context can be seen from 2 different perspectives: 1) the beneficial nutritional aspects of MD, apporting micronutrients and bio-active compounds, preventing microelements deficiency and malnutrition, which is a growing problem in developed and developing countries; 2) the social/economic dimension of MD, which, by having a close link with food production and land, allows the economic development and access to food nutritional quality by the potentially vulnerable groups.

When looking to the first point, the increased availability of food, mainly low-cost, high-energy and nutrient-poor, not only results in obesity development, as previously discussed, but is also a cause of malnutrition caused by the lack of micronutrient (Kaidar-Person et al., 2008). Deficiencies in Vitamin D, chromium, biotin, thiamine and different antioxidant vitamins can be associated with the development of several diseases, including diabetes (Via, 2012). Evidence supports that a high adherence to MD increases the probability of fulfilling the nutrient recommendations (Castro-Quezada et al., 2020).

Taking a look to the second point, to promote MD will also to develop rural zones. One of the characteristics of MD is the consumption of foods locally produced, through production methods mostly environment-friend (not intensive) and respecting seasonality. As such, MD is greatly dependent on small producers. An increase in adherence to MD would have the effect of increased demand of these products, which would give economic viability, and access to high quality foods, to rural populations, usually poorer. As it was recently concluded by some authors "*Promotion of the MD as a sustainable diet will help reduce the prevalence of food insecurity and provide the foundations of food security for posterity*" (Grammatikopoulou et al., 2020).

Finally, it should also be noted that the MD and the global characteristics of this lifestyle are associated with well-being, which is a broad multi-parameter concept that includes the individual's subjective assessment of their own well-being, an assessment that can vary in sociodemographic terms, economics and lifestyles.

SUSTAINABLE PRODUCTION AND CONSUMPTION OF MD

Sustainable consumption and production is identified as one of the essential requirements for sustainable development (Wang et al., 2018). Aspects such as seasonal consumption, food production respecting environment characteristics, the use of spontaneous plants and the predominance of foods from vegetable are some of the reasons for that sustainability. The propose of sustainable production is to efficiently manage scarce resources with respect to socio-cultural factors in production (Ghadimi et al., 2013; Pallaro et al., 2015), while in sustainable consumption is to increase consumer awareness toward pursuing sustainable purchasing behaviours (Liu et al., 2016).

In general, previous work related to sustainable consumption and production include "top-down" efforts by policymakers, with an economical intervention by government, and "bottom-up" activities by firm's commitment into their business activities (Akenji and Bengtsson, 2014; Tseng et al., 2013). However, the focus on sustainable consumption and production practices varies based on economic conditions. Thus, the economy growth is always a top priority in developing countries, where the sustainability not always has been pursuit and, generally the efforts have been made by companies on "bottom-up" activities. The "top-down" efforts are mostly neglected by government authorities. The opposite occurs in developed economies, where the coexistence of "top-down" and "bottom-up" efforts with diverse interventions at both the supply and demand side (Wang et al., 2018).

There is increasing interest in promoting local economically viable production (Sonnino, 2013) as well as an increasing global demand for local foods and concerns about its implications for sustainable development goals (Hinrichs and Charles, 2012). In this sense the Mediterranean diet can be a lever for sustainable development. It leads local production with small-scale producers, incites empower women and creating opportunities for small and medium-sized enterprises, cooperatives and producer organisations of the agro-food sector. Consequently, improve local economic growth and job creation, and the development of new products, services and experiences (gastronomic, medical tourism between others), and new markets (Hachem et al., 2016). Producers are also involved in post-production activities including small-scale processing of traditional and local food products selling into markets. Through their agricultural work women into organisations and family farms actively contribute to sustainable development and play an essential role in knowledge transfer on the MD and safeguarding of its know-how and gastronomic techniques and cooking practices. Also, they preserve the knowledge of indigenous plants as edible foods and stock the abundant seasonal produce for later use during the year, hence safeguard bio-diversity and in expanding availability and accessibility of nutritional food on a sustainable basis. These traditional food-saving practices, produced to make use of food surplus reduce the food wasted (Hachem et al., 2016) and highlight the environmental-friendly characteristics of MD context.

Other point is related to the environmental footprint of MD. Several studies demonstrate that MD pattern demands less soil, water and energy compared to the other diet patterns due the different varieties of crops used and the native animals breeding locally, hence increasing the genetic diversity, biodiversity and ecosystem services (Sáez-Almendros et al., 2013). Furthermore, the dependence on local resources and food stocks, induce less pressure on fishing and hunting on local populations. According FAO (2014), through the implementation of suitable policies and provision of incentives to farmers, it is possible to achieve sustainable food production that meets present and future needs.

Socio-cultural, economic, and demographic factors are drivers of sustainable or unsustainable consumption. The consumers, when choosing and eating mainly plant, diverse and local foods produced through sustainable production systems, also contribute to the sustainable nature of the MD (Hachem et al., 2016).

The wandering away from MD patterns to adopt less healthy lifestyles moving away of the social and familiar way of consuming foods is related to the acceleration of modernisation and urban growth due to the rural exodus towards cities, to the changes in demographics, lifestyles and food consumption patterns and economic activities related to them. The urbanisation has largely contributed to farming and natural land degradation and loss, affecting the traditional agricultural livelihoods and local food production, favours the mass production of low-cost food and the necessity for the transfer and storage of these products in urban centres, increasing the amount of spoiled and wasted food (Hachem et al., 2016).

Thus, the promotion of MD sustainable consumption needs to be accompanied by parallel initiatives to communicate their benefits in terms of sustainable diet, agrobiodiversity, high quality local and typical products and, nutrient content. Sustainable food and agriculture policies should, simultaneously (Hachem et al., 2016): i) improve the efficiency of food production, preserving their diverse ecosystem services; ii) minimise the adverse effects on the environment of inputs use,; iii) optimise the use of energy and water in all operations (transport, storage, food processing, retail, consumption); iv) recognize and treated food losses and waste; v) improve food handling systems; vi) ensure food quality and safety; vii) deliver more education to all stakeholders of the agri-food chain, including farmers and consumers; ix) develop better infrastructures (storage facilities and marketing systems), x) make consumers aware of the need to adopt sustainable consumption choices and change their food choices and behaviours; xi) improve research and development capacity, particularly of small-scale producers.

MD GEOGRAHICAL VARIATION AND ADHERENCE

The geographical variations in the characteristics and adherence to MD, namely the particularities of different countries in the type of foods, habits and adherence levels, drivers and barriers and, the socio-demographics variations are presented in this section.

Adherence Levels, Drivers and Barriers

Adherence

Despite the benefits of MD, discussed previously, the Mediterranean region and the countries where MD was tradition, are now facing different problems of food security (Berry, 2019), with under nutrition co-existing with overweight and obesity. The type of problem varies between European and non-European countries, but all seem to be related with the decrease in adherence to the Mediterranean sustainable and healthy lifestyle pattern. According to World Health Organization (WHO), the countries integrating the East of Mediterranean are facing the most marked changes in demographic, socio-economic and healthy conditions of the last 30 years. These changes include a strong nutritional transition characterized by the replacement of traditional and ancestral recipes by foods rich in fat, refined sugars and highly processed, with the consequent rise in the prevalence of chronic nutrition-related diseases, such as diabetes, hypertension, cardiovascular diseases and obesity (Musaiger et al., 2012). This decrease in the adherence to MD pattern has been recognized in different scientific documents. In 2009, a study about the adherence level to MD in different 41 countries compared the situation between 1961-1965 with the one in 2000-2003 (Da Silva et al., 2009). Despite the evidence that countries considered with Mediterranean characteristics had high adherence levels than non-Mediterranean countries, in both cases this adherence considerably decreased in the 40 years evaluated, with the decrease being even higher in Mediterranean countries. This situation can be done to the increased availability, in these countries, of foreign foods, more processed and with higher amounts of conservatives (Garcia-Closas et al., 2006), as well as to the increase in meat, sugar and fats (others that not olive-oil) (Helsing, 1995). At the same time, during this period, the increased availability of fruits and vegetables in North Europe allowed this region to get closer of MD pattern (Balanza et al., 2007).

In Sicilia, Italy, the adult population presents only 18.9% of the individuals with high adherence to MD, with the majority having a medium adherence (Grosso et al., 2014). The reduced adherence raises even more concerns when seen in children. A study from 2020 shows that, in Spain, 74.4% of the evaluated children presented only a medium or lower level of adherence. Considering the KIDMED score, which is based on 16 questions, this sample presented a mean of 5.9 points (Da Rocha et al., 2020). Adolescents also have limited adherence, with high adherences observed only in 9,1%, in Italy (Mistretta et al., 2017), and low adherence in 73.7% of Greek adolescents (Theodoridis et al., 2018). In Portugal, although data numbers are limited, it is recognized that, since 1960, the dietary patterns have been diverging from MD (Durão et al., 2008; Pinho et al., 2016). In a study performed with 10,153 adult participants, published in 2017, it was identified high levels of meat consumption, together with lower consumption of fresh fruits, vegetables, fish and dairy (Gregório et al., 2017).

It is interesting to note that socio-economic conditions can affect the adherence levels, with the rural population showing higher adherence levels, comparatively to urban population, possibly due to the proximity to agriculture (Grosso et al., 2014; Mohtadi et al., 2020). Due to different economic conditions and socio-demographic factors, MD also requires a dissimilar focus in developing and developed economies (Wang et al., 2018).

Drivers and Barriers

The decrease in adherence of a food and lifestyle pattern with positive effects for health and sustainability makes more urgent the understanding of the factors that contribute to the acceptance to this. Among the different factors that contribute to food choices, palatability is essential, since an unpalatable food is not easily accepted to be consumed, particularly at medium-long term. The fact of MD being a plant-based diet, rich in polyphenols (Louro et al., 2021b), results in foods that may have sensory properties that are usualy aversive, namely astringency and bitterness. Several studies show that bitter taste and astringency act as a barrier for the acceptance of diverse foods of vegetable origin (e.g. Drewnowski and Gomez-Carneros, 2000). It is important to note that the way different individuals perceive sensory characteristics, i.e., the different sensitivity to those characteristics, varies from one individual to another. So, the polyphenolic composition of MD foods can have more deterrent effects for some individuals than another. Despite the lack of information about the effect of taste and astringency perception to MD adherence, a recent study showed how these can be important in the preferences for fruits and vegetables (Louro et al., 2021a).

Neophobia, i.e., the fear for new, unknown foods, also appears as a barrier to MD adherence. Higher neophobia levels results in less diverse diet. In students of different school levels, it was observed that higher neophobia was associated with lower MD adherence (Rodríguez-Tadeo et al., 2018). Similar results were also observed in adults (Predieri et al., 2020). Socio-economic factors need also to be considered, when discussing divers and barriers for MD consumption. For some countries there is the perception that plant-based diets, as MD, can be more expensive, and that can constitute a barrier to a shift for this dietary type. A study performed with the United Kingdom population evidenced that, in that countries, to follow MD is more expensive that to follow a Wester-type data (Tong et al., 2018). Not only price, but also school level and social position need to be considered, since these are determinant factors for the adoption of healthy dietary patterns (Darmon and Drewnowski, 2008), which were also observed to affect MD, with higher adherence levels for higher school levels (Maugeri et al., 2019). Finally, there is an important point to be considered, in terms of barriers to MD adherence, that is convenience. In actual societies of developed countries, people buy a high proportion of their foods in supermarkets and search for the ones that are ready-to-go, ready-to-cook, and ready-to-eat. This results in a high demand of ultra-processed, pre-cooked foods (Lane et al., 2021; Schnabel et al., 2019; Wolfson and Bleich, 2015). This tendency goes against the principles of MD, where the time spend cooking, as well as the social aspect of eating, is a major pilar.

If, on one hand, the actual lifestyle can be seen as a barrier to MD adherence, the way some groups of consumers start looking to food can be a driver. In fact, there is a growing concern about sustainability and animal well-being that is resulting in higher expression of "veggie" consumption, which includes vegan, vegetarian and flexitarian regimens. The first study about the Portuguese reality, in terms of veggie, was carried on by a food consultancy agency, which showed that, similarly to other European countries, also in Portugal food habits are changing, with flexitarian segment growing, with a total of 7.4% (mainly women, adolescents and young adults). The passage to flexitarian regimen, in which MD can fit, is higher than to more strict regimens, because most of the people do not really want to stop consuming animal products, due to the high palatability of those (Burkert et al., 2014), but only to reduce consumption due to health, environment or animal wellbeing motivations (Pohjolainen et al., 2015).

MEDITERRANEAN DIET AND COVID-19 PANDEMIC

This section presents MD and COVID-19 pandemic, where the known effects of the actual pandemic in adherence levels, on one hand, and the potential effects of adherence levels in the severity of symptoms, in the other, will be reported.

MD, by its known health benefits, already reported in previous sections, has been suggested has having benefits in the context of COVID-19. Some studies observed that higher levels of adherence to MD were associated with lower severity of symptoms of individuals infected with the virus Sars-Cov2 (Perez-Araluce et al., 2021). Similarly, a study made with participants from 23 different countries also reported lower number of COVID-19 deaths associated with higher adherence levels to MD, hypothesizing that this may be related with the anti-inflammatory properties of this dietary pattern, namely the polyphenol content of the diet (Greene et al., 2021). The positive effect of this dietary pattern, in alleviating the symptoms of disease was also suggested by other authors (Angelidi et al., 2021)

But, besides considering the relationship that MD can have with the disease, it is important to consider the influence that the actual pandemic had in the way people changed dietary habits and food-related behaviors, including their adherence to MD. The restrictions imposed by the pandemic affected the way people relate with food. Different studies emerged, during this period, showing that confinement resulted in changes in the intake frequency and amount of diverse food products. One of the negative effects, transversal to several different countries, was the average increase in the total amount of food consumed, due to higher time at home and the consequent "snacking" (Lamy et al., 2020). In fact, individuals referred to feel more motivated to choose comfort foods, and increased the intake of cookies and cakes and other sweet snacks (Lamy et al., 2020). But, pandemic also had positive consequences in food habits changes, namely through an increase in vegetable and fruit intake frequency (Lamy et al., 2020). Since meals were mostly made and consumed at home, together with the increased time most of the individuals had to dedicate to cooking, the interest for new recipes and new ways of preparing foods increased considerably, in different countries, including Portugal, with the enormous advantage of decreasing the consumption of fast-food and processed foods (Ben Hassen et al., 2020; Castellini et al., 2021; Chenarides et al., 2021; Poelman et al., 2021). This change, by a considerable amount of population, demonstrates that people are available to make changes in their food-related habits, when conditions are favorable. This tendency for changes in habits, observed during lock-down, namely the increase in ingestion frequency, family meals and the concern with the type of production and local of acquisition, goes in line with the principles of MD, suggesting that, if having conditions to that, people accept and appreciate this pattern.

The limited number of studies that evaluated the direct effect of the pandemic at the level of MD adherence showed increases during this period. One study, from Croatia, show that the increases were mainly in women with ages between 20 and 50 years old and with higher school-levels. Concerning food habits, these improved, with the increase in the consumption of plant-based foods and the decrease in the intake of processed foods. The negative point was the physical activity, which was observed to decrease during this period (Pfeifer et al., 2021). As such, with the pandemic, it emerged the opportunity to promote healthy and sustainable dietary patterns, as the case of MD.

FUTURE RESEARCH DIRECTIONS

The MD encourages local and seasonal food production and consumption, thereby minimizing the environmental impact and preserving biodiversity. Future research should consider the MD sustainability issue in a holistic perspective, with the lens of TBL ("Triple Bottom Line") framework. To date, few studies have been made to systematically compare the status of sustainable consumption and production and its direction from the perspective of sustainable development and no one with TBL focus. Also, the direct and indirect relationships between sustainable consumption and production can be regarded as a dynamic environment that requires future investigation. In addition, future research should also addresse the adherence to the MD (in all dimensions, health, socio-culture, economics, and environment), showing that it is a sustainable diet model for the Mediterranean food production systems, minimizing food loss and waste (from production to consumption).

The work also highlighted the need to research the way to keep the costs of MD healthy foods, renowned for their beneficial nutritional features and taste, accessible to all people. In this way, more research is also needed to be conducted to precisely measure the associations between MD and public health.

Despite the widely recognition of MD as being associated with lower incidence and lower death rates of cardiovascular diseases, cancer and other non-communicable diseases, this traditional diet is also recognize as being worsen due global industrialization and commerce, massive urbanization, increase in the cost of plant foods leading to their substitution by less wealthy people with less nutritious but more energy dense lower quality foods. It means decline in the quality of diet, which, together with the increased sedentariness has produced a caloric imbalance responsible for the obesity diseases. Additionally, some food systems have contributed to generate social inequalities, with a few actors in the system each day more impoverished. This negative trend requires a strong research commitment in MD issues related to nutrition, health and food industry sustainable practices. Also, research on the indicators for MD sustainability assessment and on local biodiversity to analyze the nutrient content of the local species including wild plants, should also be addressed.

In future other important topic of research is technological developments, which may help to overcome some of the problems faced in the application of a healthy diet and increase knowledge about new technologies that may help farmers to produce healthy foods sticking local traditions but a fair cost.

As well, it is worthwhile to analyze the awareness of people (especially young people) to become aware of the current problems inherent to MD and lead them to move to responsible and sustainable consumption practices. This can be done with improved consumer behavior research in broad sense and, particularly, in topics such as changing attitudes, lifestyles and decision-making processes and drivers. Moreover, it is necessary to generate international actions and develop synergic initiatives and research networks for the diffusion of the MD as a possibly universal model of healthy diet worldwide. But careful is needed since due to different economic conditions and socio-demographic factors, MD consumption analysis requires a diverse focus in developing and developed economies and in their respective population segments. This means deeply study the complex lifestyle model which MD suppose, characterized by biodiversity, seasonality and local food production, involving a set of skills, knowledge, processing, cooking and traditional recipes, the sense of hospitality, constructive relationships, intercultural dialogue and respect for diversity.

CONCLUSION

This work pursued to present MD as a sustainable food system, essential to promote food security supported in food production and consumption practices that respect environment, maintaining biodiversity and providing economic valorisation of the societies.

After an introductory presentation of MD characteristics, the chapter offered a critical review of the benefits of MD pattern, both at health, sustainable production and consumption levels, as well as the geographical variations in the characteristics and adherence to MD.

MD has been the first traditional food practice to receive the award recognition as Intangible Cultural Heritage of Humanity by the UNESCO Intergovernmental Committee. This marked a milestone in the history of nutrition. MD contributes to beneficial health status and a decreased risk of many chronic diseases. Its adherence can vary with sociodemographic conditions and lifestyles, with the rural population showing higher adherence levels, comparatively to urban population, possibly due to the proximity to agriculture production systems. But, it is necessary to have in mind that MD requires also a dissimilar focus in developing and developed economies due the different economic conditions and socio-demographic factors. Moreover, MD is a valuable lifestyle and food pattern for Mediterranean regions, but other plant-based food patterns may perform better in other geographical parts of the globe.

Maintaining the traditional MD pattern, in Mediterranean countries, is crucial for public health. Concerning MD and COVID-19 pandemic, in one hand a potential protection against COVID-19 by a high-quality dietary pattern such as MD is to be expected given the biological plausibility supporting the beneficial effects of an adequate dietary intake on the immune system. By the other hand, knowledge on the relationship between long-term maintained healthy dietary patterns, such as the MD, and the risk of SARS-CoV-2 infection is still sparse. Moreover, the actual pandemic situation also had effects that can be considered at MD level. More precisely, COVID-19 pandemic highlighted that consumers are willing to adopt healthy and sustainable habits, if the message is transmitted in the right way and conditions are met.

ACKNOWLEDGMENT

This research was supported by the Fundação para a Ciência e a Tecnologia [grants number UID/04007/2020 and UIDB/05183/2020]. Funding was additionally provided by the FCT– Portuguese Foundation for Science and Technology under the research contract CEECIND/04397/2017 to Elsa Lamy.

REFERENCES

Akenji, L., & Bengtsson, M. (2014). Making sustainable consumption and production the core of sustainable development goals. *Sustainability*, *6*(2), 513–529. doi:10.3390u6020513

Álvarez-Álvarez, I., Martínez-González, M. Á., Sánchez-Tainta, A., Corella, D., Díaz-López, A., Fitó, M., Vioque, J., Romaguera, D., Martínez, J. A., Wärnberg, J., López-Miranda, J., Estruch, R., Bueno-Cavanillas, A., Arós, F., Tur, J. A., Tinahones, F. J., Serra-Majem, L., Martín, V., Lapetra, J., ... Toledo, E. (2019). Adherence to an Energy-restricted Mediterranean Diet Score and Prevalence of Cardiovascular Risk Factors in the PREDIMED-Plus: A Cross-sectional Study. *Revista Española de Cardiología (English Ed.)*, *72*(11), 925–934. Advance online publication. doi:10.1016/j.rec.2018.08.010 PMID:30287240

Angelidi, A. M., Kokkinos, A., Katechaki, E., Ros, E., & Mantzoros, C. S. (2021). Mediterranean diet as a nutritional approach for COVID-19. Metabolism: Clinical and Experimental, 114. doi:10.1016/j.metabol.2020.154407

Bach-Faig, A., Berry, E. M., Lairon, D., Reguant, J., Trichopoulou, A., Dernini, S., Medina, F. X., Battino, M., Belahsen, R., Miranda, G., & Serra-Majem, L. (2011). Pirámide de dieta mediterránea hoy. Ciencia y actualizaciones culturales. *Public Health Nutrition*, *14*(12A).

Balanza, R., García-Lorda, P., Pérez-Rodrigo, C., Aranceta, J., Bonet, M. B., & Salas-Salvadó, J. (2007). Trends in food availability determined by the Food and Agriculture Organization's food balance sheets in Mediterranean Europe in comparison with other European areas. *Public Health Nutrition*, *10*(2), 168–176. Advance online publication. doi:10.1017/S1368980007246592 PMID:17261226

Ben Hassen, T., El Bilali, H., & Allahyari, M. S. (2020). Impact of covid-19 on food behavior and consumption in qatar. *Sustainability (Switzerland)*, *12*(17), 319. Advance online publication. doi:10.29117/quarfe.2020.0296

Bennich, T., Weitz, N., & Carlsen, H. (2020). Deciphering the scientific literature on SDG interactions: A review and reading guide. *The Science of the Total Environment*, *728*, 138405. doi:10.1016/j.scitotenv.2020.138405 PMID:32388023

Berry, E. M. (2019). Sustainable food systems and the mediterranean diet. *Nutrients, 11*(9), 2229. Advance online publication. doi:10.3390/nu11092229 PMID:31527411

Bosetti, C., Gallus, S., Trichopoulou, A., Talamini, R., Franceschi, S., Negri, E., & La Vecchia, C. (2003). Influence of the Mediterranean Diet on the Risk of Cancers of the Upper Aerodigestive Tract. *Cancer Epidemiology, Biomarkers & Prevention, 12*(10). PMID:14578148

Burkert, N. T., Muckenhuber, J., Großschädl, F., Rásky, É., & Freidl, W. (2014). Nutrition and health - The association between eating behavior and various health parameters: A matched sample study. *PLoS One, 9*(2), e88278. Advance online publication. doi:10.1371/journal.pone.0088278 PMID:24516625

Burlingame, B., & Dernini, S. (2011, December 13). Burlingame, Barbara, & Dernini, S. (2011). Sustainable diets: The Mediterranean diet as an example. *Public Health Nutrition, 14*(12A), 2285–2287. Advance online publication. doi:10.1017/S1368980011002527

Caiado, R. G., Filho, W. L., Quelhas, O. L., Nascimento, D. L., & Ávila, L. V. (2018). A literature-based review on potentials and constraints in the implementation of the sustainable development goals. *Journal of Cleaner Production, 198*, 1276–1288. doi:10.1016/j.jclepro.2018.07.102

Castellini, G., Savarese, M., & Graffigna, G. (2021). The Impact of COVID-19 Outbreak in Italy on the Sustainable Food Consumption Intention From a "One Health" Perspective. *Frontiers in Nutrition, 8*, 622122. Advance online publication. doi:10.3389/fnut.2021.622122 PMID:33791331

Castro-Quezada, I., Román-Viñas, B., & Serra-Majem, L. (2020). Nutritional adequacy of the Mediterranean diet. In The Mediterranean Diet. doi:10.1016/B978-0-12-818649-7.00013-8

Chakraborty, C., & Das, S. (2016). Dynamics of Diabetes and Obesity: An Alarming Situation in the Developing Countries in Asia. *Mini-Reviews in Medicinal Chemistry, 16*(15), 1258–1268. Advance online publication. doi:10.2174/1389557516666160505115512 PMID:27145852

Chenarides, L., Grebitus, C., Lusk, J. L., & Printezis, I. (2021). Food consumption behavior during the COVID-19 pandemic. *Agribusiness, 37*(1), 44–81. Advance online publication. doi:10.1002/agr.21679 PMID:33362340

D'Alessandro, A., & De Pergola, G. (2018). The Mediterranean Diet: Its definition and evaluation of a priori dietary indexes in primary cardiovascular prevention. *International Journal of Food Sciences and Nutrition, 69*(6), 647–659. Advance online publication. doi:10.1080/09637486.2017.1417978 PMID:29347867

Da Rocha, B. R. S., Rico-Campà, A., Romanos-Nanclares, A., Ciriza, E., Barbosa, K. B. F., Martínez-González, M. Á., & Martín-Calvo, N. (2020). Adherence to Mediterranean diet is inversely associated with the consumption of ultra-processed foods among Spanish children: The SENDO project. *Public Health Nutrition*. Advance online publication. doi:10.1017/S1368980020001524 PMID:32698921

Da Silva, R., Bach-Faig, A., Raidó Quintana, B., Buckland, G., Vaz De Almeida, M. D., & Serra-Majem, L. (2009). Worldwide variation of adherence to the Mediterranean diet, in 1961-1965 and 2000-2003. *Public Health Nutrition, 12*(9). doi:10.1017/S1368980009990541

Darmon, N., & Drewnowski, A. (2008). Does social class predict diet quality? *The American Journal of Clinical Nutrition, 87*(5), 1107–1117. Advance online publication. doi:10.1093/ajcn/87.5.1107 PMID:18469226

Dernini, S., & Berry, E. M. (2015). Mediterranean Diet: From a Healthy Diet to a Sustainable Dietary Pattern. *Frontiers in Nutrition, 2*. Advance online publication. doi:10.3389/fnut.2015.00015 PMID:26284249

Drewnowski, A., & Gomez-Carneros, C. (2000). Bitter taste, phytonutrients, and the consumer: A review. *The American Journal of Clinical Nutrition, 72*(6), 1424–1435. Advance online publication. doi:10.1093/ajcn/72.6.1424 PMID:11101467

Durão, C. R., Oliveira, J. F. S., & De Almeida, M. D. V. (2008). Portugal e o padrão alimentar mediterrânico. *Revista de Alimentação Humana, 14*(3).

Esposito, K., Chiodini, P., Maiorino, M. I., Bellastella, G., Panagiotakos, D., & Giugliano, D. (2014). Which diet for prevention of type 2 diabetes? A meta-analysis of prospective studies. *Endocrine, 47*(1), 107–116. Advance online publication. doi:10.100712020-014-0264-4 PMID:24744219

Estruch, R., Ros, E., Salas-Salvadó, J., Covas, M.-I., Corella, D., Arós, F., Gómez-Gracia, E., Ruiz-Gutiérrez, V., Fiol, M., Lapetra, J., Lamuela-Raventos, R. M., Serra-Majem, L., Pintó, X., Basora, J., Muñoz, M. A., Sorlí, J. V., Martínez, J. A., Fitó, M., Gea, A., ... Martínez-González, M. A. (2018). Primary Prevention of Cardiovascular Disease with a Mediterranean Diet Supplemented with Extra-Virgin Olive Oil or Nuts. *The New England Journal of Medicine, 378*(25), e34. Advance online publication. doi:10.1056/NEJMoa1800389 PMID:29897866

Fang, Z., & Cao, C. (2019). The state of food security and nutrition in the world 2019. Building climate resilience for food security and nutrition. Food and Agriculture Organization of the United Nations.

FAO. (2006). https://www.fao.org/fileadmin/templates/faoitaly/documents/pdf/pdf_Food_Security_Cocept_Note.pdf

FAO. (2014). *Building a common vision for sustainable food and agriculture. Principles and approaches*. FAO. www.fao.org/3/a-i3940e.pdf

Fasanelli, F., Giraudo, M. T., Vineis, P., Fiano, V., Fiorito, G., Grasso, C., Polidoro, S., Trevisan, M., Grioni, S., Krogh, V., Mattiello, A., Panico, S., Giurdanella, M. C., Tumino, R., De Marco, L., Ricceri, F., & Sacerdote, C. (2019). DNA methylation, colon cancer and Mediterranean diet: Results from the EPIC-Italy cohort. *Epigenetics, 14*(10), 977–988. Advance online publication. doi:10.1080/15592294.2019.1629230 PMID:31179817

Garcia-Closas, R., Berenguer, A., & González, C. A. (2006). Changes in food supply in Mediterranean countries from 1961 to 2001. *Public Health Nutrition, 9*(1), 53–60. Advance online publication. doi:10.1079/PHN2005757 PMID:16480534

Ghadimi, P., & Heavey, C. (2013). A Review of applications of agent-based modelling and simulation in supplier selection problem. In *Modelling and Simulation (EUROSIM), 2013 8th EUROSIM Congress on*. IEEE.

Giraldi, L., Panic, N., Cadoni, G., Boccia, S., & Leoncini, E. (2017). Association between Mediterranean diet and head and neck cancer: Results of a large case-control study in Italy. *European Journal of Cancer Prevention*, *26*(5), 418–423. Advance online publication. doi:10.1097/CEJ.0000000000000277 PMID:27483412

Grammatikopoulou, M. G., Gkiouras, K., Tranidou, A., & Goulis, D. G. (2020). Food security and adherence to the Mediterranean diet: An interplay of socio-demographic characteristics. In The Mediterranean Diet. doi:10.1016/B978-0-12-818649-7.00008-4

Greene, M. W., Roberts, A. P., & Frugé, A. D. (2021). Negative Association Between Mediterranean Diet Adherence and COVID-19 Cases and Related Deaths in Spain and 23 OECD Countries: An Ecological Study. *Frontiers in Nutrition*, *8*, 591964. Advance online publication. doi:10.3389/fnut.2021.591964 PMID:33748170

Gregório, M. J., Rodrigues, A. M., Eusébio, M., Sousa, R. D., Dias, S., André, B., Grønning, K., Coelho, P. S., Mendes, J. M., Graça, P., Espnes, G. A., Branco, J. C., & Canhão, H. (2017). Dietary Patterns Characterized by High Meat Consumption Are Associated with Other Unhealthy Life Styles and Depression Symptoms. *Frontiers in Nutrition*, *4*, 25. Advance online publication. doi:10.3389/fnut.2017.00025 PMID:28660194

Grosso, G., Marventano, S., Giorgianni, G., Raciti, T., Galvano, F., & Mistretta, A. (2014). Mediterranean diet adherence rates in Sicily, southern Italy. *Public Health Nutrition*, *17*(9), 2001–2009. Advance online publication. doi:10.1017/S1368980013002188 PMID:23941897

Hachem, F., Capone, R., Yannakoulia, M., Dernini, S., Hwalla, N., & Kalaitzidis, C. (2016). The Mediterranean Diet: A Sustainable Consumption Pattern. In Mediterra 2016. Zero Waste in the Mediterranean. Natural Resources, Food and Knowledge. International Centre for Advanced Mediterranean Agronomic Studies (CIHEAM) and Food and Agriculture Organization of the United Nations (FAO), Paris: Presses deSciences Po.

Helsing, E. (1995). Traditional diets and disease patterns of the Mediterranean, circa 1960. *The American Journal of Clinical Nutrition*, *61*(6), 1329S–1337S. Advance online publication. doi:10.1093/ajcn/61.6.1329S PMID:7754984

Hinrichs, C. C., & Charles, L. (2012). Local Food Systems and Networks in the US and the UK: Community Development Considerations for Rural Areas. In M. Shuck-smith, D. L. Brown, S. Shortall, M. Warner, & J. Vergunst (Eds.), *Rural transformations and rural policies in the UK and US* (pp. 156–176). Routledge.

Kaidar-Person, O., Person, B., Szomstein, S., & Rosenthal, R. J. (2008). Nutritional deficiencies in morbidly obese patients: A new form of malnutrition? Part A: Vitamins. *Obesity Surgery*, *18*(7), 870–876. Advance online publication. doi:10.100711695-007-9349-y PMID:18465178

Keys, A., Mienotti, A., Karvonen, M. J., Aravanis, C., Blackburn, H., Buzina, R., Djordjevic, B. S., Dontas, A. S., Fidanza, F., Keys, M. H., Kromhout, D., Nedeljkovic, S., Punsar, S., Seccareccia, F., & Toshima, H. (1986). The diet and 15-year death rate in the seven countries study. *American Journal of Epidemiology*, *124*(6), 903–915. Advance online publication. doi:10.1093/oxfordjournals.aje.a114480 PMID:3776973

Kosaka, S., Suda, K., Gunawan, B., Raksanagara, A., Watanabe, C., & Umezaki, M. (2018). Urban-rural difference in the determinants of dietary and energy intake patterns: A case study in west Java, Indonesia. *PLoS One*, *13*(5), e0197626. Advance online publication. doi:10.1371/journal.pone.0197626 PMID:29768478

Lamy, E., Viegas, C., Rocha, A., Lucas, M. R., Tavares, S., Capela e Silva, F., Guedes, D., Laureati, M., Zian, Z., Salles Machado, A., Ellssel, P., Freyer, B., González-Rodrigo, E., Calzadilla, J., Majewski, E., Prazeres, I., Silva, V., Juračak, J., Platilová Vorlíčková, L., . . . Anzman-Frasca, S. (2020). Changes in food behavior during COVID-19 pandemic: a cross-country study about changes in eating habits, motivations and food-related. Academic Press.

Liu, W., Oosterveer, P., & Spaargaren, G. (2016). Promoting sustainable consumption in China: A conceptual framework and research review. *Journal of Cleaner Production*, *134*, 13–21. doi:10.1016/j.jclepro.2015.10.124

Louro, T., Simões, C., Castelo, P. M., Capela e Silva, F., Luis, H., Moreira, P., & Lamy, E. (2021a). How Individual Variations in the Perception of Basic Tastes and Astringency Relate with Dietary Intake and Preferences for Fruits and Vegetables. *Foods*, *10*(8), 1961. Advance online publication. doi:10.3390/foods10081961 PMID:34441738

Louro, T., Simões, C., Penetra, M. J., Carreira, L., Castelo, P. M., Luis, H., Moreira, P., & Lamy, E. (2021b). Relationship between mediterranean diet adherence and saliva composition. *Nutrients*, *13*(4), 1246. Advance online publication. doi:10.3390/nu13041246 PMID:33920127

Lucas, M. R. (2006). Handbook of Consumer Behaviour. Universidade de Varsóvia, Polónia.

Mall, R. K., Gupta, A., & Sonkar, G. (2017). Effect of Climate Change on Agricultural Crops. In Current Developments in Biotechnology and Bioengineering: Crop Modification, Nutrition, and Food Production. doi:10.1016/B978-0-444-63661-4.00002-5

Maugeri, A., Barchitta, M., Fiore, V., Rosta, G., Favara, G., La Mastra, C., La Rosa, M. C., San Lio, R. M., & Agodi, A. (2019). Determinants of adherence to the mediterranean diet: Findings from a cross-sectional study in women from Southern Italy. *International Journal of Environmental Research and Public Health*, *16*(16), 2963. Advance online publication. doi:10.3390/ijerph16162963 PMID:31426512

Mayén, A. L., Marques-Vidal, P., Paccaud, F., Bovet, P., & Stringhini, S. (2014). Socioeconomic determinants of dietary patterns in low- and middle-income countries: A systematic review. *The American Journal of Clinical Nutrition*, *100*(6), 1520–1531. Advance online publication. doi:10.3945/ajcn.114.089029 PMID:25411287

Mistretta, A., Marventano, S., Antoci, M., Cagnetti, A., Giogianni, G., Nolfo, F., Rametta, S., Pecora, G., & Marranzano, M. (2017). Mediterranean diet adherence and body composition among Southern Italian adolescents. *Obesity Research & Clinical Practice*, *11*(2), 215–226. Advance online publication. doi:10.1016/j.orcp.2016.05.007 PMID:27269367

Mohtadi, K., Msaad, R., Benalioua, N., Jafri, A., Meftah, H., Elkardi, Y., Lebrazi, H., Kettani, A., Derouiche, A., Taki, H., & Saïle, R. (2020). Sociodemographic and Lifestyle Factors Associated with Adherence to Mediterranean Diet in Representative Adult Population in Casablanca City, Morocco: A Cross-Sectional Study. *Journal of Nutrition and Metabolism, 3105271*, 1–9. Advance online publication. doi:10.1155/2020/3105271 PMID:32280541

Musaiger, A. O., Al-Hazzaa, H. M., Takruri, H. R., & Mokhatar, N. (2012). Change in nutrition and lifestyle in the eastern mediterranean region: Health impact. Journal of Nutrition and Metabolism. doi:10.1155/2012/436762

Nani, A., Murtaza, B., Khan, A. S., Khan, N. A., & Hichami, A. (2021). Antioxidant and anti-inflammatory potential of polyphenols contained in Mediterranean diet in obesity: Molecular mechanisms. *Molecules (Basel, Switzerland)*, *26*(4), 985. Advance online publication. doi:10.3390/molecules26040985 PMID:33673390

Nardone, A., Ronchi, B., Lacetera, N., Ranieri, M. S., & Bernabucci, U. (2010). Effects of climate changes on animal production and sustainability of livestock systems. *Livestock Science*, *130*(1–3), 57–69. doi:10.1016/j.livsci.2010.02.011

Pallaro, E., Subramanian, N., Abdulrahman, M. D., & Liu, C. (2015). Sustainable production and consumption in the automotive sector: Integrated review framework and research directions. *Sustainable Production & Consumption*, *4*(2), 47–61. doi:10.1016/j.spc.2015.07.002

Panagiotakos, D. B., Tzima, N., Pitsavos, C., Chrysohoou, C., Zampelas, A., Toussoulis, D., & Stefanadis, C. (2007). The association between adherence to the mediterranean diet and fasting indices of glucose homoeostasis: The attica study. *Journal of the American College of Nutrition*, *26*(1), 32–38. Advance online publication. doi:10.1080/07315724.2007.10719583 PMID:17353581

Perez-Araluce, R., Martinez-Gonzalez, M. A., Fernández-Lázaro, C. I., Bes-Rastrollo, M., Gea, A., & Carlos, S. (2021). Mediterranean diet and the risk of COVID-19 in the 'Seguimiento Universidad de Navarra' cohort. *Clinical Nutrition (Edinburgh, Lothian)*. Advance online publication. doi:10.1016/j.clnu.2021.04.001 PMID:33934925

Pfeifer, D., Rešetar, J., Gajdoš Kljusurić, J., Panjkota Krbavčić, I., Vranešić Bender, D., Rodríguez-Pérez, C., Ruíz-López, M. D., & Šatalić, Z. (2021). Cooking at Home and Adherence to the Mediterranean Diet During the COVID-19 Confinement: The Experience From the Croatian COVIDiet Study. *Frontiers in Nutrition*, *8*, 617721. Advance online publication. doi:10.3389/fnut.2021.617721 PMID:33869262

Pinho, I., Rodrigues, S., Franchini, B., & Graça, P. (2016). Padrão Alimentar Mediterrânico: Promotor De Saúde. *Direção-Geral Da Saúde Programa Nacional Para a Promoção Da Alimentação Saudável.*

Poelman, M. P., Gillebaart, M., Schlinkert, C., Dijkstra, S. C., Derksen, E., Mensink, F., Hermans, R. C. J., Aardening, P., de Ridder, D., & de Vet, E. (2021). Eating behavior and food purchases during the COVID-19 lockdown: A cross-sectional study among adults in the Netherlands. *Appetite*, *157*, 105002. Advance online publication. doi:10.1016/j.appet.2020.105002 PMID:33068668

Pohjolainen, P., Vinnari, M., & Jokinen, P. (2015). Consumers' perceived barriers to following a plant-based diet. *British Food Journal, 117*(3), 1150–1167. Advance online publication. doi:10.1108/BFJ-09-2013-0252

Pradhan, P., Costa, L., Rybski, D., Lucht, W., & Kropp, J. P. (2017). A Systematic Study of Sustainable Development Goal (SDG) Interactions. *Earth's Future, 5*(11), 1169–1179. doi:10.1002/2017EF000632

Predieri, S., Sinesio, F., Monteleone, E., Spinelli, S., Cianciabella, M., Daniele, G. M., Dinnella, C., Gasperi, F., Endrizzi, I., Torri, L., Toschi, T. G., Bendini, A., Pagliarini, E., Cattaneo, C., Di Monaco, R., Vitaglione, P., Condelli, N., & Laureati, M. (2020). Gender, age, geographical area, food neophobia and their relationships with the adherence to the mediterranean diet: New insights from a large population cross-sectional study. *Nutrients, 12*(6), 1778. Advance online publication. doi:10.3390/nu12061778 PMID:32549267

Robertson, A., Tirado, C., Lobstein, T., Jermini, M., Knai, C., Jensen, J. H., Ferro-Luzzi, A., & James, W. P. (2004). Food and health in Europe: a new basis for action. WHO Regional Publications. European Series (Issue 96).

Rodríguez-Tadeo, A., Patiño-Villena, B., González Martínez-La Cuesta, E., Urquídez-Romero, R., & Ros Berruezo, G. (2018). Food neophobia, mediterranean diet adherence and acceptance of healthy foods prepared in gastronomic workshops by spanish students. *Nutrición Hospitalaria, 35*(3). Advance online publication. doi:10.20960/nh.1337 PMID:29974774

Rojas-Downing, M. M., Nejadhashemi, A. P., Harrigan, T., & Woznicki, S. A. (2017). Climate change and livestock: Impacts, adaptation, and mitigation. In Climate Risk Management (Vol. 16). doi:10.1016/j.crm.2017.02.001

Sáez-Almendros, S., Obrador, B., Bach-Faig, A., & Serra-Majem, L. (2013). Environmental Footprints of Mediterranean versus Western Dietary Patterns: Beyond the Health Benefits of the Mediterranean Diet. *Environmental Health, 12*(118), 1–8. doi:10.1186/1476-069X-12-118 PMID:24378069

Salas-Salvadó, J., Becerra-Tomás, N., García-Gavilán, J. F., Bulló, M., & Barrubés, L. (2018). Mediterranean Diet and Cardiovascular Disease Prevention: What Do We Know? *Progress in Cardiovascular Diseases, 61*(1), 62–67. Advance online publication. doi:10.1016/j.pcad.2018.04.006 PMID:29678447

Santangelo, C., Filesi, C., Varì, R., Scazzocchio, B., Filardi, T., Fogliano, V., D'Archivio, M., Giovannini, C., Lenzi, A., Morano, S., & Masella, R. (2016). Consumption of extra-virgin olive oil rich in phenolic compounds improves metabolic control in patients with type 2 diabetes mellitus: A possible involvement of reduced levels of circulating visfatin. *Journal of Endocrinological Investigation, 39*(11), 1295–1301. Advance online publication. doi:10.100740618-016-0506-9 PMID:27344308

Schnabel, L., Kesse-Guyot, E., Allès, B., Touvier, M., Srour, B., Hercberg, S., Buscail, C., & Julia, C. (2019). Association between Ultraprocessed Food Consumption and Risk of Mortality among Middle-aged Adults in France. *JAMA Internal Medicine, 179*(4), 490. Advance online publication. doi:10.1001/jamainternmed.2018.7289 PMID:30742202

Schröder, H. (2007). Protective mechanisms of the Mediterranean diet in obesity and type 2 diabetes. Journal of Nutritional Biochemistry, 18(3). doi:10.1016/j.jnutbio.2006.05.006

Smith, D. L., & Almaraz, J. J. (2004). Climate change and crop production: Contributions, impacts, and adaptations. *Canadian Journal of Plant Pathology*, *26*(3), 253–266. Advance online publication. doi:10.1080/07060660409507142

Sonnino, R. (2013). Local Foodscapes: Place and Power in the Agri-food System. *Acta Agriculturæ Scandinavica. Section B, Soil and Plant Science*, *63*(1), 2–7. doi:10.1080/09064710.2013.800130

Springmann, M., Clark, M., Mason-D'Croz, D., Wiebe, K., Bodirsky, B. L., Lassaletta, L., de Vries, W., Vermeulen, S. J., Herrero, M., Carlson, K. M., Jonell, M., Troell, M., DeClerck, F., Gordon, L. J., Zurayk, R., Scarborough, P., Rayner, M., Loken, B., Fanzo, J., ... Willett, W. (2018). Options for keeping the food system within environmental limits. *Nature*, *562*(7728), 519–525. Advance online publication. doi:10.103841586-018-0594-0 PMID:30305731

Theodoridis, X., Grammatikopoulou, M. G., Gkiouras, K., Papadopoulou, S. E., Agorastou, T., Gkika, I., Maraki, M. I., Dardavessis, T., & Chourdakis, M. (2018). Food insecurity and Mediterranean diet adherence among Greek university students. *Nutrition, Metabolism, and Cardiovascular Diseases*, *28*(5), 477–485. Advance online publication. doi:10.1016/j.numecd.2018.02.007 PMID:29655531

Toledo, E., Salas-Salvado, J., Donat-Vargas, C., Buil-Cosiales, P., Estruch, R., Ros, E., Corella, D., Fito, M., Hu, F. B., Aros, F., Gomez-Gracia, E., Romaguera, D., Ortega-Calvo, M., Serra-Majem, L., Pinto, X., Schroder, H., Basora, J., Sorli, J. V., Bullo, M., ... Martinez-Gonzalez, M. A. (2015). Mediterranean diet and invasive breast cancer risk among women at high cardiovascular risk in the predimed trial a randomized clinical trial. *JAMA Internal Medicine*, *175*(11), 1752. Advance online publication. doi:10.1001/jamainternmed.2015.4838 PMID:26365989

Tong, T. Y. N., Imamura, F., Monsivais, P., Brage, S., Griffin, S. J., Wareham, N. J., & Forouhi, N. G. (2018). Dietary cost associated with adherence to the Mediterranean diet, and its variation by socio-economic factors in the UK Fenland Study. *British Journal of Nutrition*, *119*(6), 685–694. Advance online publication. doi:10.1017/S0007114517003993 PMID:29553031

Tsakiraki, M., Grammatikopoulou, M. G., Stylianou, C., & Tsigga, M. (2010). Nutrition transition and health status of Cretan women: Evidence from two generations. *Public Health Nutrition*, *14*(5), 793–800. Advance online publication. doi:10.1017/S1368980010003010 PMID:21272421

Tseng, M. L., Chiu, A. S. F., Tan, R. R., & Siriban-Manalang, A. B. (2013). Sustainable consumption and production for Asia: Sustainability through green design and practice. *Journal of Cleaner Production*, *40*(2), 1–5. doi:10.1016/j.jclepro.2012.07.015

Turati, F., Trichopoulos, D., Polesel, J., Bravi, F., Rossi, M., Talamini, R., Franceschi, S., Montella, M., Trichopoulou, A., La Vecchia, C., & Lagiou, P. (2014). Mediterranean diet and hepatocellular carcinoma. *Journal of Hepatology*, *60*(3), 606–611. Advance online publication. doi:10.1016/j.jhep.2013.10.034 PMID:24240052

UN. (n.d.). *The 17 Goals*. Retrieved September 26, 2021, from https://sdgs.un.org/goals

UNESCO. (2010). *Evaluation of nominations for inscription in 2010 on the representative list of the intangible cultural heritage of humanity*. UNESCO ICH.

UNESCO. (n.d.). *Redesigning lifestyles, consumption and production through Education for Sustainable Development.* Retrieved September 26, 2021, from https://en.unesco.org/news/redesigning-lifestyles-consumption-and-production-through-education-sustainable-development

Verberne, L., Bach-Faig, A., Buckland, G., & Serra-Majem, L. (2010). Association between the mediterranean diet and cancer risk: A review of observational studies. *Nutrition and Cancer*, *62*(7), 860–870. Advance online publication. doi:10.1080/01635581.2010.509834 PMID:20924961

Via, M. (2012). The Malnutrition of Obesity: Micronutrient Deficiencies That Promote Diabetes. *ISRN Endocrinology*, *2012*, 1–8. Advance online publication. doi:10.5402/2012/103472 PMID:22462011

Wang, C., Ghadimi, P., Lim, M. K., & Tseng, M. (2018). A literature review of sustainable consumption and production: A comparative analysis in developed and developing economies. *Journal of Cleaner Production*. . doi:10.1016/j.jclepro.2018.09.172

WHO. (2018). *Childhood Obesity Surveillance Initiative (COSI). Highlights 2015-17.* World Health Organization.

Widmer, R. J., Flammer, A. J., Lerman, L. O., & Lerman, A. (2015). The Mediterranean diet, its components, and cardiovascular disease. *The American Journal of Medicine*, *128*(3), 229–238. Advance online publication. doi:10.1016/j.amjmed.2014.10.014 PMID:25447615

Willett, W. C., Sacks, F., Trichopoulou, A., Drescher, G., Ferro-Luzzi, A., Helsing, E., & Trichopoulos, D. (1995). Mediterranean diet pyramid: A cultural model for healthy eating. *The American Journal of Clinical Nutrition*, *61*(6), 1402S–1406S. Advance online publication. doi:10.1093/ajcn/61.6.1402S PMID:7754995

Wolfson, J. A., & Bleich, S. N. (2015). Is cooking at home associated with better diet quality or weight-loss intention? *Public Health Nutrition*, *18*(8), 1397–1406. Advance online publication. doi:10.1017/S1368980014001943 PMID:25399031

ADDITIONAL READING

Garcia-Burgos, D., & Zamora, M. C. (2013). Facial affective reactions to bitter-tasting foods and body mass index in adults. *Appetite*, *71*, 178–186. doi:10.1016/j.appet.2013.08.013 PMID:23994505

Pasquet, P., Oberti, B., El Ati, J., & Hladik, C. M. (2002). Relationships between threshold-based PROP sensitivity and food preferences of Tunisians. *Appetite*, *39*(2), 167–173. Advance online publication. doi:10.1006/appe.2002.0503 PMID:12354685

Springmann, M., Clark, M., Mason-D'Croz, D., Wiebe, K., Bodirsky, B. L., Lassaletta, L., de Vries, W., Vermeulen, S. J., Herrero, M., Carlson, K. M., Jonell, M., Troell, M., DeClerck, F., Gordon, L. J., Zurayk, R., Scarborough, P., Rayner, M., Loken, B., Fanzo, J., ... Willett, W. (2018). Options for keeping the food system within environmental limits. *Nature*, *562*(7728), 519–525. Advance online publication. doi:10.103841586-018-0594-0 PMID:30305731

Veček, N. N., Mucalo, L., Dragun, R., Miličević, T., Pribisalić, A., Patarčić, I., Hayward, C., Polašek, O., & Kolčić, I. (2020). The association between salt taste perception, mediterranean diet and metabolic syndrome: A cross-sectional study. *Nutrients*, *12*(4), 1164. Advance online publication. doi:10.3390/nu12041164 PMID:32331287

KEY TERMS AND DEFINITIONS

COVID-19 Pandemic: It is an infectious disease caused by the SARS-CoV-2 virus which spread in an ongoing global pandemic.

Food Diseases: Diseases associated with food, including obesity, cardiovascular disease, diabetes, hypertension, and some types of cancer and also hunger, which affect more than 820 million people in the world.

Food Security: As defined by the United Nations' Committee on World Food Security, means that all people, at all times, have physical, social, and economic access to sufficient, safe, and nutritious food that meets their food preferences and dietary needs for an active and healthy life.

Healthy Food: A food that satisfactorily to affect beneficially body functions, beyond adequate nutritional effects, in a way which is relevant to either the state of well-being and health or the reduction of the risk of a disease.

Local Foods: Foods that have a specific geographical proximity provenance and must be market in shorts circuits, contributing for the sustainable development.

Mediterranean Diet: It can be described as a dietary pattern characterized by a) olive oil as the main source of dietary fat, b) high intake of plant foods (vegetables, fruits, whole cereals, potatoes, legumes, nuts and seeds), c) low to moderate amounts of animal foods (dairy products, fish, poultry, red meat and eggs), and d) wine in moderation within meals. Although this definition is broadly accepted, it is noteworthy that the MD has variants depending on the characteristics of each Mediterranean population, which actually makes it difficult to establish a unique universal definition.

Sustainable Production and Consumption: Is about doing more and better with less. It is also about decoupling economic growth from environmental degradation, increasing resource efficiency, and promoting sustainable lifestyles. They can also contribute substantially to poverty alleviation and the transition towards low-carbon and green economies.

Chapter 11
New Nutritional Perspectives in the Context of Chronic Disease Patient Management

Irina M. Matran
University of Medicine, Pharmacy, Science, and Technology, Targu Mures, Romania

Tuan Quoc Le
Hoa Sen University, Vietnam

Monica Tarcea
University of Medicine, Pharmacy, Science, and Technology, Targu Mures, Romania

ABSTRACT

As living standards change with the development of modern industry and social encounters, people tend to change their lifestyle and environment exposure along with their psychophysiological factors, leading to an imbalance of homeostasis and increasing the risk for chronic diseases. In addition to ingredients, methods, and food conditions storage and processing, the use of additives and certain new foods have facilitated the increased occurrence of chronic diseases in children or adults. The interaction of some components of the food system with enzymes that metabolize different types of drugs can affect the body's clearance and therapeutic index.The objective of this chapter was to present the general principles of food development for special nutritional conditions, also the adjuvants used for chronic disease status improvement, under the condition of nutritional nutrivigilence and food safety standards, and specific to introduce an adjuvant food for atopic dermatitis management.

INTRODUCTION

As living standards change along with the development of modern industry *and social encounters*, people are tending to change their lifestyle and psycho-physiological factors exposure, leading to an imbalance of homeostasis and increasing the chronic diseases frequency. In addition, as people's living standards have improved and people are increasingly busy and interested by modern industrial lifestyles, this has

DOI: 10.4018/978-1-7998-9557-2.ch011

led to a shift in the food industry following the requirements to create safe, nutritious and functional foods (for example, ready-to-eat or processed foods with stronger sensory properties and long shelf life), at competitive prices. Today's foods do not succed only sensory aspects but the trend of food supplements with nutritional and functional values is constantly evolving. However, not all companies in the food market sector have a team of multidisciplinary research and development specialists, which is a combination of nutritionists, physicians, pharmacists, economists, and food production professionals. Many companies in the food industry are limited in terms of information, measurable objectives, or qualitative verification for the nutritional composition of manufactured products. Due to these shortcomings, but also to the existing legislative regulations in each country, the food market for special nutritional conditions, preclinically or clinically verified, is very small.

In addition to ingredients, food processing methods and storage conditions, the use of additives have facilitated the chronic diseases high frequency. The interaction of some components of the food system with enzymes that metabolize different types of drugs, can affect the body clearance and therapeutic index (e.g. in vitro it has been shown that aloe vera juice and many other plant sources inhibit the enzyme CYP3A4 and may affect the metabolism of dermatological drugs metabolized by it – like terbinafinum or itraconazole). Also, the interaction between certain foods consumed at the same time can cause important side affects (like grapefruit juice which increases the effect of citritin and can induce various side effects of different severity depending on individual sensitivity).

In the production of food products for special medical purposes, the quality control (from physical, chemical, microbiological and toxicological point of view) it is necessary in order to prevent the use of fraudulent ingredients or toxic contaminants, and brings a proper quality of life to consumers with chronic diseases.

According to the WHO classification, the most frequent chronic diseases are the following: cardiovascular diseases, diabetes, mental and behavioral disorders, obesity and other disorders due to excessive food consumption, neoplasms and chronic digestive disorders. Other chronic diseases are: respiratory diseases, diseases of the osteoarticular system, muscles and connective tissue. We can add chronic kidney disease, chronic hematological disorders, unspecified nutritional anemia (simple chronic anemia), or dermatological diseases such as dermatitis (atopic, contact), eczema and psoriasis (inflammatory diseases with an increasing prevalence).

According to the Regulation of the European Union no. 128 of 2016, the formula of foods intended for special medical purposes must start from the correct medical and nutritional principles. Foods evaluated and developed for any chronic disease, for research – development, production or marketing, must be undertaken by an interdisciplinary team. This team must have reached the level of knowledge required for medicine and pharmacology (pathogenicity, complications, treatment, food-drug interaction), also for food production and protection of industrial property (prescription patenting). After establishing the new food product formulation and validating the technological data in terms of food safety and hygiene, in accordance with HACCP and nutrivigilence standards, the next important step is to verify in vivo efficacy by establishing the design of the preclinical or clinical study and statistical methods to verify the accuracy of the claimed results.

Because the number of chronic diseases is large, to understand the summary of the presentation of the principles of research - development and manufacture of organic food for special nutritional conditions, wie took atopic dermatitis as an example in our chapter.

Although food is known to influence people's well-being and health, this chapter will reveal the importance of interdisciplinary team effort that tackles the development of new organic food products for special medical purposes, aimed to increase the quality of life of patients with chronic diseases and is a useful tool for research directions in the food industry companies, other research centers, and for educational purposes as well. Both students and professionals readers will be introduced to food development, nutrition, food safety, nutrient monitoring and analysis of food components and the relevant documentation process for preclinical and clinical studies, in order to verify the efficacy of newly developed foods and the interactions of foods and components with drug-metabolizing enzymes (essential in the respective after-sales phase).

At the end, readers of this chapter will be familiar with the qualitative composition of an adjuvant food in the treatment of atopic dermatitis from the perspectives presented above.

Organising and attending the interdisciplinary scientific events which further involve decision-makers in food law and practice, nutrition and nutrivigilence or pharmacovigilence, and by the mass-media, will lead to better information on proper diet for chronic morbidity, significantly increasing the quality of life of consumers and lowering the costs of health systems as well.

BACKGROUND

Regardless of the pathogenesis, the effectiveness of drugs used in chronic diseases depends on several factors such as the type of chronic disease (inflammatory or systemical), severity (mild, moderate, severe), existing comorbidities, and related psychiatric conditions (depression or anxiety) (Whan et al., 2017). The absorption of nutrients from food is necessary for physiological processes in our body, also to produce the effectiveness of drugs for the treatment or improvement of chronic diseases status. Components from diet can induce the risk of contamination with certain toxic substances (heavy metals or free radicals obtained from different food sources), by intensive methods using pesticides, insecticides and chemical fertilizers (nitrogen, phosphorus, calcium or organic components), based on the way of preparing food (heat treatment applied during the preparation or reheating), or by the way the food is consumed or perceived (Matran, Dumitrascu, 2018).

Due to the soil pollution and water environment, plant sources (grains, vegetables and fruits) or fish (Hu et al., 2017; Chanpiwat et al.,2016; Lars, 2003), can accumulate heavy metals which can cause metabolic degradation, attention deficit, autism, neurotoxic effects (Rather et al., 2013; Heusinkveld and Westerink, 2017), myocardial damage (Maheshwari and Chaudhary, 2017), cancer (Kigen et al., 2017; Amizadeh et al, 2017; Abdollahi et al., 2004; Lioi et al., 1998; Banerjee et al., 2001; Medda et al., 2021), brain tumor (Fallahi et al., 2017), syndrome metabolism (Rosenbauma et al., 2017). Arsenic (As2O3) accumulation in the body stimulates reactive oxygen species (ROS), causing mitochondrial damage and DNA modification, by stimulating the GAD D45 and p-JNK1/2 genes and stimulating Activate nuclear factor NF-kB, leading to cell proliferation and tumor invasion (Medda et al., 2021).

In recent years, product manufacturers had to come up with various vegetable products to bring both quality and convenience to consumers due to the changed lifestyle. Pre- and post-harvest handling along with food processing is one of the important factors for obtaining foods of low or high biological and nutritional value, depending on the processing conditions and parameters applied. For example, the microwave food heat treatment of salmon fillets will reduce the lipid content, antioxidant activity of coconut water, and increase the breakdown of phytosterols, leading to an impact on food quality and safety. (2, 17, 18, 19, 20) Processed fruits and vegetables may result in lower nutritional value than fresh varieties (Dewanto et al., 2002; Al-Juhaimi et al., 2018). Due to spoilage and quality deterioration factors such as microbial growth, enzymatic degradation and chemical changes that occur after harvest, various technologies are applied to food processing and eliminate the toxicity and loss of food quality such as thermal processing with high and low temperature, chemical treatment, and non-thermal treatment such as high hydrostatic pressure and microwave. Heat treatment may be detrimental to bioactive compounds when compared to non-heat processes, which may not significantly degrade important phytochemicals with health benefits. Heat treatment also improves the digestibility and availability of sulfur amino acids of vegetable proteins (James R. Kirk, 1984); however, structure and vitamins are lost, especially vitamin C (Dewanto et al., 2002).

Regarding meat, a variety of substances of safety concern can be created in meat products depending on the form of processing (Flores et al., 2019). Using nitrite as a preservative for smoked products can produce nitrosamines. Heterocyclic aromatic amines (HAAs) are formed in the crusts of cooked meat products. Most HAAs were carcinogenic in long-term animal studies. Besides precursors in raw materials, important factors are temperature and processed time (Gibis et al. 2016). Free radicals released from lipid and protein oxidation, and/or reducing sugars generated from the Maillard reaction may be present in meat products depending on the presence of additional compounds and conditions handling and treatment conditions such as temperature, time, water activity and pH. Spices may have an impact on the mechanism of formation of these compounds due to their ability to neutralize free radicals produced as intermediates in these reactions (Flores et al., 2019). Packaging conditions (MAP, vacuum) and then storage conditions such as temperature and time are related to lipid or protein oxidation (Lund et al., 2007). Biochemical and nutritional compounds change not only during food processing but also during storage, especially under inappropriate conditions. Bioaccessibility, bioavailability, and overall nutritional value may be unexpectedly degraded during processing and subsequent storage, negatively affecting the nutritional value of foods and its physiological effects (Orlien, V., & Bolumar, T. 2019).

A stressful lifestyle, demographic factors and influence of the media, through advertisements for food and beverages, are increasing hedonism for their consumption, which can change internal homeostasis and lead to chronic diseases in time. Thus, high coffee consumption results in decreased serum vitamin D levels (Lim et al., 2017), which can lead to osteomacy (decreased density with pseudofactures of the spine, tibia, muscle weakness) and osteoporosis (decreased bone density induced by age by decreasing levels of estrogen). Other chronic diseases that can be caused by the above mentioned factors are obesity (Aryeetey et al., 2017; Marabelli et al, 2017; Puia and Leucuta, 2017), increased cholesterol and gastroesophageal reflux (Alkhathami et al, 2017), diabetes and cardiovascular risks (Lai et al, 2017, Oh et al., 2017), dementia (Chaaya et al., 2018), irritable bowel syndrome (Adeniyi et. al, 2017), alcoholism (Paulsson-Do et al., 2017), anxiety (Liu et al, 2020), psoriasis, dermatitis and eczema (Rousset and Halioua, 2018; Petersm et al., 2013). Stress in the prenatal and postnatal period, affects the formation of neural networks causing changes in emotional and cognitive function on the hypothalamic-pituitary-adrenal (HPA) axis (Nishi, 2020). Due to stress factors, through the paraventricular nucleus (PVN), the hypothalamus releases corticopin (CRH) and arginine vasopressin (AVP) and activates the anterior pituitary gland, which induces adenocorticotropic hormone (ACTH). It activates the adrenal cortex by releasing cortisol, which is the end product of the HPA axis (Nishi, 2020). In the long run, stressors, through the hypersecretion of ACTH, cause endocrine diseases, such as Cushing's disease (Pivonello et al., 2020). Unlike Cushing's disease which is produced on the HPA axis, atopic dermatitis (AD), which is another chronic (inflammatory) disease, is caused by HPA axis dysfunction and the synaptic axis (SA) (Petersm et. al.,2020), stress being the main cause. The imbalance of the two axes (HPA and SA) stimulates the synthesis of pro-inflammatory cytokines - interleukins 4 and 5 (IL-4, IL-5) (Petersm et.

al.,2020), which lead to humoral and eosonophilic inflammation, skin damage and cellular inflammation. adaptive - tumor factor alpha (TNF-α) and interferon gamma (IFN-γ), acetylcholine release (ACh) and associated receptor activation (AChR), with a role in the imbalance of the two axes). A new mechanism discovered in the production of AD is the cholinergic system in non-neuronal cells (NNCS).

Except stress, the causes of AD also include genetic (Bogusław Nedoszytko et al., 2020) and dietary factors (allergies to histamine-rich foods like eggs, dairy, fish, nuts, peanuts, and others to food additives like butylhydroxytoluene, p-Hydroxy of methyl (E 218), n-propyl p-hydroxybenzoate - (E216)). Food additives or highly processed ingredients used as food additives are compounds used to improve the palatability, texture, and shelf life of foods. However, these compounds have been found to be associated with cells of the immune system. In obese individuals, cells of the immune system contribute to both the maintenance of "skinny homeostasis" and dysregulation of metabolism. Food additives can also be important contributors to metabolism.

Food additives have related effects on cells of the immune system, which may contribute to immune-mediated metabolic dysregulation (Paula Neto et al., 2017). The widespread and frequent use of highly processed foods or emulsifiers may contribute to the increased societal prevalence of obesity/metabolic syndrome and other chronic inflammatory diseases (Chassaing et al., 2016)). These agents disrupt the interaction between mucus and bacteria, damage intestinal cells, and potentially promote diseases associated with inflammatory bowel disease (Chassaing et al., 2016; Aaron Lerner and Torsten Matthias, 2016). In a mouse model of diabetes, a high-fiber diet with resistant starch maintains intestinal barrier integrity and reduces the severity of kidney injury through complement inhibition (Snelson et al. 2021). These results demonstrate the mechanism by which processed, additive-rich foods can induce inflammation that leads to chronic disease. In the present life, due to the change in lifestyle has led to a marked increase in the consumption of highly processed foods. Overexposure to food additives, (eg, colours, emulsifiers, sweeteners), the cumulative effects of food chemicals may contribute to an increase in chronic diseases. However, there is currently a lack of data to reach this conclusion as safety assessments have mainly been performed on individual additives rather than additive complexes and most have only studied in-vitro/-vivo and exposure simulation without studying any potential effects of multiple additives at once on human health.

With lax regulations in the inspection of preservatives used in production, import and export of raw materials can also lead to the existence of many toxic compounds that cause many chronic diseases and dangerous diseases for humans. European countries are facing multiple recalls related to ethylene oxide after it was detected in food additives (Joe Whitworth, 2021). Ethylene oxide is used in many products such as sesame seeds, cereals, chocolates, cookies, bread, crackers, condiments, cakes and noodles. It is used as a preservative in many countries to reduce or eliminate microbial contamination (Salmonella). Ethylene oxide has also recently been found in locust bean gum additives, mainly as thickeners or stabilizers. It is used in foods including ice cream, breakfast cereals, meat products, confectionery, fermented dairy products, and cheeses. European Commission experts say there is no safe level of consumer exposure to products containing the additive known to be contaminated with ethylene oxide, and to any extent, any people may face potential risks in the case of cancer. Import recalls related to this compound are ongoing for thousands of conventional and organic items with long shelf lives.

AD is a recurrent disease, the pharmacological treatments recommended being tablets, injectable solutions or topics for application. Most pharmacological treatments for AD's treatment contain corticosteroids, that can have side effects, such as: contact dermatitis, allergic reactions, skin irritation and others. Many dermatological treatments with corticosteroids have multiple contraindications such as tuberculous or syphilitic lesions, primary infections, bacterial, rosacea or vulgar acne, perioral dermatitis or atrophic skin conditions. To this is added the decrease in work efficiency and daily activities, as well as the increase in health costs. Both the quality of life of patients with atopic dermatitis and their family members is negatively affected.

MAIN FOCUS OF THE CHAPTER

Issues, Controversies, Problems

To reduce the symptoms of AD and improve the quality of life, many patients resort to alternative adjuvant treatments, such as the administration of food supplements conditioned in various forms (capsules, tablets or tea). Whether they are recommended by healthcare professionals, or purchased online or in stores, a real danger are the side effects they may cause or their interaction with the enzymes involved in the metabolism of certain drugs, including those recommended for oral treatment AD. Thus, the plant resources that may cause skin adverse reactions are: *Hypericum perforatum* (Asher et al., 2017) (*Pemphigus foliaceus* (crust-covered ulcers) (Imbernón-Moya et al., 2016), *Agaricus blazei Murrill* (swelling of the lips and cheilitis).

Asimina triloba (allergy), D-mandelonitrile-bD-glucosido-6-bD-glucoside (amygdalin) (dermatitis, inflammation and redness of the skin, bluish discoloration of the skin (at high doses/ toxicity), *Andrographis paniculata* (allergic reactions, hypersensitivity, rash), *Sanguinaria canadensis* (skin irritation, burns), *Boswellia serrata* (allergic skin reactions), *Arctium lappa* (allergic reactions, contact dermatitis), *Matricaria recutita* (contact dermatitis), *Larrea tridentate* (rash and yellowing of the skin), *Chrysanthemum morifolium* (redness, swelling and itching of the skin), *Vitex agnus castus* (acne, itching, redness/ rash), *Salvia hispanica* (eczema and lesions, itching), *Ta raxacum mongolicum* (contact dermatitis) (38), *Echinacea purpurea* (dermatitis), *Tanacetum parthenium* (contact dermatitis). Redness, swelling and itching of the skin may cause *Zingiber officinale*, *Commiphora molmol* and *Turmeric longa*. Contact dermatitis can also cause the following plant resources: *Crataegus monogyna*, *Lavandula angustifolia*, *Centella asiatica*, *Azadirachta indica* and *Nigella sativa*. Other plant resources that can cause skin side effects are: *Mentha piperita* L, *Staurocucumis liouvillei*, *Glycine max*, *Stillingia sylvatica* (About Herbs, Botanicals & Other Products).

The interaction of plant resources with enzymes that metabolize dermatological drugs, influence their clearance (intracellular concentration), and is extremely important in the effectiveness of treatments, as well as in improving of AD patient's quality of life. In detail:

- Terbinafine (Davis et al., 2019) is metabolized by the enzyme CYP3A4. The clearance (intracellular concentration) of this medicine may be influenced by food/ food supplements containing *Agaricus blazei Murrill*, *Aloe vera*, *Artemisia annua*, *Andrographis paniculata*, *Matricaria recutita*, *Chrysanthemum morifolium*, *Vitex agnus castus*, *Taraxacum parthen*, *Zingiber officinale*, *Allium sativum*, *Hydrastis Canadensis*, *Centella asiatica*, *Commiphora mukul*, *Crataegus monogyna*, *Cymbopogon citratus*, *Artemisia annua*, *Urtica dioica*, *Mentha piperita L.*, *Monascus purpureus* (Red Yeast Rice), *Trifolium pratense*, 3,5,4 ‘ -trihydroxystilbene (resveratrol), *Rhodiola rosea*, *Serenoa repens*, *Glycine max* (Soy) and *Turmeric longa* (About Herbs, Botanicals & Other Products).

- CYP2C9 enzyme metabolizes dermatological drugs like voriconazole, and can be inhibited by food supplements/ other products containing *Andrographis paniculata, Matricaria recutita, Harpagophytum procumbens, Tanacetum parthenium, Allium sativum, Centella asiatica, Cymbopogon citratus, Uremolia, Lemolia Mentha piperita L., Trifolium pratense*, 3,5,4'-trihydroxystilbene (resveratrol), *Serenoa repens, Eleutherococcus senticosus* (About Herbs, Botanicals & Other Products).

Aloe vera - also inhibits the enzyme CYP2D6, *Andrographis paniculata* - also metabolizes drugs metabolized by the cytochrome P450 isoenzyme - CYP1A1, *Matricaria recutita* - also inhibits the enzymes CYP1A2 and CYP2D6, *Harpagophytum procumbens* - inhibits the enzymes C2P2, CYP2, CYP2. *Echinacea purpurea* - also inhibits CYP2CB, *Centella asiatica* - also inhibits the enzymes CYP1A2 and CYP2D6; *Magnolia officinalis* - inhibits the enzymes CYP1A2, CYP2C8, *Trifolium pratense* - also inhibits the enzymes CYP1A2, CYP2C8, CYP2D6, 3,5,4'-trihydroxystilbene (resveratrol) - inhibits the enzymes CYP2D1 and induces CYP2D6 and *Rhodiola rosea* inhibits enzyme CYP2C9 and may affect drugs metabolised by this enzyme, especially those with a limited therapeutic index, such as phenytoin and warfarin. *Serenoa repens* - also influences drug clearance metabolized by CYP2D6, inhibits UGT enzymes (Uridine 5'-diphospho-glucuronosyltransferase), may increase the effect of nonsteroidal anti-inflammatory drugs and may influence anticoagulant drugs (clopidogrel), in vitro, and in vitro, CYP1A2 and anticoagulant drugs. *Staurocucumis liouvillei*/ Mensamaria intercedens Lampert - may potentiate the effect of anticoagulant drugs, *Turmeric longa* (Turmeric) - also inhibits the enzymes CYP1A2, CYP2A6, CYP2D6. *Hypericum perforatum* also reduces the effectiveness of drugs such as protease inhibitors, irinotecan, theophylline digoxin, venlafaxine and oral contraceptives (Asher et al, 2017). In addition to the CYP3A4 enzyme, terbinafine is also metabolised by the CYP2C19 enzyme (Davis et al., 2017).

One of the major risks is the interaction between drugs and the plant/animal sources from food supplements, nutraceuticals, nutricosmetics or food, represented by over-the-counter (OTC) medicines. Given the side effects (including dermatological nature) and interactions with many drugs, the need to legislate nutrivigilence and its implementation by health professionals (doctors, pharmacists) and specialists in the food industry and food supplements, is extremely important for consumer's safety and to reduce the costs of worlwide health systems.

Adverse/Side effects (SE) to vitamin food supplements have been observed and reported: doses greater than 500 mg/day pyridoxine (vitamin B6) - photosensitivity, neurotoxicity (Ronis et al., 2018), chronic sensory polyneuropathy associated with pyridoxine (Ronis et al., 2018), bleeding associated with action antiplatelet therapy at doses of 800 - 1200 mg/ day (Ronis et al., 2018), diarrhea, weakness, blurred vision and gonadal dysfunction at doses higher than 1200 mg/ day α-tocopherol (vitamin E) (Ronis et al., 2018) and increased recurrence of early cancer 3.5 years (Ronis et al., 2018). Retinol Efficacy Trial and ATBC studies have demonstrated the toxicity of β-carotene (vitamin A) supplements by significantly increasing the risk of lung cancer in male smokers (Ronis et al., 2018). Following the study "Teratogenicity of high vitamin A intake", researchers Rothman KJ, Moore LL, Singer MR et al. showed the association between dietary supplements containing vitamin A with adverse effects on bone health and decreased bone density, thus increasing the risk of fracture (Ronis et al., 2018). Adverse effects have also been found in the case of dietary supplements with minerals, omega 3/ fish oil, soy protein, soy protein, plant nutrients, antioxidants, anti-inflammatory, weight loss or bodybuilding supplements, or various botanical supplements (Ronis et al., 2018).

With the exception of "classic" foods (peanuts, nuts, eggs, etc.) known to cause allergies, the development of the food industry has led to the specific foods consumed especially by teenagers, such as energy drinks. Frequent consumption of this type of drink (≥7 times a week) was significantly associated with asthma, allergic rhinitis and atopic dermatitis (Wee et al., 2020), high stress, lack of sleep, poor performance at school, and even suicide attempts in Korean adolescents (Kim et al., 2017) .

Also with regard to possible adverse effects, special attention should be paid to novel food ingredients monitoring, authorized by the European Food Safety Authority (EFSA). Following the Scientific Opinions on certain plant resources authorized by EFSA, or information from the Food and Drug Administration (FDA), the following adverse reactions, including chronic diseases, to the following plant resources or substances authorized for addition to food have been identified:

- (6S)-5-methyltetrahydrofolic acid, glucosamine salt, as a source of folate (side effects reported for patients taking 15 mg folic acid for 1 month: gastrointestinal - nausea, abdominal distension and flatulence; nervous system - bitter taste, changes in sleep, difficulty concentrating, hyperactivity, impaired thinking, metabolic - anorexia, others - decreased serum diphenylhydantoin in patients taking 5 or 15 mg of folic acid daily (European Commission, 2015; FDA., 2020);
- oil extracted from microalgae *Schizochytrium sp.* (ATCC PTA-9695): belching, slight increase in cholesterol; pregnant women – dose > 3g/day slowing blood clotting and increasing the chances of bleeding; respiratory sensitivity, in the case of people sensitive to aspirin, diabetes: increased blood sugar before meals in people with type 2 diabetes, blood pressure - low blood pressure, respiratory infection and gastroesophageal reflux (European Commission, 2015; FDA., 2018);
- xylo-oligosaccharides: the following side effects have been reported - transient gastrointestinal disorders: diarrhea. Allergy has been reported as a possible adverse reaction (European Commission, 2018; Turck et al., 2018).

With the exception of (6S) -5-methyltetrahydrofolic acid, the glucosamine salt, as a source of folate, the oil extracted from the microalga *Schizochytrium sp.* (ATCC PTA-9695) and xyloligosaccharides are also authorized for addition to food, as follows:

- oil extracted from microalgae *Schizochytrium sp.* (ATCC PTA-9695): may be added to dairy products, excluding milk-based beverages - maximum 200 mg/100 g or, for cheeses, maximum 600 mg/100 g, dairy products, excluding beverages - maximum 200 mg/100 g or, for cheese-like products, maximum 600 mg/100 g, spreadable fat and salad dressings - maximum 600 mg/100 g, bakery products (bread and rolls), sweet biscuits - maximum 200 mg/100 g, cereal bars - maximum 500 mg/ 100 g, cooking fats - maximum 360 mg/100 g, non-alcoholic beverages (including similar dairy products and milk-based drinks) - maximum 80 mg/100 ml, infant formulas and preparations young children - used in accordance with Directive 2006/141/EC, cereal-based preparations and baby foods for infants and young children, including those used in accordance with Directive 2006/125/EC - maximum 200 mg / 100 g (European Commission, 2015);
- xylo-oligosaccharides are also accepted in white bread - maximum 14 g/kg, wholemeal bread - maximum 14 g/kg, biscuits - maximum 14 g kg, yogurt - maximum 3.5 g/kg (when used in products dairy products, xyloligosaccharides may not replace, in whole or in part, any constituent of milk), fruit spreads - maximum 30 g/kg, soy drink - maximum 3.5 g/kg (TEuropean Commission, 2018);

Constipation is another example of a chronic disease in which phytotherapeutic monotherapy has adverse reactions induce by: *Peganum harmala* (intoxication with severe vomiting, neuropsychiatric symptoms, and in greater quantities even death) (Sadr et al., 2016), *Gymnema sylvestre* (hypoglycemia, weakness, tremors, sweating, muscular dystrophy) (Tiwari et al, 2014), or kiwi fruit (allergies) (Chang et al.,2010).

SOLUTIONS AND RECOMMENDATIONS

Analyzing the plant resources/ substances used in food, and food supplements which may cause adverse reactions or adversely affect the treatment of the aforementioned chronic diseases, we notice that it is necessary to introduce nutrivigilence, both in health and food industry.

Cytochrome P450 enzymes convert certain plant components into toxic or mutagenic substances and reactive, carcinogenic metabolites by 1'-hydroxylation to allyl side chains, and bioactivation of alkylbenzenes, such as CYP1A1 and CY enzymes. CYP3A4 contributed to the metabolic activation of elemicin, by its high catalytic capacity, and the formation of 1'-hydroxyelemicine (Wang et al., 2019). Thus, excessive intake of foods and spices containing elemicin can lead to cellular toxicity. This is an argument for which the authors of this chapter recommend legislation in the food industry and food supplements, dietetics and nutrition, also the pharmaceutical industry and medicine. Another argument is the potential adverse reactions caused by foods containing plant resources or novel food ingredients.

The proposed solution for improving the chronic diseases consumer's quality of life is research development and manufacture of organic foods for special nutritional conditions, in conditions of nutritional vigilence and food safety. We propose these foods as adjuvants in the relief of chronic diseases. This means that food for the atopic dermatitis treatment does not contain preservatives, because the European Regulation No. 848 of 2018 on organic production and labeling of organic products, prohibits the introduction of preservatives. An organic food is proposed, because preservatives and stress can have a synergistic action in causing atopic dermatitis.

We will present next the related steps and a food proposal for the special nutritional status for chronic disease - atopic dermatitis (subject to industrial property), the algorithm for identifying potential adverse reactions and its analysis in terms of food safety and criteria for obtaining industrial protection (obtaining the patent). The proposed new food is in line with protecting the environment, reducing resource consumption and reducing the carbon footprint.

Stages of research - development and manufacture of new foods proposed:

physiology of stress (or other provocative factor) → etiopathology of chronic diseases (eg atopic dermatitis) → pharmacological treatments (mechanisms, side effects, contraindications, interaction with other drugs or foods) and plant or animal resources (ditto "pharmacological treatments) (nutrition surveillance) → Hazard analysis and critical control points (HACCP) → zero (pilot) batch manufacturing → preclinical verification of the efficacy of the new food → verification prospective food clinic → protection of industrial property → approval as food for special nutritional status by national authorities (or EFSA, in case of processors from European Union).

The qualitative recipe of the adjuvant food in the relief of atopic dermatitis, proposed by the authors of this chapter, is: bovine lactoferrin, coconut sugar, pectin, tryptophan, powdered fruit flavor, water-soluble -optional, and anti-caking agent phosphate (E341) (Nurhadi et al., 2020).

The physiology of stress, the etiopathology of atopic dermatitis, pharmacological treatments and herbal resources were presented in the content of the chapter.

The following are the side effects, contraindications and drug interactions of the ingredients of the new food presented as an example of the proposed solution. This information will be the basis for the application of the Naranjo algorithm (Kazuki et al., 2015), in case of adverse reactions in the after-sales period.

Table 1. Naranjo algorithm (Kazuki et al., 2015)

No.	The question	Yes*	No*	I do not know*
1.	Is there a notification about the reaction, on the label or in the package leaflet food supplement?			
2.	Did the side effect occur after taking the food supplement?			
3.	Did the adverse reaction improve when the suspected food supplement was discontinued?			
4.	Did the adverse event recur when the food supplement was given?			
5.	Are there any causes other than the dietary supplement that could have caused side effects?			
6.	Was the reaction more severe when the dose was increased or less severe when the dose was decreased?			
7.	Did the consumer have a similar adverse reaction to the similar food supplement in the previous exposure?			
8.	Has the adverse event been confirmed by any objective evidence?			

* Scores applied: ≥ 9 most likely, 5 - 8 likely, 3 - 4 very likely, 1 - 2 possible, ≤ 0 unlikely.

No adverse reactions and drug interactions have been reported for coconut sugar; consumption in very large quantities is contraindicated.

In the case of pectin: no side effects have been reported, but there are warnings about the interaction with medicines: with tetracycline antibiotics, as it may change their clearance and therapeutic index. For this reason, the novel food should be consumed two hours before the administration of antibiotics with tetracycline, or after 4 hours after administration. Other drugs whose clearance may be altered are digoxin and lovastatin (WebMD, 2020).

Lactoferrin is considered safe, with no side effects reported (Cutone et.al, 2020).

Regarding Tryptophan, special attention should be paid to the amount and preclinical and clinical monitoring of the following adverse reactions: serotonin or gastrointestinal syndrome, dose-dependent vomiting, nausea (58). Tryptophan precautions are for patients diagnosed with kidney and liver disease, and for pregnant women. The possible interaction with psychiatric drugs must also be considered (Rx-List, 2020).

For tricalcium phosphate, the risk for chronic insufficiency, vascular calcification and left ventricular hypertrophy has been demonstrated. The problem is the accumulated amount of phosphate in various foods (meat, frozen pastries) (Stenvinkel, 2014).

The benefits of the proposed ingredients are as follows: coconut sugar - low glycemic index - 35 (as opposed to palm sugar which has a glycemic index of 42 and cane sugar with a glycemic index of 58 to 82) (Asghar et al., 2019), pectin: lowers/ prevents the risk of inflammatory bowel disease (IBD) by prebiotic effect (increasing the concentration of propionic acid) and anti-inflammatory (decreasing IL-1β and IL-6) in host intestinal cells, in a microbiota-independent manner (Ishisono et al., 2019). Lactoferrin prevents and inhibits cancer by stimulating the adaptive immune response, has high oral availability and very good tolerability.

Preclinical research (Xu et al., 2019) has shown that lactoferrin improves dopaminergic neurodegeneration and motor deficits and neuroprotective effects through its anti-inflammatory and antioxidant properties (van de Looij et al., 2019). Tryptophan or L-5-hydroxytryptophan (5-HTP) by decarboxylating serotonin (5-hydroxytryptamine, 5-HT) which is converted to melatonin (N-acetyl-5-methoxytryptamine), being effective in the treatment of endogenous depression, unipolar depression and bipolar disorder, treatment of anxiety and panic syndrome, improvement of sleep quality, treatment of migraine and ataxia, relief of symptoms in Alzheimer's disease and antidepressant symptoms in patients with Parkinson's. Tryptophan also has antioxidant, anti-inflammatory effects (decreases lipopolysaccharide (LPO) activity and IL-6 synthesis, activating protein kinase (ERK) and cyclooxygenation - 2 (COX-2) and analgesics (Maffei et al., 2020). Tricalcium phosphate: phosphorus is part of DNA, RNA and ATP, have important tasks for cell signaling, nucleic acid synthesis and energy production. Phosphate in the form of phosphate also acts as an important buffer and is needed to maintain acid-base balance and to form bones. and teeth (Stenvinkel et al., 2014).

The stages of the technological flow for obtaining the new food for the treatment of chronic disease are the following: evaluation of possible suppliers for the necessary ingredients and packaging (including from the existing procedures for identifying ingredient fraud) → ordering and supplying ingredients and packaging → quantitative and qualitative reception of ingredients → their storage (depending on the indications on the labels or in the related documents (eg technical specification, working and hygiene instructions) → dosing of the ingredients (according to the working order (manufacturing recipe in relation to the quantity to be manufactured), the instruction and hygiene) → homogenization of ingredients → qualitative verification of the manufactured product (according to the Technical Data Sheet or the Company Standard) → packaging of the finished product (according to the working order) → closing of the packaging → labeling of the packaging (if they are not marked by to the supplier) → marking of the lot or batch (for traceability) → collective packaging (according to the work order or sales orders received from customers) → final storage (depending on the conditions mentioned on the label, or in the Work Order, in compliance Working and hygiene instructions) → dispatch of products (preparation of products for transport, depending on sales orders received from customers and work and hygiene instructions) → delivery of products to customers (delivery of products and accompanying documents, specific to transport).

Given the nutrivigilence analysis performed on adverse reactions, contraindications and drug interactions, following the HACCP analysis, two critical control points were identified.

The first critical control point is the qualitative reception of the ingredients (chemically), with an emphasis on preventing their fraud, or the presence of impurities (eg for tryptophan - 4,5 tryptophandione, L-tryptophan, 3-anilinoalamine) (Maffei, 2020).

The second critical control point is the quantitative dosing of tryptophan to prevent serotonin syndrome (Maffei, 2020).

The evaluation of the preclinical efficacy of the new adjuvant food in managing atopic dermatitis is based on the Opinion of the Ethics Commission of the University where this study was conducted. The preclinical study can be performed on rodents (mice or rats), whose pathology is induced. The results must be interpreted statistically.

Evaluation of clinical efficacy can be performed prospectively double-blind randomized, after prior approval of the Ethics Commission Opinion and Patient Consent. As with preclinical evaluation, results should be interpreted statistically.

Industrial protection or obtaining the Patent is made in accordance with the legislation of each country. In order to obtain the Patent, the product for which protection is requested must meet the following criteria: it is new, has industrial applicability and has inventive step. The fulfillment of the first criterion can be achieved using the World Intellectual Property Organization (WIPO) database. The second criterion can be met by presenting the flow chart and describing its stages, and the third criterion can be met by presenting the laboratory activity and in vivo / in vitro analyzes / clinical trials. An essential role is played by the presentation of the invention.

We proposed organic food for the treatment of atopic dermatitis, because worldwide there are no validated foods (finished products) to treat atopic dermatitis at this moment. The global databases on industrial property (patents) - WIPO (World Intellectual Property) and PubMed were checked. The following keywords were used: "atopic anti-dermatitis food", "anti-dermatitis food" and "anti-eczema food". Patents resulting from search using this term consist of cosmetic compositions that can also be added to food, food supplements, cosmetics or drugs, such as plant or fruit extracts. In both WIPO and PubMed platforms used, there were no foods as a finished product, ready for consumption, for the amelioration/ treatment of atopic dermatitis. The patented products have as general mechanism the inhibition of histamine release, but do not treat the causes of atopic dermatitis, such as stress.

The approval of food for special nutritional status is done in accordance with the legislation specific to each country. At the level of the European Union, the approval of these foods is carried out by EFSA, based on Regulation (EU) no. 128/2016 - supplementing the Regulation (EU) no. 609/2013 on the specific requirements on the composition and information applicable to foods intended for special medical purposes, amended and supplemented.

In addition to the benefits of the ingredients presented above, new adjuvant food for the relief of atopic dermatitis, has the following advantages: the product is 100% natural, can be manufactured and certified organic, Halal and Kosher, does not contain synthetic substances, can be consumed by older children older than 12 years and pregnant women do not require special storage and transport conditions. Being a powdery product, by the way of packaging, the transported quantity can be large, conditions which increases the economic efficiency and decreases the carbon footprint. It also does not require heat treatment (pasteurization) and as a result, utility consumption (water, energy, gas) is low. Due to its composition, the new food has another advantage that can make it attractive to patients with atopic dermatitis, also regardless of age: it can be eaten hydrated, or in the form of jellies, for which only citrus juice should be added to the already hydrated mixture.

FUTURE RESEARCH DIRECTIONS

Although sales of food supplements or drugs in the form of tablets or capsules have increased, there is a growing interest in foods for special nutritional status. Due to their sensory characteristics, patients prefer these foods. It is important that the universities and schools with programs for the food industry, to complete the curriculum with disciplines such as clinical nutrition, general pharmacology and nutrivigilence, for the training of engineers with the skills necessary for the development of this type of food products. It is also essential that food industry processors work with healthcare professionals, and form their own multidisciplinary team.

The international trend is the development of food from plant resources or cell cultures, and the minimization of the zootechnical field and the processing of food of animal origin. We believe that the complex analysis of the benefits of strictly plant-based foods from the perspective of intensive agriculture versus organic farming is appropriate, due to population growth and climate change.

Marine resources are another possible resource for ingredients for the food and pharmaceutical industries. This resources safety and possible drug interactions, precautions and side effects also, must be well documented.

As future research directions, the authors of this chapter propose that in case of certain chronic diseases (eg atopic dermatitis or psoriasis), the conduct of preclinical research has to verify the effectiveness of new adjuvant foods for disease relief, to be performed starting with the preconception period followed by pregnancy and lactation (postnatal). In addition to visual analysis, the authors also propose immunological, histological, or biochemical analyses.

Also as a future direction of research, the authors propose the clinical evaluation of the synergy of newly developed foods and drugs administered for the amelioration or treatment of a chronic disease, also studies on over-exposure to a mixture of food additives, (e.g. color, emulsifier, sweetener), the cumulative effects of food chemicals contributing to increase chronic diseases, and the long-term health effects in the use of many additives at the same time.

Future studies should follow the direction of using new technologies in limiting the loss of biochemical and nutritional compounds from food, using mild processing conditions, food additives of plant origin, and smart packaging technology along with a stable shelf life of the product.

CONCLUSION

Plant or animal resources, food or food supplements, can cause multiple side effects (sometimes leading even to death), also they can alter the clearance of drugs (interacting with enzymes that metabolize drugs) and negatively influencing the effectiveness of treatments. For the quality of life of patients and the decrease of health systems costs, it is vital to legislate and implement nutrivigilence and specialists and population education.

The development of food products for special nutritional conditions, like adjuvants in chronic diseases management, is a solution and a challenge for both the food industry and health professionals, focused on patient's benefits.

REFERENCES

Abdollahi, M., Ranjbar, A., Shadnia, S., Shekoufeh Nikfar, S., & Rezaie, A. (2004). Pesticides and oxidative stress: A review. *Medical Science Monitor*, *10*(6), 141–147.

About Herbs, Botanicals & Other Products. (2021). https://www.mskcc.org/cancer-care/treatments/symptom-management/integrative-medicine/herbs

Adeniyi, O. F., Lesi, O. A., Olatona, F. A., Esezobor, C. I., & Ikobah, J. M. (2017). Irritable bowel syndrome in adolescents in Lagos. *The Pan African Medical Journal, 28*, 1–7. doi:10.11604/pamj.2017.28.93.11512

Alkhathami, A. M., Alzahrani, A. A., Alzhrani, M. A., Alsuwat, O. B., & Mahfouz, M. E. M. (2017). Risk Factors for Gastroesophageal Reflux Disease in Saudi Arabia. *Gastroenterology Research, 10*(5), 294–300. doi:10.14740/gr906w

Amizadeh, M., Safari-Kamalabadi, M., Askari-Saryazdi, G., Amizadeh, M., & Reihani-Kermani, H. (2017). Pesticide Exposure and Head and Neck Cancers: A Case-Control Study in an Agricultural Region. *Iranian Journal of Otorhinolaryngology, 29*(94), 275–285.

Aryeetey, B., Lartey, A., Marquis, G. S., Nti, H., Colecraft, E., & Brown, P. (2017). Prevalence and predictors of overweight and obesity among school-aged children in urban Ghana. *BMC Obesity, 4*(38), 1–8. doi:10.118640608-017-0174-0

Asghar, M.T., Yusof, A.Y., Mokhtar, M.N., Ya'acob, M.E., Ghazali, H.M., & Lee Sin Chang, L. S., & Manaf, Y.N. (2019). Coconut (Cocos nucifera L.) sap as a potential source of sugar: Antioxidant and nutritional properties. *Food Science & Nutrition, 8*(4), 1777–1787. doi:10.1002/fsn3.1191

Asher, G. N., Corbett, A. H., & Hawke, R. L. (2017). Common Herbal Dietary Supplement-Drug Interactions. *American Family Physician, 96*(2), 101–107.

Banerjee, B. D., Seth, V., & Ahmed, R. S. (2001). Pesticide-induced oxidative stress: Perspectives and trends. *Reviews on Environmental Health, 16*(1), 1–40. doi:10.1515/REVEH.2001.16.1.1

Bastías, J. M., Balladares, P., Acuña, S., Quevedo, R., & Muñoz, O. (2017). Determining the effect of different cooking methods on the nutritional composition of salmon (Salmo salar) and chilean jack mackerel (Trachurus murphyi) fillets. *PLoS One, 12*(7), 1–10. doi:10.1371/journal.pone.0180993

Chaaya, M., Phung, K., Atweh, S., Asmar, K. E., Karam, G., Khoury, R. M., Ghandour, L., Ghusn, H., Assaad, S., Prince, M., & Waldemar, G. (2018). Socio-demographic and cardiovascular disease risk factors associated with dementia: Results of a cross-sectional study from Lebanon. *Preventive Medicine Reports, 9*, 1–5. doi:10.1016/j.pmedr.2017.11.006

Chang, C. C., Yi-Ting, L., Ya-Ting, Lu., Liu, Y. S., & Liu, J. F. (2010). Kiwifruit improves bowel function in patients with irritable bowel syndrome with constipation. *Asia Pacific Journal of Clinical Nutrition, 19*(4), 451–457.

Chanpiwat, P., Sthiannopkao, S., Widmer, K., Seiichiro Himeno, S., Hideki Miyataka, H., Vu, N. U., Tran, V. V., & Pham, T. T. N. (2016). Assessment of metal and bacterial contamination in cultivated fish and impact on human health for residents living in the Mekong Delta. *Chemosphere, 163*, 342–350. doi:10.1016/j.chemosphere.2016.08.003

Cutone, A., Rosa, L., Ianiro, G., Lepanto, M. S., Bonaccorsi di Patti, M. C., Piera Valenti, P., & Musci, G. (2020). Lactoferrin's Anti-Cancer Properties: Safety, Selectivity, and Wide Range of Action. *Biomolecules, 10*(3), 1–26. doi:10.3390/biom10030456

Davis, M. A., Barnette, D. A., Flynn, N. R., Pidugu, A. S., Swamidass, S. J., Boysen, G., & Miller, G. P. (2019). CYP2C19 and 3A4 Dominate Metabolic Clearance and Bioactivation of Terbinafine Based on Computational and Experimental Approaches. *Chemical Research in Toxicology*, *32*(6), 1151–1164. doi:10.1021/acs.chemrestox.9b00006

European Comission/Comisia Europeană. (2018). Regulamentul de punere în aplicare (UE) 1648 al Comisiei din 29 octombrie 2018 de autorizare a introducerii pe piață a xilo-oligozaharidelor ca aliment nou în temeiul Regulamentului (UE) 2015/2283 al Parlamentului European și al Consiliului și de modificare a Regulamentului de punere în aplicare (UE) 2017/2470 al Comisiei (Romanian). Jurnalul Oficial al Uniunii Europene. *Official Journal of the European Union, L, 275*(1), 1–6.

European Commission. (2015). Regulation (EU) no. 414 amending Directive 2002/46 / EC of the European Parliament and of the Council as regards (6S) -5-methyltetrahydrofolic acid, the glucosamine salt used in the manufacture of food supplements. *Official Journal of the European Union, L, 68*(26), 1–2.

European Commission. (2015). Implementing Decision (EU) no. Commission Decision 545 of 31 March 2015 authorizing the placing on the market of oil extracted from microalgae Schizochytrium sp. (ATCC PTA-9695) as a novel food ingredient under Regulation (EC) No Regulation (EC) No 258/97 of the European Parliament and of the Council. *Official Journal of the European Union, L, 90*(7), 1–4.

Fallahi, P., Elia, G., Foddis, R., Cristaudo, A., & Antonelli, A. (2017). High risk of brain tumors in military personnel: A case control study. *Clinical Therapeutics*, *168*(6), 376–379.

FDA. (2018). *Safety Notice Schizochytrium sp. strain FCC-3204. Clinical studies.* https://www.fda.gov/media/117827/download

FDA. (2020). *FaLessa. Prescribing information, side efects and uses.* https://www.drugs.com/pro/falessa.html

Heusinkveld, H. J., & Westerink, R. H. S. (2017). Comparison of different in vitro cell models for the assessment of pesticide-induced dopaminergic neurotoxicity. *Toxicology In Vitro*, *45*(Pt 1), 81–88. doi:10.1016/j.tiv.2017.07.030

Hu, W., Huang, B., Tian, K., Holm, E., & Zhang, Y. (2017). Heavy metals in intensive greenhouse vegetable production systems along Yellow Sea of China: Levels, transfer and health risk. *Chemosphere*, *167*, 82–90. doi:10.1016/j.chemosphere.2016.09.122

Imbernón-Moya, A., Burgos, F., Vargas-Laguna, E., Fernández-Cogolludo, E., Aguilar-Martínez, A., & Gallego-Valdés, A. M. (2016). Pemphigus foliaceus associated with Hypericum perforatum. *Journal of the American Academy of Dermatology*, *2*(4), 326–328.

Ishisono, K., Mano, T., Yabe, T., & Kitaguchi, K. (2019). Dietary Fiber Pectin Ameliorates Experimental Colitis in a Neutral Sugar Side Chain-Dependent Manner. *Frontiers in Immunology*, *10*, 1–14. doi:10.3389/fimmu.2019.02979

Kazuki, I., Hiroshi, Y., Mamoru, K., Yohei, K., Yuma, B., & Kumi, M. (2015). Methods for estimating causal relationships of adverse events with dietary, Supplements. *BMJ Open*, *5*, 1–7.

Kigen, G., Busakhala, N., Kamuren, Z., Rono, H., Kimalat, W., & Njiru, E. (2017). Factors associated with the high prevalence of oesophageal cancer in Western Kenya: A review. *Infectious Agents and Cancer*, *12*(1), 1–11. doi:10.118613027-017-0169-y

Kim, S. Y., Sim, S., & Choi, H. G. (2017). High stress, lack of sleep, low school performance, and suicide attempts are associated with high energy drink intake in adolescents. *PLoS One*, *12*(11), 1–12. doi:10.1371/journal.pone.0187759

Kim, W. B., Jerome, D., & Yeung, J. (2017). Diagnosis and management of psoriasis. *Canadian Family Physician Medecin de Famille Canadien*, *63*(4), 278–285.

Lai, S., Mastroluca, D., Matino, S., Panebianco, V., Vitarelli, A., Capotosto, L., Turinese, I., Marinelli, P., Rossetti, M., Galani, A., Baiocchi, P., D'Angelo, A. R., & Palange, P. (2017). Early Markers of Cardiovascular Risk in Autosomal Dominant Polycystic Kidney Disease. *Kidney & Blood Pressure Research*, *42*(6), 1290–1302. doi:10.1159/000486011

Lars, J. (2003). Hazards of heavy metal contamination. *British Medical Bulletin*, *68*(1), 167–182. doi:10.1093/bmb/ldg032

Leal-Castañeda, E. J., Inchingolo, R., Cardenia, V., Hernandez-Becerra, J. A., Romani, S., Rodriguez-Estrada, M. T., & García Galindo, H. S. (2015). Effect of Microwave Heating on Phytosterol Oxidation. *Journal of Agricultural and Food Chemistry*, *63*(22), 5539–5547. doi:10.1021/acs.jafc.5b00961

Lim, H. S., Lee, H. H., Byun, D. W., Lee, B., Enkhbold, T., & Kim, T. H. (2017). Serum Vitamin D Level Related to Coffee Consumption in Korean Young Adults Using the 5th Korea National Health and Nutrition Examination Survey. *Journal of Bone Metabolism*, *24*(4), 229–233. doi:10.11005/jbm.2017.24.4.229

Lioi, M. B., Scarfì, M. R., Santoro, A., Barbieri, R., Zeni, O., Di Berardino, D., & Ursini, M. V. (1998). Genotoxicity and oxidative stress induced by pesticide exposure in bovine lymphocyte cultures in vitro. *Mutation Research*, *403*(1-2), 13–20. doi:10.1016/S0027-5107(98)00010-4

Liu, W. Z., Zhang, W. H., Zheng, Z. H., Zou, J. X., Liu, X. X., Huang, S. H., You, W. J., He, Y., Zhang, J. Y., Wang, X. D., & Pan, B. X. (2020). Identification of a prefrontal cortex-to-amygdala pathway for chronic stress-induced anxiety. *Nature Communications*, *11*(1), 1–15. doi:10.103841467-020-15920-7

Maffei, E. M. (2020). 5-Hydroxytryptophan (5-HTP): Natural Occurrence, Analysis, Biosynthesis, Biotechnology, Physiology and Toxicology. *International Journal of Molecular Sciences*, *22*(1), 1–25. doi:10.3390/ijms22010181

Maheshwari, M., & Chaudhary, S. (2017). Acute Atrial Fibrillation Complicating Organophosphorus Poisoning. *Heart Views*, *18*(3), 96–99. doi:10.4103/1995-705X.217856

Marabelli, C., Munarini, E., Lina, M., Mazza, R., Boffi, R., De Marco, C., Ruprecht, A., Angellotti, G., Veronese, C., Pozzi, P., Bruno, E., Gargano, G., Cavalleri, A., Garrone, G., & Berrino, F. (2017). A pilot study with early adolescents: Dealing with diet, tobacco and air pollution using practical experiences and biological markers. *Multidisciplinary Respiratory Medicine*, *12*(30), 1–9.

Matran, I. M., & Dumitrascu, D. L. (2018). Legătura între alimente, factorii psihosociali şi suferinţă. *Revista Medicală Română*, *LXV*(1), 24–3.

Medda, N., Kumar De, S., & Maiti, S. (2021). Different mechanisms of arsenic related signaling in cellular proliferation, apoptosis and neo-plastic transformation. *Ecotoxicology and Environmental Safety*, *208*, 1–9. doi:10.1016/j.ecoenv.2020.111752

Nedoszytko, B., Reszka, E., Gutowska-Owsiak, D., Trzeciak, M., Lange, M., Jarczak, J., Niedoszytko, M., Jablonska, E., Strapagiel, J. D., Skokowski, J., Siekierzycka, A., Nowicki, R. J., Dobrucki, I. T., Zaryczańska, A., & Kalinowski, L. (2020). Genetic and Epigenetic Aspects of Atopic Dermatitis. *International Journal of Molecular Sciences*, *21*(18), 1–18. doi:10.3390/ijms21186484

Nishi, M. (2020). Effects of Early-Life Stress on the Brain and Behaviors: Implications of Early Maternal Separation in Rodents. *International Journal of Molecular Sciences*, *21*(9), 1–14. doi:10.3390/ijms21197212

Nurhadi, B., Sukri, N., Saputra, R. A., Wandhani, F. I., & Nurlita, A. I. (2020). Physical Characteristics of Amorphous and Crystalline Coconut Sugar Powder with the Addition of Tricalcium Phosphate (TCP) as an Anticaking Agent. *International Journal of Food Sciences*, *2020*, 1–10. doi:10.1155/2020/5320173

Oh, T. K., Song, I. A., Park, Y. M., Hwang, J. W., Jeon, Y. T., Do, S. H., Yoon, Y. E., Ahn, S., & Lee, J. S. (2017). Prevalence and risk factors for postoperative stress-related cardiomyopathy in adults. *PLoS One*, *12*(12), 1–11. doi:10.1371/journal.pone.0190065

Pathera, A. K., Riar, C. S., Yadav, S., & Sharma, D. P. (2017). Effect of Dietary Fiber Enrichment and Different Cooking Methods on Quality of Chicken Nuggets. *Han-gug Chugsan Sigpum Hag-hoeji*, *37*(3), 410–417. doi:10.5851/kosfa.2017.37.3.410

Paulsson-Do, U., Edlund, B., Stenhammar, C., & Westerling, R. (2017). Psychosocial vulnerability underlying four common unhealthy behaviours in 15-16-year-old Swedish adolescents: A cross-sectional study. *BMC Psychology*, *5*(1), 1–11. doi:10.118640359-017-0209-9

Petersm, E. M. J., Michenko, A., Kupfer, J., Kummer, W., Wiegand, S., Niemeier, V., Potekaev, N., Lvov, A., & Gieler, U. (2014). Mental stress in atopic dermatitis—neuronal plasticity and the cholinergic system are affected in atopic dermatitis and in response to acute experimental mental stress in a randomized controlled pilot study. *PLoS One*, *9*(12), 1–17.

Pivonello, R., Ferrigno, R., De Martino, M. C., Simeoli, C., Di Paola, N., Pivonello, C., Barba, L., Negri, M., De Angelis, C., & Colao, A. (2020). Medical Treatment of Cushing's Disease: An Overview of the Current and Recent Clinical Trials. *Frontiers in Endocrinology*, *11*, 1–29. doi:10.3389/fendo.2020.00648

Puia, A., & Leucuta, D. C. (2017). Children's lifestyle behaviors in relation to anthropometric indices: A family practice study. *Clujul Medical*, *90*(4), 385–391.

Rather, I. A., Koh, W. Y., Paek, W. K., & Lim, J. (2013). The Sources of Chemical Contaminants in Food and Their Health Implications. *Frontiers in Pharmacology*, *8*, 1–8.

Ronis, M. J. J., Pedersen, K. B., & Watt, J. (2018). Adverse Effects of Nutraceuticals and Dietary Supplements. *Annual Review of Pharmacology and Toxicology*, *58*(1), 583–601. doi:10.1146/annurev-pharmtox-010617-052844

Rosenbauma, P. F., Weinstockb, R. S., Silverstonec, A. E., Sjödind, A., & Pavuk, M. (2017). Metabolic syndrome is associated with exposure to organochlorine pesticides in Anniston, AL, United State. *Environment International*, *108*, 11–21. doi:10.1016/j.envint.2017.07.017

Rousset, L., & Halioua, B. (2018). Stress and psoriasis. *International Journal of Dermatology*, *57*(10), 1165–1172. doi:10.1111/ijd.14032

RxList. (2020). *L-tryptophan*. https://www.rxlist.com/l-tryptophan/supplements.htm

Sadr, M. R., Reza, B., Fazlollah, M., Yazdi, S. N. M., & Anari, P. Y. (2016). Peganum Harmala (Aspand) Intoxication; a Case Report. *Emergency (Tehran, Iran)*, *4*(2), 106–107.

Stenvinkel, P., Irving, G. F., Wegener, S., Eyre, S., Rippe, B., Segelmark, M., Barany, P., & Samuelsson, O. (2014). Phosphate additives in food a potential public health risk. High phosphate levels can lead to cardiovascular disease. *Lakartidningen*, *15*, 27–28.

Tiwari, P., Mishra, B. N., & Sangwan, N. S. (2014). Phytochemical and Pharmacological Properties of Gymnema sylvestre: An Important Medicinal Plant. *BioMed Research International*, *2014*, 1–18.

Turck, D., Bresson, J. L., Burlingame, B., Dean, T., Fairweather-Tait, S., Heinonen, M., Ildico Hirsch-Ernst, K., Mangelsdorf, I., McArdle, J. M., Naska, A., Monika Neuhauser-Berthold, M., Nowicka, G., Pentieva, K., Sanz, Y., Siani, A., Sjodin, A., Stern, M., Tome, D., Vinceti, M., ... Loveren, H. (2018). Scientific opinion on the safety of xylo-oligosaccharides (XOS) as a novel food pursuant to Regulation (EU) 2015/2283. EFSA NDA Panel (EFSA Panel on Dietetic Products, Nutrition and Allergies). *EFSA Journal*, *16*(7), 5361–5381.

Valadez-Carmona, L., Cortez-García, R. M., Plazola-Jacinto, C. P., Necoechea-Mondragón, H., & Ortiz-Moreno, A. (2016). Effect of microwave drying and oven drying on the water activity, color, phenolic compounds content and antioxidant activity of coconut husk (Cocos nucifera L.). *Journal of Food Science and Technology*, *53*(9), 3495–3501. doi:10.100713197-016-2324-7

van de Looij, Y., Larpin, C., Cabungcal, J. H., Sanches, E. F., Toulotte, A., Do, K. Q., & Sizonenko, S. V. (2019). Nutritional Intervention for Developmental Brain Damage: Effects of Lactoferrin Supplementation in Hypocaloric Induced Intrauterine Growth Restriction Rat Pups. *Frontiers in Endocrinology*, *10*, 1–14. doi:10.3389/fendo.2019.00046

Wang, Y. K., Yang, X. N., Zhu, X., Xiao, X. R., Yang, X. W., Qin, H. B., Gonzalez, F. J., & Li, F. (2019). Role of Metabolic Activation in Elemicin-Induced Cellular Toxicity. *Journal of Agricultural and Food Chemistry*, *67*(29), 8243–8252. doi:10.1021/acs.jafc.9b02137

Web, M. D. (2020). *AlgaOil. Side Effects and precautions*. Available at https://www.webmd.com/vitamins/ai/ingredientmono-1565/algal-oil

WebM. D. (2020). *Pectin*. https://www.webmd.com/vitamins/ai/ingredientmono-500/pectin

Wee, J. H., Min, C., Park, M. W., Park, I. S., Park, B., & Choi, H. G. (2020). Energy-drink consumption is associated with asthma, allergic rhinitis, and atopic dermatitis in Korean adolescents. *European Journal of Clinical Nutrition*, *75*(7), 1077–1087. doi:10.103841430-020-00812-2

Xu, S. F., Zhang, Y. H., Wang, S., Pang, Z. Q., Fan, Y. G., Li, J. Y., Wang, Z. Y., & Guo, C. (2019). Lactoferrin ameliorates dopaminergic neurodegeneration and motor deficits in MPTP-treated mice. *Redox Biology*, *21*, 1–13. doi:10.1016/j.redox.2018.101090

ADDITIONAL READING

Matran, I. M., & Boţan, E. (2017). Applicability of Sericulture in the Food Industry and Medicine. *Lucrări Ştiinţifice seria. Agronomie*, *60*(1), 183–186.

Matran, I. M., Farcas, A., Bojită, M., & Dumitrascu, D. L. (2017). Alternative Therapies in Gastric Hypersecretion. *Farmacia*, *65*(3), 323–328.

Rotaru, M., Iancu, G. M., & Matran, I. (2020). Importance of food in the control of inflammation in atopic dermatitis. *Experimental and Therapeutic Medicine*, *20*(6), 1–9. doi:10.3892/etm.2020.9336

Rotaru, M., Matran, I. M., & Iancu, G. M. (2020). Topical Maintenance Treatments in Chronic Dermatitis. *Acta Medica Transilvanica*, *24*(1), 18–21. doi:10.2478/amtsb-2020-0042

KEY TERMS AND DEFINITIONS

Atopic Dermatitis (AD): Also known as atopic eczema, is a long-term type of inflammation of the skin (dermatitis).

Chronic Disease: A chronic condition is a human health condition or disease that is persistent or otherwise long-lasting in its effects or a disease that comes with time.

Critical Control Point: A critical control point is defined as a step at which control can be applied and is essential to prevent or eliminate a food safety hazard or reduce it to an acceptable level.

Food for Special Nutritional States: Processed or specially designed foods intended for the diet of patients, including infants, to be used under medical supervision; they are intended for the exclusive or partial feeding of patients with limited, weakened or impaired ability to take, digest, absorb, metabolise or excrete ordinary foods or nutrients they contain or their metabolites, or patients with other nutritional requirements identified on medical basis, whose diet can not be achieved only by changing the normal diet.

Interleukin: Interleukins (ILs) are a group of cytokines (secreted proteins and signal molecules).

Nutrivigilance: A nutrivigilance system is quality system used by the marketing responsible company to fulfil the tasks and responsibilities as required by the Competent Authorities and designed to monitor the safety of authorised food supplements and detect any change to their risk-benefit balance.

OCDI OCD-10-AM: The International Classification of Diseases (ICD) is the international standard diagnostic classification for all recognised diseases and related health problems. ICD codes are alphanumeric designations given to every diagnosis and description of symptoms on medical records. These classifications are developed and monitored by the World Health Organization (WHO) for the incidence and prevalence of diseases and other health problems. The ICD is revised periodically and is currently in its tenth edition, the ICD-10.

Patent: A patent is a type of intellectual property that gives its owner the legal right to exclude others from making, using, or selling an invention for a limited period of years, in exchange for publishing an enabling public disclosure of the invention.

Chapter 12
Perspectives on Climate Smart Agriculture:
A Review

Alpana Gupta
Indian Institute of Technology, Delhi, India

Athar Hussain
https://orcid.org/0000-0002-7187-5977
NSUT West Campus, Jaffarpur, New Delhi, India

ABSTRACT

Agriculture is one of India's most significant industries, serving half of the nation. Climate is a big factor for agricultural production. Agriculture is primarily rain-fed and subjected by small-scale farmers. Rain-fed crops are 48% of the overall area under food crops and 68% on non-food crops. The complete worldwide agricultural production in India accounts for 7.39%. India's total emissions from all GHGs in 2014, according to the World Resource Institutes, total about 3,200 MTCO2 eq out of 48,892 MTCO2s eq worldwide in 2014. Agriculture is ascribed 626.86 2 MTCO2 eq of the Indian greenhouse gas emissions averaged in 2014 of 3,200 MTCO2 eq. This chapter reviews the concept of sustainable agriculture, establishes the link between climate change and agriculture, the origination of climate-smart agriculture, and relevant practical approaches, case studies, and geospatial assessment methods responding to climate-smart agriculture.

INTRODUCTION

India is one of the world's main manufacturers of new fruits such as mango, guava, chickpea, chili pepper and rice oil. Agriculture is one of India's most significant industries and offers 50% of the workers with jobs. India's main agricultural exports include cereals, (mostly rice), spices, cassake, olive oil, and tobacco, tea, coffee and marine goods. For India's agriculture and rural economy, small-scale farmers are essential. Climate is the magnificent factor for agricultural production. Agriculture is primarily rain-fed

DOI: 10.4018/978-1-7998-9557-2.ch012

and subjected by small-scale farmers. Rains fed crops are gaining 48% of the overall area under food crops and 68% on non-food crops. India's agriculture industry is projected to account for only about 14 percent of the country's economy but 42 percent of total employment. Rainfall during the monsoon season is critical for economic activity because roughly 55 percent of India's fertile land is dependent on it (Economics, 2021). India's total emissions from all GHGs in 2014, (World Resource Institute, 2016) is about 3,200 $MTCO_2$ eq out of 48,892 $MTCO_2$'seq worldwide in 2014.Unpredictable variations in the rainfall, temperature patterns have pose great threats to agriculture yield posing food insecurity. In the last century, significant changes in revisions have been driven by the probable bearings of climate change on agriculture. The subjects will have insight of physical impacts such as crop yield changes on agriculture and also their financial impacts. This establishes an immediate need enabling the farmers to combat this problem by modifying the "PRESENT AGRICULTURE to "CLIMATE-SMART AGRICULTURE." This chapter reviews the concept of sustainable agriculture, establishes the methodology of problem link between climate change & agriculture, the origination of Climate-Smart agriculture, and relevant practical approaches, case studies, and geospatial assessment methods responding to climate-smart agriculture.

Research Questions

- Interconnections between climate change, agriculture and food security.
- Practical approaches to climate smart agriculture.
- Geospatial solutions in response to climate-smart agriculture.

The aim of the Agriculture and Food Safety Program is to promote modernism to make agricultural production more proficient and viable to improve food security and income and nutrition benefiting farmers (International Development Research Centre). Sustainable agriculture defines approaches to crop management that address interdependent objectives, such as enhancing or retaining output while conserving environment, stabilizing the use of resources which decelerates climate change. Modern agricultural, plant / animal and biotechnology technologies have helped feed the world, reduce adverse effects on the Environment and mitigate climate change. These outlines of application illustrate the role of advertise powers and hazard controlling in the opinion made by the farmers of which crop and soil control combinations are to be implemented (Gaffney et al., 2019). Food safety is described as food accessibility and access. Food insecurity ranges from food safety to hunger in full. Food safety was described by the World Food Summit in 1996 as "where everyone has always access to adequate, secure and nutritious food to sustain a good and active life". Food safety requires access to food that meets food needs and food preferences of individuals in physical and financial terms. Food safety involves an extent to which critical food supplies can be disrupted or unavailable in the future owing to multiple risk variables, including droughts, transport disruptions, fuel shortfalls, financial instability and war (Food Security: Definition & General Information). It can categorize as (i) **Constant Food Production and its supply (ii) Equitable affordability and accessibility to food.** However following factors that affects the food security are:

- ***Water scarcity occurring globally***— Water level is dropping owing to extensive excess pumping and Irrigation in several nations (including Northern China, the United States and Indian).
- ***Climate change***-The rise of worldwide temperatures is starting to impact crop supply, forest resources, water in-security and ecosystem balance.

- ***Degradation of Land***–Rigorous farming leads to an unusual period of soil fertility and a decrease of agricultural production.
- ***Acquisitions of land***–Throughout developing countries, Governments have the right to produce their own extended-term food for millions of hectares of agriculture.

CLIMATE CHANGE AND AGRICULTURE

Changes in soil moisture, temperature, rainfall, cloud cover, and rises in carbon dioxide levels can alter returns for distinct plants and geographical boundaries. The clear representation is shown in figure 1. (Kumar & Gautam, 2014). In many fields, especially tropical and mid-continental areas, the highest precipitation and elevated temperature may decrease the soil humidity content, decrease the water level and affect plant growth.

Figure 1. Agriculture and climate change
(Source: Created by author, 2021)

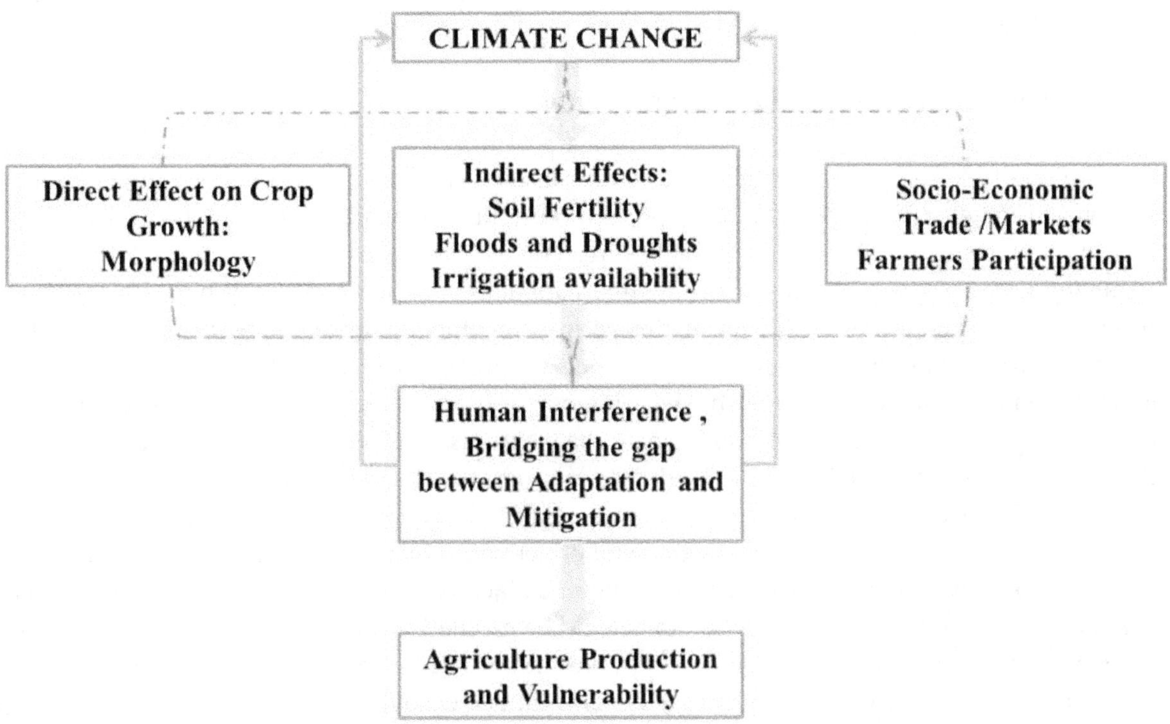

FAO statistics indicate that by intensifying food production, the global grain supply has risen by almost a factor of 2.2, outweighing the 1.3-fold development in population over the last 50 years (DeFries et al., 2016). This 2.2-fold rise was however observed at the same time as the worldwide use of fertilizers increased five-fold. Moreover, the development of biomass (food, feed, fibre, and power), as well as competition for soil and nutrients in supply areas have become the driving force for developed countries, whereas ecosystem overconsumption and eutrophication take place in import areas. Food production throughout the world does not only cause but is heavily influenced by worldwide environmental change. Agriculture is mainly at risk of climate change. The most significant drivers of food insecurity are the associated increases in natural disasters such as floods, tropical storms, prolonged periods of dryness and new pests and illnesses. (The Intergovernmental Panel on Climate Change,2015). Given climate change adverse effects in global food production, best practices for climate smart agriculture (CSA) need to be promoted to achieve food safety.Climate-smart agriculture is an methodology that provides guidance for the required measures for the transformation and transformation of agrarian structures to efficiently funding expansion, and to safeguard food security in a fluctuating environment. The CSA has three key objectives: viable productivity increases and earnings; adjustment to climate change, climate resilience and greenhouse reduction.

Adaptations will be required to counteract any negative consequences of the changing environment. Farmers must be able to alter their farming methods in response to changes. Changes in crops and crop varieties, improved water management and irrigation systems, and changes in planting schedules and tillage practices will all play a role in mitigating the negative consequences of climate change while maximising the positive advantages. Adapting to climate change on various levels and types of technology and social levels is conceivable. Crop-land suitability can be handled using geospatial tools to manage farmers' resilience to climate change, such as Land suitability index.

The Progression of Climate Change Policy

The evolution of global climate change policies over latest years has to be understood to put CSA and its disputes in context. We use (Gupta, 2010, pp. 636-653) framing, which bits the past of global climate change strategy between 1979 and 2010. He bi-furcated the five classes. This is the pre-1990 period as the time to tackle the issue, starting with The International Panel on Climate Change (IPCC) and the World Climate Conference in 1979. During this period, the global climate policy focused primarily on the global need for pollution reduction measures to achieve greenhouse gas (GHG). The publishing of the Bruntland Commission's 1987 Sustainable Development Report also resulted to the connection and the advantages of integral consideration from climate change to sustainable development. In a second phase of the international climate policy in the 1991-1996 periods, the first delivery of a global policy system defined by the Rio Treaty in 1992 and Agenda 21 was created. The establishing of the UN Climate Change Framework Convention (UNFCCC) that came into effect on 21 March 1994was an significant conclusion of the Rio Conventions. The Convention's ultimate objective is to prevent' hazardous' interference by humans with the climate scheme. This aim should be reached while ensuring that' food production is not endangered,' according to Article 2 of the Convention. The Protocol highlighted broad objectives for reducing GHGs rather than individual GHGs as regards CO2.The Clean Development Mechanism (CDM) has been created to enable advanced nations to use economic incentives to fund reductions in GHG emissions in developing nations to then take advantage of credit for their own objectives.

The emissions sequestering of carbon on soils and forestry is an essential category of emission reduction measures which are extremely essential for agricultural growth. Many possibilities were recognized through enhanced soil and forestry management for agriculture-related carbon sequestration. The Kyoto Protocol (KP) has been challenged by the need for credible and economical processes for carbon accounts, surveillance and validity, especially when carbon sequestration has been hard. (Lal,2004, pp.1-22) submitted that carbon sequestration payments could provide important additional revenue to farmers, particularly in developing nations. The increasing emphasis placed on public assistance for another source of energy affects many components of agro economy, in implementing biofuel policies across worldwide. Although the decrease in GHGs is a justification for biofuel subsidization, the need to combat increasing energy prices, enhance trade balances and boost agricultural sector revenue was perhaps even more crucial.

Agriculture, through demand for biofuels and biomass, and by supporting carbon sequestration, has been considerably affected by national and provincial plans. The soil carbon offset was a method to reduce the strain on poor farmers of developing countries, and that farmers were unlikely, but could be more likely to lose rights to their property(Actionaid,2011). At the 21st Conference, the Paris Convention was achieved of the United Nations Conference of Climate and Cooperation Parties in 2015 represents a greater worldwide commitment to addressing climate change, with nations accepting legal limitations on Greenhouse Gas Emissions that seek to limit average universal high temperature increases using a cross-market strategy which will lead to both clean power implementation and preservation.188 nations are presented to the UNFCCC on 31 March 2016.

Intended Determined National Contributions (INDCs),' including statements of actions expected for the mitigation. The adaptation is a high priority of the agriculture sector and mitigation from farming is also quite important in the applications, including sequestration. Therefore, it is more crucial than ever to consider adaptation and mitigation together and the potential synergies between them. (Skot, 2016, pp. 1-194) 31 of the INDCs unambiguously remark CSA in their attempt to achieve combined reduction in poverty and ecological merits.

The CSA concept came into being during a time of controversy over the notion and methods of sustainable agriculture and when the nature of agriculture and its position for food security were not clearly discussed in the context of action on climate change. (Scoones, 2009, pp. 547-571). The first of these was obviously expressed in discussions and controversy regarding the growth of IAASTD, which ran from 2003 to 2008. The focus of forum was Food security and agrarian mitigation in evolving countries: Capturing Choices Interactions was presented as the chief expression in the 2009 report of FAO, tossed at the Barcelona Workshop in November. In 2010, a background paper was produced by FA October 2010 on "Agriculture, Policy, Practices, and Financing for Food Security, Adaptation, and Mitigation," entitled "Climate Smart" (FAO 2010). Following on from the Joint Vision Declaration, it was decided to further improve the agriculture, food security and climate change agenda at the 17th Session of the Sustainable Development Commission(CSD-17) of May 2009.In the field of international and domestic agricultural and climatological policies, the CSA notion attracted substantial consideration &discussion and was quickly adopted as the rallying point to mobilize action on climate change and agriculture. Two parallel global political and scientific processes for CSA have been established in the aftermath of the Hague conference. In Hanoi Vietnam in 2012 and Johannesburg South Africa in 2014, the policy process involves follow-up conferences. One of the major results of these initiatives was the emerging Global Climate Smart Agriculture Alliance (GACSA) that would link policy and research in three key areas. (i) Knowledge (ii) Environmental enablement and (iii) investments. GACSA was launched at the UN Climate Summit after considerable debate in September 2014. Membership of GACSA may include governments, non-governmental /civilian society organizations, farmers, fishermen and forestry organizations, intergovernmental organizations, research / extension / training organisations, funding institutions and private sector organizations. GACSA has 122 members till January 2016 comprising 22 nations, and CSA's Climate Change and Food security initiatives have not only been implemented at international level, but have also been established in collaboration with international organizations like the FAO, the World

Bank, local and international Nguel and CGIAR, with CSA ventures launched at national and regional level. One of its key characteristics is that it aims for the accomplishment of three goals. Sustainable food safety by increased productivity and employment. Increasing sustainability and adapting to climate change. At the second global CSA policy conference held in Hanoi in 2012, the foundations of a CSA strategy and concepts are established.A number of benefits, including increased productivity, increased profitability and reduced environmental footprint, are expected from the use of innovative tools based on digital technologies and spatial planned in farming. Smart agriculture enables us to understand the application in the overall value chain, of data processing, data analyzing and automation technologies, that are jointly organized in order to increase the operation and management (analytics) of a farm in terms of standard (near-real time) operations and reuse of such data in improved chain transparency (foods) (animal-plant soil). The Internet of Things (IoT) technologies will necessarily support such capabilities. From the farmer's perspective, smart farming should add value for farmers by making better decisions or by managing and managing them more efficiently. Intelligent agriculture can also offer significant benefits in environmental issues, e.g. by improved water use or treatment optimization.

CLIMATE SMART AGRICULTURE

As demonstrated in Figure 2, climate-smart agriculture is a method to modifying and reorienting farming expansion in new realities.

Figure 2. Components of climate smart agriculture
(Source: Created by author,2021)

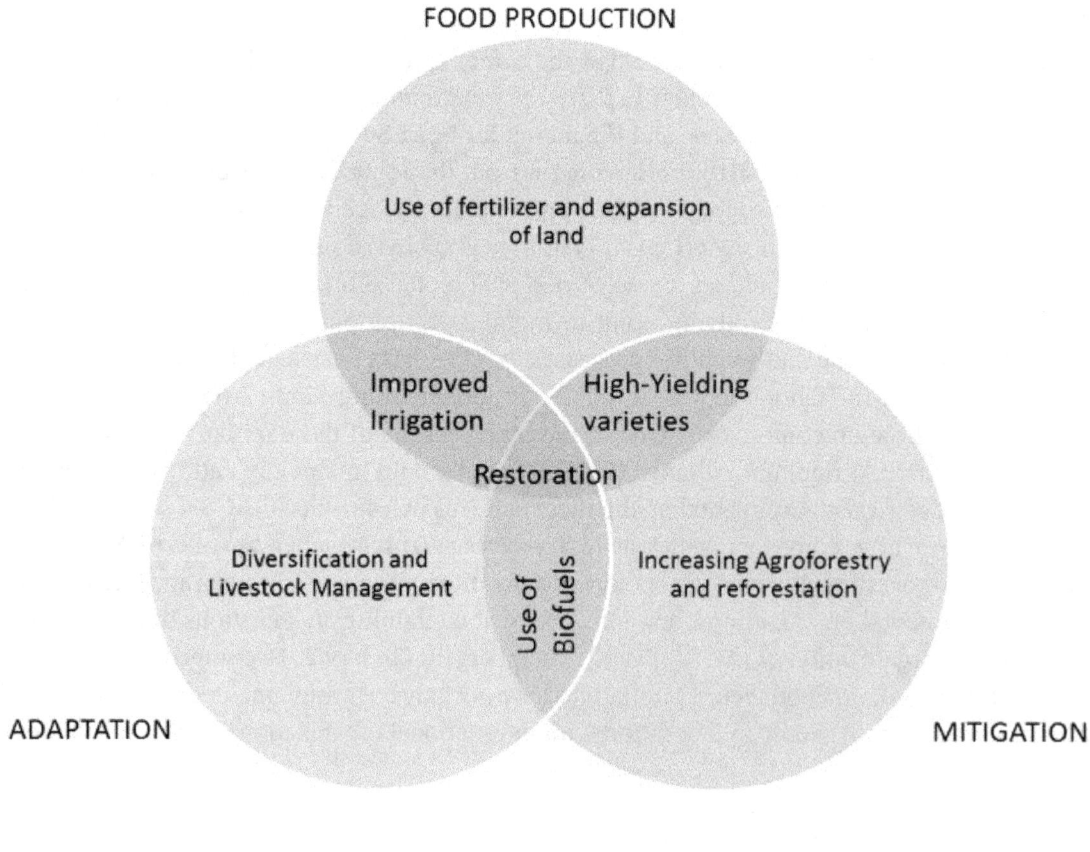

Following are the three positive outcomes of CSA:

- Increased productivity of food security for the poorest that totally rely on agriculture.
- Shock-absorber & adaptation to climate stress such as droughts, pests, & diseases.
- Reduction in emissions, avoid deforestation and promoting carbon sequestration techniques.

Chief Components of CSA

- **CSA fights climate change**

Approach to change and to reorient agricultural growth in tackling climate change can be described as climate-smart agriculture.

- **CSA combines multiple goals and maintains contracts**

CSA generates threefold benefit: higher productivity, increased resilience and decreased emissions. However, all three often cannot be achieved. Compensations must often be produced when the time comes to enforce CSA. This requires the identification of synergies and assessing the expenses and advantages of various alternatives on the basis of participatory approaches and stakeholder goals.

- **CSA maintains services to ecosystems**

CSA interventions are imperative to prevent their degradation (Food and Agriculture Organization of the United Nations, 2012). Therefore, CSA adopts an strategy based on viable farming principles and go beyond narrow sectorial methods, which lead to inept and contending land uses, cohesive preparation and supervision.

- **The CSA provides certification points at several levels**

CSA can't be considered a collection of activities or innovations. This contains a variety of entries from technologic and realistic development to progress in the dynamics and scenarios of climate change, financial schemes, the value chain and economic and political atmosphere. It involves integrating multiple initiatives at farm level through different technology into the food system, environment, value chain and policies.

- **CSA has definite framework**

Interventions need to take into account the interactions between the different elements in landscapes, inside or between ecosystems and within different political and institutional arrangements.

- **CSA involves women empowerment**

CSA methods incorporate women role and marginalized sections of society in order to attain food safety objectives and improve resilience. Often in marginal land, such as drought and flooding, these organizations are most susceptible. Consequently, climate change will most probably affect them

(Huyer,2015, pp.1-8). A further key element of CSA is gender. Women generally have less exposure to agricultural land and other economic and financial resources, which can help to cope with incidents such as droughts and floods.

Challenges in Climate Smart Agriculture

- **CSA discourses food security and undernourishment**

Despite the focus on agriculture and food security over recent centuries, around 800 million individuals worldwide are undernourished and 1 billion individuals are malnourished. It is anticipated to increase the worldwide population to over 9.7 billion by 2050.Variations were shown in food consumption pattern; increasing prosperity for meat-rich diets, for example, drives demand (Alexandratos & Bruinsma,2012). When current trends continue in the patterns of consumption and wasting it, there would be more than 70% of food production requirement by the end of 2050.

- **CSA establish the network between Poverty and Agriculture**

For many individuals residing in developing nations, agriculture is still the primary basis of nourishment, jobs & revenue. Indeed, about 76% of the poor are reported at countryside worldwide, farming is the main source of income.

- **CSA bridges the gap between climate change and agriculture**

Average temperatures around the world are already growing, and temperatures are expected not only to be warmer, but also to be more volatile, in the future. Extreme precipitation and extreme temperature will impact extensively on farming, forest and marine life (Food and Agriculture Organization of the United Nations, 2013).

- **A focus on climate change**

CSA is based on the values of enhanced efficiency and sustainability, as are others in sustainable agriculture methods. However, the focus on climate change is differentiated by addressing the problems of adaptation and mitigation explicitly and working towards food safety for all. Essentially, CSA is viable farming that includes issues of resilience while attempting to decrease emissions of greenhouse gases at the same moment.

- **Outcomes, synergies and trade-offs**

To order to concurrently tackle three issues of performance, adjustment and prevention, CSA should not only be based on strategies and processes, but also on the effects of interventions above farm level. It must take into account the collaborations between production, adaptation, and mitigation and the communications at various stages, including broader socio-ecological consequences. For example, on-farm / community interventions of CSA may impact both existing social and environmental systems and the wider landscape.

- **New funding opportunities**

Currently, the investment needed to fulfill food safety has a huge deficit. By concentrating explicitly on change of climate, CSA unlocks fresh financing possibilities for farm growth, enabling it to tap into adapting and mitigating climate finance. This involves financing from the Clean Development Mechanism and the Voluntary Carbon Market, among others: the Adaptation Facility; the LDF or the Special Climate Fund. (International Fund for Agricultural Development, 2012). The most promising of all is that the Global Environment Facility Trust Fund (GEF) and Green Climate Fund have allocated specifically to the CSA.

MAPPING OF EXISTING INITIATIVES IN THE RELEVANT AREA OF CSA

Livestock and agriculture are two of society's strategic financial operations, essential for helping and sustaining many rural regions. Currently farming is accountable for production and the environment. But technological innovation in this industry will enable for the maximum effectiveness while guaranteeing excellent quality and affordable food manufacturing results while respecting sustainable environmental requirements. However, the main tradition is agriculture, where innovation is conducted at a slower rate than in any other field (Alliance for Internet of Things Innovation, 2015). These innovation achievements have in latest years created debates on the efficiency of the innovation system in the agriculture and agriculture industries. The future of the food scheme became a subject of widespread political discussion, with a great deal of food accessible and knowledge of adverse externalities (such as the environmental and food security problems). There is therefore a need for ongoing innovation to increase the efficiency and safety of manufacturing. This section seeks to introduce the reader to current projects regarding innovation in agriculture and agriculture. It can be split into two primary kinds: -

1. **Partnerships and technological platforms**: combining and integrating current parts alternatives including surveillance, control systems, cooperative platforms, recommendations on best practices, etc.
2. **Existing goods and services**: technological developments to cover and enhance poor elements of the Farm to Fork chain in order to achieve a more effective process.

Practical Approaches to CSA

This section presents a variety of practices and techniques for CSA climate-smart (CSA) agriculture in seven areas: management of soil crop, livestock, forest, energy aquaculture management. Methods adopted such as precision agriculture, labor, and fertilization are all CSA practices as indicated in table 1.

Table 1. Practical approaches to CSA

<table>
<tr><th>Pillars of CSA</th><th>Soil Management</th><th>Crop Production</th><th>Water Management</th><th>Forestry</th></tr>
<tr><td>Introduction</td><td>Sustainable agricultural productivity would support safe land.</td><td>The vast range of methods of cultivation can now be seen as ' climatic smart ' either through adaptation or for mitigation.</td><td>The world's population is on the increase and production of more food, urban development and factories are rising, water insecurity is a rising problem.</td><td>Extreme weather conditions impacts the ecosystem services and can therefore affect the wellbeing of rural areas. One quarter of global emissions are expressed by farming, forestry and other land use industries. Afforestation would act as carbon sink.</td></tr>
<tr><td>Productivity</td><td>Increase the soil fertility and prevent the loss of nutrients.</td><td>Greater crop yields, plant nutrient management and crop choice with higher yield potential in environmental condition</td><td>Increased rainfall collection and storage and increased irrigation water management would increase plant productivity.</td><td>Integrating trees into agricultural systems will improve soil quality and increase and make crop yields more sustainable.</td></tr>
<tr><td rowspan="3">Adaptation</td><td rowspan="3">Surface harvesting and mulching, approaches to land terracing, contouring of stone and reforestation at landscape level.</td><td>Interbreeding of crop for attaining high tolerance in extreme weather conditions.</td><td rowspan="3">Supplemental irrigation and rainfall capture</td><td>Trees can serve to protect and secure crops from slippage, flooding which avalanches, and can play an important role in protecting them.</td></tr>
<tr><td>Plant reproduction is becoming increasingly necessary for resistance to drought, pest and disease as drought threats in many regions are expected to rise. (Food and Agriculture Organization of the United Nations,2008)</td><td rowspan="2">The canopy can have direct benefits: a decrease in ground temperature for underlying crops and Reduce flux rate and erosion of the soil from strong rainfall. (Leeuw et al., 2014)</td></tr>
<tr><td>Production and plantation of cultures containing thermal tolerances, drought tolerances and salinity resistant plants and switching to more temperature and hazardous plants. For example, the toughest plants are miles and sorghum of extreme, warm and dry conditions.</td></tr>
<tr><td rowspan="3">Mitigation</td><td>Modern tillage agriculture in relation to the impact on soil carbon reservoirs Agriculture (Corsil et al., 2012)</td><td rowspan="3">(Glover et al.,2007) Perennial crops can, however, stock up more CO2 than annual crops below the surface.</td><td rowspan="3">Flooded rice systems release large amounts of greenhouse gas (GHG) methane (CH_4). Alternate water cycles do not only conserve energy in these systems but also significantly reduce methane emissions. In contrast, irrigation techniques that reduce water demand will minimize pump capacity and reduce emissions.</td><td rowspan="3">Actions to improve covering tree areas (forests, reforestations, agroforestry) and reduce erosion and destruction by rising vegetation on the land and below the surface, increase carbon sequestration.</td></tr>
<tr><td>Little or no disruption of mechanical soil, permanent organic soil, diversified plant associations and rotating operations.</td></tr>
<tr><td>Leave waste as a source of carbon on the surface of the soil.</td></tr>
</table>

(Source: Created by author, 2021)

System Approaches

Practices like drought-resistant and precision-cropping are being applied beyond climate-smart agriculture. CSA needs different systems perspective in command to complete the multiple goals of production and food security, improved farmers resilience and reduced emissions of greenhouse gases. They cover landscapes, ecosystems and value chains. Landscape approaches aim to combine sustainable environment and natural resources management with conservation requirements, understanding multifunctional ecosystems and delivering services and benefits to a broad spectrum of ecosystem processes, species and social stakeholders. A wide range of land and genetic diversity throughout the entire landscape can reduce risks (pests, diseases and climate-related events). The landscape solution with increased concentration on permanent crops, grasslands, forests and wetlands is an effective way of reducing greenhouse gas emissions (GHG) and promoting carbon sequestration.

Value chains as demonstrated in table 2 shows the chain of products pass from growers (delivery of inputs such as plant, fertilizer, water, power) who cultivate and harvest them to consumers over mediators, including producer firms, haulers, traders and sellers; through the pre-farm value chain – manufacture.

Table 2. Value chain analysis

Productivity	Adaptation	Mitigation
Effective Grain Storage Project (EGSP) (CIMMYT 2014) provides storage facilities for postharvest losses that would provide economic boost to livelihoods of communities.	Reduction of poverty as the chief character of value chain will help to adapt to climate change as they develop the resources of farmers and organizational linkages. For example, through the NICADAPTA project, About 20k families build their resilience in coffee value chain.	They are incorporated to spread multiple mitigation merits across the value chain at multiple levels, such as input output, distribution, transportation, and post-harvest loss reduction. Disseminated to 600,000 farmers in Kenya, climate-smart feeding and husbandry practices will decrease the level of emission by 1.2million tCO_2e in 2018.

(Source: Created by author, 2021)

Enabling Environment

Enabling climate-smart agriculture (CSA) environments are the conditions under which climate-smart technologies and practices are implemented and encouraged. These include regulations, administrative structures, participation of investors and gender differences, infrastructure, insurance programs, climate related information and regulatory mechanism as shown in table 3.

Table 3. Enabling environment approaches

Insurance based on index	Climate related Information	Physical and social infrastructure	Policy framework	Institutional framework	Social Inclusion
Marginal farmers in low-income countries are often locked up in poverty because they cannot invest in better farming due to weather threats. Agricultural insurance, a desirable way to deal with these risks, is usually based on a clear estimate of each producer's loss or damage.	Enabling support and policy support, climate related information and advice will reduce insecurity and enable farmers to build resilience and adapt sudden shocks and take advantage of adaptation measures to combat extreme weather conditions when they occur (Cooper et al.2008)	Agriculture depends heavily on resources and physical capital, including highways, equipment and houses. In Bangladesh, for example, local innovations have created a "climate smart house" that is resilient to cyclone, food, energy and water efficient.	At national level, climate policy is usually formulated by various programs, among them national adaptation programmes, national adoption plans (NAPs), and appropriate national mitigation measures (NAMAs). Climate change policies are usually defined in national programs. Regional development and poverty reduction policies often include agricultural and food security plans; national policy reports on trade, economy, agriculture and the environment are also important.	Strengthening local institutions network. Layout of policies and programs to support farmers to implement the changes on ground (Agrawal,2008, pp.1-47).	If women have access to capital, yields on the farm will rise by between 20-30%, reducing the world's hunger populations by between 12-17% (FAO 2011).
Index-based coverage is, however, a viable choice as it uses the weather index for measuring compensation for explicitly specified threats such as precipitation (Greatrex et al.,2015).	Effective use of weather information resources can make it possible for farmers to handle weather-related risk negative effects efficiently in worse season, while also having higher average and more income than normal seasonal losses.	Climate-strengthening infrastructure investments could contribute to productivity, revenue (e.g. better market access), and livelihoods.	Some key challenges in promoting policy development and enabling environments can be described: Creating better coordination and integration across sectors. Policy should be goal oriented and transparent.		
In Eastern Africa, for nomadic pastoralists in remote areas, an inventive method is pioneered by satellite imaging of vegetation soil cover (= food supply).	Meteorological services can help alleviate the problem by encouraging more effective fertilizer utilization by improving the balance between fertilizer use and other outputs as well as inputs to the climatic conditions yearly.	Low carbon investment can help reduce emissions of GHGs.			

(Source: Created by author, 2021)

Developing CSA Plan

The first phase in the development of a CSA system is to perform contextual analysis, capture the current status and vulnerabilities of the CSA programs, threats to specific contexts and allow a multidisciplinary sector-specific and multi-tiered setting. It is important to examine the agricultural, political, social, environmental, and economic scope of the CSA approach and emphasize the points of entry for investment in priority CSA initiatives on an international scale.

Situation analysis may cover a range of subjects, but the following type of information generally includes: The first phase in designing a CSA plan is an assessment of the current status of CSA programs, of vulnerabilities and risks given specific contexts, as well as of the environment that is capable of being applied at multiple levels across industries as shown in figure 3.

Figure 3. Situation analysis
(Source: Created by author, 2021)

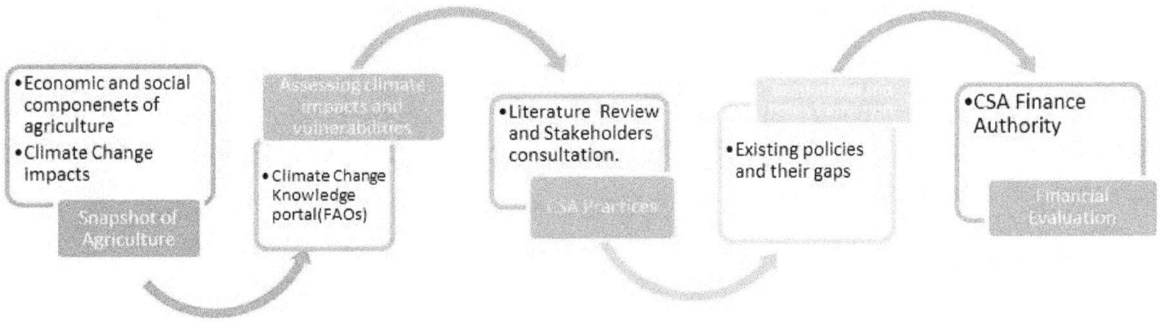

Situation analysis material is typically based on existing national or global data sources and expert feedback and surveys, including farmers and technical experts. More regional data can also be integrated, if appropriate.

Targeting and Prioritization

For climate-smart initiatives, there are a number of technical, organizational and policy choices that have specific environmental and economic implications and cost. Different CSA Prioritization Tools (CIAT / CCAFS) have been developed. The base work is divided into four stages: (i) Assessing CSA option available; (ii) Prioritizing CSA options; (iii) Cost- Benefit Analysis of Variables; and (iv) Evaluation of obstacles and alternatives as shown in figure 4.

Figure 4. Targeting and prioritization
(Source: Created by author, 2021)

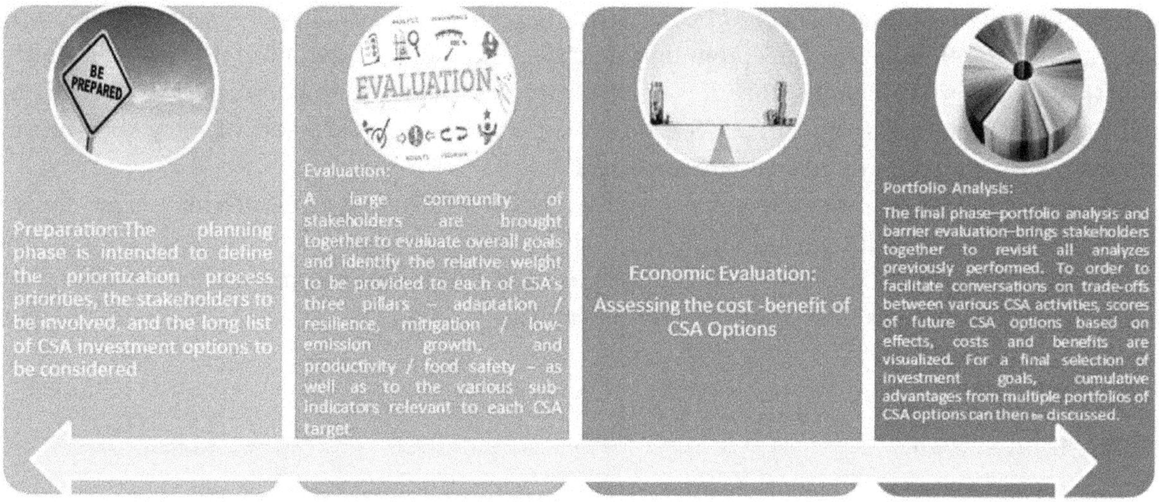

Numerous tools are available, for example the priority system of the EX-Ante Carbon Balance Tools (FAO's EX-Ante and CCAFS Mitigation Optimization Tools) and the CCAFS compendium of CSA activities (CSA Prioritization Tools).

Supporting Programme

Programme's objective is to supply tangible products driven by co-generation, business models and performing the plan on ground, etc. That allows information and services to be produced and delivered. The first principle is that the subject matter content and target market must be clearly identified. Once the "what and who" has been identified, there should be an assessment of existing resources. The goods may be modifications to existing materials, or new additions may be made. Material must meet the design criteria when user and context specific and should base on best practices and approaches from the history timeline where possible. It is important that the materials generated are properly recorded, including information of manufacturer, purpose, and time. This may help other users to save their time in duplicating the resources that has already been produced, allowing more energy for implementers to concentrate more on creativity and distribution of data.

Monitoring, Assessment and Training

The monitoring, assessment and learning (ME&L) portion of the CSA program establishes methods and resources for tracking implementation progress, evaluating effects, and encouraging iterative learning to improve preparation and implementation of the CSA.CSA Plan's indicators use-cases for the project lifecycle:

- **Readiness indicators**: Readiness is critical because CSA needs many support structures for success and the intervention often involves the building of preliminary capacity to precondition localization or target population
- **Process indicators**: Process-oriented metrics facilitate the assessment of initiatives to satisfy CSA and expected aims. Many ME&L metrics offer insights into processes such as the diversity and sex of team leadership, number, and performance of group interactions, reporting timeliness, etc.
- **Progress/impact indicators**: Typically, input indicators depend on assumptions regarding the relationships in an outcome between activities.

Measuring Productivity

Productivity shifts can be calculated in various ways. Measuring yields is the most common approach. Yield calculation methods differ among plants, ranging from weighing harvested grain from the whole field to weighing representative samples from a plot area after physiological maturity has been achieved. Definitions of measures used to calculate productivity include: yield (e.g. output per unit of property, water, power, nutrients, and labor) Profit (e.g. gross margin, net present value) Labor (e.g. individual hours, sex labor allocations). Other metrics for calculating food safety include: per capita food consumption, e.g. calories, nutrition, dietary diversity food deficits.

Mitigation Measuring

UNFCC has established platforms for mitigation measures. Emissions of methane, nitrous oxide and carbon dioxides from all agricultural sources including fuel, soil and carbon emissions from biomass and soils Modifications in land use, in particular forest and peat land use of high-continuity land.

FINANCING OF CSA PLAN

The CSA can provide a number of capital services, such as agriculture growth& food security. While the main source of climate financing is private investment, public funds remain key drivers of the climate finance system. The main stream of agricultural finance is domestic investment both from public and private sectors.

Figure 5. Financial CSA plan
(Source: Created by author, 2021)

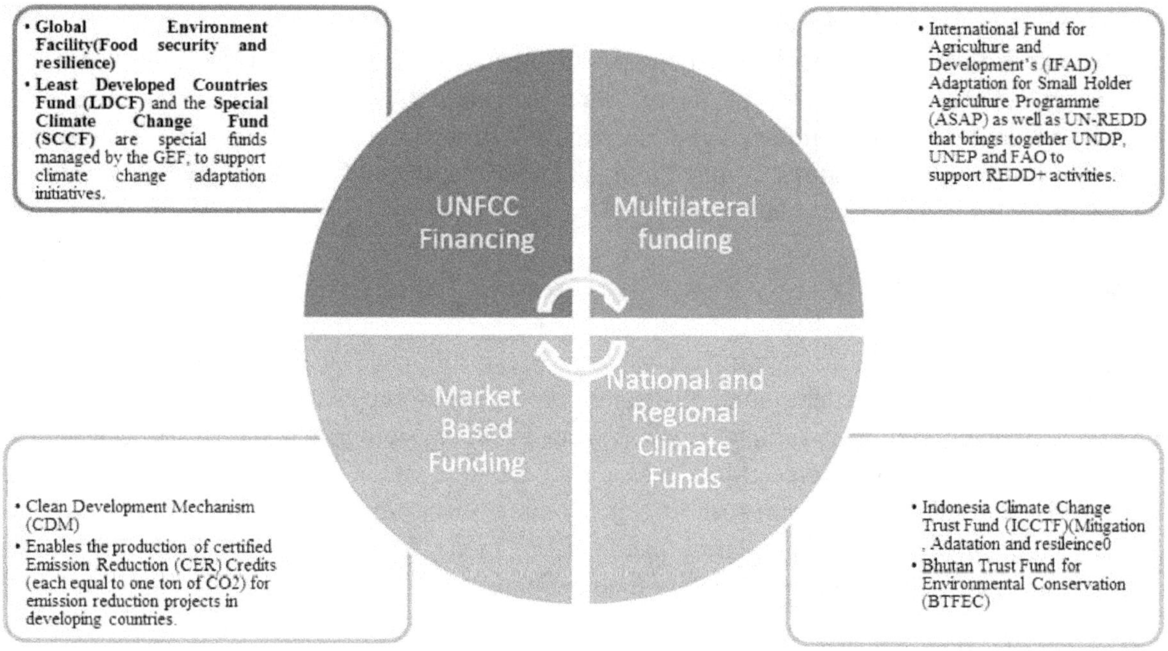

The Financial Mechanism is partly assigned to the Global Environment Facility. The GEF-6 will encourage approaches to reducing emissions of nitrous oxide (N_2O) and methane (CH_4) from crops and livestock production over the sixth program period 2014-2018 (Nakhooda et al.2016).The Green Climate Fund (GCF) is an operational body of the financial system of the UNFCCC and will invest US$ 100 billion a year by 2020, including all public and private modifications and mitigation in all sectors (FAO 2013a).The LDCF and the Special Committee on Climate Change (SCCF) are additional funds managed by the GEF as shown in figure 5. The Adaptation Fund (AF) was established in 2001 to finance in developing countries parties to the Kyoto Protocol specific adaptation schemes and initiatives, especially vulnerable to adverse consequences of climate change. CSA financing is multilateral outside the UNFCCC Financial Mechanism, including UN-REDD, UNDP, UNEP, and FAO Adaptations for REDD+ Activities (IFAD) for Agriculture and Production, the IFAD Adaptation for Small Holder Agriculture Program. Private financing through a mix of cooperative and regulatory schemes is feasible. The Clean Development Mechanography (CDM) allows for accredited EDR credits (each equal to 1 ton of CO2) in developing country EDR programs. National Climate Funds (NCF) and Regional Climate Fund (RCFs) provide support to countries in the management of their climate finance commitment through the facilitation of climate finance compilation, matching, coordination and accounting (United Nations Development Programme,2011). Most NCFs and RCFs fund projects contribute directly to agriculture and become an important source of CSA funding.

CASE STUDIES

Climate smart farming is more pronounced these days, as it is a source of enhancing the livelihood and creating employment opportunities for communities. Various approaches have been followed both in developing as well as developed nations to counteract climate change impacts on agriculture and affecting livelihood. Roughly 84% of the world's chickpea production is in South and South-East Asia. Many crops are fed to the rain and planted after the rainy season in soil humidity. Dryness and heat pressure can often be experienced. Due to extreme droughts and high temperatures, the cultivation period is as low as 90 to 155 days in two thirds of the harvesting area of chickpea, while pots are filled after raining. ICRISAT (i.e. ICCV 2) is the first extra short kabul cultivar to display resistance to fusarium and thermal sensitivity, in a span of only 85–90 days in Andhra Pradesh, India (International Crops Research Institute for the Semi-Arid Tropics,2012). This gives rise to five time increase in area (102 000 to 602 000 ha) and two time increase (583 to 1,407 kg ha-1) in output level.

Small farms in the Baramulla, Bandipora and Pulwama districts of Kashmir had to face increasingly unstable conditions and poor lands. Through replacing the yearly corn crop through hardy, annual lavender, the chance of low crop yield and full crop failure decreasing. Perennial plants over the span of 20 years may lead to the reduction of climate change by increased carbon sequestration below the roots. This ensures the crops are subject to significant losses due to repeated crop failure. Strong support by government through the federal funding has also enabled the cooperative to provide new growers with planting material and training and to link them with international markets successfully.

The climate change implications are of key importance for the Australian marine food industry, including both fisheries and aquaculture. Recent study by (Camacho et al.2014) showed that a comprehensive food system approach, which explores the entire supply of marine resources, provides chain-wide mitigation approaches such as shifting to less vulnerable and more open areas; a hierarchical distribution of operations; increased use of weather data and efficient use of resources.In the 1970's, biogas processing was implemented in Indonesia (Food and Agriculture Organization of the United Nations, 2014) even when their installation cost were higher than expected today. The use of animal waste for bioenergy decreases reliance on coal that can offset the climate change. The major grants received from State fossil fuels, and the availability of cheap wood fuel, represent other important barriers to development. Therefore, domestic biogas by-products are organic fertilizers for farm production. It also contributes to increasing rural communities ' access to food and electricity, reduces adverse health impacts and enhances their livelihoods. A low carbon farming plan of Brazil (ABC): The ABC Programme, a lending program that will provide low interest loans for farmers wanting to implement sustainable farming methods. These include degraded field regeneration, commercial tree planting, microbial nitrogen fixation, animal waste management, and crop, livestock and forest incorporation. The plan was designed to expand the area from 25 million hectares to 33 million hectares under zero tillage by 2020, in 15 million hectares on degraded pasturage. It also seeks to reduce emissions of greenhouse gasses to 160 million tons of CO_2 per year by 2020. Besides this, more than ten Climate-Smart Agriculture Investment Plans (CSAIPs) have been established by the World Bank for Bangladesh, Zimbabwe, Zambia, Lesotho, Mali, Burkina Faso, Ghana, Cote D'Ivoire, Morocco, and the Republic of Congo. The CSAIPs identify CSA initiatives worth more than US$2.5 billion that could benefit more than 80 million people in the countries covered. The Bank is supporting sustainable agricultural output in Uruguay through a number of projects, including the creation of an Agricultural Information and Decision Support System and the development of soil management plans. Since 2014, CSA has been implemented on 2,946,000 hectares, with 5,139 farmers receiving assistance to make their farms more climate-smart by improving energy efficiency and soil management (World Bank Group,2021).

The West Africa Agricultural Productivity Program (WAAP) brings together 13 countries and various partners to assist produce climate-smart rice, plantain, and maize cultivars. Farmers also gain access to technologies such as water-harvesting systems that are more efficient. By July 2019, the initiative had directly assisted over 9.6 million people and 7.6 million hectares of land in becoming more productive, resilient, and sustainable. Beneficiary yields and incomes have increased by roughly 30% on average, boosting food security for the region's 50 million people.

Geospatial technology is being used by Ravensdown, the main fertilizer manufacturer, and distributor in New Zealand, to improve farmers ' control of their production in New Zealand. With the GIS and GPS, farmers can reduce the emissions that can inflict harmful runoff on rivers and streams by specifically using fertilizers. They also reduce their total fertilizers budget by up to 10% annually. About 81percent of the population were small-scale peasants. High production with limited risk-taking ability among farmers, along with limited exposure to green technology and economical amenities results into inefficient crop yield and agricultural output.Landscape-level research to establish an action strategy for climate-smart agriculture .Kusamala, an NGO is using GIS application to spread the smart agriculture concept in 2013 for 1500 families of Dowa and Lilongwe district.

Farmers have adopted intelligent environment activities in the Indian state of Haryana, for instance, laser land raising and alternative rice wetting and drying, reduction in water use, better soil quality and economic incentives. Farmers receive agro-advisories from offices, scientists, input manufacturers, and farmers on their mobile phones, which enable them to make up time.

GIS: THE GEOGRAPHIC APPROACH FOR AGRICULTURE

ESRI's GIS software is used by all industries of the agriculture industry to share data, maximize returns, forecast performance, and optimize business practices. Agriculture operations are able to manage capital and accountability more effectively by using GIS tools for their activities, create information repositories that disseminate large amounts of agricultural and digital maps and support farming communities. Following are the merits of using GIS:

- Improved research aid for precision farming
- Better understanding of risk factoring
- Higher income production and cost recovery
- Greater efficiency by robotic processes
- Increased access to government services and results
- Increased guidance on policy making
- Easier reporting
- Increased support for decision-making.

Following are the most commonly applied approaches:

- **Remote sensing**: Pictures and data gathered by aerial or spacecraft (satellites, ships, UAVs, terrestrial sensors). Many satellite image suppliers now have images of one meter and smaller data.

- **The GPS US network.** Global Positioning system Security satellites that can provide civilian and military users with precise organized places and facilities for the receiver (note that in the next few years the similar European system called Galileo will work while the Russian system operates but is limited).
- **The GIS is a framework for the recording,** evaluation, processing, maintenance, and display of all forms of spatial and geographical information.

GEOSPATIAL TECHNOLOGY AND PRECISION FARMING

Precision Agriculture (PA) is a field management method using IT to ensure that plants and soils achieve precisely what they need for optimal health and efficiency. Data specific to site (e.g. soil characteristics, fertility and nutrients statistics, topography, irrigation characteristics, production information, yield data assembly-assembled yield sensor data, remote-conscious plant indicators) collected and stored from various sources and retained within a regional database are site specific data.

Novel Technologies for Climate-Smart Agriculture

- Climate-smart Micro Radiometer: In low waters near surface and in deep waters, climate changes can be identified.
- Sensors Enable text messages on crops for farmers to improve the Earth's health, food safety and longevity.
- Photo-catalytic solutions: Advanced product engineering not only has exciting application in the future to keep surfaces dry, but it may release germs.
- Use of robotics tractors, visualization of vegetation and digital accounting.

Agriculture Land Suitability

Major variety of crops is grown on different subsets of soil which provides nutrients to crops.

The main of GIS technology in agriculture is the agriculture land suitability to increase the productivity and crop growth which in turn boost the yield as well as farmers income. Following are the two approaches adopted for the land suitability index:

- Temporal Agriculture land Analysis: This section will include gross sown area, net cultivated area, total uncultivated land and non-agriculture land uses.
- Identification of major crops: Major crops grown in the area. Parameters such as crop area growth rate, crop production growth rate, and major crop commodity.

Figure 6. Agriculture land suitability
(Source: Created by author, 2021)

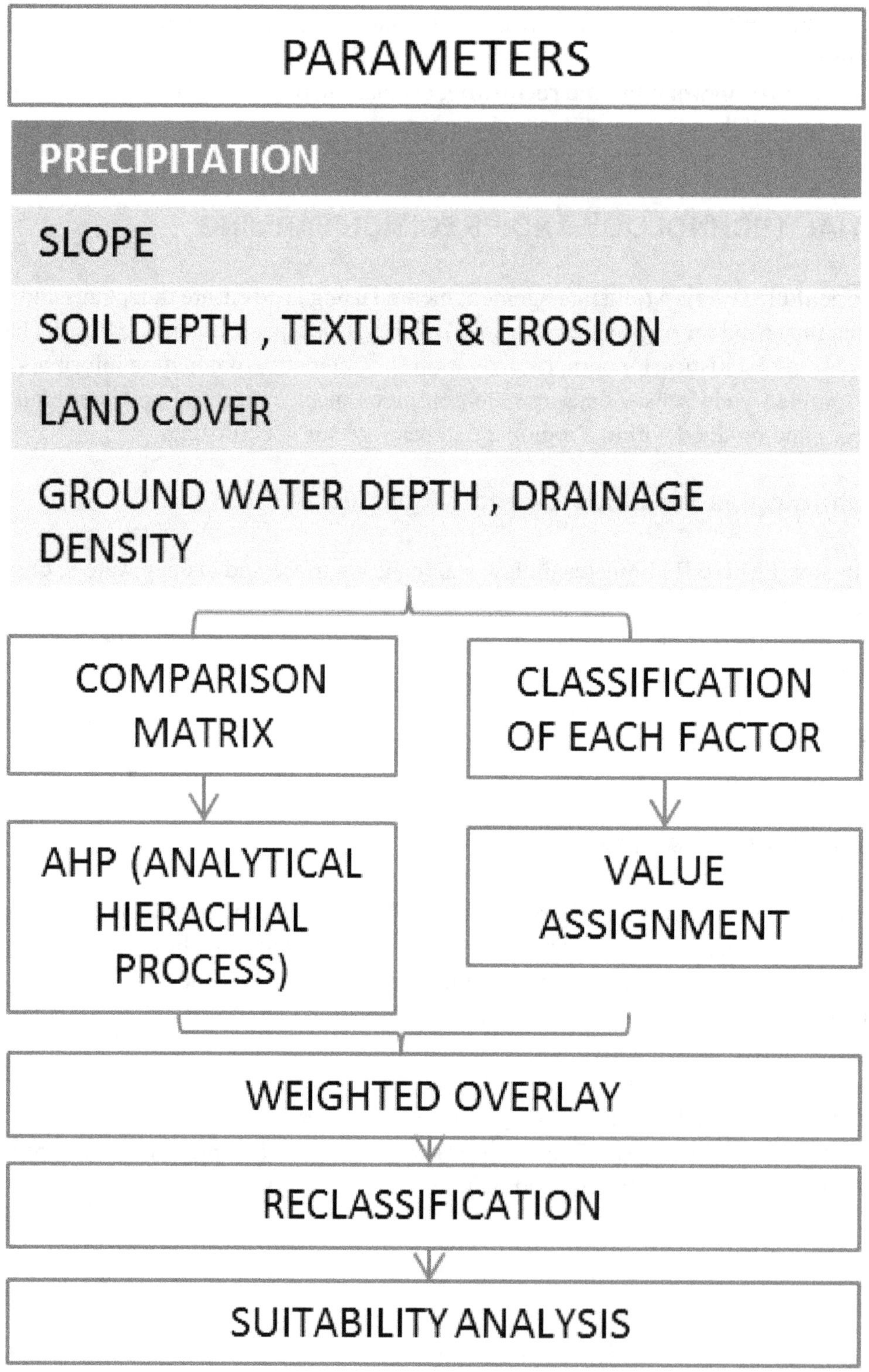

For agriculture suitability, firstly major crops are identified. Various influencing factors like: precipitation, slope, soil texture, soil depth, soil erosion, land cover, groundwater depth and drainage density are taken into account. Then a pairwise matrix comparison is done by classifying each factor. Using AHP (Analytical Hierarchical Process) Technique, weighted overlay is done (superimposing all the influencing factors) by assigning weightage to their contributing factor as shown in figure 6.

Agro Geo-Spatial Models in India

In partnership with NASA-ISRO-MoA, major National Level Programs are as follows:

- **FASAL** (Forecasting Agricultural output using Space, Agrometeorlogy& Land based observations)
- **NADAMS** (National Agricultural Drought Assessment & Monitoring System)
- **CHAMAN** (Coordinated Horticulture Assessment and Management using geoinformatics)
- **KISAN** (C[K]rop Insurance using Space Technology And geoinformatics)

CONCLUSION

Anthropology and nature coexist. Human are majorly dependent on agriculture produce in their daily life. Besides their dependency, they are freely using nature's goods without compensating which is leading to climate change i.e. Higher atmospheric carbon dioxide level. There should exist balance between supply and demand chains. Climate change is showing adverse impact on agriculture thus decreasing yield, affecting farmer's income, lending land by farmers due to crop failure. Spatial planning aspects as well as technical considerations would help in improving the conditions and determining the land suitability for other crops would cover large scale to avoid food insecurity and brings economic prosperity. Rather than mitigating the negative externalities, adaptation should bridge the gap between success and failure of crop. Adaptation strategy would include crop management, land management and afforestation... etc. Integrating IT solution for agriculture with system approaches for agriculture would enhance the livelihood of farmers as they would be ready with their adaptation strategy to combat climate change. Local participation as well as local institutions network should be set up to deliver goals and funds for smart initiatives. Decentralized planning of agriculture management would deliver the best content and subject of concern at hierarchy. Agriculture forms a supply chain from Farm land to consumers. Various phases from production to consumption are crucial for assessment in spatial as well as technical terms. Agricultural management trainings should be given to poorly deprived farmers to increased sustenance farming as well as increasing resilience through capacity building without comprising the needs of environment. For example, the programme, which is part of a strategy to help Sri Lanka's smallholder farmers develop resilience to climate change and extreme weather events, was launched by the International Water Management Institute (IWMI) in collaboration with local partners (Online,2021). Farmers receive climate risk insurance, agro-climatic consulting, and climate resilient seedlings as part of the bundled solutions package. The insurance benefits are determined by the appropriate insurance provider using a "parametric index" created by IWMI to determine agricultural production losses using satellite data. Also, there is an emerging need to account emissions inventory at each phase of Agriculture production to consumption so that best practices can be more hybrid and located at its best spatial terms in environment. In addition to research among these vocals, it would also serve or incubate models that include adaptation scenarios at farm level; capacity building approaches for agronomic challenges; also overcome the present formidable barriers for farmers during adoption of new technology and practices and an understanding of how climate affects the rural labour, land tenure thus affecting the crop productivity and food security. Effective coordination in CSA will include stakeholders, addressing uncertainties, evolving social benefits with techo-change and establishing green model of green climate fund.

REFERENCES

Actionaid. (2011). *Say no to soil carbon markets: Six reasons why soil carbon markets won't work for smallholders*. Author.

Agrawal, A. (2008). The role of local Institutions in Adaptation to climate change. Natural Resources and Environment, University of Michigan. doi:10.1596/28274

Alexandratos, N., & Bruinsma, J. (2012, June). *World Agriculture towards 2030/2050: The 2012 revision*. Agriculture Development Economics Division, FAO.

Alliance for Internet of Things Innovation. (2015). *Smart Farming and Food safety Internet of Things Applications-Challenges for large Scale Implementations*. Author.

Andrieu, N., Howland, F., Alba, I., Coq, J., Garcia, A., Baron, D., Trimin, C., Loboguerro, A., & Chia, E. (2019). Co-designing climate smart farming systems with local stakeholders: A methodological framework for achieving large-scale change. *Frontiers in Sustainable Food Systems*, *3*, 1–19. doi:10.3389/fsufs.2019.00037

Camacho, L., Hobday, A., Bustamante, R., Farmery, A., Fleming, A., Frusher, S., Green, B., Lopez, A., Plaganyi, P., Schrobback, P., Thebaud, O., Thomas, L., & Putten, I. (2014). Facing the wave of change: Stakeholder perspectives on climate adaptation for Austrailian seafood supply chains. *Environ Change*, 1-12.

Cooper, P., Dimes, J., Rao, K. P. C., Shapiro, B., Shiferaw, B., & Twomlow, S. (2008). Coping better with current climatic variability in the rain-fed farming systems of sub-Saharan Africa: An essential first step in adapting to future climate change? *Agriculture, Ecosystems & Environment*, *126*(1-2), 24–35. doi:10.1016/j.agee.2008.01.007

Corsil, S., Friedrich, T., Kassam, A., Pisante, M., & Moracs, J. (2012). Soil organic carbon accumulation and greenhouse gas emissions reductions from conservation agriculture: A literature review. *Integrated Crop Management*, *16*, 1–104.

DeFries, R., Mondal, P., Singh, D., Agrawal, I., Fanzo, J., Remans, R., & Wood, S. (2016). Synergies and trade-offs for sustainable agriculture: Nutritional yields and climate-resilience for cereal crops in Central India. *Global Food Security*, *11*, 1–10. doi:10.1016/j.gfs.2016.07.001

Economics, T. (2021). *India GDP from Agriculture*. https://tradingeconomics.com/india/gdp-from-agriculture

Food and Agriculture Organization of the United Nations. (2008). *Climate–related Transboundary Pests and Diseases*. Author.

Food and Agriculture Organization of the United Nations. (2012). Mainstreaming climate-smart agriculture into a broader landscape approach. *Second Global Conference on Agriculture, Food security and Climate Change*, 1-34.

Food and Agriculture Organization of the United Nations. (2013). *Climate Smart Agriculture Sourcebook*. Author.

Food and Agriculture Organization of the United Nations. (2014). *Small-Scale Bioenergy Initiatives in ASEAN +3*. Author.

Gaffney, J., Bing, J., Byrne, P., Cassman, K., Ciampitti, I., Delmer, D., Habbena, J., Lafitte, R., Lidstroma, U., Porter, D., Sawyer, J., Schussler, J., Settern, T., Sharp, R., Vyn, T., & Warner, D. (2019). Science based intensive agriculture: Sustainability, food security, and the role of technology. *Global Food Security*, *23*, 236–244. doi:10.1016/j.gfs.2019.08.003

Glover, J., Cox, C., & Reganold, J. (2007). Future Farming: A return to roots? *Scientific American*, *297*(2), 82–89. doi:10.1038cientificamerican0807-82 PMID:17894176

Greatrex, H., Hansen, J., Garvin, S., Diro, R., Blakeley, S., Guen, M., Rao, K., & Oagood, D. (2015). Scaling up index insurance for smallholder farmers: Recent evidence and insights. *CCAFS,* 1-32.

Gupta, J. (2010). A history of International Climate Change Policy. *Wiley Interdisciplinary Reviews: Climate Change*, *1*(5), 636–653. doi:10.1002/wcc.67

Huyer, S. (2015). Supporting women farmers in a changing climate: Five policy lessons. *CCAFS,* 1-8.

International Crops Research Institute for the Semi-Arid Tropics. (2016). *Building Climate-Smart Villages: Five approaches for helping farmers adapt to climate change*. Patancheru.

International Fund for Agricultural Development. (2012). Climate-smart smallholder agriculture: What's different? International Crops Research Institute for the Semi-Arid Tropics.

Kumar, R., & Gautam, H. (2014). Climate change and its impact on agricultural productivity in India. *Journal of Climatology & Weather Forecasting*, *2*(1), 1–4. doi:10.4172/2332-2594.1000109

Lal, R. (2004). Soil carbon sequestration to mitigate climate change. *Geoderma*, *123*(1-2), 1–22. doi:10.1016/j.geoderma.2004.01.032

Leeuw, J., Njenga, M., Wagner, B., & Iiyama, M. (2014). An assessment of the resilience provided by tress in the drylands of Eastern Africa. World Agroforestry Centre.

Nakhooda, S., Watson, C., & Schalatek, L. (2016). The Global Climate Finance Architecture. *Climate Funds Update,* 1-5.

Online, T. (2021). *Hi-tech solutions to help farmers combat climate crisis*. https://www.timesonline.lk/features/Hi-tech-solutions-to-help-farmers-combat-climate-crisis/5-1135291

Scoones, I. (2009). The politics of global assessments: The case of the International Assessment of Agricultural Knowledge, Science and Technology for Development (IAASTD). *The Journal of Peasant Studies*, *36*(3), 547–571. doi:10.1080/03066150903155008

Skot, J. (2016). *The State of Food and Agriculture: Climate Change, Agriculture and Food security*. Food and Agriculture Organization of United States.

The Intergovernmental Panel on Climate Change. (2015). *Climate change 2014 Synthesis report*. IPCC.

United Nations Development Programme. (2011). *Blending Climate Finance Through National Climate Funds*. Anvil Creatiosn Group.

World Bank Group. (2021). *Climate smart agriculture*. https://www.worldbank.org/en/topic/climate-smart-agriculture

World Resource Institute. (2016). *Climate Watch*. https://www.wri.org/initiatives/climate-watch

Chapter 13
Present and Future Land Suitability Analysis for Almond and Pistachio Crops in the Beira Baixa Region Using Spatial Multicriteria Decision Systems

Luís Quinta-Nova
https://orcid.org/0000-0002-8464-7527
Escola Superior Agrária, Instituto Politécnico de Castelo Branco, Portugal

Dora Ferreira
VALORIZA – Research Centre for Endogenous Resource Valorization, Polytechnic Institute of Portalegre, Portugal

ABSTRACT

The objective of this study is to determine the suitability for the cultivation of emerging fruit crops in the Beira Baixa region. The suitability was examined for the present time and in the face of two future emission scenarios (RCP 4.5 and 8.5). For this purpose, the biophysical criteria determining the cultivation of pistachio tree and almond tree were processed using a G. The analysis was performed by the AHP. After dividing the problem into hierarchical levels of decision making, a pairwise comparison of criteria was performed to evaluate the weights of these criteria, based on a scale of importance. In the present conditions, about 16.4% of the study area is classified as highly suitable for almond tree and 15.9% to pistachio tree. For the future scenarios, the area with high suitability will increase both for almond tree and pistachio tree. The AHP was adequate in the evaluation of the emerging fruit tree species suitability, since it allowed the integration of the several criteria studied, being a useful tool, which allows the decision making and the resolution of problems.

DOI: 10.4018/978-1-7998-9557-2.ch013

INTRODUCTION

The current trends indicate that are emerging a new consumers profile that are looking for an increase of variety, freshness and healthy options and they are also seeking a higher proportion of fresh and different produce in their eating choices (Campos & Madureira, 2019).

The concept of sustainability agriculture indicate that is necessary to promote the accountable use of resources, such water, and nutrients, and decreasing the use of pesticides in a way which considers the future needs and will not compromise the quality of the environment. Which indicates that is important to respect the agroclimatic conditions and identify and promote the appropriate adaptation strategies of the crops, especially in semiarid Mediterranean areas, where the water is the most limiting natural resource (Schaldach et al., 2021). According to the strategies of the Council of the Union on the European Innovation Partnership "Agricultural Productivity and Sustainability" (COM 2012), it is necessary to promote more sustainable use of natural resources and implement a competitive and sustainable production of food within a global compromise to reduce the impact on climate change.The climatic conditions in Beira Interior Region are characterized by low and erratic rainfall leading to water scarcity during the driest period of each year, and the high temperature during the summer period. However the by soil and climate conditions favorable to peach and cherry production (Simões et al., 2015), representing the crops with more public investment and expansion (Lopes, Alberto, Luz, & Simões, 2018). It is the most productive region of these fruits in Portugal represents 41% of the national area of peach production and 47% of the cherry production area (INE, 2021). However, in the last decade that region is seeking new patterns and configuration of agrarian landscape, through the foreign investments and benefiting from new irrigation infrastructures, expanded new areas of prunoids and introducing the new fruit crop especially almond tree and irrigated olive grove (INE, 2021).

The emerging crops that are identified in the aim of this study are the almond tree and the pistachio tree, that are distributed by different areas across the Beira Interior region (region that include Beira Baixa), that occupying around 3,800 ha (INE, 2021). In fact, the irrigation infrastructure availability made this region particularly attractive to the foreign investment that contribute to introduce new cultures and new patters of natural resources management, promoting rural innovation and showing the resilience of farmers. To these evidence joins other facts, especially the increase of market demand motivated with the health, gastronomic, and industrial properties that are linked to these fruits and their derivatives (Sottile et al., 2020).

So, these agricultural crops were selected due to their growing trend because the farmers be able to obtain more productivity by installing large-scale orchards in intensive and super-intensive mode, with higher productivity associated with the selection of varieties and greater availability of water for irrigation (Cabo, Matos & Bento, 2016).

Despite these advantages, there are associated some environmental risks that is important to prevent, especially that have been reported due to the installation of these cultures in an intensive and super-intensive mode (Malki et al., 2017). Those cultivation modes cause environmental problems as contamination of soil and water, reduction of biodiversity and soil degradation, among others, mainly derived from practices used and agrochemical products regularly used in treatments (Calatrava et al., 2021), further aggravated by the announced lack of water reserves (Sottile et al., 2020). Therefore, it is important to identify the areas with the suitability for these crops and create information to support the decision of future investments, as well prevent some consequences facing climate change scenarios and encouraging futures investments in sustainable production models and bet to native species not irrigated.

This question is especially problematic if agriculture models are dependent of irrigated cultures at the expense of species adapted to local conditions.

EMERGING FRUITS CROPS IN BEIRA BAIXA REGION: PISTACHIO AND ALMOND

According to FAO (2019) the global production of almonds in their shells is estimated to be just over 3,497,148 t in an area of 2,126,304 hectares of almond trees (*Prunus dulcis* (Mill.) D. A. Webb). The United States has the leadership (55,4%) of the world's almond production, which was followed by Spain (9,7%), Iran (5,1%), Turkey (4,3%), and Australia (4,2%). Pistachio (*Pistacia Vera* L.) which is a sub-tropical plant belongs to the Anacardiaceae family, has an estimated world production of 911,829 t in an area of 1,034,796 hectares. The main world producers are located in the Middle East with 49,9% (Iran, Turkey and Syria), United States of America (36,8%), and China (11,6%) (FAO, 2019).

The pistachio tree and almond tree are cultivated predominantly in modern orchards with high yielding varieties, intensive planting designs and farm management focused on competitiveness and economic rentability. In relation to almond production, Europe have lower productivity performance than USA, due by more limitation of irrigation level and, consequently, less productivity and competitiveness in the market. Although both plants are adapted to the poor soils and adverse climate conditions, the deficit of water compromises their profitability (García-Tejero et al., 2017). Added to this limitation, the rising global demand by dried fruits, has promoted the increase of irrigated orchards and intensive models of productions and new expansive patterns of this cultures, that tend to occupy new agricultural areas, abandoned areas or land use conversion, especially in irrigated areas traditionally occupied by other crops. However, that create more water supply demand and important water shortage problems that is important prevent and promote alternatives that don't compromise the competitiveness of the sector, such as a deficit irrigation strategy, extensive planting designs and understand the real water needs for each variety in different stages of growing. These facts lead to urgent adoption to sustainable irrigation strategies for both crops (Durán-Zuazo et al., 2020).

The profitability and prices in the market of dried fruits have influenced the decision for new investments in Portugal, where both species registered an expansion of cultivated area in the last decades, especially resulting by foreign investments or new projects supported by the new young farmers with European support.

The pistachio crop is an innovative specie in Portugal, as well the almond tree in same regions of Portugal, specifically in Beira Interior Region where predominate the cherry tree and peach tree as already mentioned. This region have technical support to the farmers installed and a close relationship between production and investigation that is supported by associations and R&D institutions that create new knowledge to solve problems related to production in traditional cultures.

But technically that region is recommended for these species due their larger areas (> 10 ha), its edaphic conditions and strong tradition in fruit growing. These factors have influenced the prospecting for the installation of pistachio and almond tree cultivation, that tripled their area between 2010 (1,050 ha) and 2020 (3,776 ha), with impact in their productivity, actually reach around 3,756 ton (2020) according to last data (INE, 2021).

However, the use of spatial planning tools proves to be extremely important in supporting the decision to set up agricultural projects, especially for young farmers and, particularly, in regions without the tradition of its cultivation and with gaps in logistical support for the farm management, technical support and knowledge transfer. Furthermore, facing the challenges to improve the economic viability and the environmental challenges, this study aims to understand the potential of sustainable development of these emerging crops and highlight the natural conditions projected to the future.

DECISION MAKING IN AGRICULTURE BASED ON LAND SUITABILITY

Facing the trends of climate change is urgent to address agricultural adaptation more coherently and promote the suitability of plant, weather, and soil conditions. Hence, that investigation identify the suitable sites for the emergent crops production using a model based on multicriteria spatial analysis AHP. The suitable areas for the emerging crops are determined by an evaluation of the climate, soil, and topographical factor and the understanding of local biophysical restraints. In this kind of situation, many variables are involved and each one should be weighted according to their relative importance on the optimal growth conditions for crops through multicriteria evaluation and Geographic Information Systems.

Multicriteria decision analysis (MCDA) has been widely applied in various studies in different fields, many of which are published and have been cited by many authors as processes of relevant decision making. Various techniques of MCDA have been used extensively in land use suitability for agricultural crops (Alkimim, Sparovek, & Clarke, 2015; Dedeoğlu & Dengiz, 2019; Elaalem, Comber, & Fisher, 2011; Mighty, 2015; Wotlolan, Wales & Glencross, 2021; Zhang, Su, Wu, & Liang, 2015). While many MCDA applications do not incorporate a spatial dimension to the analysis, MCDA lends itself well to GIS and the use of GIS within an MCDA framework is common (Malczewski, 2006; Malczewski & Rinner, 2015), particularly in forestry and agriculture (Alkimim et al., 2015; Cóndor, Scarelli, & Valentini, 2010; Kangas & Kangas, 2005; López et al., 2020; Mighty, 2015; Phua & Minowa, 2005).

An early example of the use of a GIS-MCDA application for agroforestry is that of Laskar (2003) who developed a methodology to identify whether locations in the study area were most suitable for agroforestry, economic plantation (i.e. production forestry), or energy plantation (i.e. fuel and fodder). Quinta-Nova and Roque (2018) used soil and climatic factors, complemented with socio-economic criteria within a GIS-based multicriteria analysis to spatially identify locations most suited to production forest, intensive agriculture, or multifunctional use. Others have proposed web-based multi-criteria spatial decision frameworks (Hamilton et al., 2016; Yalew, van Griensven, & van der Zaag, 2016). Modica et al. (2016) proposed a web-based multi-criteria spatial decision support system capable of performing land suitability evaluation that would be accessible to environmental managers and planners to aid in decision-making in real-time.

In the study presented in this chapter the Analytic Hierarchy Process (AHP) was used as a MCDA technique with a GIS to identify the suitable areas which can be exploited for the fruit production with new crops in the Beira Baixa region, by the integration of several criteria. AHP is an inductive, multicriteria method, effective in the estimation of the likelihood of unique events, and in the face of multiplicity, uncertainty, and the limitation of information (Saaty, 1993, 1996; Malczewski, 2006).

The suitability analysis was examined for the present time and in the face of two future emission scenarios (RCP 4.5 and 8.5), because it is essential to be aware of the suitability and resilience of new crops to meet the need to adapt to climate change.

MATERIAL AND METHODS

Study Area: Beira Baixa Region, Portugal

The Beira Baixa region is an administrative division in eastern Portugal and integrate the Beira Interior Region. The region covers an area of 4,614.6 km^2 and has a population of 84,046 inhabitants. The area includes the municipalities: Oleiros, Proença-a-Nova, Idanha-a-Nova, Penamacor, Vila Velha de Ródão and Castelo Branco. This territory is mainly occupied by forest and agroforestry uses (60.8%) and agriculture (36.2%).

The Beira Baixa is a low-relief region that is transitional between the Portuguese Central Range and the South Portugal Planation Surface. The region shows a Mediterranean climate with, predominantly, a wet cool season and a dry summer. Despite the concentration of the precipitation in winter, there is a high inter-annual variability, resulting from the latitudinal position in the south-western façade of Europe. During summer the average of higher temperatures above 29°C and more than 100 days with temperatures above 25°C and winters are cold, with an average of the lower temperatures between 1 and 4°C and 15 to 40 days with negative temperatures. The values for precipitation vary depending on the altitude, ranging between 600 and 1,200 mm (Cunha, 2008).

Figure 1. Study area location

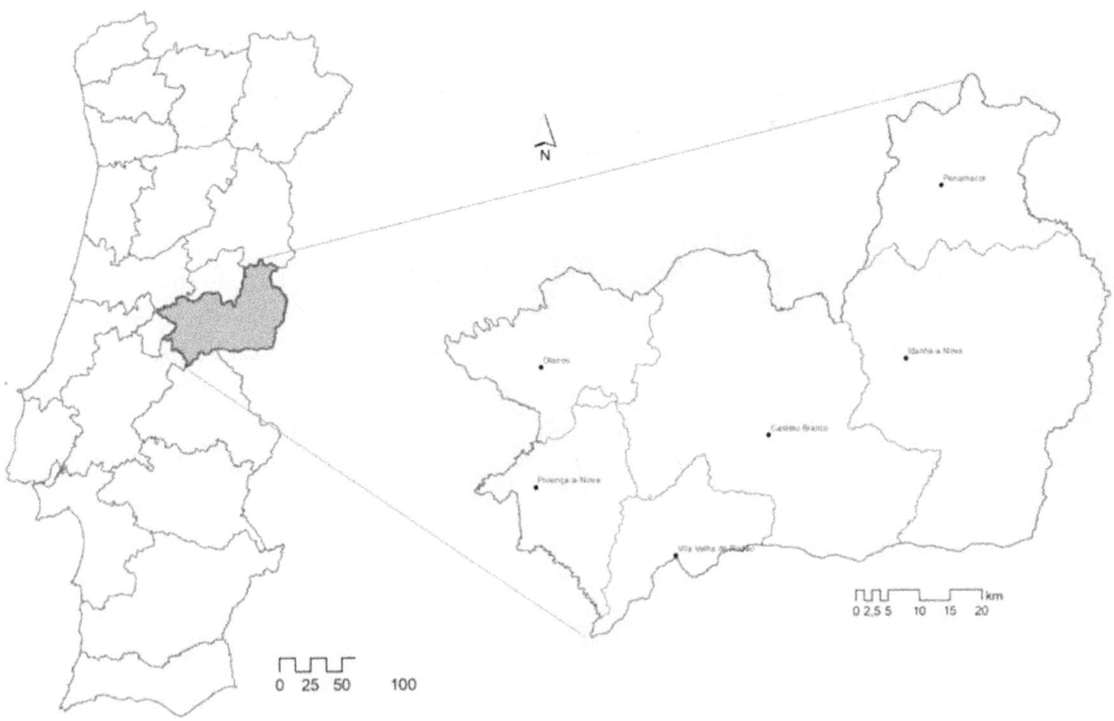

It is a territory that presents an economic weakness compared with other regions in Portugal that has been reflected in a significant population exodus in recent decades, as shown by the variation in the last decades. On the other hand, there are high levels of population aging, with a tendency towards aggravation (Carvalho, 2018).

This region is characterized by the most significant dimension of farm proprieties (around 29 ha) and where the farms with an average economic size, equal to or greater than 100,000 euros, represented more than 6.4% of the total farms in Beira Interior (Almeida, 2011). This region is characterized by the exponential growth of almond tree (INE, 2021), witnessing the installation of modern and intensive almond trees. But this region is one of the most important regions with permanent meadows and pastures with strong sheep production, olive oil production and flesh fruits production, especially peach and cherry (INE, 2021). According to the recent data (INE, 2021), Beira Interior region, especially Idanha-a-Nova and Castelo Branco municipalities, register an improve of farms certified for organic production (18%, that representing of 5% of SAU). This indicator demonstrates the relevance of new agriculture practices to protect and prevent the impact on the agriculture on the environmental.

FUTURE CLIMATIC SCENARIOS

A downscaled and bias corrected Coupled Model Intercomparison Project - CMIP5 (Taylor, Stouer and Meehl (2012). Data under representative concentration pathway (RCP) 4.5 for 2041-2060 was used for the future climate analyses, which is obtained from the WorldClim database. In the future conditions, we have considered two representative concentration pathways (RCPs) scenarios (RCP 4.5 and RCP 8.5) fitted for 2070.

Emissions in RCP 4.5 (Intermediate scenario) peak around 2040, then decline, resulting in global temperature rise between 1.1 to 2.6 °C by 2081-2100 (relative to 1986-2005). In RCP 8.5 (Worst-case climate change scenario) emissions continue to rise throughout the 21st century, resulting in global temperature rise between 2.6 to 4.8 °C by 2081-2100 (relative to 1986-2005).

ANALYTIC HIERARCHY PROCESS (AHP) APPROACH

The classification of the crops suitability resulted from the integration of a set of biophysical criteria based on the climate and soil requirements of crops and the optimal operating conditions associated with different uses. Geoprocessing and spatial analysis were performed to geographic data, namely soils, climate and elevation. In this study all the criteria (Table 1) are reflected in the corresponding GIS layers.

Table 1. Criteria considered in determining crop suitability

Criteria	Description
Mean annual temperature	Average of the 12 mean monthly temperatures (°C)
Mean total annual rainfall	Total annual depth of precipitation from a given precipitation time series (mm)
Chilling hours	Sum of hours with temperature ≤ 7.2 °C (h)
Crop heat units	Influence of temperature on a crop's growth and development (h)
Mean relative humidity	Ratio of the actual amount of water vapor present in a volume of air at a given temperature to the maximum amount that the air could hold at that temperature (%)
Elevation	Height above the Earth's sea level (m)
Soil pH	Measure of the acidity or alkalinity of a soil
Soil Organic Matter	Fraction of the soil that consists of plant or animal tissue in various stages of decomposition (%)

To characterizing the main climatic parameters that influence the crops studied in different stages and therefore its productive potential, are used the climate data of 7 meteorological stations around Beira Baixa region (Fundão, Castelo Branco, Portalegre, Coimbra, Guarda, Santarém in Portugal, and Caceres in Spain). In Table 2 are identified the stations nearest of the region and identified the main representative climatic parameters that influence crops' ecological conditions. Was observed that the mean of total annual rainfall goes by 523 mm (Cáceres) until 882 mm (Guarda). During the blossom period the rain can damage fruit production, specially combining with low temperatures (with risk of frost). The mean monthly minimum temperatures registered during the blossom period was around 3.5°C and 6,8°C and 11,2°C of minimum in Castelo Branco. Generally, the mean relative humidity during the fruit's growth with average of 62%, that parameter is important because influence the diseases risks of the crops and influence their capacity of evapotranspiration.

Table 2. Mean annual and monthly values of climatic parameters (1971-2000)

Parameters		Fundão	Castelo Branco	Portalegre	Cáceres	Coimbra	Santarém	Guarda
Mean total annual rainfall (mm)		842.9	758.3	852.4	523	905	696	882
Mean total monthly rainfall (mm)	April	78.5	58.1	78.4	39	84.8	65.7	83.1
	May	68.3	65.1	67.5	48	79.5	56.2	84.9
	Sept.	35.0	36.5	42.1	32	51.7	36.2	47.5
Mean monthly temperature (°C)	March	10.7	12.7	11.5	7.7	12.6	12.9	7.1
	April	12.0	13.1	12.3	9.6	13.9	14.1	8.0
	May	15.3	16.8	15.3	13.4	16.2	16.5	11.6
Mean monthly minimum temperatures (°C)	March	5.2	7.5	7.6	1.4	6.9	7.3	3.3
	April	6.8	8.0	8.2	3.5	8.4	8.6	4.2
	May	9.6	11.2	10.6	7	10.8	10.5	7.4
Chilling hours		1150	1050	950	950	705	750	1350
Mean relative humidity (Jun-Aug)		51.7	53.7	61.3	52.0	77.3	75	65

The different layers corresponding to each criteria were classified in two suitability levels: low to medium suitability (1) and high suitability (2) (Table 3). After creating layers resulting from the reclassification in suitability levels, the general suitability for each crop was performed using a multicriteria decision analysis - the Analytic Hierarchy Process - AHP (Saaty, 1980).

Table 3. Sub-criteria suitability thresholds (requirements) and restrictions for the crops in the Beira Baixa region.

Criteria	Pistachio tree		Almond tree	
	High suitability (2)	**Low to medium suitability (1)**	**High suitability (2)**	**Low to medium suitability (1)**
	Climatological factors			
Mean annual temperature (°C)	*	*	>16; <35	<16; >35
Mean total annual rainfall (mm)	<600	>600	>300; <1600	<300; >1600
Chilling hours (h)	<1200	>1200	<500	>500
Crop heat units (h)	>3550	<3550	*	*
Mean relative humidity (%)	<50	>50	*	*
	Edaphological and physiographic factors			
Soil pH	>6.0	<6.0	>5.5	<5.5
Soil organic matter	>1.5	<1.5	>1.5	<1.5
Elevation (m)	<1000	>1000	<1000	>1000

* Not a limiting factor for the crop. Data taken from: López et al. (2013), AJAP (2017), AJAP (2017a).

The Analytic Hierarchy Process (AHP) consists of four essential phases: criteria generating and spatial analysis, standardization, AHP, and suitability assessment.

First, a spatial database was created to include all vector and rater layers and data models. All spatial layers were prepared, and consistency of coordinates was maintained in ArcGIS 10.8 software. All criteria included in the analysis had to be standardized. Standardization makes all spatial layers constant and in the same measurement units' format (Saaty, 1980). Hence, all vector layers were converted into raster format and the reclassify tool in ArcGIS was used to standardize and assign values for each criterion.

The AHP decomposes a problem, question, or decision, in all the variables that constitute it, in a scheme of criteria and subcriteria and then makes pairwise comparisons between them (Antunes, 2012). The comparison between criteria is made using a scale of 1 to 9, wherein 1 is equally preferred and 9 is highly preferred (Saaty, 1980). The AHP reverts comparisons on numerical values that can be processed and compared to the full extent of the problem. The AHP calculates the weight value for each criterion (w_i) by taking the eigenvector corresponding to the largest eigenvalue of the matrix, and then normalizing the sum of the components to a unity. Pairwise comparison matrices were used with AHP software to value the selected factors and their classes.

The consistency of the matrix after obtaining the weight values is judged based on a consistency ratio CR. If $CR < 0.10$, the pairwise comparison matrix is considered to have acceptable consistency and the weight values calculated are considered valid and can be utilized. Finally, the spatial data was superimposed to integrate all the factors in a single layer - the suitability map for each crop.

Figure 2. Flow chart of the methodology used for the crop's suitability analysis

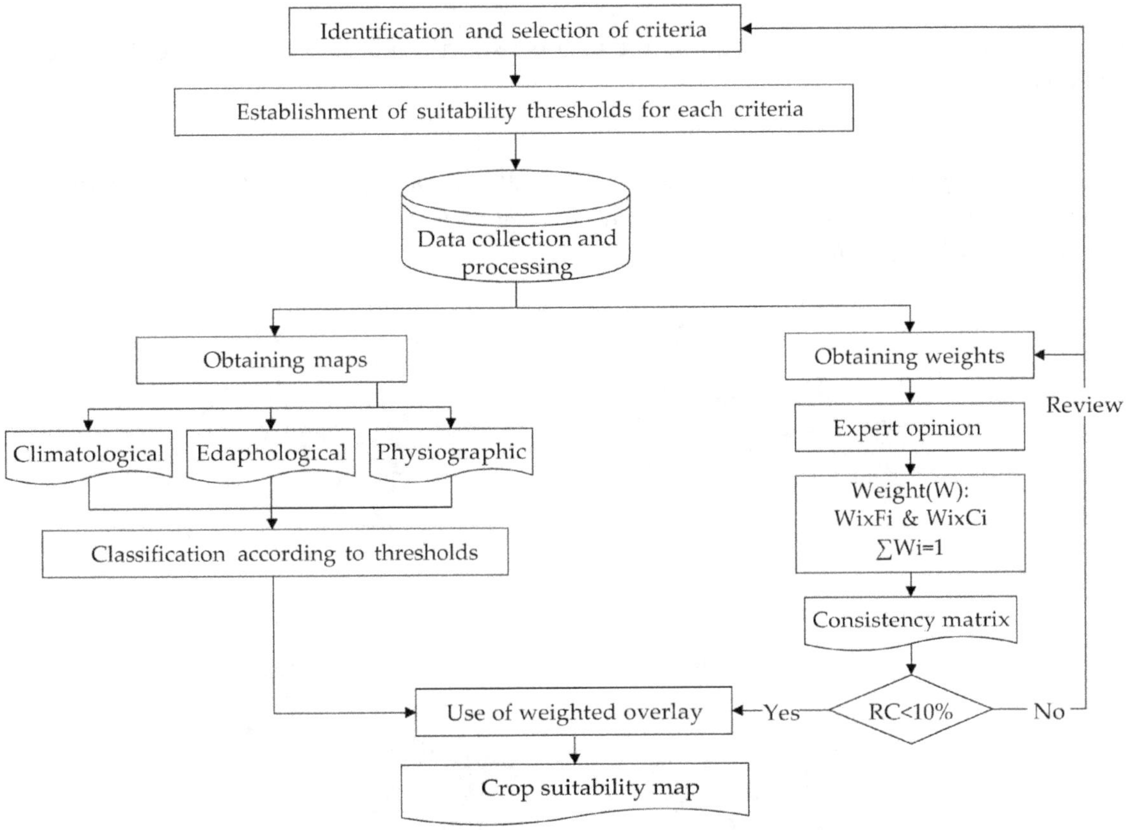

RESULTS AND DISCUSSION

The AHP analysis of the selected crops indicates that the climatic influence is determinant in the development of these species since its weight in the AHP analysis varies between 79.44% for Almond tree and 86.21% for Pistachio tree. The analysis of the consistency ratio (CR) of the AHP showed that there was consistency in the pairwise comparison matrix, and thus, the weight values calculated could be considered valid (Table 3).

Table 3. Criteria weights and consistency ratios - present time

Crop	Criteria weights	Consistency ratio
Pistachio tree	Mean total annual rainfall (25.08%); Chilling hours (25.08%); Crop heat units (25.08%); Mean relative humidity (10.97%); Elevation (4.60%); Soil Organic Matter (4.60%); Soil pH (4.60%)	0.008
Almond tree	Mean total annual rainfall (26.48%); Mean annual temperature (26.48%); Chilling hours (26.48%); Elevation (10.94%); Soil Organic Matter (4.81%); Soil pH (4.81%)	0.009

CROP SUITABILITY - PRESENT TIME

The Figure 3 shows the maps representing the factors used as criteria to perform the Analytical Hierarchy Process.

Figure 3. Base maps for the biophysical factors: a) DEM; b) Soil organic matter; c) Mean annual temperature; d) Mean total annual rainfall; e) Soil Organic matter; f) Chilling hours; g) Crop heat units and h) Mean relative humidity

Process.

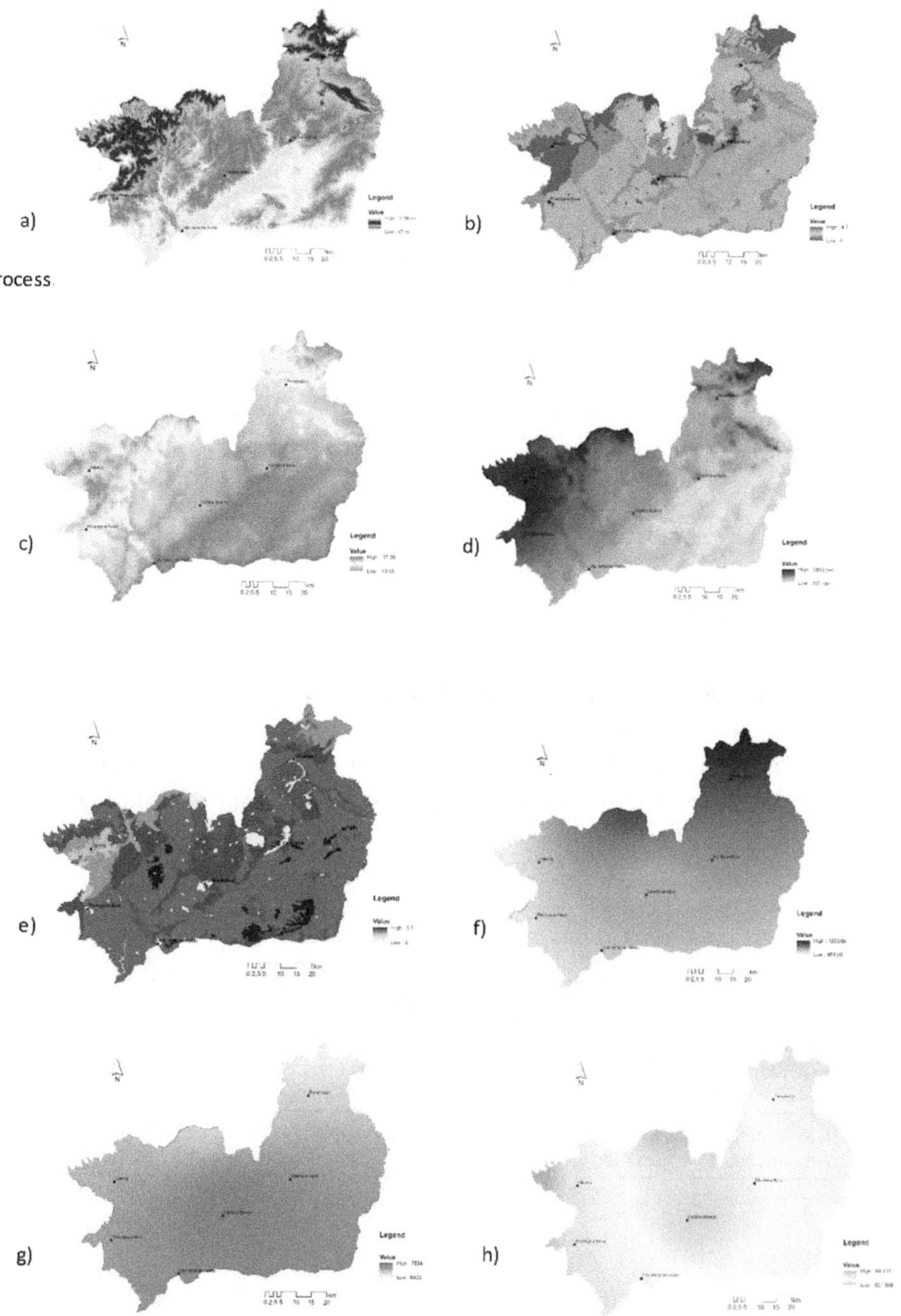

Based on the map analysis, about 752.35 km^2, corresponding to 19.7% of the total area available, are classified as highly suitable for almond tree, especially in the southeast part of the region. Low winter and spring temperatures and very high summer temperatures inhibit growth and fruit set. On the other hand, very high summer temperatures when accompanied with low soil moisture can result in the shrinkage of almond. The needs of almond in cold (to break the dormancy of the buds) are 250-350 (and in some cases over 500) hours of exposure to temperature less than 7 °C (Alonso, et al. 2005).

The area with higher potential to pistachio tree is 726.79 km^2, corresponding to 18.8% of the total area available and overlaps the potential almond tree crop area, resulting from natural conditions, especially the climatic influence. These plants are known as drought tolerant and can survive and even produce fairly yield with very little water (Ferguson et al., 2002). An annual rainfall of at least 300 to 450 mm has been reported as the optimum amount of precipitation for this crop (Goldhamer, 2005).
In Figure 4 we present the crop suitability maps resulting from the Analytical Hierarchy Process.

Figure 4. Suitability maps: a) Almond tree; b) Pistachio tree

a)

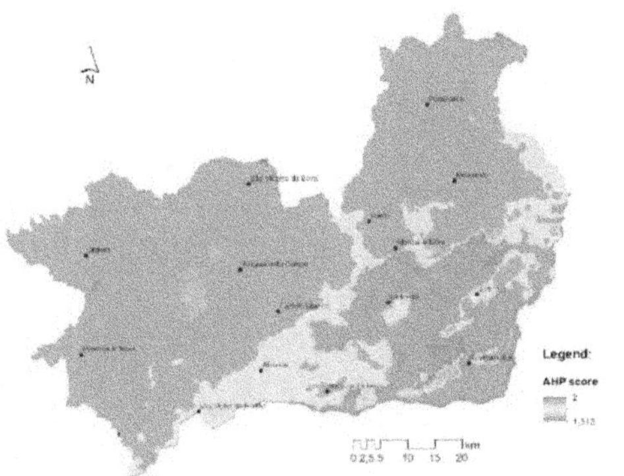

b)

CROP SUITABILITY - FUTURE CLIMATIC SCENARIOS

The future different scenarios outputs - RCP 4.5 and the RCP 8.5 for 2070, show an increase of suitable area for pistachio tree and almond tree compared to the present time (Fig 5). For the Intermediate scenario (RCP 4.5), the total area with high suitability for pistachio tree is identical to the area with high suitability for almond tree, corresponding to 1492.83 km^2 (49.3% of the total area). For the worst-case climate change scenario (RCP 8.5) the total area with high suitability for pistachio tree and almond tree as well is the highest with 1719.78 km^2 (61.7% of the total area).

Figure 5. Suitability maps: a) Pistachio tree - RCP 4.5; b) Pistachio tree - RCP 8.5; c) Almond tree - RCP 4.5; d) Almond tree - RCP 8.5

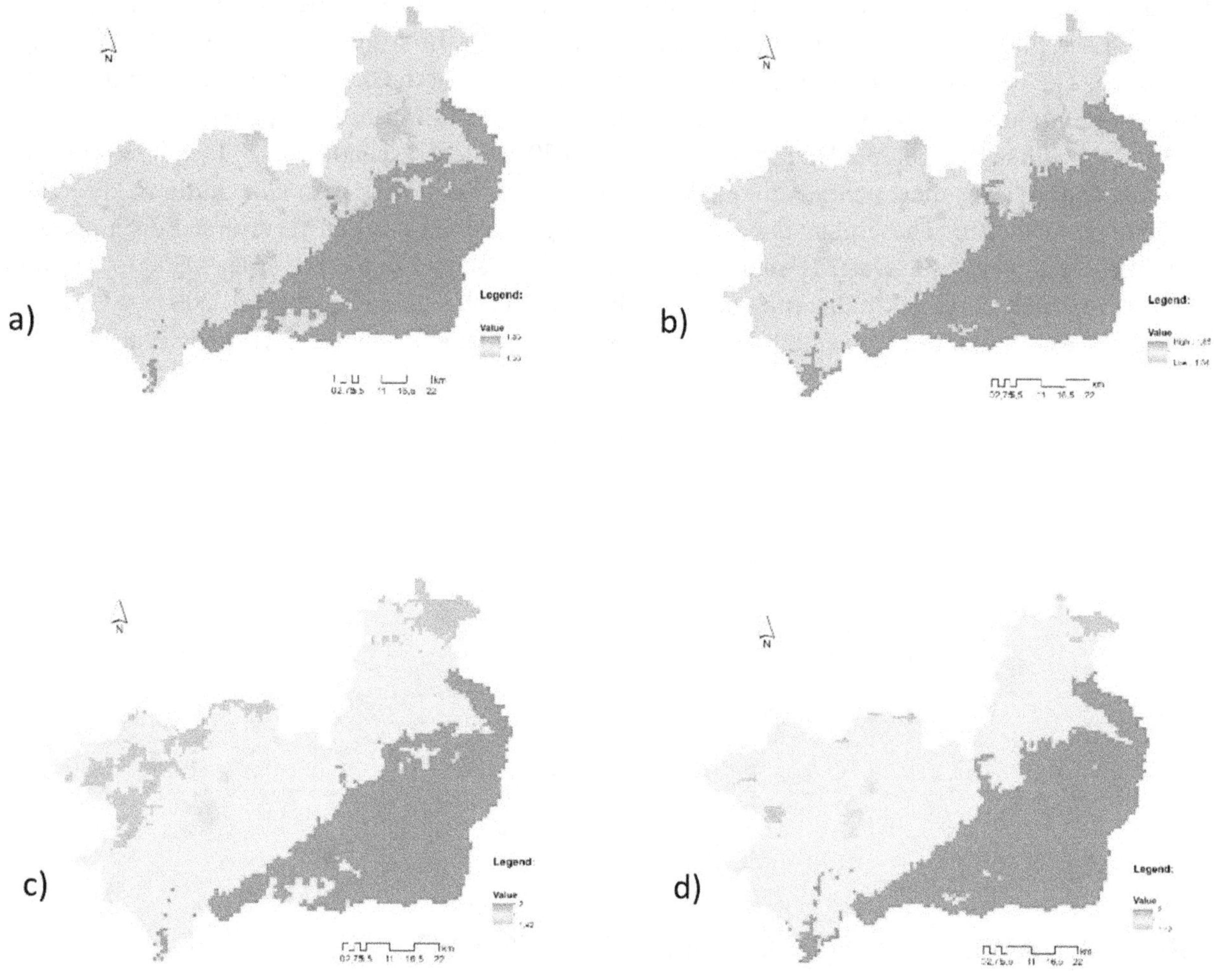

Based on the map analysis, about 16% of the total area of Beira Baixa is available with highly suitable for almond tree, especially in the southeast part of the region. The area with higher potential to pistachio tree corresponding to 16% of the total area available and overlaps the potential almond tree crop area.

Effectively, all different scenarios for 2070, shows an increase of suitable area for pistachio tree and almond tree compared to the present time resulting the climatic changing.

So, despite the increase of the area for production and, consequently, the competitiveness of the sector, it is important to reinforce that in a context of global climatic changes, the water scarcity is expected. Thus, the trend analysis and projection change of climatic factors revealed how the crop production and water requirement will become affected in future. therefore, its affect dramatically the current production models based on intensive production.

Considering all scenarios projected, a significant increasing trend for maximum and minimum temperature will lead the agricultural practices to require more water. So, it's important to know the phenological needs of these new crops in this region and test agricultural practices, such as deficit irrigation.

Therefore, the production of these crops should focus on extensive production models and bet on sustainable agriculture practices to protect ecosystems and soil heritage.

Finally, it's important to privilege the conservation of traditional landscapes in these regions, characterized by the mosaic of agrosilvopastoral systems with high biodiversity and preserve agricultural heritage betting on innovative but preventive agriculture.

CONCLUSION

The carrying out of the study presented on suitable areas which can be exploited for the fruit production with new crops in the Beira Baixa region, using GIS and AHP as multicriteria decision analysis (MCDA) technique, allows the following conclusions to be drawn:

- This methodological approach allowed us to assess the suitability of two emerging fruit crops (Almond tree and Pistachio tree) in the Beira Baixa region of Portugal. The AHP was based on a set of criteria contributing to a reflection on the adequacy of those crops for the climatic and soil characteristics of the region.
- The crops suitability maps thus obtained enable decision making. The main results obtained indicate that the methodology used, using AHP and GIS, could provide a guide map for decision makers to achieve better agriculture productions facing their ecological limitations.
- The results put on evidence the biophysical evaluation of territory and provide information at a local level that could be used by farmers to choose their crops.
- It's important the local authorities that support the farmers create tools to raise awareness and guide the farmers about the global climatic changes and their implications.
- According to the results obtained, that revels the potential of almond and pistachio in Beira Interior region, its urgent to plan the new investments in agriculture considering the emergence of climatic changes.
- Considering all climatic conditions that predict the increase of temperatures and limitations of water reserves, its important privilege traditional cultures, and techniques that contribute of biodiversity conservation and Mediterranean heritage. This is the key of sustainability of rural areas and the agriculture sector.
- For further study is recommend selecting other factors, like irrigation facilities and socio-economic factors, and other parameters which influence the sustainable land use. However, further investigation is needed to integrate the impact of climate change in crops planning to assist in supporting future national strategies for agriculture.

REFERENCES

AJAP. (2017). Manual Boas Práticas para Culturas Emergentes. A Cultura da Amêndoa. Associação dos Jovens Agricultores de Portugal, Lisboa.

AJAP. (2017a). Manual Boas Práticas para Culturas Emergentes. A Cultura do Pistácio. Associação dos Jovens Agricultores de Portugal, Lisboa.

Alkimim, A., Sparovek, G., & Clarke, K. C. (2015). Converting Brazil's pastures to cropland: Na alternative way to meet sugarcane demand and to spare forestlands. *Applied Geography (Sevenoaks, England)*, *62*, 75–84. doi:10.1016/j.apgeog.2015.04.008

Almeida, V. (2011). *A importância e as alterações recentes na agricultura da região centro*. CCDRC.

Alonso, J., Ansón, J., Espiau, M., & Company, R. S. (2005). Determination of endormancy break in almond flower buds by a correlation model using the average temperature of different day intervals and its application to the estimation of chill and heat requirements and blooming date. *Journal of American Horticultural Science*, *130*(3), 308–318. doi:10.21273/JASHS.130.3.308

Antunes, O. E. D. (2012). *Multicriteria GIS analysis for determination of a spacialized index of coastal anthropogenic pressure. Cases of Espinho, Caparica and Faro*. Dissertação de Mestrado em Gestão do Território. Área de Especialização em Deteção Remota e Sistemas de Informação Geográfica. Universidade Nova de Lisboa. Faculdade de Ciências Sociais e Humanas. Lisbon.

Cabo, P., Matos, A., & Bento, A. (2016). *Da produção ao consumo. Breve análise do mercado nacional da amêndoa*. ESADR.

Campos, S., & Madureira, L. (2019). Can Healthier Food Demand be Linked to Farming Systems' Sustainability? The Case of the Mediterranean Diet. *International Journal on Food System Dynamics*, *10*(3), 262–277.

Calatrava, J., Martínez-Granados, D., Zornoza, R., González-Rosado, M., Lozano-García, B., Vega-Zamora, M., & Gómez-López, M. D. (2021). Barriers and opportunities for the implementation of sustainable farming practices in mediterranean tree orchards. *Agronomy (Basel)*, *11*(5), 1–24. doi:10.3390/agronomy11050821

Carvalho, P. (2018). Dynamics of rural low density spaces in Portugal. *Méditerranée*, *130*(130), 1–17. doi:10.4000/mediterranee.10516

COM. (2012). *Comunicación de la Comisión Al Parlamento Europeo y al Consejo sobre la cooperación de innovación europea. 'Productividad y sostenibilidad agrícolas'*. Bruselas, 29.2.2012.

Cóndor, R. D., Scarelli, A., & Valentini, R. (2010). Multicriteria Decision Aid to support Multilateral Environmental Agreements in assessing international forestry projects. *International Environmental Agreement: Politics, Law and Economics*, *11*(2), 117–137. doi:10.100710784-010-9125-7

Cunha, L. (2008). A Beira Interior - Portugal, caracterização física. In *PINHEIRO, Elisa Calado (coord.). Rota da Lã TRANSLANA percursos e marcas de um território de fronteira: Beira Interior (Portugal) e Comarca Tajo-Salor-Almonte (Espanha)* (Vol. I). Museu de Lanifícios da Universidade da Beira Interior.

Dedeoğlu, M., & Dengiz, O. (2019). Generating of land suitability index for wheat with hybrid system aproach using AHP and GIS. *Computers and Electronics in Agriculture*, *167*, 105062. Advance online publication. doi:10.1016/j.compag.2019.105062

Durán-Zuazo, V. H., Rodriguez, B. C., Gutiérrez-gordillo, S., Bilbao, M., Sacristán, P. C., Parra, J. J. P., & García-Tejero, I. F. (2020). Rethinking irrigated almond and pistachio intensification: A shift towards a more sustainable water management paradigm. *Revista de Ciências Agrárias (Belém)*, *43*(2), 24–49.

Elaalem, M., Comber, A., & Fisher, P. (2011). A Comparison of Fuzzy AHP and Ideal Point Methods for Evaluating Land Suitability. *Transactions in GIS*, *15*(3), 329–346.

FAO. (2019). *Statistics*. FAO.

Ferguson, L., Poss, J., Grattan, S., Grieve, C., Wang, D., Wilson, C., Donovan, T., & Chao, C.-T. (2002). Pistachio rootstocks influence scion growth and ion relations under salinity and boron stress. *Journal of the American Society for Horticultural Science*, *127*(2), 194–199. doi:10.21273/JASHS.127.2.194

Ferreira, D., Simões, M. P., Martins, C., & Gaspar, P. D. (2016). Triple Helix in Agriculture Context: The case of Prunus Network in Beira Interior Region. In *Regional HELIX Conference 2016 - International Conference on Regional Triple Helix Dynamics* (pp. 1-10). Academic Press.

García-Tejero, I. F., Gutiérrez Gordillo, S., Souza, L., Cuadros-Tavira, S., & Durán Zuazo, V. H. (2019). Fostering Sustainable Water Use in Almond (*Prunus dulcis* Mill.) Orchards in a Semiarid Mediterranean Environment. *Archives of Agronomy and Soil Science*, *65*, 164–181.

Gijón, M. C. (2013). *Relaciones hídricas y manejo del riego en pistachero* (*Pistacia vera* L.) (PhD Thesis). Universidad de Córdoba.

Goldhamer, D.A. (2005). Tree water requirements and regulated deficit irrigation. *Pistachio Production Manual*, 103-116.

Hamilton, M. C., Nedza, J. A., Doody, P., Bates, M. E., Bauer, N. L., Voyadgis, D. E., & Fox-Lent, C. (2016). Web-based geospatial multiple criteria decision analysis using open software and standards. *International Journal of Geographical Information Science*, *30*(8), 1667–1686. doi:10.1080/13658816.2016.1155214

INE. (2021). Recenseamento Agrícola - Análise dos principais resultados - 2019. Instituto Nacional de Estatística.

Magrach, A., & Sanz, M. J. (2020). Environmental and social consequences of the increase in the demand for 'superfoods' world-wide. *People Nat.*, *2*(2), 267–278. doi:10.1002/pan3.10085

Malki, M., Bouchaou, L., Hirich, A., Ait Brahim, Y., & Choukr-Allah, R. (2017). Impact of agricultural practices on groundwater quality in intensive irrigated area of Chtouka-Massa, Morocco. *The Science of the Total Environment*, *574*, 760–770. doi:10.1016/j.scitotenv.2016.09.145 PMID:27664763

Kangas, J., & Kangas, A. (2005). Multiple criteria decision support in forest management – the approach, methods applied, and experiences gained. *Forest Ecology and Management*, *207*(1-2), 133–143. doi:10.1016/j.foreco.2004.10.023

Kangas, J., Store, R. L., Leskinen, P., & Mehtatalo, L. (2000). Improving the Quality of Landscape Ecological Forest Planning by Utilizing Advanced Decision-Support Tools. *Forest Ecology and Management, Amsterdam*, *132*(2-3), 157–171. doi:10.1016/S0378-1127(99)00221-2

Laskar, A. (2003). *Integrating GIS and Multicriteria Decision Making Techniques for Land Resource Planning* (MSc Thesis). University of Twente, Enschede, The Netherlands.

Lopes, C., Alberto, D., Luz, J. P., & Simões, M. P. (2018). As Políticas Agrícolas Aplicadas à Fruticultura em Produção Integrada na Beira Interior : O Estudo de Caso do Setor das Prunóideas na Cova da Beira [Agricultural Policies Applied to Integrated Fruit Production in Beira Interior: The Case Study of The Stone Fruits Sector in Cova da Beira]. *Revista Portuguesa de Estudos Regionais*, *49*, 98–101.

López, J., Villaseñor, J., López, M., Elvira, A., López, D., & Francisco, M. (2013). *El cultivo del Pistacho*. Edição Mundo-Prensa.

Malczewski, J. (2006). GIS-based multicriteria decision analysis: A survey of the literature. *International Journal of Geographical Information Science*, *20*(7), 703–726. doi:10.1080/13658810600661508

Malczewski, J., & Rinner, C. (2015). *Multicriteria Decision Analysis in Geographic Information Science*. Springer. doi:10.1007/978-3-540-74757-4

Mighty, M. A. (2015). Site suitability and the analytic hierarchy process: How GIS analysis can improve the competitive advantage of the Jamaican coffee industry. *Applied Geography (Sevenoaks, England)*, *58*, 84–93. doi:10.1016/j.apgeog.2015.01.010

Modica, G., Pollino, M., Lanucara, S., La Porta, L., Pellicone, G., Di Fazio, S., & Fichera, C. R. (2016). Land suitability evaluation for agro-forestry: Definition of a Web-Based Multi-Criteria Spatial Decision Support System (MC-SDSS): Preliminary results. In Computational Science and Its Applications - ICCSA 2016 (pp. 399-413). Academic Press.

Phua, M.-H., & Minowa, M. (2005). A GIS-based multi-criteria decision making approach to forest conservation planning at a landscape scale: A case study in the Kinabalu Area, Sabah, Malaysia. *Landscape and Urban Planning*, *71*(2-4), 207–222. doi:10.1016/j.landurbplan.2004.03.004

Quinta-Nova, L., & Roque, N. (2018). An integrated agroforestal suitability model using a GIS-Based Multicriteria Analysis Method: A case study of Portugal. The Eurasia Proceedings of Science, Technology. *Engineering & Mathematics*, *3*, 11–20.

Saaty, T. L. (1980). *The Analytic Hierarchy Process*. McGraw-Hill.

Saaty, T. L. (1993). The analytic hierarchy process: A 1993 overview. *Central European Journal of Operation Research and Economics*, *2*(2), 119–137.

López, S., Rolando, D. G. F., Silva López, J. O., Rojas Briceño, N. B., Oliva, M., Terrones Murga, R. E., Iliquín Trigoso, D., Barboza Castillo, E., & Barrena Gurbillón, M. Á. (2020). Land Suitability for Coffee (*Coffea arabica*) Growing in Amazonas, Peru: Integrated Use of AHP, GIS and RS. *ISPRS International Journal of Geo-Information*, *9*(11), 673. doi:10.3390/ijgi9110673

Schaldach, R., Koch, J., Aus der Beek, T., Kynast, E., & Flörke, M. (2012). Current and future irrigation water requirements in pan-Europe: An integrated analysis of socioeconomic and climate scenarios. *Global and Planetary Change*, *94*, 33–45. doi:10.1016/j.gloplacha.2012.06.004

Simões, M. P., Alberto, D., Dias, C., Duarte, A. C., Ferreira, D., Gavinhos, C., & Castanheira, I. (2015). O projeto + pêssego - inovação e desenvolvimento na cultura do pessegueiro na região da beira ["+Peach" Project: Innovation and development of peach production in Beira Interior Region of Portugal]. In III Jornadas Potencial Técnico e Científico do IPCB (pp. 25-27). Academic Press.

Sottile, F., Massaglia, S., & Peano, C. (2020). Ecological and Economic Indicators for the Evaluation of Almond (*Prunus dulcis* L.) Orchard Renewal in Sicily. *Agriculture*, *10*(7), 301. doi:10.3390/agriculture10070301

Taylor, K. E., Stouer, R. J., & Meehl, G. A. (2012). An Overview of CMIP5 and the Experiment Design. *Bulletin of the American Meteorological Society*, *93*(4), 485–498. doi:10.1175/BAMS-D-11-00094.1

Wotlolan, D. L., Lowry, J. H., Wales, N. A., & Glencross, K. (2021). Land suitability evaluation for multiple crop agroforestry planning using GIS and multi-criteria decision analysis: A case study in Fiji. *Agroforestry Systems*, *95*(8), 1519–1532. doi:10.100710457-021-00661-3

Yalew, S. G., van Griensven, A., & van der Zaag, P. (2016). AgriSuit: A web-based GIS-MCDA framework for agricultural land suitability assessment. *Computers and Electronics in Agriculture*, *128*, 1–8. doi:10.1016/j.compag.2016.08.008

Zhang, J., Su, Y., Wu, J., & Liang, H. (2015). GIS based land suitability assessment for tobacco production using AHP and fuzzy set in Shandong province of China. *Computers and Electronics in Agriculture*, *114*, 202–211. doi:10.1016/j.compag.2015.04.004

Chapter 14
Sustainability of Rice Production at Baixo Mondego, Portugal:
Drivers, Risks, and System Improvements

Maria de Fátima Lorena Oliveira
https://orcid.org/0000-0002-3055-0885
Polytechnic Institute of Coimbra, Portugal

Sergio Oliveira
Polytechnic Institute of Coimbra, Portugal

António Terrão Russo
Lower Mondego Water Users Association, Portugal

kiril bahcevandziev
Polytechnic Institute of Coimbra, Portugal

Ana Bela M. Lopes
Polytechnic Institute Coimbra, Portugal

António José Jordão
Regional Directorate of Agriculture and Fisheries of the Center, Portugal

José Manuel Gonçalves
Polytechnic Institute of Coimbra, Portugal

Manuel Nunes
Polytechnic Institute of Coimbra, Portugal

Pedro Arnaldo Reis
https://orcid.org/0000-0002-0233-953X
National Institute of Agrarian and Veterinary Research, Portugal

ABSTRACT

This chapter aims to analyze the rice production system at the Baixo Mondego Valley to understand the main concerns. Field research and field trials were carried out to analyze rice production, marketing systems, and different irrigation alternatives. An analysis on the worries was made, and a correlational attempt was done. The results show a production system oriented by agri-environmental policies. The problems related with rice irrigation are water scarcity, environmental impacts on water quality, agro-ecosystems, and methane emissions. To reduce water demand, the alternate wetting and drying flooding method, and the improvement of the precise land levelling were studied on the scope of MEDWATERICE Project. About 12-14% of water saving was observed, with impact on production lower than 3.5%, allowing period of 11-19 days of dry soil, expecting positive implications for greenhouse gas emissions. Innovation in the irrigation system may help to reduce some of the farmers' concerns and help to better adapt this crop to the new needs of agriculture in terms of environmental competitiveness.

DOI: 10.4018/978-1-7998-9557-2.ch014

INTRODUCTION

Agricultural Risks

The agricultural business always involves risks associated with the nature of its production, since it is prone to climatic and biotic factors (climate, soil, and pests). In addition to issues related to agricultural production, the agricultural enterprise also must address market issues, such as price volatility, labor issues, seasonality, and changes in agri-food policies. According to literature, research on risk, risk perception and risk management strategies are increasingly analyzed. Climate change, globalization of markets and different consumer perceptions of food safety have raised concerns about risk.

Duong *et al.* (2019) carried out a systematic review of the literature and concluded that the amount of research on risk perception in agriculture and risk management has increased substantially since 1985. The author states that market risk is considered as the most significant followed by biosecurity, which highlights the gaps between the risks mentioned by farmers and the research on socioeconomic factors that explain the perception of worries. However, there are papers discussing both risks and adaptation strategies (Crane *et al.*, 2013; Ahsan & Roth, 2010; Harwood *et al.*, 1999).

Girdžiūtė (2012) and Komarek *et al.* (2020) examine the risks in agriculture and identify that research work focuses on one risk, particularly on production risk, and that there are a limited number of studies exploring the various sources of risk, also showing the importance of risk assessment methods and the importance of studying the interrelationship and interaction between risks. The results of this work show that different types of risk are relevant at different levels and that it is necessary to interconnect risks to understand risk relevance to farmers who must struggle with many risk causes, from natural risks to economic ones (price volatility or flow production).

Market issues are the most important areas of concern followed by biosecurity which is defined by Waage & Mumford (2008) as:

means the protection of countries against alien pests (insects, vertebrates, etc.) and diseases. (p. 863).

There are other risks, such as climate risk, financial risk, governing restraints, and new technologies. Crane et al., (2013) defines risk as the chance of damage or a negative outcome linked to an action and uncertainty falls within these definitions. Harwood *et al.* (1999) reveals that risk affects an individual's welfare and is linked with loss. Risk perception influences farmers behavior and future business decisions such as the continuation in the market. Risk perception and behavior are linked with the resilience of the agricultural sector itself. According to Keil *et al.* (2000) loss level is the most important factor in shaping risk perception, and there is a significant relationship between risk perception and decision making. Sjöberg (1998) says that cultural biases are not major factors in risk perception and the variability within the public in a country is probably due to factors such as trust, beliefs, and human concepts.

Risk perception encompasses the mental processing of information and the skill mechanisms that people use to deal with uncertain events. In addition to definition issues and risk perception, one of the issues in literature is risk categorization. Literature presents different risk categories according to the sector and within agriculture. Categorization varies according to the subject matter focused by the researcher.

Climate change, globalization of markets and different consumer perceptions of food safety have raised concerns about risk. Doung *et al.* (2019) carried out a systematic review of the literature and concluded that the volume of research on risk perception in agriculture and risk management has increased substantially since 1985.

Crane *et al.*, (2013) and Harwood *et al.*, (1999) identified production, marketing, economics, human resources, and the respective legal framework as the main five risk sources. According to Hardaker *et al.* (1997) business risks are those affecting farm business performance such as production, market, institutional and personal risks, and financial risks, connected to the company's financing. The OECD (2009) uses a holistic approach that identifies three risk layers requiring different responses, and this categorization is linked with risk management. The first layer concerns the normal variation in production, process, and weather; it does not require policy reply, and can be managed by farmers as business strategy. The second layer is marketable risk which can be handled through market tools. The third layer, considered as of low probability but reaching catastrophic levels, leads to high and irreversible losses affecting many or all farmers. Under these circumstances, resilience is beyond farmers or markets, and government intervention may be required. In the first layer, risk probability is high, and losses are low, the second layer covers low frequency risk and medium losses; finally, the third layer includes very low frequency, causing very high losses and requiring risk mitigation and risk transfer (Tedesco, 2017).

RICE PRODUCTION

Rice is the world's most important food crop as it is a staple food for more than half of the world's population and the world demand for rice will increase by approximately 24% over the next 20 years (Nguyen & Ferrero, 2012). Rice is cultivated over about 1.3 Mha in Mediterranean countries (FAOSTAT, 2016). Although in the Mediterranean region it is concentrated in specific areas, rice production has great socio-economic and environmental importance. Since it is a fundamental staple food for some countries due to its high quality, and to its role in preserving biodiversity (many important rice areas are in river deltas, estuaries, coastal wetlands or are part of protected ecosystems such as the EU Natura-2000 network). The most important rice-producing countries in the Mediterranean region are Italy and Spain in Europe (72% of the EU production; 345,000 ha), and Egypt and Turkey among the non-EU countries (practically all the production; 789,000 ha).

Traditionally, rice is grown in paddies flooded from pre-sowing to pre-harvest, thus requiring much more irrigation water than non-ponded crops (Cesari *et al.*, 2016). This practice is highly water demanding, in comparison with most methods applied in irrigation of other crops, due to significant deep percolation, and the need for surface drainage of water from the basin. The main problems related with continuous flooding refer to water scarcity, environmental impacts on water quality and agroecosystems, and soil methane emissions into the atmosphere (Katoh *et al.*, 2004). Rice paddies are one of the most important sources of atmospheric methane (CH4), producing about 5-20% of the total emission from anthropogenic sources (USEPA, 2006) and approximately 30% of global agriculture CH4 emissions. Moreover, many important rice growing areas in the Mediterranean region are in environments where soil salinity is an important constraining production factor.

Although the common use of laser land leveling has allowed a significant reduction of water use, there are still common water management problems to deal with, to face climate global changes and the raising of social emergent consensus. The issue of rice irrigation water saving is challenging and intensively studied, demonstrating several irrigation management systems, such as zero-grade fields, alternate wetting and drying, multiple-inlet, furrow, pivot irrigation and drip irrigation (Datta *et al.*, 2017; Vories, 2017).

Mediterranean rice agroecosystems are nowadays facing numerous problems, such as the need to match irrigation demand with the availability of the resource, environmental protection, the need to ensure an adequate income for rice producers, the impossibility of introducing the crop in farmlands characterized by limited water availability despite the increase of rice consumption in the Mediterranean basin, and the lack of specific studies conducted in Mediterranean countries addressing environmental and socio-economic peculiarities of these areas. Due to these problems, the introduction of water management practices alternative to continuous flooding is imperative to enhance water use efficiency and safeguard environmental quality in Mediterranean rice agroecosystems.

Rice Crop in Portugal

Rice was introduced by the Arabs in southern Iberian Peninsula in the eighth century. Thus, rice growing was a key factor in the economic development of originally very underprivileged areas and for the emergence of social and cultural traditions contributing to the reputation of these regions still today. In 2020, rice accounted for about 830 thousand hectares in the 27 countries of the European Union (EU) with an average annual production of 3.4 million tons (Agri-Food Data Portal, 2021), and in 2019/2020 market year the EU import 1.4 million tons of rice (EU, 2021). Today, rice cultivation plays an essential role in maintaining the ecological balance and biological wealth of the ecosystems mentioned above.

The rice value chain begins with the production of paddy rice, which can be dried either on the farm, on the cooperative (farmers' associations) or by the industry. In both cooperative and industrial facilities, dry paddy rice is subject to husking, bleaching, polishing, and packaging. Imported rice may be imported as paddy rice for the rice industry or already packed for final consumption.

Rice plays a role of great social and economic importance in Portugal. We must bear in mind that rice cultivation occupies areas that could not or could hardly be allocated to other productions, thus playing an important role in the conservation of these specific ecosystems. Due to cultural issues rice is of utmost important in the Portuguese diet.

The rice producer is entitled to the Basic Payment Scheme (BPS) of the Common Agricultural Policy (CAP) whose payments are allocated to land rights. Rice production also benefits from voluntary coupled support (VCS). VCS is applied voluntarily by Member States to support certain sectors or types of agriculture facing difficulties and that are economically, socially, or environmentally significant. This support has two goals: to ensure a stable supply to the local processing industry, enabling it to maintain a certain level of production and to avoid disruptive situations in the sector leading to the abandonment of the activity. In Portugal the VCS grant for rice is 194€/hectares (ha) *per* year. The support given to integrated rice production amounts to around 376 €/hectare (ha) for areas under 30 ha and 75 € for areas over 120 ha (Table 1). In the *Baixo Mondego* Valley (case study), a significant part of rice production done is an integrated production system. In some cases, farmers can apply to *Greening* measures, but the set of supports may not exceed 600 €/hectare.

Table 1. Support amounts per hectare per year

Hectares	30 ha	> 30 and ≤ 60 ha	> 60 and ≤ 120 ha	> 120 ha
Grant (Euros)	376	301	188	75

Source: (https://www.ifap.pt/, accessed in 20/10/2020)

Portuguese rice growers are involved in the Integrated Production System for sustainable production, in compliance with agri-environmental CAP measures. To accomplish this measure, farmers are required to use 120 kilograms of selected and certified rice per hectare of cultivated area. Depending on seed availability of *Carolino* rice in international and domestic markets, the Government may alter the quantity of seed to be used.

Table 2 shows the evolution occurred between 2004 and 2018 (triannual average). It is worth noting that Portuguese exports increased more than imports, which led to an improvement in the trade balance. The national representativeness of *Beira Litoral* rice (Mondego representing the main area of production) decreased both in terms of relative area and production, despite the increase of rice production. In 2004/2006, the *Beira Litoral* region accounted for about 27% of the surface area (6,617 ha) and 23% of production (31,723 tons). In 2016/2018, these numbers fell to 22% (6,359 ha) in surface area and to 19% in production (32,186 tons). Rice productivity in Portugal has slightly increased from 5.7 tons/ha in 2004/2006 to 5.8 tons/ha in 2016/2018. In *Beira Litoral*, the productivity was lower but the increase in the analyzed period was higher than the national average; it rose from 4.7 in 2004/2006 to 5.1 tons/ha in 2016/2018 (INE, several years).

Table 2. Variables of rice production and portuguese milled and semi milled rice market

Average	**Rice Surface**	**Rice Prod.**	**Yield**	**Market of milled and semi-milled rice**					
	1,000 Ha	1,000 Tons	Tons/ha	Prod.	Imp.	Exp.	Human Consumption	Consumption *per capita*	Degree self-sufficiency
				1,000 tons				kg *per capita*	%
2004- 2006	24	139	5.7	157	21	5	166	15.8	92
2010-2012	31	180	5.9	164	14	8	169	16.0	96
2016-2018	29	170	5.8	173	26	45	156	15.2	109
Annual Average Growth Rate	2%	2%	0%	1%	2%	20%	-1%	0%	1%
Growth Rate	20%	22%	2%	10%	20%	800%	-6%	-4%	19%

Note: Annual Average Growth Rate and Growth Rate between 2004/06-2016/18).
Source: (INE, several years)

Objectives

The aim of this paper is to provide empirical analyses of how farmers grow rice in the *Baixo*-Mondego Valley in Portugal, how they prioritize their concerns, how they relate their concerns with risk perception, and how these concerns interact with rice production systems and sociodemographic characteristics. The focus was not to directly address the idea of risk, but rather on worries or concerns. In OECD (2019), the word "worries" was used instead of "risk" and "concerns" words was also applied in Sjoberg's work (Sjöberg, 1998) to avoid manipulating the dialog.

Medwaterice project (prima-section 2-2018; www.medwaterice.org) is studying these problems aiming to explore the sustainability of innovative rice irrigation methods and technologies in the mediterranean basin, to reduce rice water use and environmental impacts, and to extend rice cultivation outside traditional paddy areas to meet growing demand. Studies are carried out at the farm scale, including the *baixo* mondego valley, to support the selection of the most appropriate irrigation management options to be tested and demonstrated. In turn, data collected at the farm scale is upscaled at the irrigation district level to support management and policy making decisions.

METHODOLOGY

Baixo-Mondego Valley Scheme

Field work was performed at the *Baixo* Mondego Valley (Lower Mondego Valley), located in the Beira Litoral region, covering an area of about 12,000 hectares down river. Water used for irrigation comes from a weir on the Mondego river located in Coimbra city. The concrete embankment along the river is a multipurpose canal and the left bank is served by a pipeline installed along the irrigation fields. In the main valley, the predominant crops are maize and rice, occupying more than 90% of the area. In the tributary valleys, these are also the leading crops, but rice keeps the top position.

In the Baixo Mondego Valley, rice cultivation started in the second half of the 18th century and was done by the Coimbra friars in swamp areas. Rice was a marginal crop until the 19th century but during the 19th and 20th centuries was supported by the state, which allowed higher producer prices while consumption was bound to urban areas. However, the culture was negatively seen at different times, because of the effects land flooding, required by cultivation, had on the populations, notably the emergence of diseases such as malaria. With the help of technology and soil drainage, the negative effects have diminished, and rice cultivation has become an important income source for local populations. Moreover, rice became part of the gastronomic culture and rice fields modelled part of the Mondego heritage.

Rice Farmers Survey

A survey was conducted to grasp the full range of farming enterprises typology. This methodology provided indicators to address the actual farm conditions and represents current agronomic managing practices as noted by Dantsis *et al.* (2010). The survey was conducted in person, face to face, at the farms site. The questionnaire was divided into four groups of questions designed to provide information about:

1. the farmers' socioeconomic profile,
2. agricultural cultural practices and its production inputs,
3. production costs optimization,
4. production mode according to its environmental sustainability, mainly related with the application of fertilizers and phytopharmaceuticals.

The questions presented were closed single or multiple-choice questions with a maximum of two possible choices.

To assess the degree of fulfillment in relation to certification instruments, policy support and marketing channels, an analysis of the degree of satisfaction was done. The five points of the Likert scale were used (Vagias, 2006) to perform the satisfaction analysis. According to Hardaker *et al.* (1997) the source of risk may be production, human or personal, financial, and business risks. Chartier *et al.* (2019) conduct risk analysis *per* type of risk classified as: price output, yield and income risks, environmental and climate risk, animal, and plant health and the financial, institutional, legal and policy risks. In the current analysis, the categorization of Crane *et al.* (2013) which considers marketing risk (price and market channel) the production risk (production costs and yield risks) and legal and financial risks was used.

The Chartier *et al.* (2019) categorization, namely environmental risks perceptions were also considered.

Nonparametric correlations the Spearman coefficient (rho) were applied, using the SPSS 25 software. For the Regression analysis the SPSS 25 and STATA15 software were used.

The 34 validated surveys were carried out in June 2019 at the central region of the *Baixo-Mondego* Valley. The 34 questionnaires represent 20% of rice farmers at this location (table 3). The interviewed farmers were chosen by opportunity due to the difficulty to have personal contacts within their available time. To determine the sample size, Taro Yamane formula (Yamane, 1973) was applied:

$n = N/[1+N(e^2)]$ (1)

where, n: sample size; N: population size; and *e:* acceptable error.

Alternate Wetting and Drying Flooding Irrigation

Alternate wetting and drying (AWD) consists of intermittent flooding, through a sequence of irrigation cycles with flooding for a certain period, followed by an interruption of supply, until the soil dries up, creating conditions of non-saturation in the surface layer (Tuong & Bouman, 2003). In this way, the volume of water used to irrigation is reduced, compared to continuous flooding (CF), as well as the reduction of greenhouse gas emissions (Runkle *et al.*, 2019) and lower arsenic accumulation in the grain (Linquist *et al.*, 2019), due to soil aeration conditions during non-saturation periods. AWD has been successfully used in several countries, such as India (Jalindar *et al.*, 2019), the Philippines (Lampayan *et al.*, 2015), and the Mi-South of USA (Reba & Massey, 2020).

The experimental farm scale plots were in Montemor-o-Velho (Bico da Barca) and Quinta do Canal, with the work carried out in 2020, on the scope of MEDWATERICE Project. In each location, two parts were considered, to evaluate the practices of CF and AWD. The measurements and evaluations carried out in these plots allowed daily data according to the following procedure: irrigation and drainage flows, by spillways or volumetric counters; stored surface water through water pipes, equipped with automatic level sensors; evaporation from free water, based on a class A evaporimeter can; reference evapotranspiration, by the Penman-Monteith method, based on data from a local meteorological station; cultural evapotranspiration (ETc) through the cultural coefficients 1.25, in flooded soil, and 1.0 to 1.10, in dry soil; deep percolation determined by the method of hydrological balance in the site, applying the previous data. Conventional cultivation techniques were used in soil preparation fertilization, and health control operations. Ariete was the selected crop variety and to assess productivity, samples of the crop were collected at representative points

The AWD essay applied the following procedure based on Bouman *et al.* (2007) and Gonçalves *et al.* (2021), with the adjustments to local agronomic practices:

1. Initial flooding allowing wet sowing, like the traditional practice (follows an initial drying event to favor the emergence),
2. Shallow ponding during the vegetative phase, considering the drying periods required for herbicide application, usually twice, with particular attention during the flowering period because it is very sensitive to water stress,
3. AWD technique during all stages after flowering, until last irrigation; the target was a flood water depth not higher than 5-7 cm; the irrigation schedule considered was an interval between 10 to 14 days of irrigation events; ensuring that the water level should not fall to 15 cm below the soil surface, measured in a water tube,
4. The last irrigation should be about 20 days before the harvest.

OUTCOMES

Characterization of the Lower Mondego Farmer Sample

Due to the high number of very small agricultural properties and the heterogeneity of properties in terms of cultivated area and the importance of properties with medium or large size as decision makers in the rice chain and in production systems, the cultivated area was adopted as a basis of the sample size. If we apply as the population size the number of hectares of rice cultivated area, and for an acceptable error of 5%, the sample size will be 328 hectares. ´

The sample size is displayed in table 3. The surveyed farm holders own 764 hectares (representing about 42% of the rice-cultivated area in the lower Mondego valley), that is, the sample is larger than the minimum required. The decision to use all the data obtained in the research fieldwork allows a better image of the production systems and avoids data loss. Four categories were considered according to the production area (Table 3).

Table 3. Sample categories and number of producer's holders according to production areas

Hectares	Farmers Survey		Rice Farmer		Representative sample
	Number	%	Number	%	%
≤6.30	5	14.7	103	59.2	4.9
≥6.31 and ≤ 22.60	16	47.1	52	9.2	30.8
≥22,61 and ≤ 38.80	5	23.5	8	4.6	62.5
≥38.81	5	14.7	11	6.3	45.5
Total	34	100	174	100	19.5

Source: Oliveira et al. (2019:1215)

The study conducted with rice farmers shows that 82.4% of respondents are full-time farmers, 2.9% are part-time farmers and 14.7% are agricultural pensioners. About 32.4% of respondents: are between the ages of 20 and 40; 26.6% are older than 60 years old. Approximately 62% of the interviewees have nine years of schooling and 26.5% of the respondents have university-level education. When compared to national results, these figures are higher in terms of both the farmers' age and educational level. The average literacy level in Portugal shows that 46.3% of farmers have elementary education and only 8.0% attended university on a nonagricultural sector and 1.3% attends university on the agricultural sector. In terms of age group, about 52.5% of Portuguese farmers are over 65 years (INE, 2021).

All farmers apply integrated production systems. A weak correlation was identified (significant at 5% and 1% level of significance), between the importance of agricultural activity in the farmer's professional life (full-time; part-time, retirees) and farmer's level of education and age, respectively. In the case of young rice producers, agriculture is their main activity compared to the other age groups. At the same time, young producers hold higher education degrees. No correlation between the farmer's characteristics and the size of the farm they manage was found. The farmers' education level is positively related with the turnover generation level (Table 4).

Table 4. Socioeconomic correlation between economic variables (Spearman coefficient)

Variables	Education Level	Age Class	Size famer Category (ha)
Farmer relevance activity	-.364*	.517**	-.013
Sig. Level	*.034*	*.002*	*.942*
Education Level	-	-.748**	.088
Sig. Level	-	*.000*	*.622*
Age Class	-.748**	-	.027
Sig. Level	*.000*	-	*.880*
Size famer Category (ha)	.088	.027	-
Sig. Level	*.622*	*.880*	-

Note: * significance level of 5% (p=0.05); ** significance level of 1% (p=0.01)

The *Carolino* rice variety, *Ariete*, (Oryza sativa L. subspecies Japonica), represents approximately 77% of the rice cultivated area (considering the area analyzed in this work). About 56% of the farmers produce one only rice variety, 19% grow two varieties and 6% produce more than two rice varieties. Variety diversification can reduce the production risk. Around 91%, produce *Carolino Ariete*, followed by *Euro* and *Opale* varieties.

Fixed Assets and Costs

In agriculture, fixed assets refer to farmers' technical equipment. The fixed assets to total assets ratio are the main factor distinguishing farms. Farmers' competitiveness depends on the use of technical capacity. Property, plants, and equipment are the physical and technical basis of production capacity. An analysis of a farmer's ability to invest in equipment and his dependence on equipment from others can explain and help understanding the farmer´s notion or perception of risk in relation to long-term production or investment. All respondents own their tractors (88% respondents have three or four tractors) which are used with different types of equipment needed for rice cultivation. In general, most farmers own their equipment (Table 5).

Table 5. Owner of fixed assets

	Plowing and harrowing		**Levelling land Machine**		**Rice Seed Sowing Machine**		**Pulverizer**		**Rice Harvester Machine**	
	Nº	%	Nº	%	Nº	%	Nº	%	Nº	%
Owner	34	100	30	88.2	31	91.2	32	94.1	28	82.4
Rent	0	0	4	11.8	3	8.8	2	5.9	6	17.6

Only two farmers rent rice dryers and only one owns peeling and bleaching equipment. Land ownership may be important for risk perception. Not knowing if land may be accessed in the future may be a concern, because some lands are rent. All respondents are landowners, but about 47% also use leased land. When controlling weeds, plagues and diseases farmers are subject to integrated production rules regarding the quantity and quality of herbicides and pesticides applied. All farmers claim to apply herbicide one to four times. Regarding other pesticide uses (insecticides and fungicides), the number of applications is less frequent: 5.9% do not apply, 50% apply once and 44% apply twice.

Production costs are the most important variable on farm profitability, and they can affect risk perception. Production risk involves both the quantity and the quality of the output, as noted by Chartier *et al.* (2017).

The relationship between farm income and underlying factors such as yields, and prices is not as straightforward as one might think. (p. 28).

Income variability depends on the correlation between price and income risks, the farm's cost structure (fixed and variable costs) and specially the variable costs that constitute the activity's crop account and bring up to date the farmer´s efficiency and even their ability to support themselves and the risk awareness brought to family life by their activity. Duong *et al.* (2019) include in production risks: climate problems, biosecurity, technology change and yields; and Komarek *et al.* (2020) refer that literature focus on production risks is understandable.

Production risks arise from natural processes and are connected to climate variability and to pests and diseases outbreaks. Other yield limiting factors, also acting as production risks, (e.g., excessive heavy metals in the soil or soil salinity) are very important factors in rice crop. It is important to know the cost system of rice production since it depends a lot on climate variability and other factors such as weed and pest emergence.

Respondents were asked about rice production and processing costs in the last three years. Taxes, interest, amortization of fixed assets and certification costs were not included. It is important to note that water costs include a fixed component (conservation fees) and a variable component of water use, that is, exploitation fees (0.00262 €/m^3; one hectare uses about 16,390 m^3 of water).

Table 6 shows that five variables represent 70.4% of rice production costs. To understand the relationship between costs and the production costs for the rice activity a regression analysis was performed.

The method of Ordinary Least Squares (OLS) was applied. The presence of multicollinearity, heteroscedasticity, and residual autocorrelation tests (Durbin-Watson) was searched for. The heteroscedasticity test was performed by analyzing regression residues in SPSS25 and later confirmed by regression performed by STATA15. According to table 6, we can safely conclude that there is no collinearity and the absence of multicollinearity since the *Tolerance* (Tol.) is greater than 0.1 and the *Variance Inflation Factor* (VIF) is less than 10.

Table 6. OLS regression and weight of costs components in costs rice production

OLS Regression (exogenous variable: Costs for Rice production; Euros/hectare). Endogenous Variables. Costs in Euros/hectare; The variables Surface are in hectares and Production are in tonnes						Costs (Euros/ha)			
Explanatory Variables	**Beta**	**t-test**	**p-value**	**Tol.**	**VIF**	**Costs**	**Average**	**Std Desv**	**Weight**
Intercept	6857	1.22	.24			Total	1731.3	299,7	100%
Costs Labor	.23	1.65	.11	.18	5.60	Labor	287.0	116.1	16.6%
Costs Herbicides	.23	2.59	.02	.43	2.32	Herbicides	305.9	54.8	17.7%
Costs Fungicide/ insecticide	.12	1.39	.18	.46	2.16	Fungicides/ insecticides	111.4	52.7	6.4%
Costs Seeds	.20	2.49	.02	.50	1.98	Seeds	191.1	24.3	11.0%
Costs Machine & fuel	.20	2.13	.05	.39	2.56	Machines & fuel	225.0	43.1	13.0%
Costs Fertilizers	19	2.21	.04	.46	2.16	Fertilizers	209.9	55.8	12.1%
Costs Land Rent	.59	7.84	.00	.58	1.71	Rent	188.4	193.0	10.9%
Costs Insurance	.01	0.07	.95	.44	2.29	Insurance	2.7	1.2	0.2%
Costs Water/ Irrigation	-.10	-1.20	.24	.48	2.09	Water/ Irrigation	88.1	0.3	5.1%
Costs Dryer Rice	.19	2.25	.04	.45	2.25	Rice dryer	117.6	46.5	6.8%
Costs Peeler/ Bleaching	.34	2.85	.01	.24	4.15	Peeler/ Bleaching	4.4	25.5	0.3%
Surface	-.43	-3.32	.00	.20	5.07				
Production	. 45	4.22	.00	.30	3.39				
Adj R^2 = .889	F(13, 20) = 21.50			Breusch-Pagan chi2(1) = .00					
Durbin-Watson = 1.890	Prob > F = .00			Prob > chi2 =.98					

The production and area variables introduced and improved the value of the Durbin-Watson statistic (DW). According to DW test, we can reject the *autocorrelation* in residuals. The model is fit and explains the independent variable (Table 6).

Production is negatively correlated with costs, that is, the higher the production, the lower the costs and the larger the cultivated area the higher the costs. The rental cost variable is significant and has the highest partial regression coefficient (the increase of one unit in the rent costs implies an increase of 0.59 units in the total production costs).

Labor costs represent 16.6% of the total cost, although, according to regression, this variable does not help explaining production costs. It is interesting to note that the weight of each component of total costs is not directly related to the weight of this component in explaining the costs of producing rice. It is important to emphasize that despite the questions about total costs of rice production and the costs per cost component in rice production have been carried out in different parts of the questionnaire, there were no significant differences between the total costs calculated by component and the total production costs indicated by the interviewees. This means that farmers have a good perception of both the costs *per* cost component and the rice activity cost as a whole.

We checked whether there is a correlation between cost components and education level, age group and activity relevance to farmers. To measure the degree of association between variables we can apply for different criterions. For Cohen (1988), rho coefficient between between 0.10 and 0.29 represent a small relationship; between 0.30 and 0.49 are moderate; values between 0.50 and 1.0 signified a large association. However, other authors such as Pestana & Gageiro (2014) point to a slightly different classification. For Pestana & Gageiro (2014), values of $|rho|<0.20$ represent a very weak association; $0.2 \leq |rho| < 0.4$ weak; $0.4 \leq |rho| < 0.7$ moderate; $0.7 \leq |rho| < 0.9$ higher and values of $|rho| \geq 0.9$ very high association. According to Pestana & Gagueiro (2014) differences between the scales.is due to the domain of science in which this classification is applied. For this work, we will apply the classification of Jacob Cohen (1988).There is a large negative relationship, significant at the 0.05 level between herbicide costs and farmer relevance to activity (Spearman's coefficient: rho = -0.525; p-value = 0.001).

Channel Marketing

Market risk is considered in literature as the most significant risk and is related to the market process, price volatility and the marketing channel efficiency. Hardaker *et al.* (1997) refers that output prices matter but so do inputs prices, and Harwood *et al.* (1999) gives greater importance to price aspects, output price, inputs prices, the vertical integration and market contracts as well.

Market risks are related to the possibility of market loss or revenue loss due to a lower price than expected. Lower sales and lower prices caused by the increasing numbers of competing producers or by consumers' preference changes are common sources of market risk. Market risks can also arise from the loss of market access due to the loss of one element of the value chain, either the cooperative or the industry capable of turning rice into a finished product. To analyze market behavior, respondents were asked about the channel to market used to sell their rice, the reason for their choice and their degree of satisfaction.

With respect to processing capacity, only one interviewee is able to transform and sell under his own brand. All other respondents depend on either the cooperative or the industry for processing paddy rice. Considering the channels to market, one of the farmers sells 80% of his yield for retail, a unique case only possible because he has his own brand. The remaining 20% are sold to industry. About 20 of the rice farmers (59% of the respondents) sell all their harvest straight to manufacturing and 11 (32%) of the interviewees sell their entire crop to the Cooperative of *Montemor-o-Velho*. There are currently three cooperatives in Portugal's Center Region capable of processing paddy rice. Only three producers use two channels to market simultaneously. The degree of satisfaction is higher for the industry then to the cooperative and it has been noticed that they are satisfied (average value: 2,93) with the channel to market used (Table 7).

Table 7. Degree channel to market satisfaction

Point	Level of satisfaction	Industry		Cooperative Montemor-o-Velho		Retail with Brand		Total	
		Nº	%	Nº	%	Nº	%	Nº	%
1	Not at all satisfied	1	4.3	0	0	0	0	1	2.7
2	Slightly satisfied	4	17.4	6	46.2	0	0	10	27.0
3	Satisfied	10	43.5	5	38.5	0	0	15	40.5
4	Very satisfied	8	34.8	2	15.4	1	100	10	27.0
5	Extremely satisfied	0	0	0	0	0	0	0	0
	Total	23	67.6	13	38.2	1	100	36	100
	Average Degree *Std. Deviation*	3.09 .85		2.69 .75		4.00		2.93 .80	

These results match the results found by Székely & Pálinkás (2009) who state that selling agricultural products through contracts with industry or cooperatives is less risky due to contractual factors. Selling products individually is probably the most profitable way to market them, especially when there is greater competition, and the farmer has no bargaining power.

Out of the 34 farmers interviewed, 32 respondents (representing 94%) of the sample answered the question concerning the reason to choose the channel to market in which each respondent could choose two options. Since respondents could choose two options, a set of 55 responses was obtained (Table 8). About 38% respondents (out 32 respondents) choose the channel to market because of "Flow ease production" and 34% because they have no processing capacity (out of 32 respondents). Regarding quality certification standards applied by the farmer, all respondents answered to the question and they only have the certification of integrated production. Although "*Arroz Carolino do Baixo Mondego*" has been certified since 2015 as a product with "Protected Geographical Indication, (PGI) none of the interviewed producers claimed to PGI certification

Table 8. Reason to choose the channel marketing and standard certification

	Reason to choose the channel marketing (% in total of answers)							
Reason	**Best price**	**Easy flow production**	**No processing capacity**	**Price stability**	**Guaranteed Production flow**	**Fastest payment**	**Support production**	**Total**
Nº answers	7	12	11	8	9	6	2	55
% in total	12.7	21.8	20.0	14.5	16.4	10.9	3.6	100
	Reason to choose the standard quality certification (% in total of answers)							
Reason	**Best price**	**Easy flow production**	**Because other did**	**Price stability**	**Guaranteed Production flow**	**Option to choose the channel**		**Total**
Nº	12	20	0	8	15	2		57
% in total	21.1	35.1	0	14.0	26.3	3.5		100

All respondents answered the question on the reason for choosing the certification system and about 59% respondents (% in total respondents) chose this type of certification due to the "ease production flow", and 44% due to the "guarantee production flow", that imply that ease and ensuring flow production are key factors for choosing the channel to market and influenced the quality certification system (Table 8).

Main Concerns

The producer's concern analysis provides an important explanation for the degree of risk perception. A list of "concerns" was presented and farmers were asked to respond to their degree of concern for each question on a Likert scale from one to five points (Table 9). There were 33 answers to all items, except for the question "Flow production" (32 responses). Respondents indicated seven most important worries (values above the average: 3.31). The most important concern was the issue of weeds, followed by seed costs which is easily understandable.

Because farmers produce in an integrated production approach, the application and the type of herbicides used by farmers must comply with the regulation and they must apply certified seeds, as mentioned above. These two variables are significant in the regression (Table 6) accounting for the highest estimated regression coefficients. Five out of the seven concerns are about the costs of paddy rice produced; market and environmental concerns, mainly climate change, complete the list of issues.

Table 9. Degree of concerns by worries

Degree	1 -No worry		2 - Little worry		3 - Worry		4- Pretty worried		5 - Very worried		Average	*Std, Desv* Value
Answers	Nº	%	Nº	%	Nº	%	Nº	%	Nº	%		
Weeds issues	0	0.0	0	0.0	0	0.0	4	12.1	29	87.9	4.88	*0.33*
Seed cost	0	0.0	1	0.0	2	6.1	9	27.3	22	66.7	4.61	*0.61*
Climate issues	0	0.0	0	0.0	3	9.1	11	33.3	19	57.6	4.48	*0.67*
Plagues and Diseases issues	1	3.0	0	0.0	2	6.1	11	33.3	19	57.6	4.42	*0.87*
Rice price producer	0	0.0	3	9.1	4	12.1	9	27.3	17	51.5	4.21	*0.99*
Fertilization issues	0	0.0	5	15.6	13	40.6	10	31.3	4	12.5	3.41	*0.91*
Machine and fuel costs	0	0.0	5	15.2	17	51.5	7	21.2	4	12.1	3.30	*0.88*
Rice dryer issues	1	3.0	10	30.3	14	42.4	6	18.2	2	6.1	2.94	*0.93*
Labor availability	4	12.1	13	39.4	7	21.2	3	9.1	6	18.2	2.82	*1.31*
Labor cost	2	6.1	19	57.6	5	15.2	4	12.1	3	9.1	2.61	*1.09*
Machine availability issue	10	30.3	13	39.4	2	6.1	6	18.2	2	6.1	2.30	*1.26*
Seed quality issues	0	0.0	1	3.0	3	9.1	10	30.3	19	57.6	2.30	*0.79*
Water issues	14	42.4	5	15.2	12	36.4	2	6.1	0	0.0	2.06	*1.03*
Flow production issue	8	24.2	21	63.6	1	3.0	1	3.0	2	6.1	2.03	*0.98*
Soil salinity	23	69.7	3	9.1	0	0.0	4	12.1	3	9.1	1.82	*1.42*

If we apply for Cohen rank (Cohen, 1988), we found a moderate positive correlation (at 0.05 level of significant) between education level and water issues, (rho = 0.375), a moderate negative correlation (at 0.05 level of significant) between class age level and water issues (rho= -0.412). A moderate or medium negative correlation (at 0.05 level of significant) between class age level and soil salinity (rho= - 0,351). Several moderate correlations between production area and different worries were found (at 0.05 level of significant) such as: water issues (rho = 0.388); soil salinity (rho = 0.395); seed quality (rho = -0.414). We found (at 0.01 level of significant) a large negative correlation between rice price producer and production area (rho = -0.520). This finding is interesting to explore in future works because it goes against the economic theory of the supply model. Small farmers are more sensible to rice price.

Flooding Irrigation Water Savings

The number of days with wetland and dry soil, with CF and AWD, in the three experimental sites, is shown in table 10. The differences result from the practice of AWD after flowering starts, since the irrigation management was similar in the vegetative stages and even beginning of flowering. The increase in time with dry soil due to AWD was 10, and 25 days, corresponding to a period with dry soil related to the cultural cycle of 47%, and 57%, for Quinta do Canal, Bico da Barca, respectively (Gonçalves *et al.* 2021; *Nunes et al.* 2021).

Table 10. Number of days with flooded and dry soil, in continuous and AWD flooding trials

Plot	Soil condition	Complete crop season period (days)		After vegetative stage (days)	
		CF	AWD	CF	AWD
Quinta do Canal	Flooded	88	78	40	29
	Dry soil	59	69	33	44
	Total	147	147	73	73
Bico da Barca	Flooded	83	58	40	21
	Dry soil	52	77	39	58
	Total	135	135	79	79

CF - Continuous (traditional) flooding; AWD - alternate wetting and drying flooding.

Water use values in the experimental plots, with CF and AWD, are shown in Table 11. The irrigation allocations with CF were 1588 mm at Quinta do Canal, and 1725 mm at Bico da Barca, with the relative savings of AWD water was 12.6%, and 11.8%, with reductions in cultural evapotranspiration of 1.6%, and 3.4%, and in deep percolation of 22.1%, and 15.0%, for Quinta do Canal, and Bico da Barca, respectively.

Table 11. Water use in continuous and AWD flooding trials

Plot	Water use (mm)	Complete crop season period		After vegetative stage	
		CF	AWD	CF	AWD
Quinta do Canal	Evapotranspiration	696.3	685.1	298.9	287.7
	Irrigation	1588	1388	651.5	425.1
	Precipitation	130.4	130.4	77.6	77.6
	Deep Percolation	538.5	419.7	261.4	152.6
	Surface Drainage	516.4	460.8	211.6	117.0
Bico da Barca	Evapotranspiration	588.0	568.1	282.0	263.4
	Irrigation	1725	1522	742.1	537.5
	Precipitation	99.6	99.6	87.8	87.8
	Deep Percolation [1]	1264	1075	651.4	494.2

[1]Includes a small fraction of surface drainage

Production values in the experimental plots, with CF and AWD, are shown in Table 12. Production was 9.58 t/ha at Quinta do Canal, and 8.10 t/ha at Bico da Barca, with production decreasing in AWD by 3.4% at Quinta do Canal and increased by 0.3% at Bico da Barca. In turn, water productivity increased in both locations, 10.6%, and 13.6%, in the same order, reaching the highest at Quinta do Canal with 0.667 kg/m^3, and 0.543 kg/m^3 at Bico da Barca.

Table 12. Yield and water productivity of continuous and AWD flooding trials. Local

Plot	Técnica	Y (t/ha)	WP (kg/m^3)	WG (g)	SY (t/ha)
Quinta do Canal	CF	9.582±1.230	0.603	28.9±1.42	5.49±0.70
	AWD	9.252±6.120	0.667	28.9±0.74	5.62±0.53
Bico da Barca	CF	8.101±0.987	0.470	31.0±1.68	4.45±0.39
	AWD	8.124±0.920	0.534	31.0±0.53	5.28±0.77

CF - Continuous (traditional) flooding; AWD - alternate wetting and drying flooding; Y- Rice grain yield at 14% humidity (t/ha); WP - Water Productivity=grain yield at 14% humidity/ (irrigation + precipitation) (m^3/ha); WG - Weight of 1000 grains at 14% humidity (g); [6]SY - Straw yield, dry matter (t/ha).

Regarding the AWD technique, the results seem to be in accordance with several published studies (Tuong & Bouman, 2003; Jalinda, 2019; Lampayan, 2015), showing that there is a relative potential for saving water. In table 12, we can see water savings, with an impact on production of less than 3.5%. AWD allowed a period of 11 to 25 days of dry soil, expecting positive implications for greenhouse gas emissions and the arsenic content of the rice grain. The improvement of precise land levelling is considered a priority to optimize the water level above soil surface, aiming water saving. On the other hand, the need to carry out frequent and planned irrigation events in the AWD period (after mid-July), makes inflow control devices more demanding, making place for its automation.

FUTURE RESEARCH DIRECTIONS

Business diversification is a strategy used by farmers as risk protection. Mishra & El-Osta (2002) suggests that diversification and farm size may be negatively correlated according to economic theory, but the study provides evidence that farms receiving government payments are more diversified than others. Diversification is the most widely used risk avoidance instrument and is based on risk spreading. The motivation for diversification is based on the principle that when one activity generates low revenue, other activities may be profitable.

The multiple activity agricultural enterprise significantly reduces the possibility for local natural disasters to have a simultaneous negative impact on all activities and allows diverse income source on and off the farm. From an environmental point of view, diversifying agricultural production on a farm or agricultural region is one of the main tactics adopted by farmers to restrain the long-term changes induce by environmental changes. However according to Lancaster &Torres (2019) the diversification of agricultural operations would result in more steady ecosystems over time, enabling quicker responses to climate and social changes.

About 50% of respondents said rice farming is their main source of income (≥ 50% of farm income) and for 24% it represents 100% of income. About 47% of respondents have two and 29% have at least three agricultural activities, including rice and 18% have a non-agricultural income (Table 13).

Table 13. Income source by agricultural activity and percentage of rice in agricultural income

Source of income	Frist		Second		Third		Rice income by % in total of farm income		
Activity	Nº	%	Nº	%	Nº	%	Class	Nº	%
Rice	17	50.0	13	38.2	4	11.8	< 25%	4	11.8
Beef	1	2.9	2	5.9	1	2.9	25 – 39%	4	11.8
Fruit			1	2.9	1	2.9	40- 49%	9	26.5
Maize	15	44.1	8	23.5			50- 79%	5	14.7
Horticultural					4	11.8	80- 99%	4	11.8
Others	1	2.9	2	5.9			100%	8	23.5

According to the responses, all farmers surveyed received support from agricultural policies, namely the Rural Development Program (RDP), over the past decade. However, no support demands were submitted to support policies for organic farming, for diversification and for commercialization and processing. Farmers were satisfied or more than satisfied with the progress of the measures on their farms, and all respondents stated that they want this production system in the near future, at least during the next couple of years (Table 14).

Table 14. Submission for RDP policy aid over the past decade and satisfaction grade

	Aids	Advice Service		Farms Modernization		Environmental		Integrated farming		Young Farmer		BPS include VCS rice	
	Application	Nº	%	Nº	%	Nº	%	Nº	%	Nº	%	Nº	%
		33	97.1	7	20.6	34	100	31	91.2	8	23.5	33	97,1
Point	**Level**	**Degree of Satisfaction**											
1	Not at all satisfied	0	0	0	0	0	0	0	0	0	0	0	0
2	Slightly satisfied	3	9.1	0	0	13	38.2	0	0	0	0	1	3.0
3	Satisfied	22	66.7	2	28.6	15	44.1	10	32.3	1	1	14	42.4
4	Very satisfied	7	21.2	5	71.4	6	17.6	15	48,4	6	6	14	42,4
5	Extremely satisfied	1	3.0	0	0	0	0	6	19,4	1	1	4	12,1
Average Degree *Std, Deviation*		3.0 *0.6*		3.7 *0.5*		3.8 *0.7*		3.9 *0.7*		4.0 *0.5*		3.6 *0.7*	

In this question, respondents could justify their answer by choosing at most two main motives to maintain the agricultural activity. The financial reasons and family income balance represented 24.3% of the responses, respectively, but 21.6% stated that agricultural activity would only continue due of the nonexistence of profitable alternatives.

Regarding rice crop, only one farmer replied he/she does not intend to keep producing rice due to the crop´s absence of economic sustainability. Approximately 89,2% of the farmers want to maintain rice production, and 45.5% said they preserve this activity since there are no suitable agronomic systems available for their land, given the current agronomic and technological systems' availability for farmers in the region. Around 24.2% of the sampled producers consider that knowledge about rice is an important reason to keep the crop. About of 21.2% of the farmers in this work believe that the economic viability justifies the option to continue cultivating rice

Rice farmers are reasonably satisfied with the functioning of this farming system. The fact that they use marginal lands that cannot be used for other cultural systems is an important factor and agri-environmental policies support their production decisions. The lack of development of own brands is highlighted, but this fact can be justified by the high level of satisfaction they have with industry, and also with supporting cooperatives. Crop diversification contributes to reduce the economic risk in the most complex agronomic systems particularly in places with high pest incidence.

Rice industry policies, together with the rice food chain, aim to promote rice, through the promotion of its quality attributes and through the European rice agri-environmental attributes This promotion of Portuguese rice involves encouraging the consumption of national rice in Portugal and increasing the consumption of rice in Northern Europe countries. The attribute of consuming European rice set to reduce the carbon footprint, and how rice is grown choosing areas of the EU's wetlands, which are home to migratory birds, stimulate environmental choices. On the other hand, if European rice gains a share in the domestic market, we will help to reduce imports of rice from third countries that do not pay customs duties entering the European Union.

The main issues related to water management in rice cultivation that determine future research are the following: 1) Water saving in irrigation, to reduce impacts on water resources and to better adapt to situations of water scarcity; in this sense, improving the practice of AWD is of great importance. 2)

Automation of irrigation supply to paddies, for better control of the water applied and the feasibility of AWD practice. 3) Reuse of drainage water, whenever water quality conditions allow it. 4) Improvement of agronomic practices to reduce the use of pesticides and facilitate weed control, namely by mechanical control or the use of crop rotation. 5) Reduction of methane emissions to the atmosphere because of flooded soil conditions; to this end, the practice of AWD in flooding irrigation, or more significant changes in irrigation techniques, such as drip irrigation, are envisaged as possible solutions, when soil and economic conditions allow it. 6) Improve water management at the irrigation district level, to ensure more reliable and consistent solutions that ensure spatial equity of water savings and farmers' income. 7) Optimize the positive role of paddy rice in preserving biodiversity of agroecosystems, making it possible to value the crop in terms of environmental services.

CONCLUSION

The work carried out an analysis of the rice production and marketing system in the Mondego valley. This system is supported by CAP measures of which farmers are aware, benefit and are satisfied with. The concern with production costs is clear, as well as the concern about rice price. Production risk is important and is related to the price obtained for rice. Market risk is spotted by the producer despite the reduction in volatility and price growth in the last five years.

The choice of channels to market has to do with rice flow output guarantees, but this is not a relevant concern. In terms of market risk, there is a difference between what farmers consider to be a price risk and safest choice. Farmers are aware of the problems and try to find risk mitigation strategies, diversifying the activity, choosing the channel, controlling production costs, and applying specific Portuguese policies as well as agri-environmental measures.

The issues of climate change and how these changes may affect production, such as rice that uses sensitive ecosystems, are important issues to consider in European governance policies and in agricultural policy discussions in Europe. Farmer's perception of risk involving important crops in terms of food sovereignty and the guarantee of national and European food self-sufficiency, namely cereals, are issues to consider when developing agricultural policies for an environmentally and economically sustainable development of agri-food production.

The need to deal with reduced availability of water for irrigation and the requirements to reduce negative environmental impacts, call for very effective actions in agronomic changes and in irrigation techniques for rice crops, to ensure economic income for farmers and rice production in quantity and quality, to satisfy consumption. These changes are generally complex, due to crop sensitivity and the soil system therefore, increased efforts are needed in research and experimentation, as well as in supporting farmers and other interested parties.

ACKNOWLEDGMENT

This research was supported by the FCT - Foundation for Science and Technology, [MEDWATERICE - PRIMA/0005/2018]; FCT - Foundation for Science and Technology [CERNAS: UIDB/00681/2020]

REFERENCES

Agri-Food Data Portal. (2021). *DG AGRI Rice Production*. https://agridata.ec.europa.eu/extensions/DashboardRice/RiceProduction.html

Ahsan, D., & Roth, E. (2010). Farmers' Perceived Risks and Risk Management Strategies in an Emerging Mussel Aquaculture Industry in Denmark. *Marine Resource Economics*, *25*(3), 309–323. doi:10.5950/0738-1360-25.3.309

Bouman, B. A. M., Humphreys, E., Tuong, T. P., & Barker, R. (2007). Rice and water. *Advances in Agronomy*, *92*, 187–237. doi:10.1016/S0065-2113(04)92004-4

Cesari de Maria, S., Rienzner, M., Facchi, A., Chiaradia, E. A., Romani, M., & Gandolfi, C. (2016). Water balance implications of switching from continuous submergence to flush irrigation I n a rice-growing district. *Agricultural Water Management*, *171*, 108–119. doi:10.1016/j.agwat.2016.03.018

Chartier, O., & Cronin, E. (Eds). (2017). *Study on risk management in EU agriculture: final report*, European Commission. https://op.europa.eu/pt/publication-detail/-/publication/5a935010-af78-11e8-99ee-01aa75ed71a1

Cohen, J. (1988). *Statistical power analysis for the behavioral sciences*. Erlbaum.

Crane, L., Gantz, G., Isaacs, S., Jose, D., & Sharp, R. (2013). *Introduction to Risk Management: Understanding Agricultural Risks*. Extension Risk Management Education and Risk Management Agency.

Dantsis, T., Douma, C., Giourga, C., Loumou, A., & Polychronaki, E. (2010). A methodological approach to assess and compare the sustainability level of agricultural plant production systems. *Ecological Indicators*, *10*(2), 256–263. doi:10.1016/j.ecolind.2009.05.007

Datta, A., Ullah, H., & Ferdous, Z. (2017). Water management in rice. In B. S. Chauhan (Ed.), *Rice Production Worldwide* (pp. 255–277). Springer Int. Pub. AG. doi:10.1007/978-3-319-47516-5_11

Duong, T., Brewer, T., Luck, J., & Zander, K. (2019). A Global Review of Farmers' Perceptions of Agricultural Risks and Risk Management Strategies. *Agriculture*, *9*(1), 10. doi:10.3390/agriculture9010010

EU. (2021). *Rice Market Situation Committee for the Common Organisation of Agricultural Markets*. European Union. https://ec.europa.eu/info/sites/default/files/food-farming-fisheries/plants_and_plant_products/presentations/rice-market-situation.pdf

FAOSTAT. (2016). *FAOSTAT Online Database*. Food and Agriculture Organization. http://faostat3. fao.org

Girdžiūtė, L. (2012). Risks in Agriculture and Opportunities of their Integrated Evaluation. *Procedia: Social and Behavioral Sciences*, *62*, 783–790. doi:10.1016/j.sbspro.2012.09.132

Gonçalves, J. M., Nunes, M., Jordão, A., Ferreira, S., Eugénio, R., Bigeriego, J., Duarte, I., Amador, P., Filipe, O., Damásio, H., & Bahcevandziev, K. (2021). The Challenges of Water Saving in Rice Irrigation: Field Assessment of Alternate Wetting and Drying Flooding and Drip Irrigation Techniques in the Lis Valley, Portugal. *Proceedings of the 1st International Conference on Water Energy Food and Sustainability (ICoWEFS 2021)*, 85–94. 10.1007/978-3-030-75315-3_11

Hardaker, J., Huirne, R., & Anderson, J. (1997). *Coping with Risk in Agriculture*. Cab International.

Harwood, J., Heifner, R., Coble, K., Perry, J., & Somwaru, A. (1999). Managing Risk in Farming: Concepts, Research, and Analysis. Market and Trade Economics Division and Resource Economics, Division, Economic Research Service, U.S. Department of Agriculture, Agricultural Economic Report, No. 774.

INE. (2021). *Agricultural Census - 2019* [Recenseamento Agrícola – 2019]. INE.

INE. (n.d.). *Agricultural Statistics* [Estatísticas Agrícolas]. INE.

Jalindar, M. K., Rao, V. P., Ramulu, V., Kumar, K. A., & Devi, M. U. (2019). Effectiveness of field water tube for standardization of alternate wetting and drying (AWD) method of water management in lowland rice (Oryza Sativa L.). *Irrigation and Drainage*, *68*(4), 679–689. doi:10.1002/ird.2354

Katoh, M., Murase, J., Hayashi, M., Matsuya, K., & Kimura, M. (2004). Nutrient leaching from the plow layer by water percolation and accumulation in the subsoil in an irrigated paddy field. *Soil Science and Plant Nutrition*, *50*(5), 721–729. doi:10.1080/00380768.2004.10408528

Keil, M., Wallace, L., Turk, D., Dixon-Randall, G., & Nulden, U. (2000). An Investigation of Risk Perception and Risk Propensity on the Decision to Continue a Software Development Project. *Journal of Systems and Software*, *53*(2), 145–157. doi:10.1016/S0164-1212(00)00010-8

Komarek, A., De Pinto, A., & Smith, V. (2020). A review of types of risks in agriculture: What we know and what we need to know. *Agricultural Systems*, •••, 178.

Lampayan, R. M., Rejesus, R. M., Singleton, G. R., & Bouman, B. A. M. (2015). Adoption and economics of alternate wetting and drying management for irrigated lowland rice. *Field Crops Research*, *170*, 95–108.

Lampayan, R. M., Rejesus, R. M., Singleton, G. R., & Bouman, B. A. M. (2015). Adoption and economics of alternate wetting and drying management for irrigated lowland rice. *Field Crops Research*, *170*, 95–108. doi:10.1016/j.fcr.2014.10.013

Lancaster, A., & Torres, P. (2019). Investigating the Drivers of Farm Diversification Among US, Fruit and Vegetable Operations. *Sustainability*, *2019*(11), 3380. doi:10.3390u11123380

Linquist, M. M., Adviento-Borbe, M. A. A., Chaney, R., Da Roda, E. F. F., & Van Kessel, C. (2014). Reducing greenhouse gas emissions, water use, and grain arsenic levels in rice systems. *Global Change Biology*, *21*(1), 407–417. doi:10.1111/gcb.12701 PMID:25099317

MEDWATERICE Homepage. (n.d.). https://www.medwaterice.org/

Mishra, A., & El-Osta, H. (2002). Risk Management Through Enterprise Diversification: A Farm-Level Analysis. *American Agricultural Economics Association*.

Nguyen, N. V., & Ferrero, A. (2012). Meeting the challenges of global rice production. *Paddy and Water Environment*, *4*(1), 1–9. doi:10.100710333-005-0031-5

Nunes, M., Jordão, A., Paixão, J., Russo, A., Eugénio, R., Damásio, H., & Gonçalves, J. M. (2021). Experimentation on water saving in rice irrigation, in Beira Litoral (Portugal). *Proceedings of 15th Portuguese Water Congress*. www.aprh.pt/congressoagua2021/

OECD. (2009). *Managing Risks in Agriculture: a Holistic Approach.* Organization for Economic Co-operation and Development.

OECD. (2019). *Risks that Matter: Main Findings from the 2018 OECD Risks that Matter Survey.* https://www.oecd.org/social/risks-that-matter.htm

Oliveira, F., Reis, P., Oliveira, S., Russo, A., Damásio, H., Eugénio, R., Paixão, J., Melo, F., Nunes, M., Teixeira, M., Bahcevandziev, K., & Gonçalves, J. (2019). Sustainability of Rice Production Systems: Agro-Economic Analysis of Baixo Mondego and Lis Irrigation Districts, Portugal. In *X International Scientific Agriculture Symposium "Agrosym 2019"* (pp. 1210-1215). Faculty of Agriculture East Sarajevo.

Pestana, M. H., & Gagueiro, J. (2014). Data Analysis for Social Sciences: The Complementarity of SPSS [Análise de dados para as Ciências Sociais: A complementaridade do SPSS] (6th ed.). Sílabos Eds.

Reba, M. L., & Massey, J. H. (2020), Surface irrigation in the Lower Mississippi River Basin: Trends and Innovations. Trans. of the ASABE, 63, 1305–1314.

Runkle, B. R. K., Suvoscarev, K., Reba, M. L., Reavis, C. W., Smith, S. F., Chiu, Y. L., & Fong, B. (2019). Methane emission reductions from the alternate wetting and drying of rice fields detected using the eddy covariance method. *Environmental Science & Technology, 53*(2), 671–681. doi:10.1021/acs.est.8b05535 PMID:30566833

Sjöberg, L. (1998). World views, political attitudes and risk perception. *Risk–Health. Safety and Environment, 9*, 137–152.

Székely, C., & Pálinkás, P. (2009). Agricultural Risk Management in the European Union and in the USA. *Studies in Agricultural Economics (Budapest)*, (109), 55–72.

Tedesco, I. (2017) A holistic approach to agricultural risk management for improving resilience. *2nd International Workshop on Modelling of Physical, Economic and Social Systems for Resilience Assessment*, 14-16.

Tuong, T. P., & Bouman, B. A. M. (2003). Rice production in water-scarce environments. In Water Productivity: Limits and Opportunities for Improvement. CABI Pub. doi:10.1079/9780851996691.0053

USEPA. (2006). Global anthropogenic non-CO2 greenhouse gas emissions: 1990–2020. United States Environmental Protection Agency, EPA 430-R-06-003.

Vagias, W. (2006). *Likert-type scale response anchors.* Clemson International Institute for Management, Clemson University.

Vories, E., Stevens, W. G., Rhine, M., & Straatmann, Z. (2017). Investigating irrigation schedul-ing for rice using variable rate irrigation. *Agricultural Water Management, 179*, 314–323. doi:10.1016/j.agwat.2016.05.032

Waage, J., & Mumford, J. (2008). Agricultural biosecurity. *Philos Trans R Soc Lond B Biol Sci, 363*(1492), 863–876. doi:10.1098/rstb.2007.2188

Yamane, T. (1973). *Statistics: An Introductory Analysis* (3rd ed.). Harper and Row.

Chapter 15
Sustainable Cocoa Value Chain:
A Review and Critical Analysis of "Triple Bottom Line" Scenarios

Ibrahim Cravid dos Prazeres
https://orcid.org/0000-0001-6833-5155
CEGAFE, University of Évora, Portugal & CETRAD, University of Trás-os-Montes and Alto Douro, Portugal

Maria Raquel Lucas
Universidade de Évora, Portugal

Ana Alexandra Marta-Costa
https://orcid.org/0000-0001-9247-9167
University of Trás-os-Montes and Alto Douro, Portugal

ABSTRACT

In the last decades, the focus of studies on cocoa value chain (CVC) has changed from the low income of farmers and the shortcomings of the educational and financial systems to the incorporation of innovations, supported on sustainability principles. However, classical theories based on economics are insufficient to understand sustainability phenomenon, and the investigation in the field is still dispersed. This study represents one first attempt to synthesize findings on the topic, in line with the triple bottom line (TBL) scenarios. TBL provides a useful framework to understand the social, economic, and environmental aspects along the CVC. This chapter performs a systematic literature review on sustainability scenarios applied to CVC, each one representing one of the three dimensions of sustainability. At the final, an agenda for future research on the topic is suggested, uncovering a set of future study propositions.

INTRODUCTION

Sustainability has emerged as a factor of growing importance among society and businesses, following the imperative matter of sustainable development raised by the World Council of Economic Develop-

DOI: 10.4018/978-1-7998-9557-2.ch015

ment (WCED, 1987). Hence, many countries are pursuing it in their diverse value chains, aiming to adopt strong measures which target the growing environmental pressure and its long-term risks such as flooding, drought, political instability, social unrest and depletion of natural resources (Giddens, 2015). Nevertheless, assessing social and environmental impacts is not new (Norman and MacDonald, 2004; Richardson, 2004) and TBL (triple bottom line) has greatly contributed to incorporating sustainability into the business agenda (McDonough and Braungart, 2002).

Although the TBL construct is explicitly based on the integration of the social, environmental and economic dimensions of sustainability, this concept is inconsistent in its literature. Several studies used sustainability to primarily refer to the environmental dimension (Yan et al., 2009). Others used the concept to refer to its social dimension (Bibri, 2008), while some used it to refer to all three (Marcus and Fremeth, 2009). Conversely, TBL places equal levels of importance on each of the three dimensions, which brings greater balance and coherence to the construct of sustainability (Elkington, 1997; Epstein, 2008; Harmon et al., 2009; Savitz and Weber, 2006). Other sustainability-related studies showed an imbalance in the level of importance distributed among the three dimensions. The importance of the economic dimension was limited (Collins et al., 2007), though many studies included it when referring to sustainability (Collins et al., 2007).

Additionally, some authors argue that the economic, environmental, and social aspects do not sufficiently cover the entire concept of the TBL related-sustainability construct (Wu et al., 2016). They proposed four additional aspects, to further promote discussion on the issue: operations, resilience, long-term and stakeholders. Tseng (2017) emphasized that sustainability issues are characterized by their high complexity and uncertainty and Isil and Hernke (2017) contributed with a critical evaluation of the TBL paradigm. To promote the quality and success of the solutions, it is essential to integrate qualitative information, quantitative data, and social media into the TBL discussion (Wu et al., 2018). The TBL can evolve in different ways independently from the object of study (Tseng et al., 2020). In each case, balancing the TBL toward sustainability, involves viewing it as an essentially dynamic and multidimensional concept (Alvarez et al. 2016).

To some extent, agriculture, agribusiness and food markets have become areas that prioritize the sustainable development, in many countries. Such is due to their responsibility for over 30% of greenhouse gas emissions, 70% of global water use (Pradhan et al., 2013) and approximately 50% of available land use worldwide, being the greatest driver of the biodiversity decline (Chiarolla, 2017), accounting for 70% of the projected loss of terrestrial biodiversity (CBD, 2014; 2016). This loss is placing global food security in serious danger by undermining the resilience of many agricultural systems to threats such as pests, pathogens and climate change (IPBES, 2019; Belanger and Pilling, 2019). The cocoa value chain sustainability is an increasing global concern, as large part of its used land is attained through deforestation, which has a large impact among the local biodiversity and social cohesion (Recanati et al, 2017). Additionally, the over-aged tree stocks, the repercussions of disease and pest infestation, the political instability in West Africa, a lack of agricultural professionalism, an absence of infrastructure, low farmer income, the shortcomings of the educational and financial systems in the cocoa-growing regions and child labour exploitation and child trafficking, prove to be sustainability problems (Matissek, et al., 2012; Ingram et al., 2018). Thus, the integration of the TBL dimensions in the cocoa chain represents a relevant issue for its adequate management.

Cocoa is one of the main goods in the global agricultural trade markets, occupying the third position in exports. It is a commodity that plays an important role in the food product market and it is a vital contributor to smallholder's producer income and to their developing countries' trades, which is facing climate challenges and risks (Prazeres et al., 2021). Furthermore, cocoa has an interesting value chain embedded in a complex global network that includes all the actors, from the producer, mainly located in Africa, Asia and Americas, to the chocolate manufacturer and the consumer market, primarily done in Europe and Northern America (Recanati et al., 2017). Besides, the demand for cocoa is persistently growing. In the past 50 years, the demand increased by 300% and some forecasts sustain that it has continuously risen. Therefore, sustainability improvements will be even more impact full (Recanati et al., 2017) and the TBL concept of Elkington (1997) can be a relevant possibility for a sustainable transition and/or improvement within the cocoa value chain (Blowfield, 2003).

This chapter attempts to fulfil the previously identified research gap, by performing a systematic literature review of the cocoa value chain, under the lens of TBL scenarios. TBL is an appropriate perspective to study this chain as it sheds light on the potential driver of systemic changes across all key-actors, including the increase in consumer awareness of more sustainable lifestyles and responsible consumption (Colwell, 2012). TBL is also an adequate terminology both to frame the multidisciplinary answers ands scenarios at a wider scale and identify the basic principles for driving value chains (Movilla-Pateiro *et al.*, 2021). In this endeavour, the structure and the guidelines of a literature review by Gilal et al. (2019) and Hungara and Nobre (2021) are followed.

The chapter is structured as follows. Its opening concerns the historical overview of TBL, following other authors (e.g., Alhaddi, 2015; Rashidi, et al., 2020). Successively, the review methodology is presented. It is primarily stimulated by the studies of Kahiya (2018), Gilal et al. (2019) and Hungara and Nobre (2021), and to a lesser extent, by some other studies (e.g., Paul, 2019; Paul and Benito, 2018). Following the review methodology, the bibliometric profile of the articles is exposed. Then, the selected sample is analysed in terms of publications trends and research contexts, and the methodologies and scenarios are discussed. The conclusions, to the research agenda for future studies and its emerging themes, and the set of propositions of study, are presented before the finale. The chapter ends with some key literature for additional reading which reflects the issues discussed.

HISTORICAL OVERVIEW OF TBL

In 1997, Elkington coined the expression TBL, a newly sustainability-related construct. Though there are no different interpretations to the exact historical roots of TBL, being widely accepted the 1997's expansion of the term and its growth in most agrifood chains (Colwell, 2012), its antecedents may date back to over 130 years ago, from an idea denominated as spaceship earth (George, 1879/2009).

The multidisciplinary construct of TBL gained important recognition when it provided the Brundtland Report (WCED, 1987) and instigated the concept of "sustainable development". The concept is defined as the "*development that meets the needs of the present generations without compromising the ability of the future generations to meet their own needs*" (WCED, 1987, p. 43). In this context TBL expresses the expansion of the environmental agenda, integrating the economic and social dimensions (Elkington, 1997), providing a framework for assessing three key-interrelated dimensions driven by sustainability, the economic dimension, the social dimension and the environmental one (Goel, 2010), to which the institutional can be added to (Waas *et al.*, 2012). In Elkington's original definition of TBL, these dimensions or lines were profit, people, and planet (He et al., 2019) and, for some, the dimensions of TBL, can also be denominated as "the three lines/pillars" or "the three E's" (economy, equity and ecology) (Colwell, 2012). The central idea behind the three E's is that sustainability cannot be achieved unless the chain is economically feasible, environmentally dependable and socially responsible, allowing interchangeability. Meaning, if there is a shortfalling in one of the dimensions, it can be offset by "investing" in another (Colwell, 2012). Elkington (1998) does not considered decreasing the level of negative outcomes as a solution to the environmental and societal problems of the planet, as settled and established in the Millennium Development Goals (MDGs).

The main research efforts in the field have three booms in publications, in 2003, 2011, and 2015, revealing a proliferation of academic studies and articles addressing the issue of sustainability by employing the TBL. One of these booms is related to the redefinition of Elkington's construct of TBL by Carter and Rogers (2008). They redefined the TBL as follows: "sustainability should hold economic performance, the natural environment and society at a broader level, and the intersection of social, environmental and economic activities can help organizations become engaged in activities that not only positively affect the natural environment and society but that also result in long-term economic benefits and competitive advantage for the firms." The second boom occurred after the United Nations climate change conference in Durban 2011, South Africa, aimed at developing a new agreement to limit carbon emissions. The final boom occurred after complementing TBL with four new dimensions (operations, resilience, long-term, and stakeholders) (Tseng et al., 2020).

After 25 years and after the legitimacy of TBL as an independent research field, Elkington (2018) came back to his analysis, recognising the inadequate use of the concept as an accounting tool and updated TBL to a more globally system approach. The main problem of the TBL framework, which was not evolving into practice and thus obtaining positive results, signalised the need to find other or specific conditions or characteristics that could promote its real implementation (Loviscek, 2021). Elkington (2018) also recognized that the TBL hadn't lost its credibility and was still considered a current approach for sustainable development. The academia has been using the term from 1998 until today, reaching its peak in the last five years due to the proliferation of a huge body of literature on the topic, owed to worldwide environmental and societal pressures. This reputation is both capturing the interest of top journals (Zaharia and Zaharia (2021) and increasing the number of published articles.

METHODOLOGICAL PROCEDURE

The study opted for an integrative literature review due to its ability to capture the diversity and development of more than one body of literature. The multiplicity of emerging topics is addressed through a holistic conceptualisation and synthesis of the extant literature (Torraco, 2016; Hungara and Nobre, 2021). This type of review generates knowledge as it culminates into a conceptual model or framework that offers new ways of thinking about the literature and a research agenda that delivers questions for future research (Torraco, 2005; Hungara and Nobre, 2021). This should allow for a better understanding of the phenomenon, provide meaningful knowledge on the TBL construct dimensions' scenarios for further research, and enable a discussion on the implementation of the TBL in the cocoa value chain.

Initially, in order to conduct this integrative review, a set of specific research questions that guided the study, was developed. These questions were:

1. How can sustainability or TBL contribute to the literature on the cocoa value chain?
2. What type of sustainability or TBL scenarios for the cocoa value chain are described in the literature so far?

3. What theories have been used in parallel with sustainability or TBL in the cocoa value chain?
4. What variables have been used under sustainability or TBL scenarios?
5. What contexts have been studied within the cocoa value chain?
6. What countries have been studied in regard to sustainability or under the TBL cocoa value chain literature?
7. What kind of research methods have been use?

This chapter seeks to answer these questions by linking the development and legitimacy of sustainability or TBL to the development of the cocoa value chain, and by demonstrating how these two concepts over-lap.

The study followed a theory-based review, similar to Hungara and Nobre (2021). Theory-based reviews advance the literature on a specific topic by applying a given theory to a subject area or field (Paul and Criado, 2020). The review methodology used, consisted of different steps according to Kahiya (2018), including, among others, the identification of the search terms and the databases to be accessed and, the definition of the eligibility criteria for and exclusion of articles (see figure 1).

The search terms selected were: "cocoa or chocolate" and "sustainability" and organic combined with "TBL scenarios "and value chain" or "supply chain". These search terms were applied through the search engine of Scopus to search for peer-reviewed journal articles. This database and process were also generally used in previous systematic reviews (e.g., Randhawa et al., 2016; Alcayaga, et al., 2019) and in cocoa value chain reviews (e.g., Nabhani, Daryanto, Yassin and Rifin, 2015; Ros-Tonen et al., 2019; Goel et al., 20121). Because sustainability and TBL are multidisciplinary topics that cross different fields of research (e.g., sociology, economy and environment), the Scopus database which is broad, diverse and multidisciplinary, offering a large sample of greatly trustworthy publications, was deemed adequate. However, other reviews additionally considered the Web of Science, Emerald Insight and EBSCO databases (Correia et al., 2017).

The English, Spanish and Portuguese language and the temporal period from 2000 to September 2021, were the main criteria used in selected articles. This period was chosen because it covers all the TBL life cycle, from its origin and enclosing the three booms of related publications. This goes in line with what a systematic literature reviews must encompass, meaning, 10 years of research in the field (Paul and Criado, 2020) similarly to what occurs in other studies (Paul and Mas, 2019; Rosado-Serrano et al., 2018; Alcayaga, et al., 2019).

The search results from the Scopus database contain a set of 208 scientific articles on diverse subject areas of research interest (see figure 1). In step two, to perform a selection and exclusion of peer-reviewed journal articles, the criteria were research area of the publications ("Agricultural and Biological Sciences" "Business, Management and Accounting", "Environmental Science", "Social Sciences", "Economic, Econometrics and Finance" and "Decision Sciences") and keywords. The aim was to include out-standing articles in the selection, comprising significant cocoa or chocolate-related research applied to the cocoa value chain. Upon these results, the number of articles retained was 129 from 9 countries around the world (United Kingdom, United States, Ghana, Colombia, Germany, Indonesia, Brazil, Ecuador and Italy). During step three, the 129 articles gathered were reduced down to 75, through an exclusion by title observation of the papers dedicated to agronomic, chemical and microbiological subjects; studies in product quality; and researches that included a set of tropical products.

Following Alcayaga, et al. (2019) and Paul and Criado (2020), the targeted search of representative literature enclosed specific keywords, present not only in the title, abstract, or list of keywords, but also in the full text. Thus, the procedures of the abstract review (step four) and full-text review (step five), allowed for the exclusion of 17 articles that did not contemplate the topics that are intended for this study. A total of 58 articles were retrieved (see figure 1). According to Kahiya (2018), another criterion was set for article inclusion. This criterion stated that, at least, one of the following conditions had to be verified:

1. Presence of an existing link between sustainability or TBL and cocoa or chocolate value chain;
2. Provide means to define or identify sustainability or TBL scenarios on cocoa or chocolate value chain;
3. Specifically mention sustainability or TBL dimensions or bottom line of sustainability (e.g., economics or environmental or social) and cocoa or chocolate value chain.
4. Provide information about an agenda for future research on sustainable cocoa or chocolate value chain.

Figure 1. Review methodology

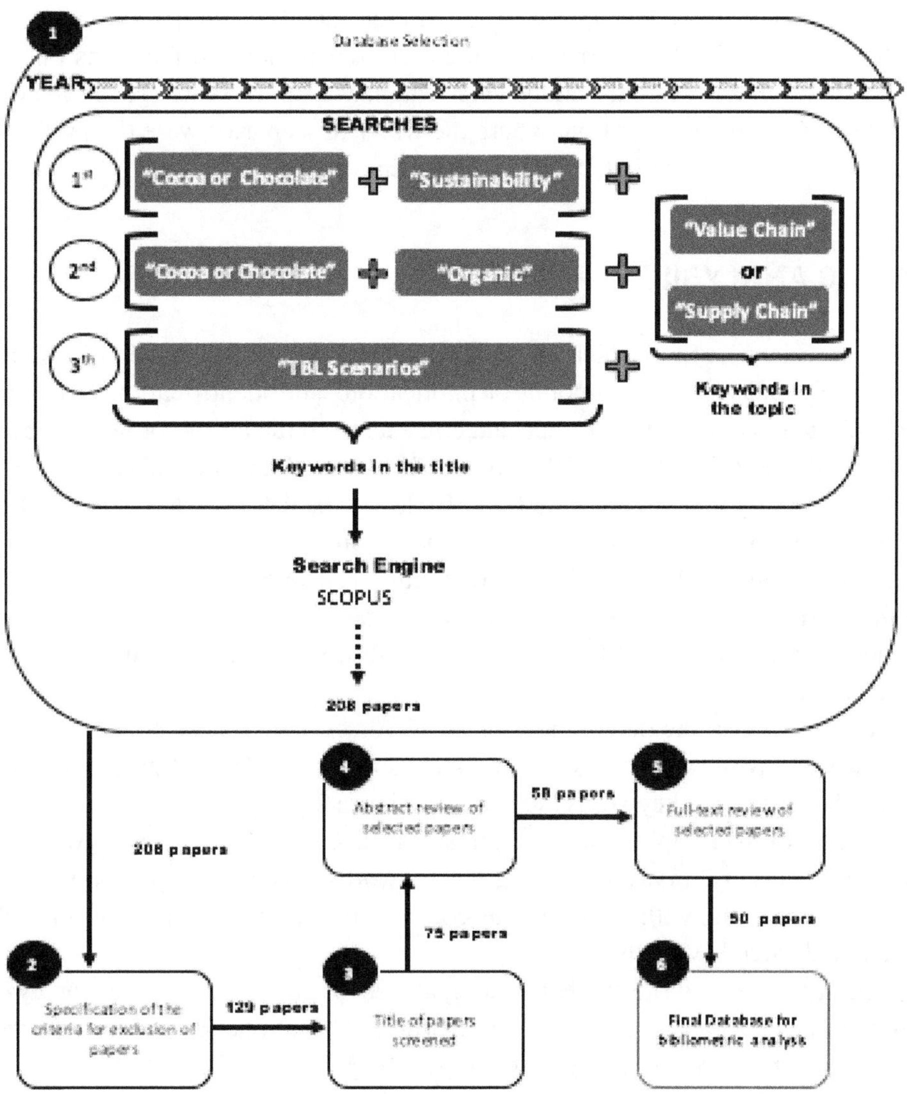

To ensure that only relevant papers were analysed, the Tranfield *et al.* (2003) recommendations on the decisions relating to the inclusion/exclusion criteria were followed. These are relatively subjective and should be carried out by more than one researcher, therefore, two researchers with expertise in sustainability and cocoa were involved in this phase. All these criteria allowed a final database of 50 scientific articles (Table 1), wich is in line with the criteria of a sample of 40-50 in a systematic literature review (Paul and Criado, 2020).

The final sample of 50 articles did not comprise articles specifically focusing on cocoa TBL scenarios, but rather those on explicit subjects of cocoa or the sustainability of the chocolate value chain, such as, sustainability practices and standards, supply chain performance and added value, supply chain on zero-deforestation ecosystems services, sustainability transition pathways, circular economy, certification, or child labour and child trafficking on the value chain or rural/territorial development with shared value. The search also permits the confirmation that there are no studies concerning TBL scenarios exclusively on the cocoa value chain and, consequently, their contribution to the literature is still commencing. Moreover, as far as known, no other systematic review has focused specifically on TBL aligned with the cocoa value chain. Thus, it wasn't possible to follow the procedure recommended by Kahiya's (2018) and compare the final sample with the samples of other systematic reviews on the topic. Notwithstanding, it should be noted that there are much more publications on TBL in a broad sense than those strictly focusing on the cocoa value chain. Multiple research scenarios were confronted with a trade-off or win-win consideration, where the win-win scenarios were the preferred options due to the increase on multiple factors in the TBL (Trienekens, 2018).

BIBLIOMETRIC ANALYSIS

Bibliometrics is the scientific field which measures publications, authors and citations. In this work, the bibliometric analysis focused on the amount of publications and identification of publication outlets, on the number of citations. publishing trends and contexts from the article sample, as evidence of their academic value. Following Hao et al. (2019), Kahiya (2018), and Hungara and Nobre (2021), a list of the distribution of articles per journal was created (see Table 2). The list, similarly to that of Rosado-Serrano et al. (2018), covers 50 articles spread across 41 journals, whilst following the recommendations of Paul and Criado (2020) of having at least 10–20 different journals to avoid biased selection criteria. With the exception of the journal *Sustainability* (n=3), eight other journals have the same number of publications (n=2), covering a large number of publications in sustainability related research. The remaining journals solely have one article.

The next step was to map the occurrences of keywords clusters related to subfields of study in the sample search, using *VOSviewer* software, a tool for constructing and visualizing bibliometric networks. When analysing the map of the keywords clusters (see figure 2), it is noted that 4 clusters can be identified as representative of the sample. These are: cluster 1 (cross), including the keywords farmer, cocoa, country and quality; cluster 2 (star) agregating the keywords stakeholder, perception, impact and, deforestation cluster 3 (triangle) with the keywords cocoa, and value chain; and, cluster 4 (rhombus) with the supply chain and value keywords.

To assess scholarly work on the subject, Hao et al. (2019) was followed, thus analyzing the global number of citations and the partials per year of each article. The number of citations per year was used to control the age of an article. According to other authors (e.g., Kahiya, 2018; Lim et al., 2021), the study

Table 1. Final database

Number	Reference	Number	Reference
(1)	Abbey et al. (2016);	(26)	Krauss and Krishnan (2021)
(2)	Aboah et al. (2021)	(27)	Lafargue et al. (2021)
(3)	Adu-Acheampong et al. (2017)	(28)	Lalwani et al. (2018)
(4)	Arias and Fromm (2019)	(29)	León Bravo et al. (2021)
(5)	Asamoah and Annan (2012)	(30)	Liliana (2020)
(6)	Astrid Fenger et al. (2017)	(31)	Mabe et al. (2020)
(7)	Barrientos (2014)	(32)	McLoughlin and Meehan (2021)
(8)	Barrientos and Asenso-Okyere (2009)	(33)	Mendoza et al. (2020)
(9)	Bonuedi et al. (2021)	(34)	Middendorp et al. (2020)
(10)	Borda et al. (2021)	(35)	Mithöfer et al. (2017)
(11)	Busquet et al. (2021)	(36)	Moreno-Miranda et al. (2020)
(12)	Carodenuto and Buluran (2021)	(37)	Nabhani et al. (2015)
(13)	Castro-Nunez et al. (2020)	(38)	Narciso et al. (2020)
(14)	de Boer et al. (2019)	(39)	Nonci et al. (2019)
(15)	Díaz-Montenegro et al. (2018)	(40)	Odijie (2018)
(16)	Doherty and Meehan (2006)	(41)	Ollivier de Leth and Ros-Tonen (2021)
(17)	Dompreh et al. (2021)	(42)	Orozco-Aguilar et al. (2021)
(18)	Escobar et al. (2020)	(43)	Ruben (2017)
(19)	Estival et al. (2016)	(44)	Rueda et al. (2018)
(20)	Glavee-Geo et al. (2020)	(45)	Schroth et al. (2016)
(21)	Grumiller (2018)	(46)	Scott et al. (2015)
(22)	Haynes et al. (2012)	(47)	Sjauw-Koen-Fa et al. (2018)
(23)	Hess (2021)	(48)	Sonwa et al. (2014)
(24)	Indah et al. (2020)	(49)	van Huellen and Abubakar (2021)
(25)	Ingram et al. (2018)	(50)	Vogel et al.(2020)

established the top 10 of the most cited articles (see Table 3). Barrientos (2014) is the most influential article, both in terms of the global number of citations (n=36) and citations per year (5,14). It is also noted that although Mithöfer et al. (2017) and Rueda et al. (2018) have a substantially lower number of citations (n=19 and n= 4, respectively) than Sonwa et al. (2014) (n=22), in terms of citations per year, they are ranked second and third. Sonwa et al. (2014) ranks second in terms of global citations (n=22), but the article itself, only ranked sixth in terms of average citations per year (3,14) and is less cited than the more recent articles of Ingram et al. (2018), which has an average citation per year of 4,0, and Lalwani et al. (2018), with an average citation per year of 3,33.

Figure 2. Keywords clusters

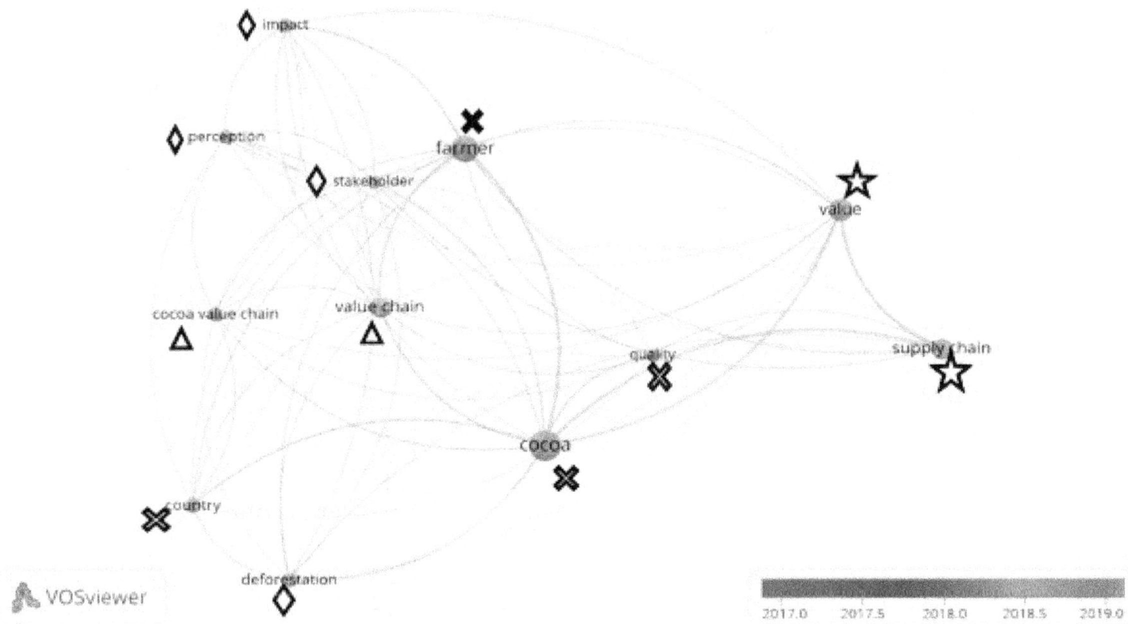

Publication Trends and Contexts

Since the research focuses on studying the cocoa chain sustainability under theTBL scenarios perspective, it began by analyzing the articles in the final sample (n=50) which specifically mention the sustainability topic in the title, abstract, and body text of the articles. According to the booms in publications mentioned by Carter and Rogers (2008), one of them related to Elkington's construct of TBL, the analysis was divided into three periods of time: 2000 – 2010, 2011 – 2015, and 2016 – 2021 that represent three publishing trends (see figure 3).

Only two articles were published between 2000-2010, one focusing on "competing on social resources" and the other on Ghana's cocoa value chain and its challenges. The fact that sustainability and the TBL related topics are present at an early research stage can explained by the scarcity in literature during this period. Between 2011- 2015, the number of articles more than doubled to 6. One of these articles is related to promoting sustainable agroforestry cocoa systems in the main production countries, and its ecosystems services. Another is focused on organic and fair trade as strategies towards environmental sustainability drivers and social equity in production and trade, particularly by small-scale actors. More recently, 42 articles (12 of them related to sustainability dimensions) reached SCOPUS academic journals. These numbers demonstrate that cocoa sustainability research topics are increasingly garnering the interest of the academic and research communities. However, it could also infer that from the 2016-2021 sample, there are still many articles that do not analyse the sustainability dimensions in light with TBL as explored in this study.

Table 2. Publication outlets

Journal	References (codes of Table 1)
Sustainability (MDPI)	(10, 22, 25)
International Journal of Agricultural Sustainability	(40, 50)
International Journal on Food System Dynamics	(4, 47)
Journal Fur Entwicklungspolitik	(8, 21)
Journal of Agribusiness in Developing and Emerging Economies	(14, 31)
Journal of Rural Studies	(1, 15)
Supply Chain Management	(29, 27)
Sustainability Science	(45, 17)
Agricultural Systems	(2)
Agriculture (MDPI)	(18)
Agroalimentaria	(30)
Agroecology and Sustainable Food Systems	(6)
Agroforestry Systems	(48)
Applied Geography	(13)
Asian Social Science	(37)
Benchmarking	(28)
Bioagro	(36)
Bulgarian Journal of Agricultural Science	(24)
Business Strategy and Development	(34)
Business Strategy and The Environment	(44)
Cahiers Agricultures	(3)
Custos e Agronegocio	(46)
Enterprise Development And Microfinance	(43)
Espacios	(19)
European Journal of Development Research	(49)
Frontiers in Sustainable Food Systems	(42)
Global Networks	(26)
International Journal of Biodiversity Science Ecosystem Services and Management	(35)
International Journal of Operations and Production Management	(32)
International Journal of Organizational Analysis	(23)
International Journal of Services and Standards	(5)
International Journal of Supply Chain Management	(39)
Journal of Agriculture and Environment For International Development	(38)
Journal of Business Ethics	(41)
Journal of Development Studies	(9)
Journal of Environmental Policy and Planning	(12)
Journal of Macromarketing	(20)
Journal of Strategic Marketing	(16)
Regional Studies	(7)
Revista de La Facultad de Agronomia	(33)
World Development	(11)

Table 3. Citation analysis

References (codes of Table 1)	Total Citations	Citations/Year
(7)	36	5,14
(48)	22	3,14
(35)	19	4,75
(16)	16	1,06
(45)	15	3,00
(44)	14	4,66
(25)	12	4,0
(22)	11	1,22
(28)	10	3,33
(6)	9	2,25

Figure 3. Publishing trends

The countries of study and the specific research contexts in which the research was applied are presented on Tables 4 and 5 and followed a similar design to Chen et al. (2021) and Hungara and Nobre (2021). Their observation shows that most studies took place in Ghana (n=17), followed by Ecuador (n=7) and Indonesia (n=5). West and Central Africa and the Ivory Coast presented the same number of studies (n=4). UK (n=2) embodies the rare exception of countries located in Europe with contributions to the field. Notwithstanding, one of UK'studies solely mentioned the field as part of a research taking place mostly in the Europe. The studies also covered a wide range of subjects related to sustainability and the cocoa value chain (see table 5) such as: sustainability initiatives and practices (n=11), value chain dynamics (n=11), value chain strategies (n=14), deforestation (n=6) and smallholder livelihoods and family farming (n=6).

Table 4. Research country

Country	References (codes of Table 1)
Ghana	(1, 2, 3, 5, 6, 7, 8, 12, 17, 20, 21, 23, 28, 31, 41, 47, 49).
Ecuador	(15, 29, 27, 33, 34, 36, 44)
Indonesia	(14, 24, 35, 37, 39)
West and Central Africa	(11, 25, 40, 48)
Ivory Coast	(12, 21, 28, 47)
Peru	(10, 42, 46)
Colombia	(13, 18)
Nicaragua	(26, 42)
UK (Europe)	(16, 32)
Brazil	(45)
Costa Rica	(22)
Cameroon	(50)
Honduras	(4)
Netherlands	(27)
Venezuela	(30)
São Tomé and Príncipe	(38)
Sierra Leone	(9)

Table 5. Research context

Research context	References (codes of Table 1)
Literature Review	(37, 43)
Sustainability (iniciatives; indicators; practices; organisation; programs; policies; management)	(10, 14, 23, 25, 26, 28, 29, 32, 35, 40, 47)
Value chain dynamics, (opportunities, risks, interventions; gender; future scenarios)	(3, 5, 7, 8, 9, 21, 22, 27, 31, 46, 49)
Value chain strategies (resilience, policies, trust relationships, performance; competitive advantage; CSV and inclusive development; certification; stakolders perceptions)	(1, 2, 4, 6, 15, 16, 17, 20, 24, 30, 36, 38, 47, 50)
Deforestation, Reforestation and Tree Cover; Circular Economy; agroforestry systems and ecosystem services)	(12, 13, 33, 42, 45, 48)
Smallholder livelihoods, family farming, child labour, child trafficking and ecosystems (weakness, improvements, challenges)	(11, 18, 19, 34, 39, 44)

REVIEW OF STUDIES

This section analyzes the main perspectives on the connection between TBL and the cocoa value chain sustainability, emphasizing the research methodologies and the sustainability scenarios.

Research Methods

The research methods used in the studies, excluding literature review articles (Ruben, 2017; Nabhani et al., 2015), are synthetized in the Table 6. The case studies that are most widely used in the cocoa value chain sustainability research are predominant in the sample (n=12). Ensuing, the predominant are the qualitative and the quantitative studies. Among the qualitative methods, the most popular were interviews (n=13). The methods used in three of the articles in the sample, combined field observation and interviews and one combined action research and interviews. Different interview types appeared, such as semi-structured, long, and in-depth interviews. Among other qualitative research methods considered were the focus group (n=3), content analysis (n=3) and participant observation (n=4) were three of them combining, respectivly, written background narratives, panel data and, remote sensing work. In the quantitative methods, the surveys (n=12) were the most represented, with only one article using supply fuctions estimation and other panel studies' methods. Thus, in regards to the research methods, it can be understood that most articles connecting sustainability to the cocoa value chain, use a widely qualitative approach, mostly through case studies and interviews.

Most of the articles report the use of primary data (n=44) which can be explained by the popularity of case studies and in-depth interviews as the main methods for data collection. Almost third of these 50 articles (n=18) present a combination of both primary and secondary data. In most cases, this combination stems from the application of case studies' methods. Finally, among the articles analyzed, excluding the literature review articles, only four of them used exclusively secondary data sources.

Table 6. Research methods

Research method	References (codes of Table 1)
Case studies	(4, 7, 10, 14, 21, 23, 28, 29, 35, 41, 42, 49)
Interview	(6, 16, 17, 19, 22, 24, 25, 27, 30, 31, 32, 46, 50)
Surveys	(1, 5, 9, 15, 18, 24, 29, 33, 36, 39, 44, 50)
Observation	(3, 8, 45, 50)
Content analysis	(12, 38, 40)
Focus group	(6, 18, 25)
Panel studies	(2, 13, 25)
Supply functions	(48)

Scenarios of TBL

Despite the absence of TBL specific scenarios in the sample, the cocoa value chain sustainability scenarios have evolved into increasingly reliable self-set voluntary commitments to address the persistent social and environmental challenges embedded in global value chains. While activities' profits continue to guide sustainability's economic dimension and value chain strategies, the onus is increasingly placed on socio-environmental awareness to pronounce measures to eliminate deforestation and other unethical practices associated with cocoa value chain operations. In this contribution, some authors analyze the novel zero-deforestation commitment that focuses mainly on the world's top cocoa producers: Côte d'Ivoire and Ghana. Others focus on the smallholder livelihoods, the family farming, the child labour, the child trafficking, the circular economy and the ecosystems'services preservation. Some transversal scenarios agregate several value chain dynamics and actions towards sustainability in specific geographic contexts and production systems. Figure 4 synthetizes the articulation of the previously identified clusters and the research context of the sustainability scenarios.

Figure 4. Clusters, scenarios and research context

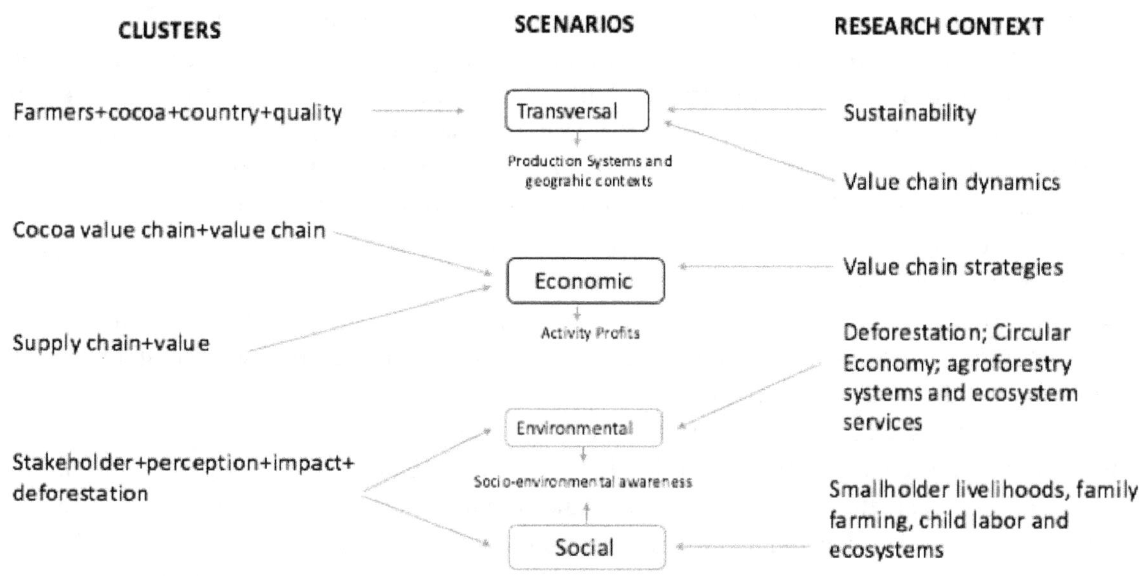

The fairtrade and organic certification are widely recognized in the literature, and are present in the cluster "farmers, cocoa, country and quality" (Haynes et al., 2012; Ingram et al., 2018). Organic certification focuses its attention on safe environmental and work practices, and provides a scenario that aligns relatively well with existing traditional shade grown cocoa production systems in several producer countries and geographic contexts (Dompreh et al., 2021). Though it may be most appropriate for larger cocoa growers, it may also assume important functions to small producers, ensuring that they maintain environmentally friendly cocoa systems and providing them and their customers with standardized quality assurance (Astrid Fenger et al., 2017). Even though smallholders lack the assets needed to join mainstream commodity markets, these certifications schemes have enabled them to capitalize on the qualities of their traditional varieties and access niche markets (Rueda et al., 2018). Nonetheless, because the value chain is not gender neutral, but rather interacts with gendered processes that affect its functioning and vice-versa, it is important that economic and social upgrading and cocoa value chain sustainability is achieved if women (farmers and workers) receive greater recognition and remuneration in both the commercial and societal spheres (Barrientos, 2014). "Sustainability" and "value chain dynamics" should reflect the interaction of local, national and international drivers (Borda et al, 2021)

In the "value chain" and "supply value chain" clusters, the traditional conceptions of value follow neoclassical economics by using price and activity profits as the appropriate unit of measurement. Still, new approaches have broken these narrow definition and have not only redefined value but, more importantly, shifted the key unit of analysis from the individual cocoa firm's value chain to the wider value constellation, assuming value as 'the utility combination of benefits delivered to the market less the total costs of acquiring the delivered benefits (De Boer et al., 2019). This is a scenario where 'price', as the measure of value, is replaced with the more cognitively grounded concept of 'delivered benefits'. It means that value occurs not in sequential chains but in complex constellations. In those scenarios, there is a departure from more vertically integrated business structures, in preference for looser 'networks', which reflects advances in flexible manufacturing technologies, ease in organisational control and co-ordination through information technology, as well as enhanced customer knowledge and relationships, further supported by information technologies (Rueda et al., 2018). These developments should result in a more sophisticated concept of value and greater understanding of the contexts (value chain networks), in which it is created. Likewise, value creation should be related to the nature of the resources from which value is created and not fail to explore the control issues (within the value chain network) and, distribution of value to all chain links, in a more ethical, fair and sustainable way.

In the context of "smallholder livelihoods, family farming, child labour, child trafficking and ecosystems preservation", the scenarios found in literature clearly reveal: i) that the payment for ecosystem services that can be provided by sustainable cocoa agroforestry systems depends upon the appropriate combination of cocoa, timber and non-timber forest trees on the same land; ii) that keeping cocoa's agroforest system is an asset for the payment of environmental services, such as carbon storage and biodiversity conservation, because of the resource potential that they can generate (Sonwa et al., 2014); iii) that the good governance and social capital can facilitate wider stakeholder participation, enhancing consensus within the cocoa value chains and socio-economic development (Díaz-Montenegro et al., 2018); iv) that it is important to mitigate the continuous disappearance of natural forests, cocoa agroforests and plantations due to the increasing demand and interest for timber production and other forest-related products (Odijie, 2018). All of these problems when linked to social and environmental awareness, are part of the third cluster. In this context, a future scenario for the cocoa, according to some authors (Aboah et al., 2021) would be the maintenance of a diversity of smallholders due to their important role in the value chains. This means a more intensified regional specialization with lower costs and higher quality, and, when possible, supported in local varieties (Scott et al., 2015). This also means an identification of agroecological zones and a subsequent adequate adjustment of varieties and production systems, offering a range of different cocoa products for different palates, markets and prices and, the adoption of good farming management practices so to improve cocoa productivity (Aboah et al., 2021). This scenario would also link cocoa production much more closely to tourism of various kinds (live-in, culinary, and environmental) and to the food/ restaurant industry, including cooking schools where cocoa can be a prominent dish ingredient.

The principal risks associated with this scenario include continued limited scale and low yields, thus, low incomes for many producers, who never reach a sufficient scale to be truly sustainable. This does not exclude an alternative large-scale scenario, with a much more expansive cocoa production development, emphasizing improved productivity to ensure future competitiveness for cocoa producers, given the relatively small volumes currently sold and the uncertainty of future demand. In this scenario, larger, more established cocoa cooperatives, of comparable scale, and private firms, would capture both production and post-harvest economies, and produce the necessary volumes to facilitate greater added-value, resultant from processing in close proximity to the growing areas. A complementary component to this would involve strategic alliances between the larger, more established and better equipped producer cooperatives and the smaller, more recently established associations. Such arrangement would help overcome the capacity utilization difficulties which larger scale cooperatives might face while giving the associated smaller cooperatives access to the knowledge, experience, and marketing contacts.

GAPS IN THE LITERATURE AND TBL PROPOSITIONS

An important contribution and originality of the present study relates to the application of TBL to critically analyze research on the cocoa value chain. Despite the lack of studies, TBL provides a useful framework to understand the social, economic and environmental aspects of this chain. In this endeavor, it suggests a research agenda on the topic, entailing its emerging themes and propositions to future testing studies. Following, the main literature gaps concerning the cocoa value chain sustainability and the TBL propositions for the future, are presented.

The gathered results of this work sustain that, contrary to what was to be expected in the face of the current challenges, sustainability scenarios are inducing less stringent environmental and social commitments when compared to the economic dimension and added-value counterparts. The review also isolates key gaps in the TBL scenarios, including the lack of precision surrounding all its dimensions, and clarity of its role and relevance under the cocoa value chain geospatial context.

Despite its highten profits, the global cocoa value chain faces numerous challenges at its different links, especially at the farm production level. Cocoa is largely grown by smallholder farmers, who experience poverty, declining productivity, volatile prices and lack of investment funding, in addition to other issues relating to human rights and the environment. Contract-farming is a value chain model that can be thought of as a future emerging trend that addresses some of the challenges faced by cocoa farmers, including disputes and inequities in the cocoa market access (Callahan, 2019). Contract-farming can offer cocoa farmers the opportunity to receive a higher price point for cocoa and can be more beneficial than some sustainability programmes. However, due to the significant gaps in the research on contract-farming, future research should adress its potential advantages for both farmers and buyers as well as its impacts on sustainability in the longer term, even when farmers obtain a higher price point for cocoa. Likewise, the methods to mitigate environmental and social risks in contract-farming must be examined, as well as the increased risks associated with land tenure insecurity, marginalization of women and crop monoculture.

Recent literature seeks to understand the gap between international price and the farmer's price and the existence of a power asymmetry, especially on supply side. Consequently, it is important to understand how social and ethical credibility are viable bases of differentiation and competitive positioning in the mainstream cocoa value chain and markets. In this perspective, new propositions should pursuit the study of the social sustainability dimension to understand how social resources may be used to improve weelbeing and, develop a more equitable market access for cocoa growers. This can provide contributions to amenity the growing imbalance between commercially sophisticated buyers and fragmented small-scale farmers who supply them, as well as, lessen their potentially adverse consequences for the sustainability.

The gradual reduction of natural forest, from which timber and non-timber products are gathered, has aroused the need to research this topic and its consequences on the sustainable integration of the natural forest in cocoa agroforestry systems. Thus, future research should aim to study the gradual reduction of natural forest and its consequences for sustainability, and understand the integration of forests' species in cocoa agroforestry systems, whilst evaluating the resulting environmental services, such as carbon storage and potential biodiversity conservation.

Given the significance of good governance and social capital for sustainable development, which is recognised by organisations such as the World Bank and the United Nations, this is an important and possibly fruitful line for research. Future studies might seek to explore the impact of the social capital on governance in the cocoa value chain (and possibly emerging industries) in some developing producing countries.

CONCLUSION

The cocoa value chains are complex, assymetric and mainly concentrated in the tropics, particularly in Africa, which represents two thirds of world production, though Côte d'Ivoire is the main producing country. The work drew on seven specific research questions in order to address the literature on TBL scenarios and sustainability related themes. The findings, supported in the literature revision of the TBL concept and its connections to sustainability dimensions and scenarios, showed that TBL remains an open and ongoing research topic.

The Bibliometrics analysis allowed for an identification of the main contexts of research dedicated to each theme, such as countries, used methods and their integration within the TBL scenarios. The articulation of these topics led to the conclusion that the pillars associated with the TBL assume different importance according to the contexts where they are analysed. The main research contexts of the selected studies are the countries located in Africa, Latin America and Asia, with more limiting structural conditions and that raise the dimension of socioeconomic problems in contemporary concerns, being environmental parameters still easily circumvented.

The iniciatives towards cocoa sustainability are hindered by the complex nature of their value chains in low economy countries where many basic social problems still dominate. Lower yields, child labour, unfair working conditions and gendered patterns are among the issues that need to be addressed in order to successfully manage the value chain towards sustainability in these production countries.

A critical observation of the consulted literature is that no research showed a clear strategy to end the problem of child labour or to the cease in the influx of children at borders and track down the traffickers. Despite the fact that raising awareness on child labour is relevant in order to restrain it, it does not make any meaningfull impact within the regions from which the children are trafficked. Poverty is a likely reason to why those children to work in cocoa farms. Likewise, it might not be economically feasible for global cocoa companies to implement any measures that aim at working with the governments in those countries or improving cocoa farmers' yield. In fact, we can conclude that the value chain reflects the problems existing in the geographical area where it is located.

Research on the sustainability of the cocoa value chain is very recent and qualitative approaches are dominant. These elements are also articulated with the smaller knowledge inputs that have been promoted at the level of the value chain, capable of originating greater added value upstream of the chain. This has been compensated by the increasing scrutiny on the value chain practices by consumers, media and other influential stakeholders that have increased the need to define, implement and effectively communicate sustainable practices and management strategies.

Notwithstanding the circumstances mentioned that impose improvement and development of governance practices that promote a more equitable production chain, the originality of the production systems stands out. Despite some value chain dynamics and the implementation of several initiatives, environmental impacts are more harmless and confidence in certification may be the most prominent feature.

Certified procurement in different schemes such as organic production and fair trade is considered to be an effective strategy thought there are still many loopholes in this system, which can raise other serious issues, such as corruption. In this context, certification decisions, which are made at top management level – considered at policy or strategy levels, should be accompanied by clear procedures at an operational level and should be scrutinised by a different group of stakeholders (third-party certification).

Partnerships with different associations, contract-farming and the value chain networks are measures and key performance indicators that were provided in literature. These help monitoring the progress within each link of cocoa value chain and improve the conditions of farmers. They are efficient for chocolate companies and for cocoa farmers and other stakeholders. Additionally, improvements in the TBL dimensions are advocated as a key result of these iniciatives.

In this follow-up and aiming to overcome a limitation found in the research, it is suggested a holistic approach to tackling wider sustainability issues and its interactions in the cocoa value chain. Highly complex social issues in cocoa value chain sustainability are particularly scarce in literature and its linkages should be analysed in the light of the TBL. In sum, there is still a long way to go to expand the existing programmes and iniciatives towards cocoa value chain sustainability, under the lens of TBL scenarios.

ACKNOWLEDGMENT

This research was supported by the Fundação para a Ciência e a Tecnologia [grants number UID/04007/2020 and UIDB/04011/2020].

REFERENCES

Abbey, P., Tomlinson, P. R., & Branston, J. R. (2016). Perceptions of governance and social capital in Ghana's cocoa industry. *Journal of Rural Studies*, *44*, 153–163. doi:10.1016/j.jrurstud.2016.01.015

Aboah, J., Wilson, M. M. J., Bicknell, K., & Rich, K. M. (2021). Ex-ante impact of on-farm diversification and forward integration on agricultural value chain resilience: A system dynamics approach. *Agricultural Systems*, *189*, 189. doi:10.1016/j.agsy.2020.103043

Adu-Acheampong, R., Jiggins, J., Quartey, E. T., Karikari, N. M., Jonfia-Essien, W., Quarshie, E., Osei-Fosu, P., Amuzu, M., Afari-Mintah, C., Ofori-Frimpong, K., Owusu-Manu, M., Eku, N., Sakyi-Dawson, O., Quarmine, W., & Acquah, F. O. (2017). An innovation platform for institutional change in Ghana's cocoa sector. *Cahiers Agricultures*, *26*(3), 35002. doi:10.1051/cagri/2017016

Alcayaga, A., Hansen, E. G., & Wiener, M. (2019). Towards a Framework of Smart-Circular Systems: An Integrative Literature Review. *Journal of Cleaner Production*, *221*, 622–634. doi:10.1016/j.jclepro.2019.02.085

Alhaddi, H. (2015). Triple Bottom Line and Sustainability: A Literature Review. *Business and Management Studies*, *1*(2), 6–10. doi:10.11114/bms.v1i2.752

Alvarez, M.E.P., Barcena, M.M., & Gonzalez, F.A. (2016). A review of sustainable machining engineering: optimization process through triple bottom line. *Manufacturing Science and Engineering, Transactions/ASME, 138*(10), 16.

Arias, R. C., & Fromm, I. (2019). From cocoa producers to chocolatiers? Developing an entrepreneurial model for small-scale producers in Honduras. *International Journal on Food System Dynamics, 10*(1), 38–54.

Asamoah, D., & Annan, J. (2012). Analysis of Ghana's cocoa value chain towards services and standards for stakeholders. *Int. J. Services and Standards, 8*(2), 116–132. doi:10.1504/IJSS.2012.049423

Astrid Fenger, N., Skovmand Bosselmann, A., Asare, R., & de Neergaard, A. (2017). The impact of certification on the natural and financial capitals of Ghanaian cocoa farmers. *Agroecology and Sustainable Food Systems, 41*(2), 143–166. doi:10.1080/21683565.2016.1258606

Banerjee, S. B. (2003). Who sustains whose development? Sustainable development and the reinvention of nature. *Organization Studies, 24*(1), 143–180. doi:10.1177/0170840603024001341

Barrientos, S. (2014). Gendered global production networks: Analysis of cocoa-chocolate sourcing. *Regional Studies, 48*(5), 791–803. doi:10.1080/00343404.2013.878799

Barrientos, S., & Asenso-Okyere, K. (2009). Cocoa value chain: Challenges facing Ghana in a changing global confectionary market. *Journal für Entwicklungspolitik, 25*(2), 88–107. doi:10.20446/JEP-2414-3197-25-2-88

Belanger, J., & Pilling, D. (Eds.). (2019). *The State of the World's Biodiversity for Food and Agriculture*. FAO.

Bibri, M. (2008). *Corporate sustainability/CSR communications and value creation: A marketing approach* [Unpublished Master's thesis]. Blekinge Institute of Technology, School of Management, University of Karlshamn.

Blowfield, M. (2003). Ethical supply chains in the cocoa, coffee and tea industries. *Greener Management International, 43*(43), 15–24. doi:10.9774/GLEAF.3062.2003.au.00004

Bonuedi, I., Gerber, N., & Kornher, L. (2021). Intervening in cash crop value chains for improved nutrition: Evidence from rural Sierra Leone. *The Journal of Development Studies*, 1–17. doi:10.1080/0022 0388.2021.1945043

Borda, A., Morales, O., Teegen, H., Rees, G. H., & Gonzalez-Perez, M. A. (2021). Addressing sustainable rural development with shared value: A Peruvian model from the cacao industry. *Sustainability, 13*(14), 8028. doi:10.3390u13148028

Busquet, M., Bosma, N., & Hummels, H. (2021). A multidimensional perspective on child labor in the value chain: The case of the cocoa value chain in West Africa. *World Development, 146*, 146. doi:10.1016/j.worlddev.2021.105601

Callahan, L. (2019). Contract-Farming in Cocoa Value Chains in Africa: Possibilities and Challenges. In Ethiopian Yearbook of International Law 2018. Springer.

Carodenuto, S., & Buluran, M. (2021). The effect of supply chain position on zero-deforestation commitments: Evidence from the cocoa industry. *Journal of Environmental Policy and Planning*, *23*(6), 716–731. doi:10.1080/1523908X.2021.1910020

Castro-Nunez, A., Charry, A., Castro-Llanos, F., Sylvester, J., & Bax, V. (2020). Reducing deforestation through value chain interventions in countries emerging from conflict: The case of the Colombian cocoa sector. *Applied Geography (Sevenoaks, England)*, *123*, 123. doi:10.1016/j.apgeog.2020.102280

Cavagnaro, E., & Curiel, G. (2012). *The Three Levels of Sustainability* (1st ed.). Routledge.

CBD - Convention on Biological Diversity. (2014). *Global Biodiversity Outlook 4*. Secretariat of the Convention on Biological Diversity.

CBD - Convention on Biological Diversity (2016). *Decision XIII/28. Indicators for the Strategic Plan for Biodiversity 2011–2020 and the Aichi Biodiversity Targets*. Author.

Chiarolla, C. (2017). Agriculture and biodiversity conservation. In Biodiversity and Nature protection law. Edward Elgar Publishing. doi:10.4337/9781783474257.III.18

Collins, C., Steg, L., & Koning, M. (2007). Customers' values, beliefs on sustainable corporate performance, and buying behavior. *Psychology and Marketing*, *24*(6), 555–577. doi:10.1002/mar.20173

Colwell, K. (2012). *Is it still green: The impacts of corporate takeovers on business sustainability?* [Unpublished doctoral dissertation].

Correia, E., Carvalho, H., Azevedo, S. G., & Govindan, K. (2017). Maturity Models in Supply Chain Sustainability: A Systematic Literature Review. *Sustainability*, *9*(1), 64. doi:10.3390u9010064

de Boer, D., Limpens, G., Rifin, A., & Kusnadi, N. (2019). Inclusive productive value chains, an overview of Indonesia's cocoa industry. *Journal of Agribusiness in Developing and Emerging Economies*, *9*(5), 439–456. doi:10.1108/JADEE-09-2018-0131

Díaz-Montenegro, J., Varela, E., & Gil, J. M. (2018). Livelihood strategies of cacao producers in Ecuador: Effects of national policies to support cacao farmers and specialty cacao landraces. *Journal of Rural Studies*, *63*, 141–156. doi:10.1016/j.jrurstud.2018.08.004

Doherty, B., & Meehan, J. (2006). Competing on social resources: The case of the day chocolate company in the UK confectionery sector. *Journal of Strategic Marketing*, *14*(4), 299–313. doi:10.1080/09652540600947847

Dompreh, E. B., Asare, R., & Gasparatos, A. (2021). Stakeholder perceptions about the drivers, impacts and barriers of certification in the Ghanaian cocoa and oil palm sectors. *Sustainability Science*, *16*(6), 2101–2122. doi:10.100711625-021-01027-5

Elkington, J. (1997). *Cannibals with Forks. The triple bottom line of 21st century business*. Capstone, Oxford.

Elkington, J. (1998). Partnerships from cannibals with forks: The triple bottom line of 21st-century business. *Environmental Quality Management*, *8*(1), 37–51. doi:10.1002/tqem.3310080106

Elkington, J. (2018). 25 years ago I coined the phrase "triple bottom line." Here's why it's time to rethink it. *Harvard Business Review*.

Epstein, M. (2008). *Making sustainability work: Best practices in managing and measuring corporate social, environmental, and economic impacts*. Greenleaf.

Escobar, S., Santander, M., Useche, P., Contreras, C., & Rodríguez, J. (2020). Aligning strategic objectives with research and development activities in a soft commodity sector: A technological plan for Colombian cocoa producers. *Agriculture*, *10*(5), 141. doi:10.3390/agriculture10050141

Estival, K.G.S., Corrêa, S.R.S., Mariani, M.A.P., & Benini, E.G. (2016). Analysis of the participation of the producers in the farming of settlements and rural communities in the south of Bahia, Brazil, in the chain of value of cocoa (chocolate). *Espacios*, *37*(17).

George, H. (2009). *Progress and poverty*. Cambridge University Press. (Original work published 1879) doi:10.1017/CBO9780511693687

Giddens, A. (2015). The politics of climate change. *Policy and Politics*, *43*(2), 155–162. doi:10.1332/030557315X14290856538163

Gilal, F. G., Zhang, J., Paul, J., & Gilal, N. G. (2019). The role of self-determination theory in marketing science: An integrative review. *Journal of Interactive Marketing*, *51*, 1–8.

Glavee-Geo, R., Burki, U., & Buvik, A. (2020). Building trustworthy relationships with smallholder (small-scale) agro-commodity suppliers: Insights from the ghana cocoa industry. *Journal of Macromarketing*, *40*(1), 110–127. doi:10.1177/0276146719900370

Goel, A., Ganesh, L. S., & Kaur, A. (2021). Sustainability assessment of construction practices in India using inductive content analysis of research literature. *Int. J. Construction Management*, *21*(8), 802–817. doi:10.1080/15623599.2019.1583851

Goel, P. (2010). Triple bottomline reporting: An analytical approach for corporate sustainability. *J. of Finance, Accounting, and Management*, *1*(1), 27–42.

Grumiller, J. (2018). Upgrading potentials and challenges in commodity-based value chains: The Ivorian and Ghanaian cocoa processing sectors. *Journal für Entwicklungspolitik*, *34*(3-4), 15–45. doi:10.20446/JEP-2414-3197-34-3-15

Hahn, T., Figge, F., Pinkse, J., & Preuss, L. (2010). Trade-offs in corporate sustainability: You can't have your cake and eat it. *Business Strategy and the Environment*, *19*(4), 217–229. doi:10.1002/bse.674

Hao, A. W., Paul, J., Trott, S., Guo, C., & Wu, H. H. (2019). Two decades of research on nation branding: A review and future research agenda. *International Marketing Review*, *38*(1), 46–69. doi:10.1108/IMR-01-2019-0028

Harmon, J., Bucy, F., Nickbarg, S., Rao, G., & Wirtenber, J. (2009). In J. Wirtenberg, W. Russell, & D. Lipsky (Eds.), *The Sustainable Enterprise Fieldbook* (pp. 89–115). Greenleaf Publishing.

Haynes, J., Cubbage, F., Mercer, E., & Sills, E. (2012). The search for value and meaning in the cocoa supply chain in Costa Rica. *Sustainability*, *4*(7), 1466–1487. doi:10.3390u4071466

He, Q., Gallear, D., Ghobadian, A., & Ramanathan, R. (2019). Managing knowledge in supply chains: A catalyst to triple bottom line sustainability. *Production Planning and Control*, *30*(5-6), 5–6, 448–463. doi:10.1080/09537287.2018.1501814

Hess, J. P. (2021). A multi-level analysis of sustainability practices in Ghana: Examining the timber, cocoa, and gold mining industries. *The International Journal of Organizational Analysis*. Advance online publication. doi:10.1108/IJOA-01-2020-2011

Hungara, A., & Nobre, H. (2021). A consumer culture theory perspective of the marketplace: An integrative review and agenda for research. *International Journal of Consumer Studies*, *45*(4), 805–823. doi:10.1111/ijcs.12670

Indah, P. N., Setiawan, R. F., Hendrarini, H., Yektingsih, E., & Sunarsono, R. J. (2021). Agriculture supply chain performance and added value of cocoa: A study in kare village, Indonesia. *Bulgarian Journal of Agricultural Science*, *27*(3), 487–497.

Ingram, V., van Rijn, F., Waarts, Y., & Gilhuis, H. (2018). The Impacts of Cocoa Sustainability Initiatives in West Africa. *Sustainability*, *10*(11), 4249. doi:10.3390u10114249

IPBES. (2019). *Global assessment report on biodiversity and ecosystem services*. IPBES.

Isil, O., & Hernke, M. T. (2017). The Triple Bottom Line: A Critical Review from a Transdisciplinary Perspective. *Business Strategy and the Environment*, *26*(8), 1235–1251. doi:10.1002/bse.1982

Kahiya, E. T. (2018). Five decades of research on export barriers: Review and future directions. *International Business Review*, *27*(6), 1172–1188. doi:10.1016/j.ibusrev.2018.04.008

Krauss, J. E., & Krishnan, A. (2021). Global decisions versus local realities: Sustainability standards, priorities and upgrading dynamics in agricultural global production networks. *Global Networks*, glob.12325. doi:10.1111/glob.12325

Lafargue, P., Rogerson, M., Parry, G.C., & Allainguillaume, J. (2021). Broken chocolate: Biomarkers as a method for delivering cocoa supply chain visibility. *Supply Chain Management*.

Lalwani, S. K., Nunes, B., Chicksand, D., & Boojihawon, D. K. (2018). Benchmarking self-declared social sustainability initiatives in cocoa sourcing. *Benchmarking*, *25*(9), 3986–4008. doi:10.1108/BIJ-07-2017-0186

León Bravo, V., Jaramillo Villacrés, M., & Silva, M.E. (2021). Analysing competing logics towards sustainable supplier management. *Supply Chain Management*.

Liliana, Q. R. M. (2020). The cocoa agri-food chain in Venezuela: Towards territorial development. *Agroalimentaria*, *26*(51), 213–238.

Lim, W. M., Yap, S. F., & Makkar, M. (2021). Home sharing in marketing and tourism at a tipping point: What do we know, how do we know, and where should we be heading? *Journal of Business Research*, *122*, 534–566. doi:10.1016/j.jbusres.2020.08.051 PMID:33012896

Loviscek, V. (2021). Triple Bottom Line toward a Holistic Framework for Sustainability: A Systematic Review. *J. Contemporary Administrator*, *25*(3), e-200017. doi:10.1590/1982-7849rac2021200017.en

Mabe, F. N., Danso-Abbeam, G., Azumah, S. B., Amoh Boateng, N., Mensah, K. B., & Boateng, E. (2020). Drivers of youth in cocoa value chain activities in Ghana. *Journal of Agribusiness in Developing and Emerging Economies*, *11*(4), 366–378. doi:10.1108/JADEE-10-2019-0177

Marcus, A., & Fremeth, A. (2009). Green management matters regardless. *The Academy of Management Perspectives*, *23*(3), 17–26. doi:10.5465/amp.2009.43479261

Matissek, R., Reinecke, J., Von Hagen, O., & Manning, S. (2012). *Sustainability in the Cocoa Sector-Review*. Challenges and Approaches.

McDonough, W., & Braungart, M. (2002). Design for the triple top line: New tools for sustainable commerce. *Corporate Environmental Strategy*, *9*(3), 251–258. doi:10.1016/S1066-7938(02)00069-6

McLoughlin, K., & Meehan, J. (2021). The institutional logic of the sustainable organisation: The case of a chocolate supply network. *International Journal of Operations & Production Management*, *41*(3), 251–274. doi:10.1108/IJOPM-11-2020-0773

Mendoza, M. L. B., Cedeño, M. R., & Cossío, N. S. (2020). Prospective of the circular economy in the agri-food chain of fine organic aromatic cocoa in the province of manabi. *Revista de la Facultad de Agronomía*, *37*(1), 95–110.

Middendorp, R. S., Boever, O., Rueda, X., & Lambin, E. F. (2020). Improving smallholder livelihoods and ecosystems through direct trade relations: High-quality cocoa producers in Ecuador. *Business Strategy and Development*, *3*(2), 165–184. doi:10.1002/bsd2.86

Mithöfer, D., Roshetko, J. M., Donovan, J. A., Nathalie, E., Robiglio, V., Wau, D., Sonwa, D., & Blare, T. (2017). Unpacking 'sustainable' cocoa: Do sustainability standards, development projects and policies address producer concerns in Indonesia, Cameroon and Peru? *The International Journal of Biodiversity Science, Ecosystem Services & Management*, *13*(1), 444–469. doi:10.1080/21513732.2018.1432691

Moreno-Miranda, C., Molina, I., Miranda, Z., Moreno, R., & Moreno, P. (2020). The cocoa value chain in Ecuador: Proposal of strategies to support sustainability. *Bioagro-*, *32*(3), 205–214.

Movilla-Pateiro, L., Mahou-Lago, X. M., Doval, M. I., & Simal-Gandara, J. (2021). Toward a sustainable metric and indicators for the goal of sustainability in agriculturaland food production. *Critical Reviews in Food Science and Nutrition*, *61*(7), 1108–1129. doi:10.1080/10408398.2020.1754161 PMID:32319311

Nabhani, I., Daryanto, A., Yassin, M., & Rifin, A. (2015). Can Indonesia cocoa farmers get benefit on global value chain inclusion? A literature review. *Asian Social Science*, *11*(18), 288–294. doi:10.5539/ass.v11n18p288

Narciso, A., Barzini, S. A., & Di Nuzzo, A. (2020). Discovering neverland: São tomé and príncipe and the development of the agricultural heritage of a multi-ethnic population. *Journal of Agriculture and Environment for International Development*, *114*(2), 63–84.

Nonci, N., Nonci, N., & Halim, H., & Muliani. (2019). Role of supply chain in cocoa blue print model performance in national development program. *Int. J. Supply Chain Managemet*, *8*(6), 198–210.

Norman, W., & MacDonald, C. (2004). Getting to the bottom of 'triple bottom line'. *Business Ethics Quarterly*, *14*(02), 243–262. doi:10.5840/beq200414211

Odijie, M. E. (2018). Sustainability winners and losers in business-biased cocoa sustainability programmes in West Africa. *International Journal of Agricultural Sustainability*, *16*(2), 214–227. doi:10.1080/14735903.2018.1445408

Ollivier de Leth, D., & Ros-Tonen, M. A. F. (2021). Creating shared value through an inclusive development lens: A case study of a CSV strategy in Ghana's cocoa sector. *Journal of Business Ethics*. Advance online publication. doi:10.100710551-021-04808-1

Orozco-Aguilar, L., López-Sampson, A., Leandro-Muñoz, M. E., Robiglio, V., Reyes, M., Bordeaux, M., Sepúlveda, N., & Somarriba, E. (2021). Elucidating pathways and discourses linking cocoa cultivation to deforestation, reforestation, and tree cover change in Nicaragua and Peru. *Frontiers in Sustainable Food Systems*, *5*, 5. doi:10.3389/fsufs.2021.635779

Paul, J. (2019). Marketing in emerging markets: A review, theoretical synthesis and extension. *Int. J Emerging Markets*, *15*(3), 446–468. doi:10.1108/IJOEM-04-2017-0130

Paul, J., & Benito, G. R. G. (2018). A review of research on outward foreign direct investment from emerging countries, including China: What do we know, how do we know and where should we be heading? *Asia Pacific Business Review*, *24*(1), 90–115. doi:10.1080/13602381.2017.1357316

Paul, J., & Criado, A. R. (2020). The art of writing literature review: What do we know and what do we need to know? *International Business Review*, *29*(4), 101717. doi:10.1016/j.ibusrev.2020.101717

Pradhan, P., Reusser, D. E., & Kropp, J. P. (2013). Embodied greenhouse gas emissions in diets. *PLoS One*, *8*(5), e62228. doi:10.1371/journal.pone.0062228 PMID:23700408

Prazeres, I., Lucas, M. R., & Marta-Costa, A. (2021). Cocoa Markets and Value Chains: Dynamics and Challenges for Sao Tome and Principe Organic Smallholders. *Int. J. Innovation and Economic Development*, *7*(2), 64–77. doi:10.18775/ijied.1849-7551-7020.2015.72.2005

Randhawa, K., Wilden, R., & Hohberger, J. (2016). A bibliometric review of open innovation: Setting a research agenda. *Journal of Product Innovation Management*, *33*(6), 750–772. doi:10.1111/jpim.12312

Rashidi, K., Noorizadeh, A., Kannan, D., & Cullinane, K. (2020). Applying the triple bottom line in sustainable supplier selection: A meta-review of the state-of-the-art. *Journal of Cleaner Production*, *269*, 122001. doi:10.1016/j.jclepro.2020.122001

Recanati, F., Marveggio, D., & Dotelli, G. (2017). From beans to bar: A life cycle assessment towards sustainable chocolate supplychain. *The Science of the Total Environment*, *613614*, 1013–1023. PMID:28946374

Richardson, J. (2004). Accounting for sustainability: measuring quantities or enhancing qualities. In A. Henriques & J. Richardson (Eds.), *The Triple Bottom Line: Does it All Add Up?* (pp. 34–44). Routledge.

Ros-Tonen, M. A., Bitzer, V., Laven, A., Ollivier de Leth, D., Van Leynseele, Y., & Vos, A. (2019). Conceptualizing inclusiveness of smallholder value chain integration. *Current Opinion in Environmental Sustainability*, *41*, 10–17. doi:10.1016/j.cosust.2019.08.006

Rosado-Serrano, A., Paul, J., & Dikova, D. (2018). International fran-chising: A literature review and research agenda. *Journal of Business Research*, *85*, 238–257. doi:10.1016/j.jbusres.2017.12.049

Ruben, R. (2017). Impact assessment of commodity standards: Towards inclusive value chains. *Enterprise Development & Microfinance*, *28*(1-2), 82–97. doi:10.3362/1755-1986.16-00020

Rueda, X., Paz, A., Gibbs-Plessl, T., Leon, R., Moyano, B., & Lambin, E. F. (2018). Smallholders at a crossroad: Intensify or fall behind? Exploring alternative livelihood strategies in a globalized world. *Business Strategy and the Environment*, *27*(2), 215–229. doi:10.1002/bse.2011

Savitz, A., & Weber, K. (2006). *The triple bottom line: How today's best-run organizations are achieving economic, social and environmental success -and how you can too*. Jossey Bass.

Schroth, G., Garcia, E., Griscom, B. W., Teixeira, W. G., & Barros, L. P. (2016). Commodity production as restoration driver in the Brazilian amazon? Pasture re-agro-forestation with cocoa (theobroma cacao) in southern pará. *Sustainability Science*, *11*(2), 277–293. doi:10.100711625-015-0330-8 PMID:30174731

Scott, G. J., Donovan, J., & Higuchi, A. (2015). Costs, quality, and competition in the cocoa value chain in Peru: An exploratory assessment. *Custos e Agronegocio*, *11*(4), 324–358.

Sjauw-Koen-Fa, A. R., Omta, S. W. F., & Blok, V. (2018). Constructing a multinationals' inclusive sourcing indicator for impacting farmer business models: Application in cocoa cases. *International Journal on Food System Dynamics*, *9*(3), 207–225.

Sonwa, D. J., Weise, S. F., Schroth, G., Janssens, M. J. J., & Howard-Yana Shapiro. (2014). Plant diversity management in cocoa agroforestry systems in west and central Africa—Effects of markets and household needs. *Agroforestry Systems*, *88*(6), 1021–1034. doi:10.100710457-014-9714-5

Torraco, R. J. (2005). Writing integrative literature reviews: Guidelines and exemples. *HR Develop. RE:view*, *4*(3), 356–367.

Tranfield, D., Denyer, D., & Smart, P. (2003). Towards a methodology for developing evidence-informed management knowledge by means of systematic review. *British Journal of Management*, *14*(3), 207–222. doi:10.1111/1467-8551.00375

Trienekens, J. H. (2018). *A search for sustainability in cocoa supply chain business models* [Unpublished Bachelor thesis dissertation]. Wageningen University.

Tseng, M. L. (2017). Using social media and qualitative and quantitative information scales to benchmark corporate sustainability. *Journal of Cleaner Production*, *142*, 727–738. doi:10.1016/j.jclepro.2016.05.062

Tseng, M. L., Chang, C. H., Lin, C. W. R., Wu, K. J., Chen, Q., Xia, L., & Xue, B. (2020). Future trends and guidance for the triple bottom line and sustainability: A data driven bibliometric analysis. *Environmental Science and Pollution Research International*, *27*(27), 33543–33567. doi:10.100711356-020-09284-0 PMID:32572746

van Huellen, S., & Abubakar, F. M. (2021). Potential for upgrading in financialised agri-food chains: The case of Ghanaian cocoa. *European Journal of Development Research*, *33*(2), 227–252. doi:10.105741287-020-00351-3 PMID:33558793

Vogel, C., Mathé, S., Geitzenauer, M., Ndah, H. T., Sieber, S., Bonatti, M., & Lana, M. (2020). Stakeholders' perceptions on sustainability transition pathways of the cocoa value chain towards improved livelihood of small-scale farming households in Cameroon. *International Journal of Agricultural Sustainability, 18*(1), 55–69. doi:10.1080/14735903.2019.1696156

Waas, T., Hugé, J., Verbruggen, A., & Wright, T. (2012). Sustainable Development: A Bird's Eye View. *Sustainability, 3*(10), 1637–1661. doi:10.3390u3101637

WCED. (1987). *Our Common Future. Brundtland Report* (Vol. 4). Oxford University Press.

Wu, K. J., Liao, C. J., Tseng, M. L., & Chiu, K. S. (2016). Multi-attribute approach to sustainable supply chain management under uncertainty. *Industrial Management & Data Systems, 116*(4), 777–800. doi:10.1108/IMDS-08-2015-0327

Wu, K. J., Zhu, Y., Tseng, M. L., Lim, M. K., & Xue, B. (2018). Developing a hierarchical structure of the co-benefits of the triple bottom line under uncertainty. *Journal of Cleaner Production, 195*, 908–918. doi:10.1016/j.jclepro.2018.05.264

Yan, W., Chen, C., & Chang, W. (2009). An investigation into sustainable product constructualization using a design knowledge hierarchy and Hopfield network. *Computers & Industrial Engineering, 56*(4), 617–626. doi:10.1016/j.cie.2008.10.015

Zaharia, R. M., & Zaharia, R. (2021). Triple Bottom Line. In D. Crowther & S. Seifi (Eds.), *The Palgrave Handbook of Corporate Social Responsibility*. Palgrave Macmillan.

ADDITIONAL READING

Eskandarpour, M., Dejax, P., Miemczyk, J., & Péton, O. (2015). Sustainable supply chain network design: An optimization-oriented review. *Omega, 54*(C), 11–32. doi:10.1016/j.omega.2015.01.006

Sadovska, V. (2018). *Sustainable value creation in the agricultural sector. A literature review* [Paper presentation]. International Association of Agricultural Economists-IAAE, Vancouver.

Somarriba, E., & Lopez-Sampson, A. (2018). *Coffee and Cocoa Agroforestry Systems: Pathways to Deforestation, Reforestation, and Tree Cover Change*. IBRD/The World Bank.

KEY TERMS AND DEFINITIONS

Circular Economy: Is a new restorative or regenerative model which disruptly change the way the societies and business are organized. It is based on three principles, eliminate waste and pollution, circulate products and materials and, regenerate nature.

Ecosystem Services: Are the aspects of ecosystems utilised to produce human well-being. This aren't only the final services of the ecosystem, including ecosystem organization, as well as process and/or functions provided that they are consumed or utilized by humans either directly or indirectly.

Sustainability: Focuses on meeting the needs of the present without compromising the ability of future generations to meet their needs. It is composed of three dimensions: economic, environmental.

TBL: Introduced by Elkington in 1997, is a sustainability-related framework which captures its essence throught measuring the impact of an organization's activities in three dimensions, social, environmental and economic.

Value Chain: First described by Michael Porter, in 1985 is a concept which describes the full chain of a business's activities in the creation of a product or service – from the initial reception of materials all the way through its delivery to market, and everything in between.

Chapter 16
Sustainable Development in the Agrifood Rice Chain in Timor-Leste

Cristóvão Reis
Universidade Nacional de Timor Lorosa'e, Timor-Leste

Andreia Dionísio
Universidade de Évora, Portugal

Maria Raquel Lucas
Universidade de Évora, Portugal

ABSTRACT

In Timor-Leste, rice is a source of livehood and a staple food. However, it presents persistently low yield, quality, price, and value to consumers, which, allied with climate projections and pressure for higher quality and productivity, raised logistics costs, and subsidized imports, creates a need to identify drivers/inhibitors of sustainable development. This chapter investigates rice agri-food chain sustainable development by recording the main actors involved and understanding their perspectives. Interviews, questionnaires, observation, and focus group have been applied to understand how sustainable development can be triggered. Results show that actors are not accurately coordinated to find a future sustainable development. An alignment of activities, innovation, best practices, and cooperation are recommended towards a future sustainability plan as a starting point to agrifood rice development. Each element of this development should be measured and quantified in future research.

INTRODUCTION

The depletion of resources, globalization, climate change and their consequent grave environmental consequences (Kopnina, 2017; Hart, 1997), allied with the projections of world population increase, have turned the multidisciplinary concept of sustainability, which emerged in the 1960s (Ahmed, 2017), into a growing object of interest from public, private, governmental, and academic organizations (Giunipero

DOI: 10.4018/978-1-7998-9557-2.ch016

et al., 2012). The issue of optimal and sustainable food supply is among the most important goals and targets of sustainable development (SDGs). Sustainability is, simultaneously, a response to the deterioration of the environment and means of achieving economic growth and development, reducing poverty, income inequality and food security (Chabowski et al., 2011; Marsden & Morley, 2014; Chagomoka et al., 2014). Particularly, in agrifood chains resulting from complex systems of interrelated and interdependent activities, the three dimensions of sustainability -environmental, social and economic - are fundamental (Flynn & Bailey, 2014), to achieve sustainable development and creating value (Ruben et al. 2007; Trienekens 2011).

The sustainability of agrifood has become imperative for all stakeholders, particularly for policy makers and decision makers (Marsden et al., 2000; Hinrichs, 2003; Govindan, 2018), as well as industry players. The latter have displayed several changes, given greater importance to consumption and increased sustainable production (Govindan, 2018). Things have particularly shifted since June 2015, when G7, the group of the most developed countries in the world, which encompasses Japan, Germany, France, Great Britain, Italy, USA and Canada, met to discuss future environmental management strategies. They set the goal of ending extreme poverty and reducing the number of people who suffer from hunger by 500 million, by 2030. This goal was passed on by promoting ideas, attitudes, sustainable consumption and production behaviors, which contribute to the ODS (Marsden & Morley, 2014; Govindan, 2018). Although the general opinion supports the need to incorporate sustainability principles into rice agrifood chains (Ray et al., 2013; OECD/FAO, 2019; USDA/ERS, 2019), such might not be eased by the configurations of its chain actors and their respective constraints.

Rice is one of the most produced and consumed cereals in the world and the basilar diet for more than half of Asia's population. It assures the caloric consumption for more than 2 billion people (Juliano & Villareal, 1993; Muthayya et al., 2014). In Timor-Leste, where agriculture is a main source of income for more than 80% of rural households, rice is the staple food, and its production is intended entirely for domestic consumption (Loly, 2014). In order to meet domestic demand, Timor-Leste is dependent on high costs from imported rice from Vietnam, Thailand and Indonesia. These costs are added to by the complexity of customs legislation, the high tax burden on imported products, the inefficient infrastructure, the insufficiency of Timorese ports, airports and highways, and also the fluctuation of acquisition costs due to variations in exchange rates, logistics and transportation costs (Fleury et al., 2009). These costs reduce the competitiveness of companies that carry out import operations and have a direct impact on the country's economic development, by reducing job opportunities and income within the domestic market (Lopes & Gama, 2013). The export of national rice does not exist. Even if the amount of production allowed for exportation, the rice would necessarily have to be destined for low value markets due to its low quality, poor packaging, conservation and non-compliance with international standards concerning pesticide residues, among others. Thus, some consumers do not get the value and quality they are looking for in rice, while others have concerns about food safety, particularly about the unregulated use of hazardous chemicals. There is also a growing awareness of the impacts of food production and marketing systems in human health, economy and society at large (Prowse & Moyer-Lee 2013).

There are several challenges that reduce the value of rice to consumers, the profitability of the main actors (especially producers) and hinder the sector's contribution to the country's socio-economic development. The following are worth noting: the low price at which rice is sold directly to government entities, low quality control and the logistics' costs and subsidized imports, leading to very low yields and production disincentive. The rice agrifood chain also faces problems concerning overuse and degradation of resources such as soil and water as well as energy scarcity. This has an additional effect on rice productivity and production efficiency. Moreover, rice marketing does not exist in domestic production and it is inefficient for imported rice, due to the numerous intermediaries, lack of market dynamism and the absence of modern and logistical support such as cold storage, transport and other equipment. Despite the existence of modern retail formats such as supermarkets, rice is distributed through traditional channels which create little value for consumers (Cruz, 2014). Furthermore, the distribution of profits among the chain's participants is uneven, with producers receiving a relatively low share of the price paid by consumers, or none at all if the rice is imported. All these challenges mean that the rice agrifood chain in Timor-Leste does not strongly contribute to the sustainable development of the country.

Based on literature and empirical work, this chapter aimed to investigate rice's agri-food chain in Timor-Leste, mapping its main actors (producers, distributors and consumers) and understanding their perspectives concerning sustainable development. It is organized in six sections and subsections. The main sections include the agrifood chain, rice and sustainability, material and methods, the sustainable development of rice's agrifood chain, the sustainability plan for the rice agrifood chain, future research directions and a conclusion.

AGRIFOOD CHAIN, RICE AND SUSTAINABILITY

Agrifood Chain

Since Porter's (1985) initial definition, adjusted by Hobbs (Hobbs et al., 2000), there is a great amount of value chain definitions. This concept has evolved over the years and has been enriched with different researcher´s points of view and advances in its respective disciplines (Gereffi et al., 2001). These perspectives range from buying and selling to logistics and transport, marketing, industrial organization, contract theory, economics of transaction costs, institutional sociology, systems engineering, network theory, good practices, strategic management and the economics of development (Fernández, 2013). Regardless of the type of value chain, behavior or integrated processes, sharing information, risks, benefits or rewards between members and cooperating and focusing on customer service are common elements (Mentzer et al., 2001; Lambert, 2000).

The agrifood chain has different definitions. It can be seen as: i) a network of organizations that work together, in different processes and activities, to bring the products and services to the market, so to satisfy consumer demand (Christopher, 2005); ii) a commercial flow starting with production and ending at the final consumer, in which it is important to objectively identify critical points, and the various pressures to which they may be subject to, whether such are due to lack of transparency, supply failures or price changes, among others (Puelles et al., 2011); iii) the path taken by a product from "field to table"; iv) a set of activities developed in different stages, in a direct and indirect manner, with the purpose of satisfying the consumer, through various productive, financial, informative and analytical functions (Briz et al., 2010; 2012). Thus, to reach the final consumer, any agrifood product follows a specific chain with a multitude of activities, which transform the initial raw material into the final product, adding value (Fernández, 2013), from the producer (the origin), to the market (the destination). According to this author, the greater complexity of these chains, in relation to those in other sectors, has to do with the characteristics of the agri-food products, which are dependent on biological cycles, with a limited lifetime, demand and price variability and, for which food quality and safety are key elements.

The value that is added along the chain, in general, is calculated by the price that the consumer is willing to pay for the product and it can be reduced by environmental contamination and misuse of resources. The fair and equitable distribution of the value created in the agri-food chain, to all the actors involved, is important and facilitates more stable and successful relationships between stakeholders. Problems such as lack of transparency, abuse of dominant positions and excessive commercial margins for some members of the chain should be analyzed and avoided so that its functioning and performance is adequate and ensures the continuity of relationships and transactions (Briz et al., 2010; Briz, 2011; Briza t al., 2012). From a globalization perspective, quality, safety and food sustainability are also fundamental to agri-food chains (Jongen, 2000; Van der Laan et al., 1999; Fernández, 2013).

The agri-food chain is complex and multidimensional in economic, technical and sociological terms. If analyzed solely from a single perspective, the conclusions obtained can be biased, or at least incomplete (Clay & Feeney, 2018). These authors identified six different approaches concerning analysis, strategy, efficiency, sustainability, value assessment and development. The strategic approach considers the agrifood chain as a whole, as an economic unit which includes multiple actors, financial and physical flows and a final consumer, whose expectations on the final product must be prioritised. It implies that goals and common strategies are taken into account so to improve competitiveness and satisfy the final consumer, who is increasingly looking at each product with greater added value and specific characteristics of quality, traceability and brand (Clay & Feeney, 2018). The efficiency approach focuses on technical aspects and the flow of products. It seeks to understand how to improve processes within the value chain and avoid monetary and physical losses (products). The latter is particularly important due to the products' perishability and sensitivity to biological cycles, processes and other risks (climate and stocks). Sustainability is often linked to circular economy as there is a growing need from consumers to be informed on the products' impacts on health, the environment and social values. These matters are assessed with the same level of relevance as any economic values. The evaluation of value created and the way it is distributed within the different structures of the agri-food chains, from the primary stages to industry and distribution, requires a developing analysis and an understanding of the power relations' indicators, governance mechanisms and potential conflicts between actors. The latter approach is very relevant in developing countries, which often enter the chains' global agrifood products as primary producers, with scarce added value for the actors involved. In these developing countries, one of the main problems identified by La Gra, et al. (2016), is the loss of food along the agrifood chain and the need to add product value through technological improvements (Ellis et al., 2019).

Rice Agrifood Chain

Several literature studies can be found on rice's agrifood chain from the last decade (Table 1). They include themes such as profit and the role of the different actors, economic value and impact on poverty reduction, key constraints to value chain competitiveness, benefits to smallholders, export capacity, collaboration, stock management, demand supply, waste reduction, productivity increase, restrictions and opportunities, among others. In other words, sustainability principles have been incorporated into agrifood chains, with various purposes, practices and levels. They can, for instance, be used to measure the expectations of different stakeholders and society (Akhtar et al., 2016; Fischer et al., 2009; Notarnicola et al., 2017; Rueda et al., 2017; Schmutz et al., 2018; Sharma et al., 2018).

Dania et al. (2018) and Stone and Rahimifard (2018) have made respectable literature reviews on the essential factors for collaboration between different partners in the agrifood chains, so to achieve sustainability. These authors make a critical analysis on the resilience of agrifood chains. Previously, Taylor (2005) had presented a critical literature review on the management of the agrifood chains, identifying the gaps to be explored in developing countries. The way in which food is produced, processed, transported and consumed is fundamental for achieving sustainability throughout the agrifood chain, and the complexity that persists in its coordination has led to an increase in food waste in recent years (Govindan, 2018).

The focus of the research has been the behavioral factors and characteristics which contribute to the effective collaboration towards sustainability (Dania et al., 2018), governance and management (Stone and Rahimifard, 2018). The main behavioral factors are: joint efforts, sharing activities, collaboration value, adaptation, trust, commitment, power, continuous improvement, coordination and stability. The main multidisciplinary elements of the resilience are the capacities to maintain core function, providing food security to the consumer and adaptation to changing conditions.

Rice and Sustainability

The topic of sustainability linked to rice is very present in literature (Both et al., 2021). Recent studies estimate the need to increase the global production of this food by around 7% and 13% over the next decade and double this by 2050 to meet expected demand at current market prices (Ray et al., 2013, 2013; OECD/FAO, 2019; USDA/ERS, 2019). Potentially, because rice production is water intensive, with a global average water footprint of 1,325 m3 ton (Chapagain & Hoekstra, 2011) and the use of about 40% of the world's irrigation water (SRP, 2020), this increase in production can place strong pressures on limited natural resources and aggravate environmental damage (Okpiaifo et al, 2020). Moreover, rice production is a major contributor to greenhouse gas emissions and the main source of agricultural methane (CH4) and nitrous oxide (N2O), accounting for 19 percent of global CH4 emissions and 11percent of global agricultural N2O emissions (US-EPA, 2006; Smith et al., 2007). Additionally, excessive and sometimes inappropriate use of chemical inputs, especially nitrogen (an important fertilizer in rice production), can increase these emissions and result in other environmental problems such as soil acidification (Guo et al., 2010) and water pollution (Diaz & Rosenberg, 2008). Thus, the future challenge for rice is to promote the necessary growth in production to meet the growing demand in a manner compatible with the quantity and quality of existing resources and the required environmental restrictions. Some of the main studies on rice sustainability are presented in Table 2.

Literature Gaps

Some of the studies carried out on the sustainability of the rice agrifood chain do not focus on the totality of the actors within the chain and therefore present limitations and gaps. These studies tend to focus solely on one of the actors, or a few of them, and on one of the dimensions of sustainability. For example, they either analyse the technical and economic efficiency of the production (Panpluem et al., 2019) or its environmental dimension (Braun et al., 2019; Han et al., 2020). Even when the study focuses on the entire agrifood chain (Othman et al., 2016; Devkota et al., 2019, Aggarwal and Srivastava, 2019) the perceptions obtained from the different parts are hardly robust enough to be applied to other contexts if the circumstances of these contexts are very different from the investigated regions. Besides this, the studies of Matopoulos et al (2007) and Aggarwal and Srivastava (2019) attempts to capture the possibilities of collaborative practices and perceptions of buyers and suppliers towards collaboration, seeking to find out how can collaboration in agri-food supply chain lead to low wastage and higher levels of efficiency.

Thus, to strengthen the reliability and applicability of the research results on the rice agrifood chain, the analysis should be carried out in its specific context as emphasized in the literature. Furthermore, there is a need to study the entire agrifood chain in order to understand the relationship and cooperation between its stakeholders and their performance, which is considered a key factor to ensure its sustainability (Panpluem et al., 2019).

In the near future, the need for a sustainable increase in rice production requires substantial improvements in productivity and efficiency, namely in the use of land, seeds, labor and capital, which are the main production inputs, identified as investigation essentials. Improving technical efficiency is the output from which farmers are the main beneficiaries (Balcombe et al., 2008; Dhungana et al., 2004; Tipi et al., 2009; Jalilov et al., 2019). Rice yield and production have increased considerably in recent years. In several countries, rice yields in favorable environments have reached the maximum yield potential of the current generation of existing rice varieties, derived from genetic improvement, including new hybrid and transgenic varieties. It is also true that, in other contexts, technological and production management improvements should be investigated in order to considerably increase the level of rice productivity and evaluate the efficiency of the production systems, with the desired result improving income and welfare for farmers. It should be emphasized that the DEA methodology, although little used in research on the rice agrifood chain so far, is referenced as useful and with potential to increase knowledge in this matter (Dhungana et al., 2004; Balcombe et al., 2008; Tipi et al., 2009; Jalilov et al., 2019).

The effects of state policies on the sustainability of the agrifood chain is one of the topics for future research suggested in the FAO study (2018) carried out in India, namely: 1) the policy of purchasing rice directly from farmers on behalf of the central government and the need to create the respective infrastructure in the rice catchment area to increase efficiency in obtaining rice; 2) the grain storage and distribution policy and welfare schemes and their effects on controlling losses during transit and increasing their availability and supply to local markets. This issue had already been mentioned in a previous study carried out in Malaysia (Yong, 2008), where a framework of public policy choices and strategies for the development of an innovative rice agrifood chain involving farmers, processors, importers and distributors, was proposed (wholesalers and retailers). The study concluded the need for research to consider a comprehensive approach to the rice agrifood chain in order to develop robust and sustainable solutions. The literature also suggested the need to further study consumers so to understand how they value sustainably produced rice, as long as they are properly informed about the certification and origin of the product which they are buying (My et al., 2018).

The analysis of trade-offs between different production factors and the efficiency of rice production can be useful to identify sustainable rice intensification and other production strategies as well as for impact assessment and comparison of technology, region or country. For Devkota et al. (2019) a combination of agronomic, environmental, economic and social data should be studied and analyzed to evaluate if and how they can be integrated, using appropriate sustainability indicators.

Regarding the management of the agrifood chain, namely its strategic management and performance, Obura et al. (2017) refer the need to investigate the conceptual Framework. This allows for an analysis of the relationship between uncertain factors, concerning demand and supply, and the practices of rice agrifood chain, as well as the research methodologies that need to be employed. In the specific case of Kenya, where rice plays a vital role in food security, the authors mention a need to carry out a comprehensive value chain analysis to determine gaps, develop the necessary interventions to address those gaps, increase investment in research, developing and establishing policy measures that create and establish a system for producing and distributing improved seeds in the right qualities and quantities to farmers, improving extension activities, empowering producers and their organizations with a view to reducing losses and improving post-harvest quality, adopting new production and processing technologies and instilling greater cooperation and transparency in the governance of the agrifood chain. It is also important to refer Rampersan et al. (2019), which research work investigate how supply chains can become more resilient through innovation initiatives. The authors analysed the expansion and deepening of relationships between buyers and suppliers and the facilitatory role of the government in this process. In terms of management processes, it become clear the importance of of power distribution, coordination, communication, trust and commitment for innovation within these relationships.

MATERIAL AND METHODS

Taking into account the multidimensional perspective of the rice agrifood chain (Taylor, 2005), the study used a mixed (quantitative and qualitative) methodology, supported by an exploratory research approach to collect primary data, and an extensive analysis of the literature, to gather secondary data.

The exploratory phase of the research was based on an inductive approach, in which secondary sources, the researcher's point of view and qualitative interviews of the focus group were fundamental for the design of the second phase, which consisted of a conclusive-descriptive research, for which the deductive approach and the combination of quantitative and qualitative research methods were considered.

The focus group, conducted in two groups with a total of 16 participants, during the month of January 2018, was subject to an interview guide with five sections: i) the concept of sustainability and its dimensions; ii) the main bottlenecks identified and the challenges to the sustainable development of the rice agrifood chain; iii) rice production, sustainability and food security; iv) sustainable distribution, consumption and marketing; v) the group's global understanding of the "sustainability of the rice agrifood chain" After the introduction and presentation of the group participants and their objectives and evaluated the possibility of recording (audio and video), the discussion targeted the exchange and unveiled different opinions over the sustainable pathway for the rice agrifood chain in Timor-Leste. In order to guarantee its exploratory nature, the study used both, structured and open questions to probe new ideias and opinions, allowing the participants to have an exchange and a rather free discussion.

In the second phase, two structured questionnaires targeted the key stakeholders in order to analyse the vision of both the producers and consumers, within the rice value chain in Timor-Leste. At this stage of the process, qualitative distributor interviews (aimed at capturing individual perspectives) were conducted during a three-month research stay, from January to April 2018. The reason behind employing a mixed approach (quantitative and qualitative) was so the study relied on distinct and complementary instruments.

Field observation was done during the entire period when empirical data collection took place, in order to complement the information collected and, help adding information and triangulate the data generated by the other research methods in place. Table 3 summarizes the methodology used in the study.

SUSTAINABLE DEVELOPMENT OF RICE AGRIFOOD CHAIN

The results on the sustainable development of the rice agrifood chain in Timor-Leste are presented in this section. It begins with the analysis of the efficiency and characteristics of the production, followed by the vision of its distributors and consumers on the sustainability of the rice agrifood chain. Finally, the study identifies the main drivers and inhibitors of a sustainability development and proposes the basis for a sustainability plan for the rice agrifood chain in Timor-Leste.

Production Characteristics and Efficiency

In Timor-Leste, rice production is carried out in small production units owned by households, where the manager is mostly male, illiterate, aged between 26 and 70 years old and with experience in rice production. This typology is similar to that of previous studies (Loly, 2014; Cruz et al., 2016) which shows that there has not been a great variation in family structure and exploitation in recent years.

Producers use fertilizers and pesticides, irrigation is done with rainwater and spring water, and the initial capital is very small, limited to bags, cans, sickles, cutlasses and shovels. This production technology is essentially manual for most operations (pre-germination, cleaning, leveling, sowing, watering, fertilizing, protection from pests and diseases, mowing, drying, storage and marketing), with the exception of land preparation, threshing and husking of rice, referred to by Loly (2014) and by Deus (2019). The latter distinguished two rice production systems based precisely on the criterion of mechanization, namely, the lowland and highland system of the districts of Ailéu, Bobonaro and Covalima. It is called upland rice by Silva (2011) because it is based on rainwater, from the districts of Ermera, Bobonaro and Covalima. The superior productivity of one system (in the districts of Covalima and Aileu) compared to the other (in the districts of Ermera and Bobonaro), is due to mechanization, considered one of the main factors of production that, when added to technology, can improve production efficiency and end results.

The assumptions of variable returns to scale (DEA BCC) with output orientation were considered in order to obtain the technical efficiency levels and scale of rice producers in the municipalities of Boborao and Baucau in Timor-Leste. Based on Devkota et al., (2019), efficient producers were considered to be those who reached efficiency levels above 0.90, that is, at 1 and inefficient producers were those who obtained levels below of that value. Table 4 shows the efficiency levels for the sampled producers. Of the total number of rice producers, only six were technically efficient, which corresponds to 2.0% of the total sample. The average inefficiency was 48.2%.

Table 5 presents a comparison between the six most efficient and six least efficient producers, estimated by the BCC model, with output orientation. The values of the quantity produced are presented in kilograms, the area used is in square meters, the seeds in kilos, the workers in singular people and the current value of the capital in US dollars. The values of the six least efficient producers represent the values that should be reduced from the inputs for these producers to become efficient, based on benchmarking.

The first most efficient producer, reached a produced amount of 3750 kg, using an area of 15000 square meters of land, seeds 30 kg/ha, employed 189 people per hectare and 100 US dollars in capital.

Among the six least efficient producers, it was observed that, in order to increase the quantity produced by the owner, through the optimal use of inputs, some changes in producer expenses would be necessary. It was found that both the first, the second, the third and fourth producers in the classification of least efficient had increasing returns to scale, thus, an increase in inputs would lead to a more than proportional growth in production. It appears that a better use of all inputs would promote greater efficiency for rice producers in the Bobonaro and Baucau municipalities in Timor-Leste.

Improving technical efficiency is necessary as only 2% of the rice producers studied produce at the optimal production scale. In the assumption of constant income, only six of the producers are technically efficient, and the inefficiency of the others can be reduced if the use of inputs is reduced to about half, maintaining the level of production. This may mean that producers use excess intermediate consumption (seeds, fertilizers, pesticides and labor) with negative consequences for the sustainability of the production. Even when considering variable returns, despite the fact that the average technical efficiency increases and the number of efficient producers increases to seven, the technical inefficiency is due to the production scale (scale inefficiency). This shows, on the one hand, that the cultivation area (production scale) is a factor that influences the technical efficiency of the production and, on the other, that there seems to be room for rice producers to increase production using the same resources, or using less resources for the same production.

Despite the use of factors (days of work for men and women, days of mechanical traction and amount of fertilizers, seeds and pesticides), the differences in the results among producers are not substantially large, either due to the similar characteristics of the production system used or due to a combined lack of planning initiatives, resources, knowledge and technology (mechanization) to reduce and improve them. For Loly (2014), production levels influence efficiency, and higher production levels have greater efficiency. According to this author, most producers show increasing returns to scale and an increase in the area of production units has an advantage in terms of efficiency of farms. The levels of self-consumption, the type of market where rice is sold and the type of consumer, do not influence the production efficiency as rice is mostly used for self-consumption, with a small part of it being sold in local markets and to end consumers as was defended by (Loly, 2014) and in accordance with the results of this study.

Reducing production gaps and improving the productivity and efficiency of rice production can involve greater and active institutional support from the government, particularly in regard to training, in production management, planning and good practices in the use of intermediate consumptions. Such is, while taking into account the levels of environmental and socioeconomic sustainability and the reduction of climate change risk, through the provision of credit to households and villages and through stronger and more assertive extension bonds. Additionally, the government could promote closer collaboration between local authorities, non-governmental organizations (NGOs), universities and private sectors, in order to identify specific constraints and improve performance. It could also adopt appropriate technologies and solutions and take concerted actions to bridge the gap between rice production and other levels of the agrifood chain, through participatory approaches and applied research, which is essential. All of this, relies heavily on the government's willingness to support, coordinate and monitor an integrated and holistic rice sustainability program in Timor-Leste, and on international support for this government initiative towards inducing the sustainable development of rice production and the conservation of natural resources and the environment for future generations.

Consumption and Consumer Behaviour

A questionnaire was applied to rice consumers in East Timor, and 240 responses were obtained. The sample is characterized according to the information on Table 6.

Globally, the responses showed that the consumption of rice is very important in Timor-Leste, as this food is appreciated by the population in general, constituting the basis of their diet.

Cluster analysis applied to the sample of consumers led to the creation of two distinct clusters of rice consumers in Timor-Leste. Cluster I was obtained with 171 individuals and Cluster II with 69 individuals. In order to characterize these clusters, an analysis was carried out based on the sociodemographic variables obtained, as well as on issues related to taste and preference for rice consumption. Tables 7 and 8 summarize the main sociodemographic characteristics of each cluster (higher frequencies of sociodemographic variables in the two clusters of rice consumers).

In Cluster I, consumers consider all the items under evaluation as very important, namely the taste for rice, whether is a staple food, their habitual consumption or cultural dependence on this product within the diet. In Cluster II, in addition to the importance attributed to the aforementioned items, though less than that of Cluster I, the item related to the cultural dependence of rice, considered as both unimportant and important, should be highlighted.

It is understood that a sustainable rice consumption is particularly important in East Timor, for the sustainable development of the rice agrifood chain, when achieved through traditional production systems, which are also globally sustainable. This situation is not different from that of other developing countries, where rice is a cheap product, easy to access and to accept (Wang & Gao, 2017). Hence, consumer behavior is a much-researched topic, in several aspects and contexts, including the preference for local and imported rice and the attitude towards the sustainability of its production (Wang & Gao, 2017; My et al., 2018; Okpiaifo et al., 2020) and commercialization (My et al., 2018). In less developed or developing countries, the preference for domestic products tends to be weaker than in developed ones, and domestic products are less considered than imported ones (Opoku & Akorli, 2009). This situation that is not completely similar to that found in Timor-Leste, even so, many consumers consider imported rice (or both) as important for consumption.

The fact that the vast majority of consumers claims to be willing to pay an additional average amount of approximately 15 dollars per bag for the quality of the rice, allows, alike the study by My et al. (2018), on the one hand, for the emergence of a target group which produces rice sustainably and with quality and, on the other hand, for the need to communicate its environmental and health benefits with credibility and reliability. The latter will allow for a reinforcement of the consumer's beliefs and improve their appreciation of sustainably produced domestic rice. This preference and willingness to pay for quality, identical to those found in other studies (Custodio et al., 2016), can be used to induce improvements in the agrifood chain, adjusted from an economic, social and environmental point of view.

In the present investigation, as in other studies (Suwannaporn & Linnemann, 2008; Bairagi et al., 2017), quality is considered relevant when purchasing, and can influence the price of rice (Cuevas et al., 2016). The value of sustainably produced rice is strongly affected by consumer belief in the health benefits of the product and its nutritional content (De Steur et al., 2012; De Steur et al., 2017; Mondelaers et al., 2009). Consumers who are willing to pay premiums also tend to be more environmentally conscious (Abdul-Muhmin, 2007, Yadav, 2016), varying between income classes (Cuevas et al., 2016). In this investigation, these evidences were not found.

As expected, the study recognized the importance of consuming rice as the sole nutritional source or, in addition to other foods such as meat, fish and eggs, in high daily amounts per family (1-3 kilograms) and a daily consumption in every meal. This situation is common in other Asian countries, where rice is the stable staple food for 2 billion people in Asia alone, who get 80% of their energy needs from it (contains 80% carbohydrates, 7–8% protein, 3% fat and 3% fiber) and often consume it on its own, without associated harms, because its natural molecule has the best physiological compatibility and little or no toxic effect (Chaudhari et al. 2018). In Timor-Leste, there seem to be two components related to consumption, one labelled Access, with greater weight of variables related to distance and transport and availability/accessibility, and the other, named Quality, with greater weight on variables related to quality and human benefit, already discussed, and healthy life.

There is a balance concerning the different places where consumers make over half of their rice purchases, namely, stores, supermarkets, local markets, kiosks and the road trader with the lowest weight. Once again, quality, good service, price (cheaper) and short distance (close to home) prove to be the main reasons for purchasing in these places. These findings are in line with those of Bunyasiri and Sirisupluxana (2017) and Okpiaifo et al. (2020), in which consumers make their purchases mainly in stores and supermarkets. These places, in particular supermarkets, are seen as a key player in building the foundations for sustainability in the agrifood chain (Reardon et al., 2012).

Despite the identification of consumer perceptions and preferences for national rice produced in sustainable manners, and their willingness to pay for this product, and regardless of this being a determining element in the motivation of producers and distributors to produce this rice and market it (Wang & Gao, 2017), the results found in Timor-Leste appear to be somewhat inconsistent. If, on the one hand, good quality is one of the determinants of the purchase and consumption of rice in East Timor, along with the product's human benefits, accessibility and the possibility of it replacing other foods, the mention of an affordable price is yet another determinant. Purchase and consumption, seem contradictory with the willingness to pay an additional amount for the higher quality of rice. It is also worth mentioning that though consumers in Timor-Leste give importance to the quality of the rice, due to its caloric content, there is also the suggestion that this quality should be improved, controlled and certified. This result suggests the possibility, similarly to the study by My et al. (2018) to incorporate quality (and also sustainability) through certification and labeling as a viable incentive mechanism to increase sustainability in the rice agrifood chain in Timor-Leste. It may also have important implications on market organization and the adoption of certification schemes similar to those in other rice-producing countries in Southeast Asia and South Asia (My et al., 2018), provided that the changes suggested by consumers to improve rice sales are considered. Legislating and regulating, having greater organization and intervention on the market, clearly informing consumers, having an affordable price and giving incentives to national production, are some of these suggestions.

Regarding the perception of rice as sustainable, consumers emphasize the importance of defending agriculture and resources, improving the quality of life and inducing the development of rural areas as very important, and economic viability, promoting research and guaranteeing human survival. In the way rice is produced and sold, they show that it is a non-polluting cereal, that it is national and that it helps producers. These results are in line with other recent studies (Okpiaifo et al., 2020). Increasing the quality and quantity of rice, its process transparency and knowledge on marketing, improving processing and distribution, adopting technology, and having coordination between relevant institutions and partners, taking care of the environment and natural resources are aspects that consumers value as very important or important and which are improving. Also, the findings of Okpiaifo et al. (2020) identify food safety and health and safety as important sustainability attributes for rice consumers.

Distribution Channels

In regard to this section, it is worth mentioning the difficulty in discussing the results, as only eight distributors were interviewed. Even so, it is feasible to say that the rice distribution companies under study, located in the district of Dili, are relatively new, single-person, small in terms of the amount of human resources used and the average volume of business. The proportion of rice turnover in total turnover is quite high. Regarding its cost structure, personnel, energy, telephone, internet and transport stand out, which reveal economies of scale. That is, larger companies spend a smaller proportion. Some costs are specific to the size of companies, such as security and other costs, which only occur in larger companies.

The ways of distributing rice, its destinations and the promotion actions that are carried out were identified. The supply to distributors is mainly made through imports, with few direct purchases from Timorese producers. Although cooperation between different economic agents and promotion are considered important, only half actually promote rice, especially in newspapers and on the radio. The financing of companies is made with equity, and business agreements are based on the confidence of the handshake that seals them. Payments are made in cash, with companies offering discounts as a percentage of the price or even a reduction in price. The months with the highest turnover are December, August and July. Most of the distributors' rice is destined for the final consumer. Often the distributor includes transportation to customers. The most commercialized rice brand is Globos. Companies obtain price information mainly from importers, they have already received technical training in this business area, do not find cooperation between companies in the same business area important and promote their products in various ways. It was only possible to identify the costs in connection to the amount of time spent transporting rice, which amounted to around two hours. By identifying four types of distribution channel, the study noted that the frequency of transactions is higher, when these degrees of integration of higher channels exist.

Regarding the most appropriate typology, the recommendations to be made focus on the so-called short circuits, based on valuing close relationships, boosting the local economy and guaranteeing food and nutritional security for the population. The literature review highlights the potential of this strategy from the point of view of increased production, product presentation conditions, the expansion of social networks, food security and sustainable development (Chiffoleau, 2012; Lamine & Chiffoleau, 2012; Harun et al., 2021). In Timor-Leste, despite the importance of close relations in the country, there are asymmetries, with consumers having a limited role in monitoring the distribution channels. Hence the importance of short circuits, which provide greater proximity between producers, importers and consumers and are important vectors for sustainable development.

Sustainability Drivers and Inhibitors

Timor-Leste is divided into three important ecological zones (lowlands, highlands and slopes) based on their respective intensity of atmospheric precipitation and topography, and, as such, they set the edafo-climatic conditions which are necessary for the sustainable production of rice, in familial subsistence agricultural systems, the irrigated and the dryland. Though rice is the population's staple food, the reduced quality and quantity of the national rice production does not ensure the country's food sovereignty. All of the main actors in rice's agrifood chain along with their individual functions can, simultaneously, be boosters and inhibitors of sustainability, in all of their three pillars, economic, social and environmental.

In the specific case of rice production, which is sustained on natural resources that are fundamental to the planet's future (land and water), it can be the inducer of both good and bad environmental, social and economic results. Soil degradation is a problem which affects fertile and marginal land, with a growing aggravant of rising pressure on agriculture to increase its yield per soil unit. The growing lack of resources, the effects of climate change and the increasing concern regarding the environment (collaterals) sustain that the relation between sustainable development of the agrifood chain and environmental protection grows increasingly important. Good governance, security and political stability of the country are pre-requisites to rice's sustainable agrifood chain. The producers, in their individual or familial form along with other organizations that are legally responsible for the production (associations or producer agglomerates) should assure that the best production practices are in place.

The distributors can also be boosting elements, if they assume the configurations inherent to the commercialization of rice, with prices and payment methods that are fair to farmers, as well as common negotiating elements, providing consumers with quality products at an accessible price and a clear communication of their origins and benefits. Alongside producers, which assure food for the growing populations, distributors can allow products to be made available at accessible prices, contributing to both food security and political stability. In Timor-Leste, security and food sovereignty are all encompassing matters, involving all sectors at multiple levels, concerning intervenient from the familial level to policy makers.

Regarding sustainability inhibitors at a production level, it is worth emphasizing low productivity, resulting from lack of agricultural infrastructures, knowledge, good production and transformation processes, and the effects of climate change and resulting implications. Besides the lack of machinery, there is a lack of irrigation systems, quality seed provisions, fertilizers, pesticides, technological improvements, technical assistance, research, efficient extension services, access to credit and support for investments in infrastructures (roads, dams, irrigation systems, storage units, rural electricity and rice's genetical improvement). Concerning distribution, there are deficiencies in transportation and multiple ways, product loss and inefficient storage conditions, as these are usually not equipped and rice is usually not stowed properly in warehouses). Besides what has been listed, the rice agrifood chain generally lacks cooperation's, organized intersectorial connections, trust and management of the relationship between the different actors, improvement of the human capital with practical training and competence development on sustainability, transparent governance, investigation and political and institutional compromise, all of which are basal elements to the chain's sustainable development. Additionally, there is a growing populational rural/urban migration as well as the reduction of the familial workforce needed for rice production.

Consumers are also and importance link and are growing increasingly more contientious of quality and food security, regardless of whether the rice is national or imported. Consequently, it is necessary to optimize the economic, social and environmental performance of the agrifood chain so to ensure its sustainable development. In order for this to happen, better bonds among the links and better fluxes and coordination of information are necessary, so to proportionate the consumers with what they need and desire. The actors in the chain of value may not be able to achieve that alone; therefore, they need to play an active role, bettering their practices and developing collaboration among the chain. Collaboration between all the players is a key element for reach efficiency levels and to have higher performance. According to Aggarwal and Svirastava (2019) in developing countries, agriculture industry is the backbone of economy. So, the results from this kind of studies should be useful for managers in agribusiness in terms of supply chain practices that would increase profit and efficiency, and decrease waste and pollution.

In Timor-Leste, there are yet another set of inhibitors of this type of development, namely the fragmented nature of the chain, the lack of organization and the need for a better and more regulated chain. The development of the internal market is also a determining factor in the creation, maintenance and promotion of an adequate environment for the sustainable development of the country's rice agrifood chain.

Those responsible for agricultural policy may boost or inhibit the path towards a sustainable rice agrifood chain, depending on the chosen strategy and its consequent measures and planned activities, as well as the incentives to distinct operators, motorization of their compliance. Data collection, the identification of all the actors in the chain and their respective functions in the form of official statistical data should be assured by the state officials (ministries and their regional directions and dependencies), such as the training of producers and eventual and future rice certification.

SUSTAINABILITY PLAN FOR THE RICE AGRIFOOD CHAIN

A sustainability plan for the rice agrifood chain in Timor-Leste (PSCAATL) is what is being proposed, as it is the solution that allows for sustainability and seems to be more feasible in the short/medium time. In the long run, the structure that is eventually created and developed in East Timor may integrate the *Sustainable Rice Platform* (SRP). This is a multiple stakeholder global aliance of multiple members, with over 100 institutional members, led by the United Nations Environment Program (UNEP) and the Internations Rice Research Institute (IRRI). This platform connects diferent partners in the public and private sector, such as investigation centres and units, non-profit organizations, and others from the the internacional development community (SRP, 2020).

PSCAATL should aim to achieve resource efficiency and climate change resilience concerning the production, distribution and consumption (along the entire chain). For this to happen, there is a need for information on all the links of the chain as well as a search for transformations in production methods, distribution procedures, market organization and responsible consumption, which assure/ develop a sustainable pattern, supported in criteria. The purpose of PSCAATL is that of establishing the national pattern of production, distribution/commercialization and sustainable consumption, structured by a set of indicators and goals which quantify and verify the current and future evolution achieved through the implemented improvement practices, At the core of this, is the aim to minimize the environmental impacts of rice's agrifood chain, increasing income and the wellbeing of the producing households, the revenues of the distributors and consumer satisfaction, contributing to the soverignty and food security of Timor -Leste.

PSCAATL should be a voluntary initiative, promoted by the Ministry of Agriculture or one of its oficial dependencies, aimed mostly at rice producers but with impacts on transformation and distribution. Its basis relies on the definition of a sustainability strategy oriented towards cost reduction and bettering/increase of economic viability, cultural durability and related businesses, incentivating proactivity in the face of environmental pressures and climate change, answering social concerns, bettering quality and the competitiveness of the final product. Reducing waste, minimizing risks, entering other market segments, integrating producers into networks, increasing provider and implementing intelligent communication stategies are also at play. Meaning, allying competitiveness with environmental objectives, amplifying the national rice production with quality and attending to the needs of the consumers and the market opportunities. The basis of the PSCAATL' proposal, listed on Table 9, showcases the main dimensions under consideration.

PSCAATL proposal was envisioned to: 1) Support rice producers in the quest to better their environmental, social and economic performance, promoting the recognition of the sustainable development of the national rice; 2) Develop rice's agrifood chain through the incorporation of ecoefficiency principles with the objective of promoting a more efficient use of resources, incentivating the reduction and reutilization of coproducts, reducing operational costs; 3) Producing national rice with sustainable performance; and, 4) Identify the producer's perfomance, comparing their results among peers and districts, defining, areas for improvement and plans of action that change the production practices. If created, the implementation of PSCAATL should be initiated by a self-evaluation, for which the present study is a valuable contributor. This self-evaluation, should be thought of in three different fields (Production; Distribution; Production & Distribution /Agrifood Chain) as previously depicted in Table 9. This self-evaluation should respect the classic methodology for continuous improvement of a system, with a mandatory annual execution. This self-evaluation, which should be supported and even performed by technicians from the ministry of agriculture, will serve as an annual diagnosis on the sustainable practices or lack there of, performed by the elements of the agrifood chain, particularly producers and distributors, during their activities

FUTURE RESEARCH DIRECTIONS

This study presents some suggestions for future investigation, based on its findings. Firstly, a more thorough analysis should be performed on the entire agrifood chain and its three pillars of sustainability. Otherwise, the benefits of the environmental improvements on the production phase of rice's agrifood chain may be lost in the latter stages of processing or distribution and result in an increase in waste or environmental risk. This investigation, focused solely on analysing rice's agrifood chain, in the perpective of improving its sustainability, thus, a broader analysis of the chain, which explores all its actors' sustainability aspects, in all its stages, including imports, would be desirable.

Another relevant research suggestion is the identification of the agrifood chain's risks, so to manage its sustainability and identify the adequate measures for risk mitigation and decrease their frequency of impact. Thus, it is recommended that new studies adopt a quantitative approach to the identification and analysis of the risks, especially those resultants from climate change, among others, and evaluate their occurrence and practical impact on the management of the agrifood chain.

Increasing the knowledge on the different stages of the agrifood chain, investigating operations and transaction costs (whether they are at production level, post-harvest, storage and transport outside production limits, or whether they regard the commercialization of rice), understanding flow reversal inside the chain in the perspective of circular economy and innovative forms of governance are also considered important for subjects for investigation.

Given the growing interest in quality assurance and rice verification and certification, further studies can be carried out to explore stakeholder readiness in relation to the costs arising from rice which is both certified and sustainable.

Theoretical Implications

There is a lack of Timorese governmental measures which promote sustainable industries, and a lack of precisely what this study focuses on, policies which promote the sustainability of the rice agrifood chain. Evidence of this lack of policies is found in post-harvest losses, excessive and unregulated use of chemicals, waste and unsanitary conditions in wholesale and retail markets, and lack of consideration for consumer concerns. Consequently, it can be verified that the practices in the value chain are not aligned with the requisites of sustainable development.

Considering the high level of population growth, the urban development which reaches unprecedented levels, the changes in citizens' daily habits and strict international compliance requirements, government policies, in general, and agricultural policies in in particular, should be framed from the point of view of sustainable development. In addition to economic development, it is also important to promote social and environmental policies.

The sustainable development of the rice industry caught the eye of the policy-makers in Timor-Leste. The results of this study can contribute significantly to the framing of that policy. This study verified that that consumer preferences are not properly taken into account by government policies, especially those related to food. There are no measures aimed at protecting consumers from food unfit for consumption. In fact, the unregulated use of hazardous chemicals, especially those that promote faster rice ripening, generate anxiety among consumers.

Consumers are an important part of society and the effectiveness of policies depends on their compatibility with the needs expressed by consumers. Mowat and Collins (2000) believe that the development of agrifood industries is not possible if these do not understand and meet the needs of consumers. De Jonge et al. (2007) argue that effective food policies must take into account consumer preferences and address consumer concerns, particularly those related to food safety.

Another important policy implication to be drawn from this study is the need to shift the focus of development policies and interventions in Timor-Leste from disintegrated to integrated approaches, such as the rice agrifood chain development approach. Previously, the focus was on the sustainability of specific parts of the value chains and concentrated mainly on the agri-food production aspects of rice. The government needs to adopt an effective agrifood chain development approach that promotes sustainability. Similarly, government-led projects and programs are needed to promote the sustainable development of rice agrifood chains in Timor-Leste.

Practical Implications

The actors in the rice value chain operate in both traditional and modern chains. They can learn from the different requirements of the rice consumer segments and if/when concerned with safety, proceed to adapt the quality of their products. It is interesting to note that most consumers are more concerned about the value of the rice, being willing to pay more if the rice quality is as desired.

Value chain actors develop and strengthen their chains, improving practices that emphasize product quality and safety. A considerable portion of consumers prefer to buy in traditional stores, which forces the actors of the chain to redirect their efforts to meet modern retail standards. Faced with this changing scenario, value chain actors, especially producers, need to improve the quality attributes of their rice fields, aligning their practices with the specific needs of consumers (Donovan and Jason 2011; Tinsley, 2012; Gunden & Thomas 2012; Jiménez-Guerrero et al. 2012).

This study highlighted concerns about the food safety of consumers, who demand an update of practices, particularly those who are still closely linked to traditional value chains (Verbeke et al., 2007). As the market share of modern value chains gradually grows, traditional value chain actors cannot remain economically viable if they continue to ignore quality attributes related to food safety and marketing. This changing landscape makes it imperative that traditional wholesalers understand consumer preferences and provide good quality, safe and affordable rice. Beginning this process would include creating hygienic conditions in and around retail facilities.

CONCLUSION

Sustainable development in the rice agrifood chain is essential for producers, distributors, consumers and the strengthening of their respective relationships. In the particular case of Timor-Leste, the chain is related to the most important primary and secondary activities in the country and constitutes a strategy to improve competitiveness, both in terms of market leadershio and costs. From the transformation of raw materials into products for the final consumers, rice goes through a series of intermediaries (wholesalers, retailers and agents). In this context, the strategy of short agrifood chains is essential to create a greater proximity between producers and consumers, as well as to improve the connection and interactivity in the mutual construction of trusting relationships. In terms of sustainability, collaboration between partners is a fundamental factor in achieving positive results, especially on an economic level, but also at an environmental, social and operational level. Although there is an awareness of these relationships concerning the rice food industry in Timor-Leste, they are not yet fully established.

Regarding the efficiency of the production link in the rice agrifood chain, six efficient decision-making units were identified, the others being inefficient, an occurance similar to that of other studies (Linn & Maenhout, 2019). Globally, the explanatory variables affect production, while individually, only a few influences it. As for consumption, the vision and perception that rice distributors and consumers have about the agrifood chain itself is that it has undergone major recent transformations, resulting from globalization, which hinder the competitiveness of many agricultural producers who are unable to reach the required standards of quality. The anticipated changes in the socio-cultural and political environment reinforce this trend. There tends to be a loss of 'space' for informal distribution networks, which lack efficient coordination, while the leading networks will be gainning ground, strengthening themselves. The creation of a structure to ensure this coordination will be essential for all actors in the rice agrifood chain.

Consumers favor the flavor of rice, its health benefits and its sociocultural importance. As predicted, they prefer good quality rice at an affordable price. On the other hand, they show dissatisfaction when faced with rice that can cause health problems due to the high amount of chemicals used in its production. Most attach importance to clear information on the laws created for product sale regulation, market intervention and the creation of farmer subsidies. In regard to the importance given to producer safety and sustainability, consumers consider local and imported rice to be very important for the sustainable development of the rice agrifood chain in Timor-Leste, making themselves available to pay more for better quality of rice.

The drivers of sustainability in the rice agrifood chain are related to the use of soil, seeds and capital. The production, commercialization, raw material acquisition, consortium activities, low production cost and product quality processes are also important driving factors. This category also includes the improvement of infrastructure, productivity, product quality, technology in terms of agricultural equipment and the experience-based knowledge of producers. Inhibiting elements relate to variations in environmental, social, economic practices and in environmental laws and regulations. Market competitiveness, product sales prices, lack of skilled labor, climatic variations, soil fertility and agricultural policies can also be considered as inhibiting factors.

The proposal of a sustainable development plan for the rice agrifood chain in Timor-Leste is a solution towards sustainability. This should highlight various sustainable practices, such as soil management and fertility, good sowing practices and organic materials, water management and quality, herbicides, disease control, harvesting and transport. With respect to a marketing plan which underlines the segmentation and positioning of the market, the study highlights the organization/identification of consumer groups and the demand for agricultural goods in the market. The development of this plan will serve to promote better practices in agricultural producers, interconnect the different links of the rice agrifood chain in a more inclusive way and orient it to the market.

ACKNOWLEDGMENT

This research was supported by the Fundação para a Ciência e a Tecnologia [grant number UID/04007/2020].

REFERENCES

Abdul-Muhmin, A. G. (2007). Explaining consumers' willingness to be environmentally friendly. *International Journal of Consumer Studies, 31*(3), 237–247. doi:10.1111/j.1470-6431.2006.00528.x

Abebe, M. (2016). *The Contributions and Challenges of Rice Value Chain Development on the Livelihood of Small Holder Rice Farmers (In the case of Ethiopia Driving Growth Entrepreneurship and Trade (EDGET) project in Fogera and Libokmkem District)* [Unpublished master dissertation]. Addis Ababa University.

ADB – African Development Bank (2018). *The Gambia rice value chain transformation programme (RVCP).* RDGW/AHAI/PGCLDEPARTMENTS, November.

Aggarwal, S., & Srivastava, M. K. (2016). Towards a grounded view of collaboration in Indian agri-food supply chains: A qualitative investigation. *British Food Journal, 118*(5). Advance online publication. doi:10.1108/BFJ-08-2015-0274

Ahmed, K. H. H. (2017). *A Decision Support Framework for Sustainable Supply Chain Management* [Unpublished doctoral dissertation]. Loughborough University, UK.

Akhtar, P., Tse, Y. K., Khan, Z., & Rao-Nicholson, R. (2016). Data-driven and adaptive leadership contributing to sustainability: Global agri-food supply chains connected with emerging markets. *International Journal of Production Economics, 181*, 392–401. doi:10.1016/j.ijpe.2015.11.013

Ampadu-Ameyaw, R., Omari, R. & Owusu, G. E. (2017). Development and Analysis of the Rice Value Chain for the Hohoe and Jasikan Districts of the Volta Region, Ghana. *FARA research Results, 1*(1), 48.

Bach, V. N., Pham; C. H. & Hoang, N. (2016). A Critical Analysis of the Value Chain in the Rice Industry and Its Effects on the Export Rice Industry in Kien Giang Province, Vietnam. *International Journal of Financial Research, 7*(3).

Bairagi, S., Mohanty, S., Ynion, J., & Demont, M. (2017). *Determinants of Consumer Preferences for Rice Attributes: Evidence from South and Southeast Asia*. Conference: Agricultural & Applied Economics Association Annual Meeting, Chicago, IL.

Balcombe, K., Fraser, I. M., Latruffe, L., Rahman, M., & Smith, L. (2008). An Application of the DEA Double Bootstrap to Examine Sources of Efficiency in Bangladesh Rice Farming. *Applied Economics, 40*(15), 1919–1925. doi:10.1080/00036840600905282

Both, A. V. R., Saraiva, M. S. M. M. de M., & Tibério, M. L. (2021). Rice Sustainability Trend: A bibliometric review. *Research. Social Development, 10*(7), e8310716224. doi:10.33448/rsd-v10i7.16224

Briz, J. (2011). La cadena de valor como instrumento de análisis del funcionamiento y riesgo del Sistema Agroalimentario. In *La Cadena de Valor Alimentaria: Análisis Internacional de casos reales*. Editorial Agrícola Española, SA.

Briz, J., De Felipe, I., & Briz, T. (2010). La cadena de valor alimentaria: un enfoque metodológico. *Boletín Económico del ICE, 2983*.

Briz, J., De Felipe, I., & Briz, T. (2012). La sostenibilidad del sistema alimentario a través de la cadena de valor. In *Las Redes Cadenas de Valor Alimentarias en el siglo XX1: Retos y oportunidades internacionales*. Editorial Agrícola Española, SA.

Bunyasiri, I. N., & Sirisupluxana, P. (2017). Consumers behavior and rice attributes for Thai hommali rice in Sichuan province of China. *The Business and Management Review, 8*(4), 329–341.

CABRI - Collaborative Africa Budget Reform Initiative. (2019). The role of governments in developing agriculture value chains. Case Study 2. Rice and cassava value chains. Author.

Chabowski, B., Mena, J., & Gonzalez-Padron, T. (2011). The structure of sustainability research in marketing, 1958–2008: A basis for future research opportunities. *Journal of the Academy of Marketing Science, 39*(1), 55–70. doi:10.100711747-010-0212-7

Chagomoka, T., Afari-Sefa, V., & Pitoro, R. (2014). Value chain analysis of traditional vegetables from Malawi and Mozambique. *The International Food and Agribusiness Management Review, 17*(4), 59–86.

Chapagain, A., & Hoekstra, A. (2011). The green, blue and grey water foot print of rice from production and consumption perspectives. *Ecological Economics, 70*(4), 749–758. doi:10.1016/j.ecolecon.2010.11.012

Chaudhari, P. R., Tamrakar, N., Singh, L., Tandon, A., & Sharma, D. (2018). Rice nutritional and medicinal properties: A review article. *Journal of Pharmacognosy and Phytochemistry, 7*(2), 150–156.

Chiffoleau, Y. (2012). Circuits courts alimentaires, dynamiques relationnelles e lutte contre l'exclusion en agriculture. *Économie Rurale (Paris), 332*(332), 88–101. doi:10.4000/economierurale.3694

Christopher, M. (2005). *Logistics and supply chain management: creating value-adding networks*. Pearson Education.

Clay, P. M., & Feeney, R. (2018). Analyzing agribusiness value chains: A literature review. *The International Food and Agribusiness Management Review, 22*(1), 31–46. doi:10.22434/IFAMR2018.0089

Cruz, R., Lucas, M. R., & Carvalho, M. L. S. (2016). Comercialização do arroz no Distrito de Bobonaro, Sub-distrito de Maliana – Análise dos factores que a afectam e da Margem de Comercialização. *Proceedings do VIII congresso da APDEA, ESADR 2016*.

Cruz, R. S. W. (2014). *Análise da margem e dos factores que afectam a comercialização do arroz no Distrito de Bobonaro, sub-distrito de Maliana* [Unpublished master dissertation]. UNTL, Universidade Nacional de Timor Lorosa'e, Dili, Timor Leste.

Cuevas, R. P., Pede, V. O., McKinley, J., Velarde, O., & Demont, M. (2016). Rice Grain Quality and Consumer Preferences: A Case Study of Two Rural Towns in the Philippines. *PLoS One*, *11*(3), e0150345. doi:10.1371/journal.pone.0150345 PMID:26982587

Custodio, M. C., Demont, M., Laborte, A., & Ynion, J. (2016). Improving food security in Asia through consumer-focused rice breeding. *Global Food Security*, *9*, 19–28. doi:10.1016/j.gfs.2016.05.005

Dania, W. A. P., Xing, K., & Amer, Y. (2018). Collaboration behavioural factors for sustainable agri-food supply chains: A systematic review. *Journal of Cleaner Production*, *186*, 851–864. doi:10.1016/j.jclepro.2018.03.148

De Jonge, J., Van Trijp, H., Jan Renes, R., & Frewer, L. (2007). Understanding consumer confidence in the safety of food: Its two-dimensional structure and determinants. *Risk Analysis: An International Journal*, *27*(3), 729–740. doi:10.1111/j.1539-6924.2007.00917.x PMID:17640219

De Steur, H., Demont, M., Gellynck, X., & Stein, A. J. (2017). The social and economic impact of biofortification through genetic modification. *Current Opinion in Biotechnology*, *44*, 161–168. doi:10.1016/j.copbio.2017.01.012 PMID:28231514

De Steur, H., Gellynck, X., Feng, S., Rutsaert, P., & Verbeke, W. (2012). Determinants of willingness-to-pay for GM rice with health benefits in a high-risk region: Evidence from experimental auctions for folate biofortified rice in China. *Food Quality and Preference*, *25*(2), 87–94. doi:10.1016/j.foodqual.2012.02.001

Deus, C. C. (2019). *Os Sistemas de agricultura familiar de Timor-Leste:Uma abordagem com foco no bem-estar dos produtores* [Unpublished doctoral dissertation]. Universidade de Évora, Évora, Portugal.

Devkota, K. P., Pasuquin, E., Elmido-Mabilangan, A., Dikitanan, R., Singleton, G. R., Stuart, A. M., Vithoonjit, D., Vidiyangkura, L., Pustika, A. B., Afriani, R., Listyowati, C. L., Keerthisena, R. S. K., Thikieu, N., Malabayabas, A. J., Hu, R., Pan, J., & Beebout, S. E. J. (2019). Economic and environmental indicators of sustainable rice cultivation: A comparison across intensive irrigated rice cropping systems in six Asian countries. *Ecological Indicators*, *105*, 199–214. doi:10.1016/j.ecolind.2019.05.029

Dhungana, B. R., Nuthall, P. L., & Nartea, G. V. (2004). Measuring the economic inefficiency of Nepalese rice farms using data envelopment analysis. *The Australian Journal of Agricultural and Resource Economics*, *48*(2), 347–369. doi:10.1111/j.1467-8489.2004.00243.x

Diaz, R., & Rosenberg, R. (2008). Spreading dead zones and consequences for marine eco-systems. *Science*, *321*(5891), 926–929. doi:10.1126cience.1156401 PMID:18703733

Donovan, J. A. (2011). *Value chain development for addressing rural poverty: asset building by smallholder coffee producers and cooperatives in Nicaragua* [Unpublished doctoral dissertation]. SOAS (School of Oriental and African Studies).

Dzudzor, M. I. (2013). *Analysis of Rice Production and Opportunities in Ghana* [Unpublished master dissertation]. Faculté d'Ingénierie Biologique, Agronomique et Environnementale at Université catholique de Louvain, Agosto.

Ellis, E., Kwofie, E. M., & Ngadi, M. (2019). Value Beyond Price: End User Value Chain Analysis. *International Journal on Food System Dynamics*, *10*(4), 347–360.

Ellram, M. (1991). Supply-Chain Management: The Industrial Organisation Perspective. *International Journal of Physical Distribution & Logistics Management, 21*(1), 13–22. doi:10.1108/09600039110137082

FAO. (2018). Food loss analysis: Causes and solutions - Case study on the rice value chain in the Republic of India. FAO.

Fasasi, A. R. (2012). *A study on the contribution of supply chain management to Ghana's rice industry* [Unpublished master dissertation]. Universiti Teknikal Malaysia Melaka, Faculty of Technology Management and Technopreneurship.

Fernández, M. C. (2013). *La confianza como instrumento de análises em la cadena de valor alimentaria* [Unpublished doctoral dissertation]. Escola Técnica Superior de Ingenieros Agrónomos, Universidad Politécnica de Madrid.

Fischer, C., Hartmann, M., Reynolds, N., Leat, P., Revoredo-Giha, C., Henchion, M., Albisu, L. M., & Gracia, A. (2009). Factors influencing contractual choice and sustainable relationships in European agri-food supply chains. *European Review of Agriculture Economics, 36*(4), 541–569. doi:10.1093/erae/jbp041

Fleury, P. F., Wanke, P., & Figueiredo, K. (2009). Logística empresarial. coleção copped de administração. *Centro de estudos em logística São Paulo: Atlas*, 235.

Flynn, A., & Bailey, K. (2014). Sustainable food supply chains: the dynamics for change. In T. Marsden & A. Morley (Eds.), *Sustainable food systems: building a new paradigm*. Routledge.

Gereffi, G., Humphrey, J., Kaplinsky, R., & Sturgeon, T. (2001). Value chains and development. *IDS Bulletin, 32*(3), 3. doi:10.1111/j.1759-5436.2001.mp32003001.x

Giunipero, L. C., Hooker, R. E., & Denslow, D. (2012). Purchasing and supply management sustainability: Drivers and barriers. *Journal of Purchasing and Supply Management, 18*(4), 258–269. doi:10.1016/j.pursup.2012.06.003

Govindan, K. (2018). Sustainable consumption and production in the food supply chain: A conceptual framework. *International Journal of Production Economics, 195*, 419–431. doi:10.1016/j.ijpe.2017.03.003

Gunden, C., & Thomas, T. (2012). Assessing consumer attitudes towards fresh fruit and vegetable attributes. *Journal of Food Agriculture and Environment, 10*(2), 85–88.

Guo, J., Liu, X., Zhang, Y., Shen, J., Han, W., Zhang, W., Christie, P., Goulding, K., Vitousek, P., & Zhang, F. (2010). Significant acidification in major Chinese croplands. *Science, 327*(5968), 1008–1010. doi:10.1126cience.1182570 PMID:20150447

Han, X., Hu, C., Chen, Y., Qiao, Y., Liu, D., Fan, J., Li, S., & Zhang, Z. (2020). Crop yield stability and sustainability in a rice-wheat cropping system based on 34-year field experiment. *European Journal of Agronomy, 113*, 125965. doi:10.1016/j.eja.2019.125965

Hart, S. L. (1997). Beyond Greening: Strategies for a Sustainable World. *Harvard Business Review, 75*(1), 66–76.

Harun, S. N., Hanafiah, M. M., & Aziz, N. I. H. A. (2021). An LCA-Based Environmental Performance of Rice Production for Developing a Sustainable Agri-Food System in Malaysia. *Environmental Management, 67*(1), 146–161. doi:10.100700267-020-01365-7 PMID:33001258

Henriksen, L. F., Riisgaard, L., Ponte, S., Hartwich, F., & Kormawa, P. (2010). *Agro-food value chain interventions in Asia: a review and analysis of case studies.* Working Paper. United Nations Industrial Development Organization (UNIDO).

Hinrichs, C. C. (2003). The practice and politics of food system localization. *Journal of Rural Studies, 19*(1), 33–45. doi:10.1016/S0743-0167(02)00040-2

Hobbs, J., Cooney, A., & Fulton, M. (2000). *Value Chains in the Agri-Food Sector. What are they, How do they work and Are they for me?* Department of Agricultural Economics, University of Saskatchewan.

Ilu, I. Y. (2015). *Value chain analysis of rice (Oryza sativa) in Kano river irrigation project (KRIP) Kano State, Nigeria* [Unpublished doctoral dissertation]. Ahmadu Bello University, Zaria, Nigeria.

Jalilov, S.-M., Mainuddin, M., Maniruzzaman, M., Mahbubul Alam, M., Towfiqul Islam, M., & Jahangir Kabir, M. (2019). Efficiency in the Rice Farming: Evidence from Northwest Bangladesh. *Agriculture, 9*(11), 245. doi:10.3390/agriculture9110245

Jiménez-Guerrero, J. F., Gázquez-Abad, J. C., Mondéjar-Jiménez, J. A., & Huertas-García, R. (2012). *Consumer preferences for olive-oil attributes: a review of the empirical literature using a conjoint approach. Olive oil-constituents, quality, health properties and bioconversions.* InTech Europe.

Jongen, W. M. F. (2000). Food supply chains: from productivity toward quality. *Fruit and Vegetable Quality: An Integrated View*, 3-20.

Juliano, B. O., & Villareal, C. P. (1993). *Grain quality evaluation of world rices.* Int. Rice Res. Inst.

Kaplisky, R., & Morris, M. (2002). *A handbook for value chain research.* Institute of Development Studies.

Kisanga, P. S. (2015). *Perfomance of rice value chain in Kahama district, Tanzania* [Unpublished master dissertation]. Sokoine University of Agriculture. Morogoro, Tanzania.

Kopnina, H. (2017). Sustainability: New strategic thinking for business. *Environment, Development and Sustainability, 19*(1), 27–43. doi:10.100710668-015-9723-1

La Gra, J., Kitinoja, L. & Alpízar, K. (2016). *Metodología de evaluación de cadenas agroalimentarias para la identificación de problemas y proyectos: un primer paso para la disminución de pérdidas de alimentos.* IICA.

Lambert, D. M., & Cooper, M. C. (2000). Issues in supply chain management. *Industrial Marketing Management, 29*(1), 65–83. doi:10.1016/S0019-8501(99)00113-3

Lamine, C. & Chiffoleau, Y. (2012). Reconnecter agriculture e alimentation dans lês territoires: dynamiques et défis. *Pour, 3*(215-216), 85-92.

Linn, T., & Maenhout, B. (2019). Measuring the efficiency of rice production in Myanmar using data envelopment analysis. *Asian Journal of Agriculture and Development, 16*, 1-24.

Loly, J. F. R. B. (2014). *Análise de influência da utilização dos fatores da produção no aumento da produção de arroz no subdistrito de Maliana, Distrito de Bobonaro em Timor-Leste* [Unpublished master dissertation]. Universidade de Évora, Évora.

Lopes, J. M. C., & Gama, M. (2013). Comércio exterior competitive (4th ed.). São Paulo: Aduaneiras.

Makosa, D. (2015). Constraints and opportunities to upgrading Uganda's rice markets: A value chain approach. *Journal of Development and Agricultural Economics*, *7*(12), 386–399. doi:10.5897/JDAE2015.0643

Maris, C.-M. N. (2017). *Analysis of value chain in rice production systems in Ebonyi State, Nigeria* [Unpublished doctoral dissertation]. Federal University of Technology, Owerri, Nigeria.

Marsden, T., Banks, J., & Bristow, G. (2000). Food supply chain approaches: Exploring their role in rural development. *Sociologia Ruralis*, *40*(4), 424–438. doi:10.1111/1467-9523.00158

Marsden, T., & Morley, A. (2014). Current food questions and their scholarly challenges: creating and framing a sustainable food paradigma. In T. Marsden & A. Morley (Eds.), Sustainable Food Systems: Building a New Paradigm. Routledge. doi:10.4324/9780203083499-9

Matopoulos, A., Vlachopoulou, M., Manthou, V., & Manos, B. (2007). A conceptual framework for supply chain collaboration: Empirical evidence from the agri-food industry. *Supply Chain Management*, *12*(3), 177–186. doi:10.1108/13598540710742491

Mentzer, J. T., DeWitt, W., Keebler, J. S., Min, S., Nix, N. W., Smith, C. D., & Zacharia, Z. G. (2001). Defining Supply Chain Management. *Journal of Business Logistics*, *22*(2), 1–25. doi:10.1002/j.2158-1592.2001.tb00001.x

Mondelaers, K., Verbeke, W., & Van Huylenbroeck, G. (2009). Importance of health and environment as quality traits in the buying decision of organic products. *British Food Journal*, *111*(10), 1120–1139. doi:10.1108/00070700910992952

Mowat, A., & Collins, R. (2000). Consumer behaviour and fruit quality: Supply chain management in an emerging industry. *Supply Chain Management*, *5*(1), 45–54. doi:10.1108/13598540010312963

Muthayya, S., Sugimoto, J. D., Montgomery, S., & Maberly, G. F. (2014). An overview of global rice production, supply, trade, and consumption. *Annals of the New York Academy of Sciences*, *1324*(1), 7–14. doi:10.1111/nyas.12540 PMID:25224455

My, N. H. D., Demont, M., Loo, E. J. V., Guia, A., Rutsaert, P., Tuan, T. H., & Verbeke, W. (2018). What is the value of sustainably-produced rice? Consumer evidence from experimental auctions in Vietnam. *Food Policy*, *79*, 283–296. doi:10.1016/j.foodpol.2018.08.004

Ndirangu, S. N., Wilson, A., & Oyange, W. A. (2019). Analysis of Millers in Kenya's Rice Value Chain. *Journal of Agriculture and Veterinary Science*, Volume 12 (1). *Ser. III*, (January), 38–47.

Nkuba, J., Ndunguru, A., Madulu, R., Lwezaura, D., Kajiru, G., Babu, A., Chalamila, B., & Ley, G. (2016). Rice value chain analysis in Tanzania:Identification of constraints, opportunities and upgrading stategies. *African Crop Science Journal*, *24*(1), 73–87.

Notarnicola, B., Sala, S., Anton, A., McLaren, S. J., Saouter, E., & Sonesson, U. (2017). The role of life cycle assessment in supporting sustainable agri-food systems: A review of the challenges. *Journal of Cleaner Production, 140*(2), 399–409.

Obura, J., Ombok, B. O., & Omuga, G. (2017). Analysis of Rice Supply Chain in Kenya. *International Journal of Managerial Studies and Research, 5*(8), 12-17.

Odokonyero, T., Barungi, M. & Mbowa, S. (2016). *Supply of Improved Rice Seed in Eastern Uganda: The Gap and Required Investment.* Series Policy Briefs 71, Economic Policy Research Centre (EPRC).

OECD/FAO-Organization for Economic Cooperation and Development/Food and Agriculture Organization. (2019). *Agricultural Outlook 2019-2028.* OECD Publishing.

Ojo, T. O., Ogundeji, A. A., Babu, S. C., & Alimi, T. (2019). *Estimating financing gaps in rice production in southwestern Nigeria.* IFPRI Discussion Paper 1818. Washington, DC: International Food Policy Research Institute (IFPRI).

Okpiaifo, G., Durand-Morat, A., West, G. H., Nalley, L. L., Nayga, R. M. Jr, & Wailes, E. J. (2020). Consumers' preferences for sustainable rice practices in Nigeria. *Global Food Security, 24*, 100345. doi:10.1016/j.gfs.2019.100345

Omoare, O. A., & Oyeleke, O. W. (2017). Assessment of factors affecting rice (oryza spp.) value chain (RVC) in Ogun and Niger states, Nigeria. *Global Journal of Agricultural Research, 5*(4), 43–59.

Opoku, R. A., & Akorli, P. A. K. (2009). The preference gap: Ghanaian consumers' attitudes toward local and imported products. *African Journal of Business Management, 3*(8), 350–357.

Othman, S. N., Othman, Z., Yaacob, N. A., & Abhamid, K. (2016). Sustainable rice production and its impact on the rice value chain: A case study of rural paddy farm in Kedah. *Proceedings of the International Conference on Applied Science and Tecnhology 2016 (ICAST'16). Project Sustainable agriculture and value chain.* 10.1063/1.4960925

Panpluem, N., Mustafa, A., Huang, X., Wang, S., & Yin, C. (2019). Measuring the Technical Efficiency of Certified Organic Rice Producing Farms in Yasothon Province: Northeast Thailand. *Sustainability, 11*(24), 6974. doi:10.3390u11246974

Porter, M. E. (1985). *Competitive Advantage. Creating and Sustaining Superior Performance.* Free Press.

Prowse, M. (2013). A history of tobacco production and marketing in Malawi, 1890 – 2010. *Journal of Eastern African Studies: the Journal of the British Institute in Eastern Africa, 7*(4), 691–712. doi:10.1080/17531055.2013.805077

Puelles, J. A., Gómez, M., & Puelles, M. (2011). *Marcas de distribuidor, concepto, evolución, protagonistas y adaptación a los ciclos económicos.* Ediciones Pirámide.

Rampersad, G. C., Hordacre, A.-L., & Spoehr, J. (2020). Driving innovation in supply chains: An examination of advanced manufacturing and food industries. *Journal of Business and Industrial Marketing, 35*(5), 835–847. doi:10.1108/JBIM-03-2019-0101

Ray, D., Mueller, N., West, P., & Foley, J. (2013). Yield trends are insuficiente to double global crop production by 2050. *PLoS One*, *8*(6), e66428. doi:10.1371/journal.pone.0066428 PMID:23840465

Razafinjoelina, S. M. (2017). *Rice Value Chain Development and Policy Sequencing in Southern Africa* [Unpublished bachelor dissertation]. University of Arkansas, Fayetteville.

Reardon, T., Timmer, C. P., & Minten, B. (2012). Supermarket revolution in Asia and emerging development strategies to include small farmers. *Proceedings of the National Academy of Sciences of the United States of America*, *109*(31), 12332–12337. doi:10.1073/pnas.1003160108 PMID:21135250

Rerkasem, B. (2017). The Rice Value Chain: a Case Study of Thai Rice. *Journal of Social Sciences and Humanities*, *4*(1).

Ruben, R., Van Boekel, M., Van Tilburg, A., & Trienekens, J. (2007). Governance for quality in tropical food chains. Wageningen: Wageningen Acadamic Publishers.

Rueda, X., Garrett, R. D., & Lambin, E. F. (2017). Corporate investments in supply chain sustainability: Selecting instruments in the agri-food industry. *Journal of Cleaner Production*, *142*, 2480–2492. doi:10.1016/j.jclepro.2016.11.026

Schmutz, U., Kneafsey, M., Kay, C. S., Doernberg, A., & Zasada, I. (2018). Sustainability impact assessments of different urban short food supply chains: Examples from London, UK. *Renewable Agriculture and Food Systems*, *33*(6), 518–529. doi:10.1017/S1742170517000564

Sharma, V., Giri, S., & Rai, S. S. (2013, March). Supply chain management of rice in India: A rice processing company's perspective. *International Journal of Managing Value and Supply Chains*, *4*(1), 25–36. Advance online publication. doi:10.5121/ijmvsc.2013.4103

Sharma, Y. K., Mangla, S. K., Patil, P. P., & Uniyal, S. (2018). Analyzing sustainable food supply chain management challenges in India. In R. Mangey & D. J. Paulo (Eds.), *Soft computing techniques and applications in mechanical engineering* (pp. 162–180). IGI Global. doi:10.4018/978-1-5225-3035-0.ch008

Silva, J. M. F. (2011). *Desenvolvimento dos recursos humanos no sector de produção de arroz em distrito de Baucau, Timor-Leste* [Unpublished master dissertation]. Universidade de Évora, Évora.

Smith, P., Martino, D., Cai, Z., Gwary, D., Janzen, H., Kumar, P., McCarl, B., Ogle, S., O'Mara, F., Rice, C., Scholes, B., & Sirotenko, O. (2007). Agriculture. In Climate Change 2007: Mitigation. Contribution of Working Group III to the Fourth Assessment Report of the Intergovernmental Panel on Climate Change. Cambridge University Press.

Soullier, G. (2017). *Modernization of domestic food chains in developing countries: What effects on small-scale farmers? The rice value chain in Senegal* [Unpublished doctoral dissertation]. Universidade de Montpellier.

SRP - Sustainable Rice Platform. (2020). *The SRP Assurance Scheme (Version 1.3)*. Sustainable Rice Platform.

Stone, J., & Rahimifard, S. (2018). Resilience in agri-food supply chains: A critical analysis of the literature and synthesis of a novel framework. *Supply Chain Management*, *23*(3), 207–238. doi:10.1108/SCM-06-2017-0201

Stryker, J. D. (2010). *Developing competitive rice value chains*. Second Africa Rice Congress, Bamako, Mali.

Suwannaporn, P., & Linnemann, A. (2008). Consumer Preferences and Buying Criteria in Rice: A Study to Identify Market Strategy for Thailand Jasmine Rice Export. *Journal of Food Products Marketing*, *14*(4), 33–53. doi:10.1080/10454440801986348

Takele, A. (2010). *Analysis of rice profitability and marketing chain: The case of Fogera Woreda, South Gondar zone, Amhara National Regional State, Ethiopia* [Unpublished master dissertation]. Haramaya University.

Taylor, D. H. (2005). Value chain analysis: An approach to supply chain improvement in agrifood chains. *International Journal of Physical Distribution & Logistics Management*, *35*(10), 744–761. doi:10.1108/09600030510634599

Tinsley, R. (2012). *Rice Value Chain Analysis – Sokoto State Nigeria* (NIG 244). Winrock International Nigeria Farmer-to-Farmer Program, Consultant Volunteer Report.

Tipi, T., Yildiz, N., Nargeleçekenler, M., & Çetin, B. (2009). Measuring the technical efficiency and determinants of efficiency of rice (Oryza sativa) farms in Marmara region, Turkey. *New Zealand Journal of Crop and Horticultural Science*, *37*(2), 121–129. doi:10.1080/01140670909510257

Trevor, R. W. & Lewis, I. (2015). *The Rice Value Chain in Tanzania. A report from the Southern Highlands Food Systems Programme*. FAO (Food and Agriculture Organization of the United Nations).

Trienekens, J. H. (2011). Agricultural value chains in developing countries: A framework for analysis. *The International Food and Agribusiness Management Review*, *14*(2), 51–81.

US-EPA - United States/Environmental Protection Agency. (2006). *Global Anthropogenic Non-CO2 Greenhouse Gas Emissions:1990-2020*. U.S. Environmental Protection Agency, Office of Atmospheric Programs, Climate Change Division.

USDA/ERS-United States Department of Agriculture/Economic Research Service. (2019). *International Baseline Data:2019. International Long-Term Projections to 2028*. Author.

Van der Laan, H. L., H Dijkstra, T., & Tilburg, A. V. (1999). Agricultural marketing in tropical Africa: obstacles to systematic study. *Agricultural marketing in tropical Africa. Contributions from the Netherlands*, 1-17.

Verbeke, W., Frewer, L. J., Scholderer, J., & De Brabander, H. F. (2007). Why consumers behave as they do with respect to food safety and risk information. *Analytica Chimica Acta*, *586*(1-2), 2–7. doi:10.1016/j.aca.2006.07.065 PMID:17386689

Wang, E., & Gao, Z. (2017). Chinese Consumer Quality Perception and Preference of Traditional Sustainable Rice Produced by the Integrated Rice–Fish System. *Sustainability*, *9*(12), 2282. doi:10.3390u9122282

Weerabahu, S., & Nanayakkara, J. (2014). A Strategic Framework for Managing Rice Supply Chains. *Conference: Wayamba International Conference, WinC 2014*.

Yadav, P. L., Han, S. H., & Rho, J. J. (2016). Impact of environmental performance on firm value for sustainable investment: Evidence from large US firms. *Business Strategy and the Environment*, *25*(6), 402–420. doi:10.1002/bse.1883

Yong, H. N. (2008). *Supply Chain Model for Rice in Malaysia – Basics and Challenges*. ECER Regional Conference 2008 (ERC08), Kota Bharu, Kelantan, Malaysia.

ADDITIONAL READING

Costa, M. V. (2018). *Canais de Distribuição de Arroz em Díli, Timor-Leste* [Unpublished master dissertation]. UNTL, Universidade Nacional de Timor Lorosa'e, Díli, Timor Leste.

Graham-Acquaah, S., Manful, J. T., Ndindeng, S. A., & Tchatcha, D. A. (2015). Effects of soaking and steaming regimes on the quality of artisanal parboiled rice. *Journal of Food Processing and Preservation*, *39*(6), 2286–2296. doi:10.1111/jfpp.12474

Ito, S., Peterson, E. W. F., & Grant, W. R. (1989). Rice in Asia: Is it becoming an inferior good? *American Journal of Agricultural Economics*, *71*(1), 32–42. doi:10.2307/1241772

Jarma, S. E., & Seo, H. S. (2017). Effects of the type of reference scale on descriptive sensory analysis of cooked rice: Universal aromatic scale versus rice aromatic scale. *Journal of Sensory Studies*, *32*(5), e12295. doi:10.1111/joss.12295

Parwez, S. (2016). A conceptual model for integration of Indian food supply chains. *Global Business Review*, *17*(4), 834–850. doi:10.1177/0972150916645681

Ruekkasaem, L., & Sasananan, M. (2017). The Factors Affecting Decisions to Purchase Rice and Consumer Perception of Organic Rice in Bangkok, Thailand. *International Journal of Advances in Agricultural & Environmental Engineering*, *4*(1), 6–11.

Silva, J., Henriques, P., Jorge, F., & Narciso, V. (2012). *Produção, consumo e importância da cultura do arroz: da colonização portuguesa a Timor-Leste independente*. Silva.

KEY TERMS AND DEFINITIONS

Agrifood Chain: A set of activities developed in different stages, in a direct and indirect manner, with the purpose of satisfying the consumer with food and agricultural products, through various productive, financial, informative, and analytical functions.

Economic Efficiency: Results from the optimization of resource-use to best serve an economy.

Globalisation: Is the process by which the world is becoming increasingly interconnected as a result of massively increased trade and cultural exchange.

Sustainability: It focuses on meeting the needs of the present without compromising the ability of future generations to meet their needs. It is composed of three pillars: economic, environmental, and social—also known informally as profits, planet, and people.

Sustainable Development: The development that meets the needs of the present without compromising the ability of future generations to meet their own needs.

Compilation of References

Abbey, P., Tomlinson, P. R., & Branston, J. R. (2016). Perceptions of governance and social capital in Ghana's cocoa industry. *Journal of Rural Studies*, *44*, 153–163. doi:10.1016/j.jrurstud.2016.01.015

Abd El-Gawad, H. G., Mukherjee, S., Farag, R., Abd Elbar, O. H., Hikal, M., Abou El-Yazied, A., Abd Elhady, S. A., Helal, N., ElKelish, A., El Nahhas, N., Azab, E., Ismail, I. A., Mbarki, S., & Ibrahim, M. F. M. (2021). Exogenous γ-aminobutyric acid (GABA)-induced signaling events and field performance associated with mitigation of drought stress in *Phaseolus vulgaris* L. *Plant Signaling & Behavior*, *16*(2), 1853384. doi:10.1080/15592324.2020.1853384 PMID:33356834

Abdelaal, K., Attia, K. A., Niedbała, G., Wojciechowski, T., Hafez, Y., Alamery, S., Alateeq, T. K., & Arafa, S. A. (2021). Mitigation of Drought Damages by Exogenous Chitosan and Yeast Extract with Modulating the Photosynthetic Pigments, Antioxidant Defense System and Improving the Productivity of Garlic Plants. *Horticulturae*, *7*(11), 510. doi:10.3390/horticulturae7110510

AbdElgawad, H., Zinta, G., Beemster, G. T. S., Janssens, I. A., & Asard, H. (2016). Future Climate CO_2 Levels Mitigate Stress Impact on Plants: Increased Defense or Decreased Challenge? *Frontiers in Plant Science*, *7*, 556. doi:10.3389/fpls.2016.00556 PMID:27200030

Abdollahi, M., Ranjbar, A., Shadnia, S., Shekoufeh Nikfar, S., & Rezaie, A. (2004). Pesticides and oxidative stress: A review. *Medical Science Monitor*, *10*(6), 141–147.

Abdul-Muhmin, A. G. (2007). Explaining consumers' willingness to be environmentally friendly. *International Journal of Consumer Studies*, *31*(3), 237–247. doi:10.1111/j.1470-6431.2006.00528.x

Abebe, M. (2016). *The Contributions and Challenges of Rice Value Chain Development on the Livelihood of Small Holder Rice Farmers (In the case of Ethiopia Driving Growth Entrepreneurship and Trade (EDGET) project in Fogera and Libokmkem District)* [Unpublished master dissertation]. Addis Ababa University.

Aboah, J., Wilson, M. M. J., Bicknell, K., & Rich, K. M. (2021). Ex-ante impact of on-farm diversification and forward integration on agricultural value chain resilience: A system dynamics approach. *Agricultural Systems*, *189*, 189. doi:10.1016/j.agsy.2020.103043

About Herbs, Botanicals & Other Products. (2021). https://www.mskcc.org/cancer-care/treatments/symptom-management/integrative-medicine/herbs

Acquier, A., Gand, S., & Szpirglas, M. (2008). From Stakeholder to Stakeholder Management in Crisis Episodes: A Case Study in a Public Transportation Company. *Journal of Contingencies and Crisis Management*, *16*(2), 101–114. doi:10.1111/j.1468-5973.2008.00538.x

Actionaid. (2011). *Say no to soil carbon markets: Six reasons why soil carbon markets won't work for smallholders*. Author.

ADB – African Development Bank (2018). *The Gambia rice value chain transformation programme (RVCP).* RDGW/AHAI/PGCLDEPARTMENTS, November.

Adeniyi, O. F., Lesi, O. A., Olatona, F. A., Esezobor, C. I., & Ikobah, J. M. (2017). Irritable bowel syndrome in adolescents in Lagos. *The Pan African Medical Journal, 28*, 1–7. doi:10.11604/pamj.2017.28.93.11512

Adolwa, I. S., Schwarze, S., & Buerkert, A. (2019). Impacts of integrated soil fertility management on yield and household income: The case of Tamale (Ghana) and Kakamega (Kenya). *Ecological Economics, 161*, 186–192. doi:10.1016/j.ecolecon.2019.03.023

Adu-Acheampong, R., Jiggins, J., Quartey, E. T., Karikari, N. M., Jonfia-Essien, W., Quarshie, E., Osei-Fosu, P., Amuzu, M., Afari-Mintah, C., Ofori-Frimpong, K., Owusu-Manu, M., Eku, N., Sakyi-Dawson, O., Quarmine, W., & Acquah, F. O. (2017). An innovation platform for institutional change in Ghana's cocoa sector. *Cahiers Agricultures, 26*(3), 35002. doi:10.1051/cagri/2017016

Agarwal, M., Atri, P., & Kundu, S. (2017). Foreign direct investment and poverty reduction: India in regional context. *South Asia Economic Journal, 18*(2), 135–157. doi:10.1177/1391561417713129

Aggarwal, P. K. (2003). Impact of climate change on Indian agriculture. *Journal of Plant Biology, 30*(2), 189–198.

Aggarwal, S., & Srivastava, M. K. (2016). Towards a grounded view of collaboration in Indian agri-food supply chains: A qualitative investigation. *British Food Journal, 118*(5). Advance online publication. doi:10.1108/BFJ-08-2015-0274

Agrawal, A. (2008). The role of local Institutions in Adaptation to climate change. Natural Resources and Environment, University of Michigan. doi:10.1596/28274

Agri-Food Data Portal. (2021). *DG AGRI Rice Production.* https://agridata.ec.europa.eu/extensions/DashboardRice/RiceProduction.html

Ahmed, K. H. H. (2017). *A Decision Support Framework for Sustainable Supply Chain Management* [Unpublished doctoral dissertation]. Loughborough University, UK.

Ahmed, M., & Suphachalasai, S. (2014). *Assessing the Costs of Climate Change and Adaptation in South Asia.* Asian Development Bank and UKaid. Retrieved from https://www.preventionweb.net/files/38999_assessingcostsclimatechangeandadapt.pdf

Ahsan, D., & Roth, E. (2010). Farmers' Perceived Risks and Risk Management Strategies in an Emerging Mussel Aquaculture Industry in Denmark. *Marine Resource Economics, 25*(3), 309–323. doi:10.5950/0738-1360-25.3.309

Ainsworth, E. A., Lemonnier, P., & Wedow, J. M. (2019). The influence of rising tropospheric carbon dioxide and ozone on plant productivity. *Plant Biology, 22*(S1, Suppl. 1), 5–11. doi:10.1111/plb.12973 PMID:30734441

AJAP. (2017). Manual Boas Práticas para Culturas Emergentes. A Cultura da Amêndoa. Associação dos Jovens Agricultores de Portugal, Lisboa.

AJAP. (2017a). Manual Boas Práticas para Culturas Emergentes. A Cultura do Pistácio. Associação dos Jovens Agricultores de Portugal, Lisboa.

Akenji, L., & Bengtsson, M. (2014). Making sustainable consumption and production the core of sustainable development goals. *Sustainability, 6*(2), 513–529. doi:10.3390u6020513

Akhtar, P., Tse, Y. K., Khan, Z., & Rao-Nicholson, R. (2016). Data-driven and adaptive leadership contributing to sustainability: Global agri-food supply chains connected with emerging markets. *International Journal of Production Economics, 181*, 392–401. doi:10.1016/j.ijpe.2015.11.013

Akter, N., & Rafiqul Islam, M. (2017). Heat stress effects and management in wheat. A review. *Agronomy for Sustainable Development*, *37*(5), 37. doi:10.100713593-017-0443-9

Alayafi, A. A. M. (2020). Exogenous ascorbic acid induces systemic heat stress tolerance in tomato seedlings: Transcriptional regulation mechanism. *Environmental Science and Pollution Research International*, *27*(16), 19186–19199. doi:10.100711356-019-06195-7 PMID:31448379

Alcayaga, A., Hansen, E. G., & Wiener, M. (2019). Towards a Framework of Smart-Circular Systems: An Integrative Literature Review. *Journal of Cleaner Production*, *221*, 622–634. doi:10.1016/j.jclepro.2019.02.085

Alexandratos, N., & Bruinsma, J. (2012, June). *World Agriculture towards 2030/2050: The 2012 revision*. Agriculture Development Economics Division, FAO.

Alhaddi, H. (2015). Triple Bottom Line and Sustainability: A Literature Review. *Business and Management Studies*, *1*(2), 6–10. doi:10.11114/bms.v1i2.752

Ali, S., Liu, Y., Ishaq, M., Shah, T., Ilyas, A., & Din, I. U. (2017). Climate Change and Its Impact on the Yield of Major Food Crops: Evidence from Pakistan. *Foods*, *6*(6), 39. Doi:10.3390/foods6060039

Ali, M., Nishat, M., & Anwar, T. (2010). Do foreign inflows benefit Pakistan poor? *Pakistan Development Review*, *48*(4), 715–738. Doi.org/10.30541/v48i4IIpp.715-738

Ali, Z., Basra, S. M. A., Munir, H., Mahmood, A., & Yousaf, S. (2011). Mitigation of Drought Stress in Maize by Natural and Synthetic Growth Promoters. *Journal of Agriculture & Social Sciences*, *7*(2), 8.

Alkhathami, A. M., Alzahrani, A. A., Alzhrani, M. A., Alsuwat, O. B., & Mahfouz, M. E. M. (2017). Risk Factors for Gastroesophageal Reflux Disease in Saudi Arabia. *Gastroenterology Research*, *10*(5), 294–300. doi:10.14740/gr906w

Alkimim, A., Sparovek, G., & Clarke, K. C. (2015). Converting Brazil's pastures to cropland: Na alternative way to meet sugarcane demand and to spare forestlands. *Applied Geography (Sevenoaks, England)*, *62*, 75–84. doi:10.1016/j.apgeog.2015.04.008

Alliance for Internet of Things Innovation. (2015). *Smart Farming and Food safety Internet of Things Applications-Challenges for large Scale Implementations*. Author.

Almeida, V. (2011). *A importância e as alterações recentes na agricultura da região centro*. CCDRC.

Alonso Villa, P., & Juste Carrion, J. J. (2018). The automotive sector in Castile and Leon. 50 years of economic growth and productivity: 1961-2011. *Revista De Estudios Regionales*, *113*, 101–136.

Alonso, J., Ansón, J., Espiau, M., & Company, R. S. (2005). Determination of endormancy break in almond flower buds by a correlation model using the average temperature of different day intervals and its application to the estimation of chill and heat requirements and blooming date. *Journal of American Horticultural Science*, *130*(3), 308–318. doi:10.21273/JASHS.130.3.308

Alvarez, M.E.P., Barcena, M.M., & Gonzalez, F.A. (2016). A review of sustainable machining engineering: optimization process through triple bottom line. *Manufacturing Science and Engineering, Transactions/ASME*, *138*(10), 16.

Álvarez-Álvarez, I., Martínez-González, M. Á., Sánchez-Tainta, A., Corella, D., Díaz-López, A., Fitó, M., Vioque, J., Romaguera, D., Martínez, J. A., Wärnberg, J., López-Miranda, J., Estruch, R., Bueno-Cavanillas, A., Arós, F., Tur, J. A., Tinahones, F. J., Serra-Majem, L., Martín, V., Lapetra, J., ... Toledo, E. (2019). Adherence to an Energy-restricted Mediterranean Diet Score and Prevalence of Cardiovascular Risk Factors in the PREDIMED-Plus: A Cross-sectional Study. *Revista Española de Cardiología (English Ed.)*, *72*(11), 925–934. Advance online publication. doi:10.1016/j.rec.2018.08.010 PMID:30287240

Ambavaram, M. M. R., Basu, S., Krishnan, A., Ramegowda, V., Batlang, U., Rahman, L., Baisakh, N., & Pereira, A. (2014). Coordinated regulation of photosynthesis in rice increases yield and tolerance to environmental stress. *Nature Communications*, *5*(1), 5302. doi:10.1038/ncomms6302 PMID:25358745

Amizadeh, M., Safari-Kamalabadi, M., Askari-Saryazdi, G., Amizadeh, M., & Reihani-Kermani, H. (2017). Pesticide Exposure and Head and Neck Cancers: A Case-Control Study in an Agricultural Region. *Iranian Journal of Otorhinolaryngology*, *29*(94), 275–285.

Amjath-Babu, T. S., Krupnik, T. J., Thilsted, S. H., & McDonald, A. J. (2020). Key indicators for monitoring food system disruptions caused by the COVID-19 pandemic: Insights from Bangladesh towards effective response. *Food Security*, 1–8. doi:10.1007/s12571-020-01083-2

Ampadu-Ameyaw, R., Omari, R. & Owusu, G. E. (2017). Development and Analysis of the Rice Value Chain for the Hohoe and Jasikan Districts of the Volta Region, Ghana. *FARA research Results, 1*(1), 48.

Andrieu, N., Howland, F., Alba, I., Coq, J., Garcia, A., Baron, D., Trimin, C., Loboguerro, A., & Chia, E. (2019). Co-designing climate smart farming systems with local stakeholders: A methodological framework for achieving large-scale change. *Frontiers in Sustainable Food Systems*, *3*, 1–19. doi:10.3389/fsufs.2019.00037

Angelidi, A. M., Kokkinos, A., Katechaki, E., Ros, E., & Mantzoros, C. S. (2021). Mediterranean diet as a nutritional approach for COVID-19. Metabolism: Clinical and Experimental, 114. doi:10.1016/j.metabol.2020.154407

Anishchenko, A. (2019). Agriculture 4.0 as a promising model for scientific and technological development of agrarian sector in modern Russia. *Food Policy and Security*, 129-140.

Antoniou, N., Monlau, F., Sambusiti, C., Ficara, E., Barakat, A., & Zabaniotou, A. (2019). Contribution to Circular Economy options of mixed agricultural wastes management: Coupling anaerobic digestion with gasification for enhanced energy and material recovery. *Journal of Cleaner Production*, *209*, 505–514. doi:10.1016/j.jclepro.2018.10.055

Antunes, O. E. D. (2012). *Multicriteria GIS analysis for determination of a spacialized index of coastal anthropogenic pressure. Cases of Espinho, Caparica and Faro*. Dissertação de Mestrado em Gestão do Território. Área de Especialização em Deteção Remota e Sistemas de Informação Geográfica. Universidade Nova de Lisboa. Faculdade de Ciências Sociais e Humanas. Lisbon.

Aranjuelo, I., Molero, G., Erice, G., Avice, J. C., & Nogués, S. (2011). Plant physiology and proteomics reveals the leaf response to drought in alfalfa (*Medicago sativa* L.). *Journal of Experimental Botany*, *62*(1), 111–123. doi:10.1093/jxb/erq249 PMID:20797998

Arias, R. C., & Fromm, I. (2019). From cocoa producers to chocolatiers? Developing an entrepreneurial model for small-scale producers in Honduras. *International Journal on Food System Dynamics*, *10*(1), 38–54.

Aronsson, T., & Schöb, R. (2018). Climate change and psychological adaptation: A behavioral environmental economics approach. *Journal of Behavioral and Experimental Economics*, *74*, 79–84. doi:10.1016/j.socec.2018.03.005

Arrizabalaga-Arriazu, M., Gomès, E., Morales, F., Irigoyen, J. J., Pascual, I., & Hilbert, G. (2020). High Temperature and Elevated Carbon Dioxide Modify Berry Composition of Different Clones of Grapevine (*Vitis vinifera* L.) cv. Tempranillo. *Frontiers in Plant Science*, *11*, 1888. doi:10.3389/fpls.2020.603687 PMID:33335536

Arteta, C., & Kirby, P. (2019). *Country Economic Memorandum for Sao Tome and Principe - Background Note 2: Is it Sustainable for Sao Tome and Príncipe to Have a Large Current Account Deficit and a Fixed Exchange Rate?* World Bank. Retrieved from: https://openknowledge.worldbank.org/handle/10986/32141

Arya, R. L., Arya, S., Arya, R., & Kumar, J. (2020). *Fundamentals of Agriculture* (Vol. 1). Scientific Publishers.

Aryeetey, B., Lartey, A., Marquis, G. S., Nti, H., Colecraft, E., & Brown, P. (2017). Prevalence and predictors of overweight and obesity among school-aged children in urban Ghana. *BMC Obesity*, *4*(38), 1–8. doi:10.118640608-017-0174-0

Asamoah, D., & Annan, J. (2012). Analysis of Ghana's cocoa value chain towards services and standards for stakeholders. *Int. J. Services and Standards*, *8*(2), 116–132. doi:10.1504/IJSS.2012.049423

Asghar, M.T., Yusof, A.Y., Mokhtar, M.N., Ya'acob, M.E., Ghazali, H.M., & Lee Sin Chang, L. S., & Manaf, Y.N. (2019). Coconut (Cocos nucifera L.) sap as a potential source of sugar: Antioxidant and nutritional properties. *Food Science & Nutrition*, *8*(4), 1777–1787. doi:10.1002/fsn3.1191

Asher, G. N., Corbett, A. H., & Hawke, R. L. (2017). Common Herbal Dietary Supplement-Drug Interactions. *American Family Physician*, *96*(2), 101–107.

Ashour, E. K., & Al-Najar, H. (2012). The Impact of Climate Change and Soil Salinity in Irrigation Water Demand in the Gaza Strip. *Journal of Earth Science & Climatic Change*, *03*(02). Advance online publication. doi:10.4172/2157-7617.1000120

Asseng, S., Ewert, F., Martre, P., Rötter, R. P., Lobell, D. B., Cammarano, D., Kimball, B. A., Ottman, M. J., Wall, G. W., White, J. W., Reynolds, M. P., Alderman, P. D., Prasad, P. V. V., Aggarwal, P. K., Anothai, J., Basso, B., Biernath, C., Challinor, A. J., De Sanctis, G., ... Zhu, Y. (2015). Rising temperatures reduce global wheat production. *Nature Climate Change*, *5*(2), 143–147. doi:10.1038/nclimate2470

Asseng, S., Foster, I., & Turner, N. C. (2011). The impact of temperature variability on wheat yields. *Global Change Biology*, *17*(2), 997–1012. doi:10.1111/j.1365-2486.2010.02262.x

Astrid Fenger, N., Skovmand Bosselmann, A., Asare, R., & de Neergaard, A. (2017). The impact of certification on the natural and financial capitals of Ghanaian cocoa farmers. *Agroecology and Sustainable Food Systems*, *41*(2), 143–166. doi:10.1080/21683565.2016.1258606

Austin, E. K., Rich, J. L., Kiem, A. S., Handley, T., Perkins, D., & Kelly, B. J. (2020). Concerns about climate change among rural residents in Australia. *Journal of Rural Studies*, *75*, 98–109. doi:10.1016/j.jrurstud.2020.01.010

Awasthi, R., Kaushal, N., Vadez, V., Turner, N. C., Berger, J., Siddique, K. H. M., & Nayyar, H. (2014). Individual and combined effects of transient drought and heat stress on carbon assimilation and seed filling in chickpea. *Functional Plant Biology: FPB*, *41*(11), 1148–1167. doi:10.1071/FP13340 PMID:32481065

Bach, V. N., Pham; C. H. & Hoang, N. (2016). A Critical Analysis of the Value Chain in the Rice Industry and Its Effects on the Export Rice Industry in Kien Giang Province, Vietnam. *International Journal of Financial Research*, *7*(3).

Bach-Faig, A., Berry, E. M., Lairon, D., Reguant, J., Trichopoulou, A., Dernini, S., Medina, F. X., Battino, M., Belahsen, R., Miranda, G., & Serra-Majem, L. (2011). Pirámide de dieta mediterránea hoy. Ciencia y actualizaciones culturales. *Public Health Nutrition*, *14*(12A).

Baer-Nawrocka, A., & Poczta, W. (2018). Agricultural transformation in Poland: A long-term analysis. *Berichte über Landwirtschaft*, *96*(1).

Bahrami, H., De Kok, L. J., Armstrong, R., Fitzgerald, G. J., Bourgault, M., Henty, S., Tausz, M., & Tausz-Posch, S. (2017). The proportion of nitrate in leaf nitrogen, but not changes in root growth, are associated with decreased grain protein in wheat under elevated [CO_2]. *Journal of Plant Physiology*, *216*, 44–51. doi:10.1016/j.jplph.2017.05.011 PMID:28575746

Bairagi, S., Mohanty, S., Ynion, J., & Demont, M. (2017). *Determinants of Consumer Preferences for Rice Attributes: Evidence from South and Southeast Asia*. Conference: Agricultural & Applied Economics Association Annual Meeting, Chicago, IL.

Bakalis, S., Valdramidis, V., Argyropoulos, D., Ahrne, L., Chen, J., Cullen, P., . . . LeBail, A. (2020). Perspectives from CO+RE: How COVID-19 changed our food systems and food security paradigms. *Current Research in Food Science*, 166–172.

Balaman, S. Y., Wright, D. G., Scott, J., & Matopoulos, A. (2018). Network design and technology management for waste to energy production: An integrated optimization framework under the principles of circular economy. *Energy, 143*, 911–933. doi:10.1016/j.energy.2017.11.058

Balanza, R., García-Lorda, P., Pérez-Rodrigo, C., Aranceta, J., Bonet, M. B., & Salas-Salvadó, J. (2007). Trends in food availability determined by the Food and Agriculture Organization's food balance sheets in Mediterranean Europe in comparison with other European areas. *Public Health Nutrition, 10*(2), 168–176. Advance online publication. doi:10.1017/S1368980007246592 PMID:17261226

Balcombe, K., Fraser, I. M., Latruffe, L., Rahman, M., & Smith, L. (2008). An Application of the DEA Double Bootstrap to Examine Sources of Efficiency in Bangladesh Rice Farming. *Applied Economics, 40*(15), 1919–1925. doi:10.1080/00036840600905282

Bandara, J. S. & Cai, Y. (2014). The impact of climate change on food crop productivity, food prices and food security in South Asia. Economic Analysis and Policy, 44(4). doi:10.1016/j.eap.2014.09.005

Banerjee, B. D., Seth, V., & Ahmed, R. S. (2001). Pesticide-induced oxidative stress: Perspectives and trends. *Reviews on Environmental Health, 16*(1), 1–40. doi:10.1515/REVEH.2001.16.1.1

Banerjee, S. B. (2003). Who sustains whose development? Sustainable development and the reinvention of nature. *Organization Studies, 24*(1), 143–180. doi:10.1177/0170840603024001341

Barbier, E. B., & Hochard, J. P. (2018). The Impacts of Climate Change on the Poor in Disadvantaged Regions. *Review of Environmental Economics and Policy, 12*(1), 26–47. doi:10.1093/reep/rex023

Barčić, A. P., Kuzman, M. K., Haviarova, E., & Oblak, L. (2019). *Circular economy & Sharing collaborative economy principles: A case study conducted in wood-based sector*. Woodema, Ia-Int Assoc Econ & Manag Wood Processing & Furn Manuf.

Barrientos, S. (2014). Gendered global production networks: Analysis of cocoa-chocolate sourcing. *Regional Studies, 48*(5), 791–803. doi:10.1080/00343404.2013.878799

Barrientos, S., & Asenso-Okyere, K. (2009). Cocoa value chain: Challenges facing Ghana in a changing global confectionary market. *Journal für Entwicklungspolitik, 25*(2), 88–107. doi:10.20446/JEP-2414-3197-25-2-88

Barros, M. V., Salvador, R., de Francisco, A. C., & Piekarski, C. M. (2020). Mapping of research lines on circular economy practices in agriculture: From waste to energy. *Renewable & Sustainable Energy Reviews, 131*, 109958. Advance online publication. doi:10.1016/j.rser.2020.109958

Baslam, M., Mitsui, T., Hodges, M., Priesack, E., Herritt, M. T., Aranjuelo, I., & Sanz-Sáez, Á. (2020). Photosynthesis in a Changing Global Climate: Scaling Up and Scaling Down in Crops. *Frontiers in Plant Science, 11*, 882. doi:10.3389/fpls.2020.00882 PMID:32733499

Bastías, J. M., Balladares, P., Acuña, S., Quevedo, R., & Muñoz, O. (2017). Determining the effect of different cooking methods on the nutritional composition of salmon (Salmo salar) and chilean jack mackerel (Trachurus murphyi) fillets. *PLoS One, 12*(7), 1–10. doi:10.1371/journal.pone.0180993

Belanger, J., & Pilling, D. (Eds.). (2019). *The State of the World's Biodiversity for Food and Agriculture*. FAO.

Ben Hassen, T., El Bilali, H., & Allahyari, M. S. (2020). Impact of covid-19 on food behavior and consumption in qatar. *Sustainability (Switzerland)*, *12*(17), 319. Advance online publication. doi:10.29117/quarfe.2020.0296

Bénaben, F. (2016). A Formal Framework for Crisis Management Describing Information Flows and Functional Structure. *Procedia Engineering*, *159*, 353–356. doi:10.1016/j.proeng.2016.08.208

Bennich, T., Weitz, N., & Carlsen, H. (2020). Deciphering the scientific literature on SDG interactions: A review and reading guide. *The Science of the Total Environment*, *728*, 138405. doi:10.1016/j.scitotenv.2020.138405 PMID:32388023

Berry, E. M. (2019). Sustainable food systems and the mediterranean diet. *Nutrients*, *11*(9), 2229. Advance online publication. doi:10.3390/nu11092229 PMID:31527411

Besuspariene, E., & Niskanen, V. A. (2020). The Assessment of Factors Affecting Fertiliser Use on Family Farms in Lithuania. *Studies in Agricultural Economics (Budapest)*, *122*(1), 13–21. doi:10.7896/j.1923

Beyer, D., & Hinke, J. (2020). European benchmarking of determinants of profitability for companies with accrual accounting in the agricultural sector. *Agricultural Economics-Zemedelska Ekonomika*, *66*(11), 477–488. doi:10.17221/128/2020-AGRICECON

Bhaduri, D., Saha, A., Desai, D., & Meena, H. N. (2016). Restoration of carbon and microbial activity in salt-induced soil by application of peanut shell biochar during short-term incubation study. *Chemosphere*, *148*, 86–98.

Bhatt, B. P., & Krishnamoorthy, P. V. (1965). Radio broadcasting serves rural development. Radio Rural Forum spread throughout India. Report and Paper on Mass Communication UNESCO (No. 48). UNESCO.

Bibri, M. (2008). *Corporate sustainability/CSR communications and value creation: A marketing approach* [Unpublished Master's thesis]. Blekinge Institute of Technology, School of Management, University of Karlshamn.

Biegańska, J., Środa-Murawska, S., Kruzmetra, Z., & Swiaczny, F. (2018). Peri-Urban Development as a Significant Rural Development Trend. *Quaestiones Geographicae*, *37*(2), 125–140. doi:10.2478/quageo-2018-0019

Bijttebier, J., Ruysschaert, G., Hijbeek, R., Werner, M., Pronk, A. A., Zavattaro, L., Bechini, L., Grignani, C., ten Berge, H., Marchand, F., & Wauters, E. (2018). Adoption of non-inversion tillage across Europe: Use of a behavioural approach in understanding decision making of farmers. *Land Use Policy*, *78*, 460–471. doi:10.1016/j.landusepol.2018.05.044

Bjorklund, T., Mikkonen, M., Mattila, P. & Marel, F. (2020). Expanding entrepreneurial solution spaces in times of crisis: Business model experimentation amongst packaged food and beverage ventures. *Journal of Business Venturing Insights*, 1-14.

Blowfield, M. (2003). Ethical supply chains in the cocoa, coffee and tea industries. *Greener Management International*, *43*(43), 15–24. doi:10.9774/GLEAF.3062.2003.au.00004

Bonuedi, I., Gerber, N., & Kornher, L. (2021). Intervening in cash crop value chains for improved nutrition: Evidence from rural Sierra Leone. *The Journal of Development Studies*, 1–17. doi:10.1080/00220388.2021.1945043

Borawski, P., Borawski, M. B., Parzonko, A., Wicki, L., Rokicki, T., Perkowska, A., & Dunn, J. W. (2021). Development of Organic Milk Production in Poland on the Background of the EU. *Agriculture-Basel*, *11*(4), 323. doi:10.3390/agriculture11040323

Borda, A., Morales, O., Teegen, H., Rees, G. H., & Gonzalez-Perez, M. A. (2021). Addressing sustainable rural development with shared value: A Peruvian model from the cacao industry. *Sustainability*, *13*(14), 8028. doi:10.3390u13148028

Borrello, M., Pascucci, S., Caracciolo, F., Lombardi, A., & Cembalo, L. (2020). Consumers are willing to participate in circular business models: A practice theory perspective to food provisioning. *Journal of Cleaner Production, 259*, 1–14. doi:10.1016/j.jclepro.2020.121013

Borunda, A. (2020). Plunge in carbon emissions from lockdowns will not slow climate change. *National Geographic*. Retrieved from https://www.nationalgeographic.com/ science/ article/plunge-in-carbon-emissions-lockdowns-will-not-slow-climate-change

Bosetti, C., Gallus, S., Trichopoulou, A., Talamini, R., Franceschi, S., Negri, E., & La Vecchia, C. (2003). Influence of the Mediterranean Diet on the Risk of Cancers of the Upper Aerodigestive Tract. *Cancer Epidemiology, Biomarkers & Prevention, 12*(10). PMID:14578148

Both, A. V. R., Saraiva, M. S. M. M. de M., & Tibério, M. L. (2021). Rice Sustainability Trend: A bibliometric review. *Research. Social Development, 10*(7), e8310716224. doi:10.33448/rsd-v10i7.16224

Bouman, B. A. M., Humphreys, E., Tuong, T. P., & Barker, R. (2007). Rice and water. *Advances in Agronomy, 92*, 187–237. doi:10.1016/S0065-2113(04)92004-4

Breiman, L., Friedman, J., Stone, C. J., & Olshen, R. A. (1984). *Classification and Regression Trees* (1st ed.). Chapman and Hall/CRC.

Briz, J., De Felipe, I., & Briz, T. (2010). La cadena de valor alimentaria: un enfoque metodológico. *Boletín Económico del ICE, 2983*.

Briz, J. (2011). La cadena de valor como instrumento de análisis del funcionamiento y riesgo del Sistema Agroalimentario. In *La Cadena de Valor Alimentaria: Análisis Internacional de casos reales*. Editorial Agrícola Española, SA.

Briz, J., De Felipe, I., & Briz, T. (2012). La sostenibilidad del sistema alimentario a través de la cadena de valor. In *Las Redes Cadenas de Valor Alimentarias en el siglo XX1: Retos y oportunidades internacionales*. Editorial Agrícola Española, SA.

Broekema, W., Kleef, D., & Steen, T. (2017). What Factors Drive Organizational Learning From Crisis? Insights From the Dutch Food Safety Services' Response to Four Veterinary Crises. *Journal of Contingencies and Crisis Management Volume*, 326-340.

Bruelle, G., Naudin, K., Scopel, E., Domas, R., Rabeharisoa, L., & Tittonell, P. (2015). Short- to Mid-Term Impact of Conservation Agriculture on Yield Variability of Upland Rice: Evidence from Farmer's Fields in Madagascar. *Experimental Agriculture, 51*(1), 66–84. doi:10.1017/S0014479714000155

Buchmann-Duck, J., & Beazley, K. F. (2020). An urgent call for circular economy advocates to acknowledge its limitations in conserving biodiversity. *The Science of the Total Environment, 727*, 9. doi:10.1016/j.scitotenv.2020.138602 PMID:32325313

Buis, A. (2019). *A Degree of Concern: Why Global Temperatures Matter*. NASA's Global Climate Change. Retrieved from https://climate.nasa.gov/news/2865/a-degree-of-concern-why-global-temperatures-matter/

Bulegon, L. G., Guimarães, V. F., Battistus, A. G., Inagaki, A. M., & da Costa, N. V. (2019). Mitigation of drought stress effects on soybean gas exchanges induced by *Azospirillum brasilense* and plant regulators. *Pesquisa Agropecuária Tropical, 49*. Advance online publication. doi:10.1590/1983-40632019v4952807

Bunyasiri, I. N., & Sirisupluxana, P. (2017). Consumers behavior and rice attributes for Thai hommali rice in Sichuan province of China. *The Business and Management Review, 8*(4), 329–341.

Burkert, N. T., Muckenhuber, J., Großschädl, F., Rásky, É., & Freidl, W. (2014). Nutrition and health - The association between eating behavior and various health parameters: A matched sample study. *PLoS One*, *9*(2), e88278. Advance online publication. doi:10.1371/journal.pone.0088278 PMID:24516625

Burlingame, B., & Dernini, S. (2011, December 13). Burlingame, Barbara, & Dernini, S. (2011). Sustainable diets: The Mediterranean diet as an example. *Public Health Nutrition*, *14*(12A), 2285–2287. Advance online publication. doi:10.1017/S1368980011002527

Burow, K. R., Nolan, B. T., Rupert, M. G., & Dubrovsky, N. M. (2010). Nitrate in Groundwater of the United States, 1991-2003. *Environmental Science & Technology*, *44*(13), 4988–4997. doi:10.1021/es100546y PMID:20540531

Busquet, M., Bosma, N., & Hummels, H. (2021). A multidimensional perspective on child labor in the value chain: The case of the cocoa value chain in West Africa. *World Development*, *146*, 146. doi:10.1016/j.worlddev.2021.105601

Cabo, P., Matos, A., & Bento, A. (2016). *Da produção ao consumo. Breve análise do mercado nacional da amêndoa.* ESADR.

CABRI - Collaborative Africa Budget Reform Initiative. (2019). The role of governments in developing agriculture value chains. Case Study 2. Rice and cassava value chains. Author.

Caiado, R. G., Filho, W. L., Quelhas, O. L., Nascimento, D. L., & Ávila, L. V. (2018). A literature-based review on potentials and constraints in the implementation of the sustainable development goals. *Journal of Cleaner Production*, *198*, 1276–1288. doi:10.1016/j.jclepro.2018.07.102

Calatrava, J., Martínez-Granados, D., Zornoza, R., González-Rosado, M., Lozano-García, B., Vega-Zamora, M., & Gómez-López, M. D. (2021). Barriers and opportunities for the implementation of sustainable farming practices in mediterranean tree orchards. *Agronomy (Basel)*, *11*(5), 1–24. doi:10.3390/agronomy11050821

Callahan, L. (2019). Contract-Farming in Cocoa Value Chains in Africa: Possibilities and Challenges. In Ethiopian Yearbook of International Law 2018. Springer.

Camacho, L., Hobday, A., Bustamante, R., Farmery, A., Fleming, A., Frusher, S., Green, B., Lopez, A., Plaganyi, P., Schrobback, P., Thebaud, O., Thomas, L., & Putten, I. (2014). Facing the wave of change: Stakeholder perspectives on climate adaptation for Austrailian seafood supply chains. *Environ Change*,1-12.

Campos, S., & Madureira, L. (2019). Can Healthier Food Demand be Linked to Farming Systems' Sustainability? The Case of the Mediterranean Diet. *International Journal on Food System Dynamics*, *10*(3), 262–277.

Cao, D., Li, Y. Y., Liu, B. H., Kong, F. J., & Tran, L. S. P. (2018). Adaptive mechanisms of soybean grown on salt-affected soils. *Land Degradation & Development*, *29*(4), 1054–1064. doi:10.1002/ldr.2754

Carmen Cuerva, M. (2012). Determinants of Labour Productivity Convergence in the European Agricultural Sector. *Agrociencia*, *46*(6), 621–635.

Carodenuto, S., & Buluran, M. (2021). The effect of supply chain position on zero-deforestation commitments: Evidence from the cocoa industry. *Journal of Environmental Policy and Planning*, *23*(6), 716–731. doi:10.1080/1523908X.2021.1910020

Carvalho, P. (2018). Dynamics of rural low density spaces in Portugal. *Méditerranée*, *130*(130), 1–17. doi:10.4000/mediterranee.10516

Cassia, R., Nocioni, M., Aragunda, N. C., & Lamattina, L. (2018). Climate Change and the Impact of Greenhouse Gasses: CO_2 and NO, Friends and Foes of Plant Oxidative Stress. *Frontiers in Plant Science*. Retrieved from https://www.frontiersin.org/articles/ 10.3389/ fpls.2018.00273/full

Castellini, G., Savarese, M., & Graffigna, G. (2021). The Impact of COVID-19 Outbreak in Italy on the Sustainable Food Consumption Intention From a "One Health" Perspective. *Frontiers in Nutrition*, *8*, 622122. Advance online publication. doi:10.3389/fnut.2021.622122 PMID:33791331

Castro-Nunez, A., Charry, A., Castro-Llanos, F., Sylvester, J., & Bax, V. (2020). Reducing deforestation through value chain interventions in countries emerging from conflict: The case of the Colombian cocoa sector. *Applied Geography (Sevenoaks, England)*, *123*, 123. doi:10.1016/j.apgeog.2020.102280

Castro-Quezada, I., Román-Viñas, B., & Serra-Majem, L. (2020). Nutritional adequacy of the Mediterranean diet. In The Mediterranean Diet. doi:10.1016/B978-0-12-818649-7.00013-8

Catarino, R., Gaba, S., & Bretagnolle, V. (2019). Experimental and empirical evidence shows that reducing weed control in winter cereal fields is a viable strategy for farmers. *Scientific Reports*, *9*(1), 9004. doi:10.103841598-019-45315-8 PMID:31227731

Cattaneoa, A., Sáncheza, M. V., Toreroa, M., & Vosb, R. (2020). Reducing food loss and waste: Five challenges for policy and research. *Food Policy*, 1–9. PMID:33012956

Cavagnaro, E., & Curiel, G. (2012). *The Three Levels of Sustainability* (1st ed.). Routledge.

CBD - Convention on Biological Diversity (2016). *Decision XIII/28. Indicators for the Strategic Plan for Biodiversity 2011–2020 and the Aichi Biodiversity Targets*. Author.

CBD - Convention on Biological Diversity. (2014). *Global Biodiversity Outlook 4*. Secretariat of the Convention on Biological Diversity.

Cesari de Maria, S., Rienzner, M., Facchi, A., Chiaradia, E. A., Romani, M., & Gandolfi, C. (2016). Water balance implications of switching from continuous submergence to flush irrigation I n a rice-growing district. *Agricultural Water Management*, *171*, 108–119. doi:10.1016/j.agwat.2016.03.018

Chaaya, M., Phung, K., Atweh, S., Asmar, K. E., Karam, G., Khoury, R. M., Ghandour, L., Ghusn, H., Assaad, S., Prince, M., & Waldemar, G. (2018). Socio-demographic and cardiovascular disease risk factors associated with dementia: Results of a cross-sectional study from Lebanon. *Preventive Medicine Reports*, *9*, 1–5. doi:10.1016/j.pmedr.2017.11.006

Chabowski, B., Mena, J., & Gonzalez-Padron, T. (2011). The structure of sustainability research in marketing, 1958–2008: A basis for future research opportunities. *Journal of the Academy of Marketing Science*, *39*(1), 55–70. doi:10.100711747-010-0212-7

Chagomoka, T., Afari-Sefa, V., & Pitoro, R. (2014). Value chain analysis of traditional vegetables from Malawi and Mozambique. *The International Food and Agribusiness Management Review*, *17*(4), 59–86.

Chakraborty, C., & Das, S. (2016). Dynamics of Diabetes and Obesity: An Alarming Situation in the Developing Countries in Asia. *Mini-Reviews in Medicinal Chemistry*, *16*(15), 1258–1268. Advance online publication. doi:10.2174/1389557516666160505115512 PMID:27145852

Chaktrabarty, M. (2016). Climate Change and Food Security in India. *ORF Issue Brief*, 157. Retrieved from https://www.orfonline.org/wp content/uploads/2016/09/ ORF_ IssueBrief _1571.pdf

Chambers, R. (1983). *Rural Development: Putting the Last First*. Routledge.

Chang, C. C., Yi-Ting, L., Ya-Ting, Lu., Liu, Y. S., & Liu, J. F. (2010). Kiwifruit improves bowel function in patients with irritable bowel syndrome with constipation. *Asia Pacific Journal of Clinical Nutrition*, *19*(4), 451–457.

Chanpiwat, P., Sthiannopkao, S., Widmer, K., Seiichiro Himeno, S., Hideki Miyataka, H., Vu, N. U., Tran, V. V., & Pham, T. T. N. (2016). Assessment of metal and bacterial contamination in cultivated fish and impact on human health for residents living in the Mekong Delta. *Chemosphere*, *163*, 342–350. doi:10.1016/j.chemosphere.2016.08.003

Chapagain, A., & Hoekstra, A. (2011). The green, blue and grey water foot print of rice from production and consumption perspectives. *Ecological Economics*, *70*(4), 749–758. doi:10.1016/j.ecolecon.2010.11.012

Chartier, O., & Cronin, E. (Eds). (2017). *Study on risk management in EU agriculture: final report*, European Commission. https://op.europa.eu/pt/publication-detail/-/publication/5a935010-af78-11e8-99ee-01aa75ed71a1

Chattopadhyay, N. (2008). Climate Change and Food Security in India. *International Symposium on Climate Change and Food Security in South Asia*. Retrieved from http://wamis.org/agm/meetings/rsama08/S402-Chattopadhyay-Climate-change_Food-Security.pdf

Chaudhari, P. R., Tamrakar, N., Singh, L., Tandon, A., & Sharma, D. (2018). Rice nutritional and medicinal properties: A review article. *Journal of Pharmacognosy and Phytochemistry*, *7*(2), 150–156.

Chenarides, L., Grebitus, C., Lusk, J. L., & Printezis, I. (2021). Food consumption behavior during the COVID-19 pandemic. *Agribusiness*, *37*(1), 44–81. Advance online publication. doi:10.1002/agr.21679 PMID:33362340

Chen, D., & Dai, A. (2018). Dependence of estimated precipitation frequency and intensity on data resolution. *Climate Dynamics*, *50*(9), 3625–3647. doi:10.100700382-017-3830-7

Cheng, Y. F., & Pan, Y. J. (2005). The area of cultivated land may lead to the problem of food and clothing. *Well-off*, (2), 56–56.

Chen, S. Y., & Sun, J. K. (2020). Creating something out of nothing and changing lanes to be driven by innovation—Enlightenment from the investigation and study of the development of digital economy in Guizhou Province. *Yunnan Science and Technology Management*, *033*(002), 34–36.

Chen, Z. m. (1994). Discussion on rational use of pesticides to avoid resistance of pests. *Agricultural Science and Technology*, 4.

Chesbrough, H. (2020). To recover faster from Covid-19, open up: Managerial implications from an open innovation perspective. *Industrial Marketing Management*, *88*, 410–413. doi:10.1016/j.indmarman.2020.04.010

Chiarolla, C. (2017). Agriculture and biodiversity conservation. In Biodiversity and Nature protection law. Edward Elgar Publishing. doi:10.4337/9781783474257.III.18

Chiffoleau, Y. (2012). Circuits courts alimentaires, dynamiques relationnelles e lutte contre l'exclusion en agriculture. *Économie Rurale (Paris)*, *332*(332), 88–101. doi:10.4000/economierurale.3694

Chikaire, J., Nnadi, F. N., & Ajaero, C. (2010). Impacts of Climate Change on Food Security and Poverty Reduction. *International Journal of Natural and Applied Sciences*, *6*(4).

China Meterological Administration. (2016). *The latest related information of drought and flood disasters*. http://www.cma.gov.cn/

China National Knowledge Infrastructure. (2021). *Sustainable development of agriculture*. https://www.cnki.net/gycnki/gycnki.htm

China's National Bureau of Statistics. (2019). *Agricultural water consumption data*. https://data.stats.gov.cn/index.htm

China's National Bureau of Statistics. (2020). *Data on the use of agricultural chemical fertilizers and pesticides*. https://data.stats.gov.cn/index.htm

Chojnacka, K., Moustakas, K., & Witek-Krowiak, A. (2020). Bio-based fertilizers: A practical approach towards circular economy. *Bioresource Technology*, *295*, 122223. Advance online publication. doi:10.1016/j.biortech.2019.122223 PMID:31623921

Christopher, M. (2005). *Logistics and supply chain management: creating value-adding networks*. Pearson Education.

Clay, P. M., & Feeney, R. (2018). Analyzing agribusiness value chains: A literature review. *The International Food and Agribusiness Management Review*, *22*(1), 31–46. doi:10.22434/IFAMR2018.0089

Cline, W. R. (2008). Global Warming and Agriculture. Finance and Development. *International Monetary Fund, 45*(1). Retrieved from http:/www.imf.org/external/pubs/ft/fandd/2008/03/cline.htm

Cobb, C. W., & Douglas, P. H. (1928). A Theory of Production. *The American Economic Review*, *18*(1), 139–165.

Cohen, J. (1988). *Statistical power analysis for the behavioral sciences*. Erlbaum.

Collins, C., Steg, L., & Koning, M. (2007). Customers' values, beliefs on sustainable corporate performance, and buying behavior. *Psychology and Marketing*, *24*(6), 555–577. doi:10.1002/mar.20173

Colwell, K. (2012). *Is it still green: The impacts of corporate takeovers on business sustainability?* [Unpublished doctoral dissertation].

COM. (2012). *Comunicación de la Comisión Al Parlamento Europeo y al Consejo sobre la cooperación de innovación europea. 'Productividad y sostenibilidad agrícolas'*. Bruselas, 29.2.2012.

Cóndor, R. D., Scarelli, A., & Valentini, R. (2010). Multicriteria Decision Aid to support Multilateral Environmental Agreements in assessing international forestry projects. *International Environmental Agreement: Politics, Law and Economics*, *11*(2), 117–137. doi:10.100710784-010-9125-7

Constantin, M., Radulescu, I., Andrei, J., Chivu, L., Erokhin, V., & Gao, T. (2021). A Perspective on Agricultural Labor Productivity and Greenhouse Gas Emissions in Context of the Common Agricultural Policy Exigencies. *Ekonomika Poljoprivreda-Economics of Agriculture*, *68*(1), 53–67. doi:10.5937/ekoPolj2101053C

Cooper, P., Dimes, J., Rao, K. P. C., Shapiro, B., Shiferaw, B., & Twomlow, S. (2008). Coping better with current climatic variability in the rain-fed farming systems of sub-Saharan Africa: An essential first step in adapting to future climate change? *Agriculture, Ecosystems & Environment*, *126*(1-2), 24–35. doi:10.1016/j.agee.2008.01.007

Copolovici, L., Popitanu, A., & Copolovici, D. (2021). Volatile organic compound emission and residual substances from plants in light of the globally increasing CO_2 level. *Current Opinion in Environmental Science & Health*, *19*, 100216. doi:10.1016/j.coesh.2020.10.004

Copp, J. H. (1972). Rural sociology and rural development. *Rural Sociology*, *37*, 515–533.

Correia, E., Carvalho, H., Azevedo, S. G., & Govindan, K. (2017). Maturity Models in Supply Chain Sustainability: A Systematic Literature Review. *Sustainability*, *9*(1), 64. doi:10.3390u9010064

Corsil, S., Friedrich, T., Kassam, A., Pisante, M., & Moracs, J. (2012). Soil organic carbon accumulation and greenhouse gas emissions reductions from conservation agriculture: A literature review. *Integrated Crop Management*, *16*, 1–104.

Costantini, M., Bacenetti, J., Coppola, G., Orsi, L., Ganzaroli, A., & Guarino, M. (2020). Improvement of human health and environmental costs in the European Union by air scrubbers in intensive pig farming. *Journal of Cleaner Production*, *275*, 124007. doi:10.1016/j.jclepro.2020.124007

Craggs, G. (2016). Photosynthesis and its Role in Climate Change and Soil Regeneration. *Future Directions International*, 1–6.

Crandall, W. R., & Mensah, E. C. (2008). Crisis management and sustainable development: a framework and proposed research agenda. *International Journal of Sustainable Strategic Management*, 16-34.

Crandall, W. R., Parnell, J. A., & Spillan, J. E. (2014). *Crisis Management: Leading in the New Strategy Landscape*. SAGE.

Crane, L., Gantz, G., Isaacs, S., Jose, D., & Sharp, R. (2013). *Introduction to Risk Management: Understanding Agricultural Risks*. Extension Risk Management Education and Risk Management Agency.

Cristhian, V. Q., María, B., & Hugo, R. (2017). Production of Bioenergy in the Framework of Circular Economy: A Sustainable Circular System in Ecuador. In Advances in Biofeedstocks and Biofuels (Vol. 1, pp. 1-31). Wiley. doi:10.1002/9781119117322.ch1

Cruz, R. S. W. (2014). *Análise da margem e dos factores que afectam a comercialização do arroz no Distrito de Bobonaro, sub-distrito de Maliana* [Unpublished master dissertation]. UNTL, Universidade Nacional de Timor Lorosa'e, Díli, Timor Leste.

Cruz, R., Lucas, M. R., & Carvalho, M. L. S. (2016). Comercialização do arroz no Distrito de Bobonaro, Sub-distrito de Maliana – Análise dos factores que a afectam e da Margem de Comercialização. *Proceedings do VIII congresso da APDEA, ESADR 2016.*

Cuevas, R. P., Pede, V. O., McKinley, J., Velarde, O., & Demont, M. (2016). Rice Grain Quality and Consumer Preferences: A Case Study of Two Rural Towns in the Philippines. *PLoS One*, *11*(3), e0150345. doi:10.1371/journal.pone.0150345 PMID:26982587

Cunha, L. (2008). A Beira Interior - Portugal, caracterização física. In *PINHEIRO, Elisa Calado (coord.). Rota da Lã TRANSLANA percursos e marcas de um território de fronteira: Beira Interior (Portugal) e Comarca Tajo-Salor-Almonte (Espanha)* (Vol. I). Museu de Lanifícios da Universidade da Beira Interior.

Custodio, M. C., Demont, M., Laborte, A., & Ynion, J. (2016). Improving food security in Asia through consumer-focused rice breeding. *Global Food Security*, *9*, 19–28. doi:10.1016/j.gfs.2016.05.005

Cutone, A., Rosa, L., Ianiro, G., Lepanto, M. S., Bonaccorsi di Patti, M. C., Piera Valenti, P., & Musci, G. (2020). Lactoferrin's Anti-Cancer Properties: Safety, Selectivity, and Wide Range of Action. *Biomolecules*, *10*(3), 1–26. doi:10.3390/biom10030456

D'Alessandro, A., & De Pergola, G. (2018). The Mediterranean Diet: Its definition and evaluation of a priori dietary indexes in primary cardiovascular prevention. *International Journal of Food Sciences and Nutrition*, *69*(6), 647–659. Advance online publication. doi:10.1080/09637486.2017.1417978 PMID:29347867

D'Odorico, P., Davis, K. F., Rosa, L., Carr, J. A., Chiarelli, D., Dell'Angelo, J., Gephart, J., MacDonald, G. K., Seekell, D. A., Suweis, S., & Rulli, M. C. (2018). The global food-energy-water nexus. *Reviews of Geophysics*, *56*(3), 456–531. doi:10.1029/2017RG000591

Da Rocha, B. R. S., Rico-Campà, A., Romanos-Nanclares, A., Ciriza, E., Barbosa, K. B. F., Martínez-González, M. Á., & Martín-Calvo, N. (2020). Adherence to Mediterranean diet is inversely associated with the consumption of ultra-processed foods among Spanish children: The SENDO project. *Public Health Nutrition*. Advance online publication. doi:10.1017/S1368980020001524 PMID:32698921

Da Silva, R., Bach-Faig, A., Raidó Quintana, B., Buckland, G., Vaz De Almeida, M. D., & Serra-Majem, L. (2009). Worldwide variation of adherence to the Mediterranean diet, in 1961-1965 and 2000-2003. *Public Health Nutrition, 12*(9). doi:10.1017/S1368980009990541

Dada, J. T., & Akinlo, T. (2021). Foreign direct investment and poverty reduction in sub-Saharan Africa: Does environmental degradation matter? *Future Bussiness Journal*, *7*(1), 21. doi:10.118643093-021-00068-7

Dahlman, L., & Lindsey, R. (2020). *Climate Change: Ocean Heat Content*. Retrieved from https://www.climate.gov/news-features/understanding-climate/climate-change-ocean-heat-content

Dalgaard, R., Halberg, N., Kristensen, I. S., & Larsen, I. (2006). Modelling representative and coherent Danish farm types based on farm accountancy data for use in environmental assessments. *Agriculture, Ecosystems & Environment*, *117*(4), 223–237. doi:10.1016/j.agee.2006.04.002

Dania, W. A. P., Xing, K., & Amer, Y. (2018). Collaboration behavioural factors for sustainable agri-food supply chains: A systematic review. *Journal of Cleaner Production*, *186*, 851–864. doi:10.1016/j.jclepro.2018.03.148

Danish, S., Zafar-Ul-Hye, M., Hussain, S., Riaz, M., & Qayyum, M. F. (2020). Mitigation of drought stress in maize through inoculation with drought tolerant ACC deaminase containing PGPR under axenic conditions. *Pakistan Journal of Botany*, *52*(1). Advance online publication. doi:10.30848/PJB2020-1(7)

Danish, S., Zafar-ul-Hye, M., Mohsin, F., & Hussain, M. (2020). ACC-deaminase producing plant growth promoting rhizobacteria and biochar mitigate adverse effects of drought stress on maize growth. *PLoS One*, *15*(4), e0230615. doi:10.1371/journal.pone.0230615 PMID:32251430

Dantsis, T., Douma, C., Giourga, C., Loumou, A., & Polychronaki, E. (2010). A methodological approach to assess and compare the sustainability level of agricultural plant production systems. *Ecological Indicators*, *10*(2), 256–263. doi:10.1016/j.ecolind.2009.05.007

Darmon, N., & Drewnowski, A. (2008). Does social class predict diet quality? *The American Journal of Clinical Nutrition*, *87*(5), 1107–1117. Advance online publication. doi:10.1093/ajcn/87.5.1107 PMID:18469226

Datta, A., Ullah, H., & Ferdous, Z. (2017). Water management in rice. In B. S. Chauhan (Ed.), *Rice Production Worldwide* (pp. 255–277). Springer Int. Pub. AG. doi:10.1007/978-3-319-47516-5_11

Davis, M. A., Barnette, D. A., Flynn, N. R., Pidugu, A. S., Swamidass, S. J., Boysen, G., & Miller, G. P. (2019). CYP2C19 and 3A4 Dominate Metabolic Clearance and Bioactivation of Terbinafine Based on Computational and Experimental Approaches. *Chemical Research in Toxicology*, *32*(6), 1151–1164. doi:10.1021/acs.chemrestox.9b00006

de Boer, D., Limpens, G., Rifin, A., & Kusnadi, N. (2019). Inclusive productive value chains, an overview of Indonesia's cocoa industry. *Journal of Agribusiness in Developing and Emerging Economies*, *9*(5), 439–456. doi:10.1108/JADEE-09-2018-0131

De Jonge, J., Van Trijp, H., Jan Renes, R., & Frewer, L. (2007). Understanding consumer confidence in the safety of food: Its two-dimensional structure and determinants. *Risk Analysis: An International Journal*, *27*(3), 729–740. doi:10.1111/j.1539-6924.2007.00917.x PMID:17640219

De Steur, H., Demont, M., Gellynck, X., & Stein, A. J. (2017). The social and economic impact of biofortification through genetic modification. *Current Opinion in Biotechnology*, *44*, 161–168. doi:10.1016/j.copbio.2017.01.012 PMID:28231514

De Steur, H., Gellynck, X., Feng, S., Rutsaert, P., & Verbeke, W. (2012). Determinants of willingness-to-pay for GM rice with health benefits in a high-risk region: Evidence from experimental auctions for folate biofortified rice in China. *Food Quality and Preference*, *25*(2), 87–94. doi:10.1016/j.foodqual.2012.02.001

Dedeoğlu, M., & Dengiz, O. (2019). Generating of land suitability index for wheat with hybrid system aproach using AHP and GIS. *Computers and Electronics in Agriculture*, *167*, 105062. Advance online publication. doi:10.1016/j.compag.2019.105062

DeFries, R., Mondal, P., Singh, D., Agrawal, I., Fanzo, J., Remans, R., & Wood, S. (2016). Synergies and trade-offs for sustainable agriculture: Nutritional yields and climate-resilience for cereal crops in Central India. *Global Food Security*, *11*, 1–10. doi:10.1016/j.gfs.2016.07.001

Department of Economic Affairs. (2018). *Economic survey 2017-18 (Statistical Appendix).* Ministry of Finance, Government of India, Table 1.19. Retrieved from http://mofapp.nic.in:8080 /economicsurvey/appendixPDF/tab1.19.pdf

Dernini, S., & Berry, E. M. (2015). Mediterranean Diet: From a Healthy Diet to a Sustainable Dietary Pattern. *Frontiers in Nutrition*, *2*. Advance online publication. doi:10.3389/fnut.2015.00015 PMID:26284249

Detang-Dessendre, C., Guyomard, H., Requillart, V., & Soler, L.-G. (2020). Changing Agricultural Systems and Food Diets to Prevent and Mitigate Global Health Shocks. *Sustainability*, *12*(16), 6462. doi:10.3390u12166462

Deus, C. C. (2019). *Os Sistemas de agricultura familiar de Timor-Leste:Uma abordagem com foco no bem-estar dos produtores* [Unpublished doctoral dissertation]. Universidade de Évora, Évora, Portugal.

Devkota, K. P., Pasuquin, E., Elmido-Mabilangan, A., Dikitanan, R., Singleton, G. R., Stuart, A. M., Vithoonjit, D., Vidiyangkura, L., Pustika, A. B., Afriani, R., Listyowati, C. L., Keerthisena, R. S. K., Thikieu, N., Malabayabas, A. J., Hu, R., Pan, J., & Beebout, S. E. J. (2019). Economic and environmental indicators of sustainable rice cultivation: A comparison across intensive irrigated rice cropping systems in six Asian countries. *Ecological Indicators*, *105*, 199–214. doi:10.1016/j.ecolind.2019.05.029

Dhungana, B. R., Nuthall, P. L., & Nartea, G. V. (2004). Measuring the economic inefficiency of Nepalese rice farms using data envelopment analysis. *The Australian Journal of Agricultural and Resource Economics*, *48*(2), 347–369. doi:10.1111/j.1467-8489.2004.00243.x

Díaz-Montenegro, J., Varela, E., & Gil, J. M. (2018). Livelihood strategies of cacao producers in Ecuador: Effects of national policies to support cacao farmers and specialty cacao landraces. *Journal of Rural Studies*, *63*, 141–156. doi:10.1016/j.jrurstud.2018.08.004

Diaz, R., & Rosenberg, R. (2008). Spreading dead zones and consequences for marine eco-systems. *Science*, *321*(5891), 926–929. doi:10.1126cience.1156401 PMID:18703733

Djanaguiraman, M., Boyle, D. L., Welti, R., Jagadish, S. V. K., & Prasad, P. V. V. (2018). Decreased photosynthetic rate under high temperature in wheat is due to lipid desaturation, oxidation, acylation, and damage of organelles. *BMC Plant Biology*, *18*(1), 55. doi:10.118612870-018-1263-z PMID:29621997

Doherty, B., & Meehan, J. (2006). Competing on social resources: The case of the day chocolate company in the UK confectionery sector. *Journal of Strategic Marketing*, *14*(4), 299–313. doi:10.1080/09652540600947847

Dolman, M. A., Sonneveld, M. P. W., Mollenhorst, H., & de Boer, I. J. M. (2014). Benchmarking the economic, environmental and societal performance of Dutch dairy farms aiming at internal recycling of nutrients. *Journal of Cleaner Production*, *73*, 245–252. doi:10.1016/j.jclepro.2014.02.043

Dompreh, E. B., Asare, R., & Gasparatos, A. (2021). Stakeholder perceptions about the drivers, impacts and barriers of certification in the Ghanaian cocoa and oil palm sectors. *Sustainability Science*, *16*(6), 2101–2122. doi:10.100711625-021-01027-5

Donner, M., Gohier, R., & Vries, H. (2020). A new circular business model typology for creating value from agrowaste. *The Science of the Total Environment*, *716*, 137065. doi:10.1016/j.scitotenv.2020.137065 PMID:32044489

Donovan, J. A. (2011). *Value chain development for addressing rural poverty: asset building by smallholder coffee producers and cooperatives in Nicaragua* [Unpublished doctoral dissertation]. SOAS (School of Oriental and African Studies).

Doytch, N., & Eren, M. (2012). Institutional Determinants of Sectoral FDI in Eastern European and Central Asian Countries: The Role of Investment Climate and Democracy. *Emerging Markets Finance & Trade*, *48*(sup4), 14–32. doi:10.2753/REE1540-496X4806S402

Drewnowski, A., & Gomez-Carneros, C. (2000). Bitter taste, phytonutrients, and the consumer: A review. *The American Journal of Clinical Nutrition*, *72*(6), 1424–1435. Advance online publication. doi:10.1093/ajcn/72.6.1424 PMID:11101467

Duong, T., Brewer, T., Luck, J., & Zander, K. (2019). A Global Review of Farmers' Perceptions of Agricultural Risks and Risk Management Strategies. *Agriculture*, *9*(1), 10. doi:10.3390/agriculture9010010

Dupraz, P., & Latruffe, L. (2015). Trends in family labour, hired labour and contract work on French field crop farms: The role of the Common Agricultural Policy. *Food Policy*, *51*, 104–118. doi:10.1016/j.foodpol.2015.01.003

Durán-Zuazo, V. H., Rodriguez, B. C., Gutiérrez-gordillo, S., Bilbao, M., Sacristán, P. C., Parra, J. J. P., & García-Tejero, I. F. (2020). Rethinking irrigated almond and pistachio intensification: A shift towards a more sustainable water management paradigm. *Revista de Ciências Agrárias (Belém)*, *43*(2), 24–49.

Durão, C. R., Oliveira, J. F. S., & De Almeida, M. D. V. (2008). Portugal e o padrão alimentar mediterrânico. *Revista de Alimentação Humana, 14*(3).

Dzudzor, M. I. (2013). *Analysis of Rice Production and Opportunities in Ghana* [Unpublished master dissertation]. Faculté d'Ingénierie Biologique, Agronomique et Environnementale at Université catholique de Louvain, Agosto.

Eckstein, D., Hutfils, M. L. & Winges, M. (2018). *Global Climate Risk Index 2019*. Germanwatch e.V. Retrieved from https://www.germanwatch.org/sites/germanwatch.org /files/Global%20Climate%20Risk%20Index%202019_2.pdf

Economics, T. (2021). *India GDP from Agriculture*. https://tradingeconomics.com/india/gdp-from-agriculture

Elaalem, M., Comber, A., & Fisher, P. (2011). A Comparison of Fuzzy AHP and Ideal Point Methods for Evaluating Land Suitability. *Transactions in GIS*, *15*(3), 329–346.

Elkington, J. (1997). *Cannibals with Forks. The triple bottom line of 21st century business*. Capstone, Oxford.

Elkington, J. (1998). Partnerships from cannibals with forks: The triple bottom line of 21st-century business. *Environmental Quality Management*, *8*(1), 37–51. doi:10.1002/tqem.3310080106

Elkington, J. (2018). 25 years ago I coined the phrase "triple bottom line." Here's why it's time to rethink it. *Harvard Business Review*.

Ellis, E., Kwofie, E. M., & Ngadi, M. (2019). Value Beyond Price: End User Value Chain Analysis. *International Journal on Food System Dynamics*, *10*(4), 347–360.

Ellram, M. (1991). Supply-Chain Management: The Industrial Organisation Perspective. *International Journal of Physical Distribution & Logistics Management*, *21*(1), 13–22. doi:10.1108/09600039110137082

Epstein, M. (2008). *Making sustainability work: Best practices in managing and measuring corporate social, environmental, and economic impacts*. Greenleaf.

Ericksen, P. J., Ingram, J. S. I., & Liverman, D. M. (2009). Food security and global environmental change: Emerging challenges. *Environmental Science & Policy*, *12*(4), 373–377. Advance online publication. doi:10.1016/j.envsci.2009.04.007

Escobar, S., Santander, M., Useche, P., Contreras, C., & Rodríguez, J. (2020). Aligning strategic objectives with research and development activities in a soft commodity sector: A technological plan for Colombian cocoa producers. *Agriculture*, *10*(5), 141. doi:10.3390/agriculture10050141

Espírito Santo, S. N. (2008). *Programas de ajustamento estrutural, produção agrícola e segurança alimentar na África Sub-saariana: caso específico de S. Tomé e Príncipe* [Unpublished doctoral dissertation]. Instituto Superior de Agronomia, Universidade Técnica de Lisboa, Lisboa, Portugal.

Esposito, K., Chiodini, P., Maiorino, M. I., Bellastella, G., Panagiotakos, D., & Giugliano, D. (2014). Which diet for prevention of type 2 diabetes? A meta-analysis of prospective studies. *Endocrine*, *47*(1), 107–116. Advance online publication. doi:10.100712020-014-0264-4 PMID:24744219

Estival, K.G.S., Corrêa, S.R.S., Mariani, M.A.P., & Benini, E.G. (2016). Analysis of the participation of the producers in the farming of settlements and rural communities in the south of Bahia, Brazil, in the chain of value of cocoa (chocolate). *Espacios*, *37*(17).

Estruch, R., Ros, E., Salas-Salvadó, J., Covas, M.-I., Corella, D., Arós, F., Gómez-Gracia, E., Ruiz-Gutiérrez, V., Fiol, M., Lapetra, J., Lamuela-Raventos, R. M., Serra-Majem, L., Pintó, X., Basora, J., Muñoz, M. A., Sorlí, J. V., Martínez, J. A., Fitó, M., Gea, A., ... Martínez-González, M. A. (2018). Primary Prevention of Cardiovascular Disease with a Mediterranean Diet Supplemented with Extra-Virgin Olive Oil or Nuts. *The New England Journal of Medicine*, *378*(25), e34. Advance online publication. doi:10.1056/NEJMoa1800389 PMID:29897866

EU. (2021). *Rice Market Situation Committee for the Common Organisation of Agricultural Markets.* European Union. https://ec.europa.eu/info/sites/default/files/food-farming-fisheries/plants_and_plant_products/presentations/rice-market-situation.pdf

European Comission/Comisia Europeană. (2018). Regulamentul de punere în aplicare (UE) 1648 al Comisiei din 29 octombrie 2018 de autorizare a introducerii pe piață a xilo-oligozaharidelor ca aliment nou în temeiul Regulamentului (UE) 2015/2283 al Parlamentului European și al Consiliului și de modificare a Regulamentului de punere în aplicare (UE) 2017/2470 al Comisiei (Romanian). Jurnalul Oficial al Uniunii Europene. *Official Journal of the European Union, L, 275*(1), 1–6.

European Commission, DG Agriculture and Rural Development. (2019). *EU agricultural outlook for markets and income, 2019-2030.* Brussels: EU.

European Commission, DG Agriculture and Rural Development. (2020). Financial needs in the agriculture and agrifood sectors in Portugal. Brussels: European Commission - Directorate General, Agriculture and Rural Development.

European Commission. (2015). Implementing Decision (EU) no. Commission Decision 545 of 31 March 2015 authorizing the placing on the market of oil extracted from microalgae Schizochytrium sp. (ATCC PTA-9695) as a novel food ingredient under Regulation (EC) No Regulation (EC) No 258/97 of the European Parliament and of the Council. *Official Journal of the European Union, L, 90*(7), 1–4.

European Commission. (2015). Regulation (EU) no. 414 amending Directive 2002/46 / EC of the European Parliament and of the Council as regards (6S) -5-methyltetrahydrofolic acid, the glucosamine salt used in the manufacture of food supplements. *Official Journal of the European Union, L, 68*(26), 1–2.

European Environment Agency. (2015). *Agriculture and Climate Change.* Retrieved from https://www.eea.europa.eu/signals/signals-2015/articles/agriculture-and-climate-change

Eurostat. (2021). *Entrepreneurship and Small and medium-sized enterprises (SMEs).* Eurostat.

Evenson, R. E., Pray, C. E., & Rosegrant, M. W. (1999). *Agricultural Research and Productivity Growth in India.* International Food Policy Research Institute.

EY. (2020). *Managing the impact of COVID-19 on India's supply chains- Now, Next and Beyond*. Retrieved from https://assets.ey.com/content/dam/ey-sites/ey-com/en_in/topics/ government-and-public-sector/2020/09/managing-the-impact-of-covid-19-on-india-supply-chains.pdf

FADN. (2021). *European Commission | Agri-food data portal | Farm economics*. https://agridata.ec.europa.eu/extensions/FarmEconomyFocus/FADNDatabase.html

Fahad, S., Hussain, S., Saud, S., Hassan, S., Ihsan, Z., Shah, A. N., Wu, C., Yousaf, M., Nasim, W., Alharby, H., Alghabari, F., & Huang, J. (2016). Exogenously Applied Plant Growth Regulators Enhance the Morpho-Physiological Growth and Yield of Rice under High Temperature. *Frontiers in Plant Science*, *7*, 1250. doi:10.3389/fpls.2016.01250 PMID:27625658

Fahad, S., Nie, L., Chen, Y., Wu, C., Xiong, D., Saud, S., Hongyan, L., Cui, K., & Huang, J. (2015). Crop Plant Hormones and Environmental Stress. In E. Lichtfouse (Ed.), *Sustainable Agriculture Reviews* (Vol. 15, pp. 371–400). Springer International Publishing., doi:10.1007/978-3-319-09132-7_10

Falco, C., Galeotti, M., & Olper, A. (2019). Climate change and migration: Is agriculture the main channel? *Global Environmental Change*, *59*, 101995. doi:10.1016/j.gloenvcha.2019.101995

Fallahi, P., Elia, G., Foddis, R., Cristaudo, A., & Antonelli, A. (2017). High risk of brain tumors in military personnel: A case control study. *Clinical Therapeutics*, *168*(6), 376–379.

Fancello, G., Mola, F., Frigau, L., Serra, P., Mancini, S., & Fadda, P. (2017). A new management scheme to support reverse logistics processes in the agrifood distribution sector. *Transportation Research Procedia*, *25*, 695–715. doi:10.1016/j.trpro.2017.05.452

Fanelli, R. M. (2020). The Spatial and Temporal Variability of the Effects of Agricultural Practices on the Environment. *Environments*, *7*(4), 33. doi:10.3390/environments7040033

Fang, Z., & Cao, C. (2019). The state of food security and nutrition in the world 2019. Building climate resilience for food security and nutrition. Food and Agriculture Organization of the United Nations.

Fan, X., Pedroli, B., Liu, G., Liu, Q., Liu, H., & Shu, L. (2012). Soil salinity development in the Yellow River Delta in relation to groundwater dynamics. *Land Degradation & Development*, *23*(2), 175–189. doi:10.1002/ldr.1071

FAO (Food and Agriculture Organization of the United States). (2008). *Climate Change and Food Security: A Framework Document*. Retrieved from http://www.fao.org/ forestry/ 15538-079b31d45081fe9c3dbc6ff34de4807e4.pdf

FAO. (2006). https://www.fao.org/fileadmin/templates/faoitaly/documents/pdf/pdf_Food_Security_Cocept_Note.pdf

FAO. (2014). *Building a common vision for sustainable food and agriculture. Principles and approaches*. FAO. www.fao.org/3/a-i3940e.pdf

FAO. (2015). *Climate change and food security: Risks and responses*. Retrieved from https://www.fao.org/3/i5188e/i5188e.pdf

FAO. (2017). *The Future of food and agriculture: Trends and Agriculture*. Retrieved from https://www.fao.org/3/i6583e/i6583e.pdf

FAO. (2017). *Water for Sustainable Food and Agriculture: A report produced for the G20 Presidency of Germany*. Retrieved from https://www.fao.org/3/i7959e/i7959e.pdf

FAO. (2018). Food loss analysis: Causes and solutions - Case study on the rice value chain in the Republic of India. FAO.

FAO. (2019). *Statistics*. FAO.

FAO. (2020a). *COVID-19 crisis and support for agrifood: Public sector responses through the financial sector.* Rome: UN.

FAO. (2020b). *COVID-19 global economic recession: Avoiding hunger must be at the center of the economic stimulus.* United Nations.

FAO. (2020c). *Review of agri-food trade policy in post-Soviet countries 2017-2018.* FAO - The Food and Agriculture Organization.

FAOSTAT. (2016). *FAOSTAT Online Database*. Food and Agriculture Organization. http://faostat3. fao.org

Faradiba, F. (2018). Forecasting Rainfall and Area of Plant Pest Disturbance in Bogor Regency. *Pro Life*, *5*(3), 688-699. http://ejournal.uki.ac.id/index.php/prolife/article/view/847/685

Faradiba, F., & Zet, L. (2020). The Impact of Climate Factors, Disaster, and Social Community in Rural Development. *Journal of Asian Finance, Economics and Business*, *7*(9), 707–717. doi:10.13106/jafeb.2020.vol7.no9.707

Farmers' Portal. (2021a). *About Us.* Retrieved from https://farmer.gov.in/M_Aboutus.aspx

Farmers' Portal. (2021b). *About Wheat.* Retrieved from https://farmer.gov.in/m_cropstaticswheat.aspx

Fasanelli, F., Giraudo, M. T., Vineis, P., Fiano, V., Fiorito, G., Grasso, C., Polidoro, S., Trevisan, M., Grioni, S., Krogh, V., Mattiello, A., Panico, S., Giurdanella, M. C., Tumino, R., De Marco, L., Ricceri, F., & Sacerdote, C. (2019). DNA methylation, colon cancer and Mediterranean diet: Results from the EPIC-Italy cohort. *Epigenetics*, *14*(10), 977–988. Advance online publication. doi:10.1080/15592294.2019.1629230 PMID:31179817

Fasasi, A. R. (2012). *A study on the contribution of supply chain management to Ghana's rice industry* [Unpublished master dissertation]. Universiti Teknikal Malaysia Melaka, Faculty of Technology Management and Technopreneurship.

Fawzy, S., Osman, A. I., Doran, J., & Rooney, D. W. (2020). Strategies for mitigation of climate change: A review. *Environmental Chemistry Letters*, *18*(6), 2069–2094. doi:10.100710311-020-01059-w

FDA. (2018). *Safety Notice Schizochytrium sp. strain FCC-3204. Clinical studies.* https://www.fda.gov/media/117827/download

FDA. (2020). *FaLessa. Prescribing information, side efects and uses.* https://www.drugs.com/pro/falessa.html

Feng, J., Yang, T., Li, F., Zhou, X., Xu, C., & Fang, F. (2021). Impact of tillage on the spatial distribution of CH4 and N2O in the soil profile of late rice fields. *Soil & Tillage Research*, *211*, 105029. doi:10.1016/j.still.2021.105029

Ferguson, L., Poss, J., Grattan, S., Grieve, C., Wang, D., Wilson, C., Donovan, T., & Chao, C.-T. (2002). Pistachio rootstocks influence scion growth and ion relations under salinity and boron stress. *Journal of the American Society for Horticultural Science*, *127*(2), 194–199. doi:10.21273/JASHS.127.2.194

Fernández, M. C. (2013). *La confianza como instrumento de análises em la cadena de valor alimentaria* [Unpublished doctoral dissertation]. Escola Técnica Superior de Ingenieros Agrónomos, Universidad Politécnica de Madrid.

Fernando, A. L., Duarte, M. P., Almeida, J., Boleo, S., & Mendes, B. (2010). Environmental impact assessment of energy crops cultivation in Europe. *Biofuels Bioproducts & Biorefining-Biofpr*, *4*(6), 594–604. doi:10.1002/bbb.249

Ferreira, D., Simões, M. P., Martins, C., & Gaspar, P. D. (2016). Triple Helix in Agriculture Context: The case of Prunus Network in Beira Interior Region. In *Regional HELIX Conference 2016 - International Conference on Regional Triple Helix Dynamics* (pp. 1-10). Academic Press.

Firrisa, M. T., van Duren, I., & Voinov, A. (2014). Energy efficiency for rapeseed biodiesel production in different farming systems. *Energy Efficiency*, *7*(1), 79–95. doi:10.100712053-013-9201-2

Fischer, C., Hartmann, M., Reynolds, N., Leat, P., Revoredo-Giha, C., Henchion, M., Albisu, L. M., & Gracia, A. (2009). Factors influencing contractual choice and sustainable relationships in European agri-food supply chains. *European Review of Agriculture Economics, 36*(4), 541–569. doi:10.1093/erae/jbp041

Fischer-Kowalski, M., Singh, S. J., Lauk, C., Remesch, A., Ringhofer, L., & Gruenbuehel, C. M. (2011). Sociometabolic transitions in subsistence communities: Boserup revisited in four comparative case studies. *Human Ecology Review, 18*(2), 147–158.

Fleury, P. F., Wanke, P., & Figueiredo, K. (2009). Logística empresarial. coleção copped de administração. *Centro de estudos em logística São Paulo: Atlas*, 235.

Flynn, A., & Bailey, K. (2014). Sustainable food supply chains: the dynamics for change. In T. Marsden & A. Morley (Eds.), *Sustainable food systems: building a new paradigm*. Routledge.

FOA in India. (2021). *India at a Glance.* Retrieved from https://www.fao.org/india/fao-in-india/india-at-a-glance/en/

Food and Agriculture Organization of the United Nations. (2008). *Climate–related Transboundary Pests and Diseases.* Author.

Food and Agriculture Organization of the United Nations. (2012). Mainstreaming climate-smart agriculture into a broader landscape approach. *Second Global Conference on Agriculture, Food security and Climate Change*, 1-34.

Food and Agriculture Organization of the United Nations. (2013). *Climate Smart Agriculture Sourcebook.* Author.

Food and Agriculture Organization of the United Nations. (2014). *Small-Scale Bioenergy Initiatives in ASEAN +3.* Author.

Food and Agriculture Organization of the United Nations. (2016). *Climate change and food security: Risks and responses.* Retrieved from https://www.fao.org/3/a-i5188e.pdf

Food and Agriculture Organization of the United States. (2019). *The State of Food Security and Nutrition in the World: Safeguarding Against Economic Slowdowns and Downturns.* Retrieved from https://www.fao.org/3/ca5162en/ca5162en.pdf

Forward-looking industry research institute. (2020). *In 2020 the development of Chinese agricultural products import and export industry market size and trend analysis.* https://www.sohu.com/a/413530053_473133

Fotovat, R., Valizadeh, M., & Toorchi, M. (2007). Association between water-use efficiency components and total chlorophyll content (SPAD) in wheat (*Triticum aestivum* L.) under well-watered and drought stress conditions. *Journal of Food Agriculture and Environment, 225–227.* Advance online publication. doi:10.1234/4.2007.1166

Fouka, V., & Schlapfer, A. (2020). Agricultural Returns to Labour and the Origins of Work Ethics. *Economic Journal (London), 130*(628), 1081–1113. doi:10.1093/ej/ueaa029

Gaffney, J., Bing, J., Byrne, P., Cassman, K., Ciampitti, I., Delmer, D., Habbena, J., Lafitte, R., Lidstroma, U., Porter, D., Sawyer, J., Schussler, J., Settern, T., Sharp, R., Vyn, T., & Warner, D. (2019). Science based intensive agriculture: Sustainability, food security, and the role of technology. *Global Food Security, 23*, 236–244. doi:10.1016/j.gfs.2019.08.003

Galdeano-Gomez, E., Cespedes-Lorente, J., & Rodriguez-Rodriguez, M. (2006). Productivity and environmental performance in marketing cooperatives: An analysis of the Spanish horticultural sector. *Journal of Agricultural Economics, 57*(3), 479–500. doi:10.1111/j.1477-9552.2006.00061.x

Galli, A., Pires, S. M., Iha, K., Alves, A. A., Lin, D., Mancini, M. S., & Teles, F. (2020). Sustainable food transition in Portugal: Assessing the Footprint of dietary choices and gaps in national and local food policies. *The Science of the Total Environment, 749*, 1–15. doi:10.1016/j.scitotenv.2020.141307 PMID:32846345

Galluzzo, N. (2017). Efficiency Analysis in Different Typologies of Farming in Italian Fadn Dataset. *Ekonomika Poljoprivreda-Economics of Agriculture, 64*(2), 451–465. doi:10.5937/ekoPolj1702451G

Galton, F. (1886). Regression Towards Mediocrity in Hereditary Stature. *Journal of the Anthropological Institute of Great Britain and Ireland, 15*, 246–263. doi:10.2307/2841583

Ganić, M. (2019). Does foreign direct investment (FDI) contribute to poverty reduction? Empirical evidence from Central European and Western Balkan countries. *Scientific Ann Econ Bus, 66*(1), 15–27. doi:10.2478aeb-2019-0003

Gao, P., Wang, Y. M., Li, P. F., Zhao, G. J., Sun, W. Y., & Mu, X. M. (2018). Land degradation changes in the Yellow River Delta and its response to the streamflow-sediment fluxes since 1976. *Land Degradation & Development, 29*(9), 3212–3220. doi:10.1002/ldr.3049

Garcia-Closas, R., Berenguer, A., & González, C. A. (2006). Changes in food supply in Mediterranean countries from 1961 to 2001. *Public Health Nutrition, 9*(1), 53–60. Advance online publication. doi:10.1079/PHN2005757 PMID:16480534

García-Tejero, I. F., Gutiérrez Gordillo, S., Souza, L., Cuadros-Tavira, S., & Durán Zuazo, V. H. (2019). Fostering Sustainable Water Use in Almond (*Prunus dulcis* Mill.) Orchards in a Semiarid Mediterranean Environment. *Archives of Agronomy and Soil Science, 65*, 164–181.

Gardiner, B. (2020). Pollution made COVID-19 worse. Now, lockdowns are clearing the air. *National Geographic*. Retrieved from https://www.nationalgeographic.com/science/2020/04/pollution-made-the-pandemic-worse-but-lockdowns-clean-the-sky/

Garske, B., & Ekardt, F. (2021). Economic policy instruments for sustainable phosphorus management: Taking into account climate and biodiversity targets. *Environmental Sciences Europe, 33*(1), 56. doi:10.118612302-021-00499-7

Garske, B., Stubenrauch, J., & Ekardt, F. (2020). Sustainable phosphorus management in European agricultural and environmental law. *Review of European, Comparative & International Environmental Law, 29*(1), 107–117. doi:10.1111/reel.12318

George, H. (2009). *Progress and poverty*. Cambridge University Press. (Original work published 1879) doi:10.1017/CBO9780511693687

Gereffi, G., Humphrey, J., Kaplinsky, R., & Sturgeon, T. (2001). Value chains and development. *IDS Bulletin, 32*(3), 3. doi:10.1111/j.1759-5436.2001.mp32003001.x

Ghadimi, P., & Heavey, C. (2013). A Review of applications of agent-based modelling and simulation in supplier selection problem. In *Modelling and Simulation (EUROSIM), 2013 8th EUROSIM Congress on*. IEEE.

Giannakis, E., & Bruggeman, A. (2018). Exploring the labour productivity of agricultural systems across European regions: A multilevel approach. *Land Use Policy, 77*, 94–106. doi:10.1016/j.landusepol.2018.05.037

Gibson, D. V., & Naquin, H. (2011). Investing in innovation to enable global competitiveness: The case of Portugal. *Technological Forecasting and Social Change, 78*(8), 1299–1309. doi:10.1016/j.techfore.2011.04.004

Giddens, A. (2015). The politics of climate change. *Policy and Politics, 43*(2), 155–162. doi:10.1332/030557315X14290856538163

Gijón, M. C. (2013). *Relaciones hídricas y manejo del riego en pistachero* (*Pistacia vera* L.) (PhD Thesis). Universidad de Córdoba.

Gilal, F. G., Zhang, J., Paul, J., & Gilal, N. G. (2019). The role of self-determination theory in marketing science: An integrative review. *Journal of Interactive Marketing, 51*, 1–8.

Gilani, M., Danish, S., Ahmed, N., Rahi, A. A., Akrem, A., Younis, U., Irshad, I., & Iqbal, R. K. (2020). Mitigation of drought stress in spinach using individual and combined applications of salicylic acid and potassium. *Pakistan Journal of Botany*, *52*(5). Advance online publication. doi:10.30848/PJB2020-5(18)

Giraldi, L., Panic, N., Cadoni, G., Boccia, S., & Leoncini, E. (2017). Association between Mediterranean diet and head and neck cancer: Results of a large case-control study in Italy. *European Journal of Cancer Prevention*, *26*(5), 418–423. Advance online publication. doi:10.1097/CEJ.0000000000000277 PMID:27483412

Girdžiūtė, L. (2012). Risks in Agriculture and Opportunities of their Integrated Evaluation. *Procedia: Social and Behavioral Sciences*, *62*, 783–790. doi:10.1016/j.sbspro.2012.09.132

Giunipero, L. C., Hooker, R. E., & Denslow, D. (2012). Purchasing and supply management sustainability: Drivers and barriers. *Journal of Purchasing and Supply Management*, *18*(4), 258–269. doi:10.1016/j.pursup.2012.06.003

Glavee-Geo, R., Burki, U., & Buvik, A. (2020). Building trustworthy relationships with smallholder (small-scale) agro-commodity suppliers: Insights from the ghana cocoa industry. *Journal of Macromarketing*, *40*(1), 110–127. doi:10.1177/0276146719900370

Gleadow, R. M., Edwards, E. J., & Evans, J. R. (2009). Changes in Nutritional Value of Cyanogenic *Trifolium repens* Grown at Elevated Atmospheric CO_2. *Journal of Chemical Ecology*, *35*(4), 476–478. doi:10.100710886-009-9617-5 PMID:19352773

Glover, J., Cox, C., & Reganold, J. (2007). Future Farming:A return to roots? *Scientific American*, *297*(2), 82–89. doi: 10.1038cientificamerican0807-82 PMID:17894176

Goel, A., Ganesh, L. S., & Kaur, A. (2021). Sustainability assessment of construction practices in India using inductive content analysis of research literature. *Int. J. Construction Management*, *21*(8), 802–817. doi:10.1080/15623599.2019.1583851

Goel, P. (2010). Triple bottomline reporting: An analytical approach for corporate sustainability. *J. of Finance, Accounting, and Management*, *1*(1), 27–42.

Goldhamer, D.A. (2005). Tree water requirements and regulated deficit irrigation. *Pistachio Production Manual*, 103-116.

Gomes, C. B., Tenjua, E., Paquete, H., & Dória, A. (2018). *Relatório Revisão Estratégica "Fome Zero"- Horizonte 2030. Programa Alimentar Mundial*. PAM.

Gonçalves, J. M., Nunes, M., Jordão, A., Ferreira, S., Eugénio, R., Bigeriego, J., Duarte, I., Amador, P., Filipe, O., Damásio, H., & Bahcevandziev, K. (2021). The Challenges of Water Saving in Rice Irrigation: Field Assessment of Alternate Wetting and Drying Flooding and Drip Irrigation Techniques in the Lis Valley, Portugal. *Proceedings of the 1st International Conference on Water Energy Food and Sustainability (ICoWEFS 2021)*, 85–94. 10.1007/978-3-030-75315-3_11

González, C. H., & Fernández, M. P. R. (2017). Total factor productivity and its components for sheep and goat farms in 37 Southern European regions (2004-2012). *Economía Agraria y Recursos Naturales*, *17*(2), 5–27. doi:10.7201/earn.2017.02.01

Gornall, J., Betts, R., Burke, E., Clark, R., Camp, J., Willett, K., & Wiltshire, A. (2010). Implications of climate change for agriculture productivity in the early twenty-first century. *Philosophical Transactions of the Royal Society of London. Series B, Biological Sciences*, *365*(1554), 2973–2989. doi:10.1098/rstb.2010.0158 PMID:20713397

Govindan, K. (2018). Sustainable consumption and production in the food supply chain: A conceptual framework. *International Journal of Production Economics*, *195*, 419–431. doi:10.1016/j.ijpe.2017.03.003

Grammatikopoulou, M. G., Gkiouras, K., Tranidou, A., & Goulis, D. G. (2020). Food security and adherence to the Mediterranean diet: An interplay of socio-demographic characteristics. In The Mediterranean Diet. doi:10.1016/B978-0-12-818649-7.00008-4

Greatrex, H., Hansen, J., Garvin, S., Diro, R., Blakeley, S., Guen, M., Rao, K., & Oagood, D. (2015). Scaling up index insurance for smallholder farmers: Recent evidence and insights. *CCAFS,* 1-32.

Greene, M. W., Roberts, A. P., & Frugé, A. D. (2021). Negative Association Between Mediterranean Diet Adherence and COVID-19 Cases and Related Deaths in Spain and 23 OECD Countries: An Ecological Study. *Frontiers in Nutrition, 8,* 591964. Advance online publication. doi:10.3389/fnut.2021.591964 PMID:33748170

Greenwood, M. J. (2008). Family and sex-specific US immigration from Europe, 1870-1910: A panel data study of rates and composition. *Explorations in Economic History, 45*(4), 356–382. doi:10.1016/j.eeh.2008.02.001

Gregório, M. J., Rodrigues, A. M., Eusébio, M., Sousa, R. D., Dias, S., André, B., Grønning, K., Coelho, P. S., Mendes, J. M., Graça, P., Espnes, G. A., Branco, J. C., & Canhão, H. (2017). Dietary Patterns Characterized by High Meat Consumption Are Associated with Other Unhealthy Life Styles and Depression Symptoms. *Frontiers in Nutrition, 4,* 25. Advance online publication. doi:10.3389/fnut.2017.00025 PMID:28660194

Gregory, P. J., Ingram, J. S., & Brklacich, M. (2005). Climate Change and Food Security. *Philosophical Transactions of the Royal Society of London. Series B, Biological Sciences, 360*(1463), 2139–2148. doi:10.1098/rstb.2005.1745 PMID:16433099

Grosso, G., Marventano, S., Giorgianni, G., Raciti, T., Galvano, F., & Mistretta, A. (2014). Mediterranean diet adherence rates in Sicily, southern Italy. *Public Health Nutrition, 17*(9), 2001–2009. Advance online publication. doi:10.1017/S1368980013002188 PMID:23941897

Grumiller, J. (2018). Upgrading potentials and challenges in commodity-based value chains: The Ivorian and Ghanaian cocoa processing sectors. *Journal für Entwicklungspolitik, 34*(3-4), 15–45. doi:10.20446/JEP-2414-3197-34-3-15

Gunden, C., & Thomas, T. (2012). Assessing consumer attitudes towards fresh fruit and vegetable attributes. *Journal of Food Agriculture and Environment, 10*(2), 85–88.

Gunderson, C. A., Norby, R. J., & Wullschleger, S. D. (2000). Acclimation of photosynthesis and respiration to simulated climatic warming in northern and southern populations of *Acer saccharum*: Laboratory and field evidence. *Tree Physiology, 20*(2), 87–96. doi:10.1093/treephys/20.2.87 PMID:12651476

Guo, J., Liu, X., Zhang, Y., Shen, J., Han, W., Zhang, W., Christie, P., Goulding, K., Vitousek, P., & Zhang, F. (2010). Significant acidification in major Chinese croplands. *Science, 327*(5968), 1008–1010. doi:10.1126cience.1182570 PMID:20150447

Gupta, J. (2010). A history of International Climate Change Policy. *Wiley Interdisciplinary Reviews: Climate Change, 1*(5), 636–653. doi:10.1002/wcc.67

Győrffy, D. (2013). Crisis Management in the EU, Prospects for the de-politicization of Economic Policy. *Focus – Crisis Management in the EU,* 119-132.

Hachem, F., Capone, R., Yannakoulia, M., Dernini, S., Hwalla, N., & Kalaitzidis, C. (2016). The Mediterranean Diet: A Sustainable Consumption Pattern. In Mediterra 2016. Zero Waste in the Mediterranean. Natural Resources, Food and Knowledge. International Centre for Advanced Mediterranean Agronomic Studies (CIHEAM) and Food and Agriculture Organization of the United Nations (FAO), Paris: Presses deSciences Po.

Hahn, T., Figge, F., Pinkse, J., & Preuss, L. (2010). Trade-offs in corporate sustainability: You can't have your cake and eat it. *Business Strategy and the Environment*, *19*(4), 217–229. doi:10.1002/bse.674

Hamilton, M. C., Nedza, J. A., Doody, P., Bates, M. E., Bauer, N. L., Voyadgis, D. E., & Fox-Lent, C. (2016). Web-based geospatial multiple criteria decision analysis using open software and standards. *International Journal of Geographical Information Science*, *30*(8), 1667–1686. doi:10.1080/13658816.2016.1155214

Han, X., Hu, C., Chen, Y., Qiao, Y., Liu, D., Fan, J., Li, S., & Zhang, Z. (2020). Crop yield stability and sustainability in a rice-wheat cropping system based on 34-year field experiment. *European Journal of Agronomy*, *113*, 125965. doi:10.1016/j.eja.2019.125965

Han, Y. P., Qu, C., & Jia, D. D. (2019). Analysis of water footprint and water consumption structure of main crops in Hebei province. *Guangai Paishui Xuebao*, 10.

Hao, A. W., Paul, J., Trott, S., Guo, C., & Wu, H. H. (2019). Two decades of research on nation branding: A review and future research agenda. *International Marketing Review*, *38*(1), 46–69. doi:10.1108/IMR-01-2019-0028

Hardaker, J., Huirne, R., & Anderson, J. (1997). *Coping with Risk in Agriculture*. Cab International.

Harmon, J., Bucy, F., Nickbarg, S., Rao, G., & Wirtenber, J. (2009). In J. Wirtenberg, W. Russell, & D. Lipsky (Eds.), *The Sustainable Enterprise Fieldbook* (pp. 89–115). Greenleaf Publishing.

Harriss, J. (1982). *Rural Development: Theories of Peasant Economy and Agrarian Change*. Routledge.

Hartman, T., Harbour, J., Tharnish, B., Van Meter, J., & Jackson-Ziems, T. A. (2020). Agronomic Factors Associated with Bacterial Leaf Streak Development Caused by Xanthomonas vasicola pv. Vasculorum in Corn. *Phytopathology*, *110*(6), 1132–1138. doi:10.1094/PHYTO-02-20-0043-R

Hart, S. L. (1997). Beyond Greening: Strategies for a Sustainable World. *Harvard Business Review*, *75*(1), 66–76.

Harun, S. N., Hanafiah, M. M., & Aziz, N. I. H. A. (2021). An LCA-Based Environmental Performance of Rice Production for Developing a Sustainable Agri-Food System in Malaysia. *Environmental Management*, *67*(1), 146–161. doi:10.100700267-020-01365-7 PMID:33001258

Harwood, J., Heifner, R., Coble, K., Perry, J., & Somwaru, A. (1999). Managing Risk in Farming: Concepts, Research, and Analysis. Market and Trade Economics Division and Resource Economics, Division, Economic Research Service, U.S. Department of Agriculture, Agricultural Economic Report, No. 774.

Hatfield, J. L., & Prueger, J. H. (2015). Temperature extremes: Effect on plant growth and development. *Weather and Climate Extremes*, *10*, 4–10. doi:10.1016/j.wace.2015.08.001

Hawkes, C. (2020). *COVID-19 and the promise of food system and innovation. Chapter 29 in Book COVID-19 and Global Food Security by John Swinnen and McDermott*. International Food Policy Research Institute. Retrieved from https://ebrary.ifpri.org/ utils/getfile/collection /p15738coll2/id/133762/filename/133971.pdf

Haynes, J., Cubbage, F., Mercer, E., & Sills, E. (2012). The search for value and meaning in the cocoa supply chain in Costa Rica. *Sustainability*, *4*(7), 1466–1487. doi:10.3390u4071466

Helsing, E. (1995). Traditional diets and disease patterns of the Mediterranean, circa 1960. *The American Journal of Clinical Nutrition*, *61*(6), 1329S–1337S. Advance online publication. doi:10.1093/ajcn/61.6.1329S PMID:7754984

Henriksen, L. F., Riisgaard, L., Ponte, S., Hartwich, F., & Kormawa, P. (2010). *Agro-food value chain interventions in Asia: a review and analysis of case studies*. Working Paper. United Nations Industrial Development Organization (UNIDO).

Henry, R. (2020). Innovations in Agriculture and Food Supply in Response to the COVID-19 Pandemic. *Molecular Plant, 13*(8), 1095–1097. doi:10.1016/j.molp.2020.07.011 PMID:32711125

He, Q., Gallear, D., Ghobadian, A., & Ramanathan, R. (2019). Managing knowledge in supply chains: A catalyst to triple bottom line sustainability. *Production Planning and Control, 30*(5-6), 5–6, 448–463. doi:10.1080/09537287.2018.1501814

Herman, E. (2016). Productive Employment in Romania: A Major Challenge to the Integration into the European Union. *Amfiteatru Economic, 18*(42), 335–350.

Heskel, M. A., O'Sullivan, O. S., Reich, P. B., Tjoelker, M. G., Weerasinghe, L. K., Penillard, A., Egerton, J. J. G., Creek, D., Bloomfield, K. J., Xiang, J., Sinca, F., Stangl, Z. R., Martinez-de la Torre, A., Griffin, K. L., Huntingford, C., Hurry, V., Meir, P., Turnbull, M. H., & Atkin, O. K. (2016). Convergence in the temperature response of leaf respiration across biomes and plant functional types. *Proceedings of the National Academy of Sciences of the United States of America, 113*(14), 3832–3837. doi:10.1073/pnas.1520282113 PMID:27001849

Hess, J. P. (2021). A multi-level analysis of sustainability practices in Ghana: Examining the timber, cocoa, and gold mining industries. *The International Journal of Organizational Analysis*. Advance online publication. doi:10.1108/IJOA-01-2020-2011

Heusinkveld, H. J., & Westerink, R. H. S. (2017). Comparison of different in vitro cell models for the assessment of pesticide-induced dopaminergic neurotoxicity. *Toxicology In Vitro, 45*(Pt 1), 81–88. doi:10.1016/j.tiv.2017.07.030

Himabindu, Y., Chakradhar, T., Reddy, M. C., Kanygin, A., Redding, K. E., & Chandrasekhar, T. (2016). Salt-tolerant genes from halophytes are potential key players of salt tolerance in glycophytes. *Environmental and Experimental Botany, 124*, 39–63. doi:10.1016/j.envexpbot.2015.11.010

Hinrichs, C. C. (2003). The practice and politics of food system localization. *Journal of Rural Studies, 19*(1), 33–45. doi:10.1016/S0743-0167(02)00040-2

Hinrichs, C. C., & Charles, L. (2012). Local Food Systems and Networks in the US and the UK: Community Development Considerations for Rural Areas. In M. Shuck-smith, D. L. Brown, S. Shortall, M. Warner, & J. Vergunst (Eds.), *Rural transformations and rural policies in the UK and US* (pp. 156–176). Routledge.

Hobbs, J., Cooney, A., & Fulton, M. (2000). *Value Chains in the Agri-Food Sector. What are they, How do they work and Are they for me?* Department of Agricultural Economics, University of Saskatchewan.

Hobert, R., & Negra, C. (2020). *Climate Change and the Future of Food*. United Nations Foundation. Retrieved from https://unfoundation.org/blog/post/climate-change-and-the-future-of-food/

Högy, P., Wieser, H., Köhler, P., Schwadorf, K., Breuer, J., Franzaring, J., Muntifering, R., & Fangmeier, A. (2009). Effects of elevated CO_2 on grain yield and quality of wheat: Results from a 3-year free-air CO_2 enrichment experiment. *Plant Biology, 11*(s1), 60–69. doi:10.1111/j.1438-8677.2009.00230.x PMID:19778369

Hoo, P. Y., Hashim, H., & Ho, W. S. (2020). Towards circular economy: Economic feasibility of waste to biomethane injection through proposed feed-in tariff. *Journal of Cleaner Production, 270*, 122160. Advance online publication. doi:10.1016/j.jclepro.2020.122160

Horrell, S., Humphries, J., & Weisdorf, J. (2020). Malthus's missing women and children: Demography and wages in historical perspective, England 1280-1850. *European Economic Review, 129*, 103534. doi:10.1016/j.euroecorev.2020.103534

Horváth, I., Glatz, A., Nakamoto, H., Mishkind, M. L., Munnik, T., Saidi, Y., Goloubinoff, P., Harwood, J. L., & Vigh, L. (2012). Heat shock response in photosynthetic organisms: Membrane and lipid connections. *Progress in Lipid Research, 51*(3), 208–220. doi:10.1016/j.plipres.2012.02.002 PMID:22484828

Hou, D. Y., Ding, Z. Y., Guang, H., & Long, H. (2018). A sustainability assessment framework for agricultural land remediation in China. *Land Degradation & Development*, *29*(4), 1005–1018. doi:10.1002/ldr.2748

Huang, G. Q., Wang, X. X., Qian, H. Y., Zhang, T. L., & Zhao, Q. G. (2004). Negative effects of chemical fertilizer application on agricultural ecological environment and countermeasures. *Ecology & Environment.*

Huang, J., & Rozelle, S. (1996). Chinese rice production potential, consumption and trade. *Journal of Chinese Rural Economy*, (4), 21–27.

Huang, J., Wu, Y., & Rozelle, S. (2010). Moving off the farm and intensifying agricultural production in shandong: A case study of rural labor market linkages in china. *Agricultural Economics*, *40*(2), 203–218. doi:10.1111/j.1574-0862.2009.00370.x

Hungara, A., & Nobre, H. (2021). A consumer culture theory perspective of the marketplace: An integrative review and agenda for research. *International Journal of Consumer Studies*, *45*(4), 805–823. doi:10.1111/ijcs.12670

Hu, W., Huang, B., Tian, K., Holm, E., & Zhang, Y. (2017). Heavy metals in intensive greenhouse vegetable production systems along Yellow Sea of China: Levels, transfer and health risk. *Chemosphere*, *167*, 82–90. doi:10.1016/j.chemosphere.2016.09.122

Huyer, S. (2015). Supporting women farmers in a changing climate: Five policy lessons. *CCAFS,* 1-8.

IBEF. (2021). *Agriculture And Food Industry And Exports.* Retrieved from https://www.ibef.org/exports/agriculture-and-food-industry-india.aspx

IBM Corp. (2019). *IBM SPSS Statistics for Windows, Version 26.0*. IBM Corp.

IBM SPSS. (2021, June 10). *SPSS Software*. https://www.ibm.com/analytics/spss-statistics-software

Ilu, I. Y. (2015). *Value chain analysis of rice (Oryza sativa) in Kano river irrigation project (KRIP) Kano State, Nigeria* [Unpublished doctoral dissertation]. Ahmadu Bello University, Zaria, Nigeria.

Imbernón-Moya, A., Burgos, F., Vargas-Laguna, E., Fernández-Cogolludo, E., Aguilar-Martínez, A., & Gallego-Valdés, A. M. (2016). Pemphigus foliaceus associated with Hypericum perforatum. *Journal of the American Academy of Dermatology*, *2*(4), 326–328.

Indah, P. N., Setiawan, R. F., Hendrarini, H., Yektingsih, E., & Sunarsono, R. J. (2021). Agriculture supply chain performance and added value of cocoa: A study in kare village, Indonesia. *Bulgarian Journal of Agricultural Science*, *27*(3), 487–497.

Indian Council of Food And Agriculture. (2017). *Role of Media in Agriculture.* National Round Table Conference. Retrieved from https://www.icfa.org.in/assets/doc/reports/RTC_Media.pdf

Indian Meteorological Department. (2018). *Press Release*. Retrieved from https://mausam.imd .gov.inpages/press_release_view.php?ff=20160921_pr_56

INE. (2021). *Agricultural Census - 2019* [Recenseamento Agrícola – 2019]. INE.

INE. (2021). Recenseamento Agrícola - Análise dos principais resultados - 2019. Instituto Nacional de Estatística.

INE. (n.d.). *Agricultural Statistics* [Estatísticas Agrícolas]. INE.

INE-Instituto Nacional de Estatística. (2019). *Estatísticas Económicas*. www.ine.st/economia.html

Ingram, V., van Rijn, F., Waarts, Y., & Gilhuis, H. (2018). The Impacts of Cocoa Sustainability Initiatives in West Africa. *Sustainability*, *10*(11), 4249. doi:10.3390u10114249

International Crops Research Institute for the Semi-Arid Tropics. (2016). *Building Climate-Smart Villages: Five approaches for helping farmers adapt to climate change*. Patancheru.

International fruit and vegetable mesh. (2021). *Henan crops the affected area of over 15 million mu, rejection of 2.697 million mu*. https://www.sohu.com/a/480242165_617283

International Fund for Agricultural Development. (2012). Climate-smart smallholder agriculture: What's different? International Crops Research Institute for the Semi-Arid Tropics.

International Labour Organization. (2021). *Covid-19 and the world of work: Sectoral Impact, responses and recommendations*. Retrieved from https://www.ilo.org/ global/topics/ coronavirus/sectoral/lang--en/index.htm

International Monetary Fund. (2019). *The Economics of Climate. Finance and Development*. Retrieved from https://www.imf.org/external/pubs/ft/fandd/2019/12/pdf/fd1219.pdf

International Monetary Fund. (2020). *World Economic Outlook: A Long and Difficult Ascent*. Retrieved from file:///C:/Users/dell/AppData/Local/Temp/text.pdf

IPBES. (2019). *Global assessment report on biodiversity and ecosystem services*. IPBES.

IPCC (Intergovernmental Panel on Climate Change). (2019). *Climate Change and Land. IPCC Special Report on Climate Change, Desertification, Land Degradation, Sustainable Land Management, Food Security, and Greenhouse gas fluxes in Terrestrial Ecosystems*. Retrieved from https://www.ipcc.ch/site/assets/uploads/2019/08/4.-SPM_Approved _ Microsite_FINAL.pdf

IPCC. (2014). *Climate change 2014: synthesis report. Contribution of Working Groups I, II and III to the Fifth Assessment Report of the Intergovernmental Panel on Climate Change. Core Writing Team* (R. K. Pachauri & L. A. Meyer, Eds., Vol. 151). IPCC.

IPCC. (2018). Summary for Policymakers. In *Global Warming of 1.5°C. An IPCC Special Report on the impacts of global warming of 1.5°C above pre-industrial levels and related global greenhouse gas emission pathways, in the context of strengthening the global response to the threat of climate change, sustainable development, and efforts to eradicate poverty*. World Meteorological Organization.

Iqbal, B. A., Hassan, M., Yadav, A., & Shriaan, A. A. (2019). Agriculture Sector and Economic Growth of India: An Empirical Association. *Asian Profile*, *47*(1).

Ishisono, K., Mano, T., Yabe, T., & Kitaguchi, K. (2019). Dietary Fiber Pectin Ameliorates Experimental Colitis in a Neutral Sugar Side Chain-Dependent Manner. *Frontiers in Immunology*, *10*, 1–14. doi:10.3389/fimmu.2019.02979

Isil, O., & Hernke, M. T. (2017). The Triple Bottom Line: A Critical Review from a Transdisciplinary Perspective. *Business Strategy and the Environment*, *26*(8), 1235–1251. doi:10.1002/bse.1982

Jackson, G., Mcnamara, K. E., & Witt, B. (2020). "System of hunger": Understanding causal disaster vulnerability of indigenous food systems. *Journal of Rural Studies*, *7*(3), 163–175. doi:10.1016/j.jrurstud.2019.10.042

Jagadish, S. V. K., Bahuguna, R. N., Djanaguiraman, M., Gamuyao, R., Prasad, P. V. V., & Craufurd, P. Q. (2016). Implications of High Temperature and Elevated CO_2 on Flowering Time in Plants. *Frontiers in Plant Science*, *7*, 913. doi:10.3389/fpls.2016.00913 PMID:27446143

Jahan, M. S., Wang, Y., Shu, S., Zhong, M., Chen, Z., Wu, J., Sun, J., & Guo, S. (2019). Exogenous salicylic acid increases the heat tolerance in Tomato (*Solanum lycopersicum* L) by enhancing photosynthesis efficiency and improving antioxidant defense system through scavenging of reactive oxygen species. *Scientia Horticulturae*, *247*, 421–429. doi:10.1016/j.scienta.2018.12.047

Jalilov, S.-M., Mainuddin, M., Maniruzzaman, M., Mahbubul Alam, M., Towfiqul Islam, M., & Jahangir Kabir, M. (2019). Efficiency in the Rice Farming: Evidence from Northwest Bangladesh. *Agriculture*, *9*(11), 245. doi:10.3390/agriculture9110245

Jalindar, M. K., Rao, V. P., Ramulu, V., Kumar, K. A., & Devi, M. U. (2019). Effectiveness of field water tube for standardization of alternate wetting and drying (AWD) method of water management in lowland rice (Oryza Sativa L.). *Irrigation and Drainage*, *68*(4), 679–689. doi:10.1002/ird.2354

Jambor, A., Balogh, J., & Kucsera, P. (2016). Country and industry specific determinants of intra-industry agri-food trade in the Baltic Countries. *Agricultural Economics-Zemedelska Ekonomika*, *62*(6), 280–291. doi:10.17221/153/2015-AGRICECON

Jarre, M., Petit-Boix, A., Priefer, C., Meyer, R., & Leipold, S. (2020). Transforming the bio-based sector towards a circular economy - What can we learn from wood cascading? *Forest Policy and Economics*, *110*, 12. doi:10.1016/j.forpol.2019.01.017

Javed, S. A., Haider, A., & Nawaz, M. (2020). How agricultural practices managing market risk get attributed to climate change? Quasi-experiment evidence. *Journal of Rural Studies*, *73*, 46–55. doi:10.1016/j.jrurstud.2019.11.020

Jenny, H. (1980). *The soil resource: Origin and behavior*. Springer Science & Business Media.

Jensen, M. H. (1997). Food Production in Greenhouses. In E. Goto, K. Kurata, M. Hayashi, & S. Sase (Eds.), *Plant Production in Closed Ecosystems*. Springer., doi:10.1007/978-94-015-8889-8_1

Jiang, Q., Qi, Z., Xue, L., Bukovsky, M., Madramootoo, C. A., & Smith, W. (2020). Assessing climate change impacts on greenhouse gas emissions, N losses in drainage and crop production in a subsurface drained field. *The Science of the Total Environment*, *705*, 135969. doi:10.1016/j.scitotenv.2019.135969 PMID:31838422

Jiang, W., Deng, Y., Tang, Z., Cao, R., Chen, Z., & Jia, K. (2016). Adaptive capacity of mountainous rural communities under restructuring to geological disasters: The case of Yunnan Province. *Journal of Rural Studies*, *47*, 622–629. doi:10.1016/j.jrurstud.2016.05.002

Jiménez-Guerrero, J. F., Gázquez-Abad, J. C., Mondéjar-Jiménez, J. A., & Huertas-García, R. (2012). *Consumer preferences for olive-oil attributes: a review of the empirical literature using a conjoint approach. Olive oil-constituents, quality, health properties and bioconversions*. InTech Europe.

Jindal, S., Laveena, & Aggarwal, A. (2015). A comparitive study of crisis management- toyota v/s General motors. *Scholedge International Journal of Management & Development*, 1-12.

Jochum, M. D., McWilliams, K. L., Borrego, E. J., Kolomiets, M. V., Niu, G., Pierson, E. A., & Jo, Y.-K. (2019). Bioprospecting Plant Growth-Promoting Rhizobacteria That Mitigate Drought Stress in Grasses. *Frontiers in Microbiology*, *10*, 2106. doi:10.3389/fmicb.2019.02106 PMID:31552009

Jongen, W. M. F. (2000). Food supply chains: from productivity toward quality. *Fruit and Vegetable Quality: An Integrated View*, 3-20.

Juliano, B. O., & Villareal, C. P. (1993). *Grain quality evaluation of world rices*. Int. Rice Res. Inst.

Kahiya, E. T. (2018). Five decades of research on export barriers: Review and future directions. *International Business Review*, *27*(6), 1172–1188. doi:10.1016/j.ibusrev.2018.04.008

Kaidar-Person, O., Person, B., Szomstein, S., & Rosenthal, R. J. (2008). Nutritional deficiencies in morbidly obese patients: A new form of malnutrition? Part A: Vitamins. *Obesity Surgery*, *18*(7), 870–876. Advance online publication. doi:10.100711695-007-9349-y PMID:18465178

Kaldor, N. (1966). *Causes of the Slow Rate of Economic Growth of the UK, An Inaugural Lecture*. Cambridge University Press.

Kaldor, N. (1975). Economic Growth and the Verdoorn Law—A Comment on Mr Rowthorn's Article. *Economic Journal (London)*, *85*(340), 891–896. doi:10.2307/2230633

Kalkuhl, M., von Braun, J., & Torero, M. (2016). Volatile and Extreme Food Prices, Food Security, and Policy: An Overview. In M. Kalkuhl, J. von Braun, & M. Torero (Eds.), *Food Price Volatility and Its Implications for Food Security and Policy*. Springer., doi:10.1007/978-3-319-28201-5_1

Kaloxylos, A., Wolfert, J., Verwaart, T., Terol, C. M., Brewsterd, C., Robbemond, R., & Sundmaker, H. (2013). The Use of Future Internet Technologies in the Agriculture and Food Sectors: Integrating the Supply Chain. *Procedia Technology*, 51 – 60.

Kangas, J., & Kangas, A. (2005). Multiple criteria decision support in forest management – the approach, methods applied, and experiences gained. *Forest Ecology and Management*, *207*(1-2), 133–143. doi:10.1016/j.foreco.2004.10.023

Kangas, J., Store, R. L., Leskinen, P., & Mehtatalo, L. (2000). Improving the Quality of Landscape Ecological Forest Planning by Utilizing Advanced Decision-Support Tools. *Forest Ecology and Management, Amsterdam*, *132*(2-3), 157–171. doi:10.1016/S0378-1127(99)00221-2

Kaplisky, R., & Morris, M. (2002). *A handbook for value chain research*. Institute of Development Studies.

Kapoor, R., Ghosh, P., Kumar, M., Sengupta, S., Gupta, A., Kumar, S. S., Vijay, V., Kumar, V., Kumar Vijay, V., & Pant, D. (2020). Valorization of agricultural waste for biogas based circular economy in India: A research outlook. *Bioresource Technology*, *304*, 11. doi:10.1016/j.biortech.2020.123036 PMID:32107150

Kapuria, P., & Saha, R. (n.d.). *Resource use efficiency and productivity: An analysis of India's food grain sector.* ORF. Retrieved from https://www.orfonline .org/research/ resource-use-efficiency-and-productivity-an-analysis-of-indias-food-grain-sector-59894/

Karademir, C., Karademir, E., Çopur, O., & Gencer, O. (2012). Effect of Drought Stress on Leaf Area in Cotton (*Gossypium hirsutum* L.). *11th Meeting of Inter-Regional Cooperative Research Network on Cotton for the Mediterranean and Middle East Regions*.

Karpenstein-Machan, M. (2001). Sustainable cultivation concepts for domestic energy production from biomass. *Critical Reviews in Plant Sciences*, *20*(1), 1–14. doi:10.1080/20013591099164

Karydis, M., & Kitsiou, D. (2012). Eutrophication and environmental policy in the Mediterranean Sea: A review. *Environmental Monitoring and Assessment*, *184*(8), 4931–4984. doi:10.100710661-011-2313-2 PMID:21956336

Kassie, M., Jaleta, M., Shiferaw, B., Mmbando, F., & Mekuria, M. (2013). Adoption of interrelated sustainable agricultural practices in smallholder systems: Evidence from rural Tanzania. *Technological Forecasting and Social Change*, *80*(3), 525–540. doi:10.1016/j.techfore.2012.08.007

Katoh, M., Murase, J., Hayashi, M., Matsuya, K., & Kimura, M. (2004). Nutrient leaching from the plow layer by water percolation and accumulation in the subsoil in an irrigated paddy field. *Soil Science and Plant Nutrition*, *50*(5), 721–729. doi:10.1080/00380768.2004.10408528

Kaur, H. & Kaur, S. (2017). Climate Change Impact on Agriculture and Food Security in India. *Journal of Business Thought,* 7. Retrieved from file:///C:/Users/dell/AppData/Local/ Temp/21329-27060-1-SM-1.pdf

Kazuki, I., Hiroshi, Y., Mamoru, K., Yohei, K., Yuma, B., & Kumi, M. (2015). Methods for estimating causal relationships of adverse events with dietary, Supplements. *BMJ Open*, *5*, 1–7.

Keil, M., Wallace, L., Turk, D., Dixon-Randall, G., & Nulden, U. (2000). An Investigation of Risk Perception and Risk Propensity on the Decision to Continue a Software Development Project. *Journal of Systems and Software, 53*(2), 145–157. doi:10.1016/S0164-1212(00)00010-8

Keys, A., Mienotti, A., Karvonen, M. J., Aravanis, C., Blackburn, H., Buzina, R., Djordjevic, B. S., Dontas, A. S., Fidanza, F., Keys, M. H., Kromhout, D., Nedeljkovic, S., Punsar, S., Seccareccia, F., & Toshima, H. (1986). The diet and 15-year death rate in the seven countries study. *American Journal of Epidemiology, 124*(6), 903–915. Advance online publication. doi:10.1093/oxfordjournals.aje.a114480 PMID:3776973

Kigen, G., Busakhala, N., Kamuren, Z., Rono, H., Kimalat, W., & Njiru, E. (2017). Factors associated with the high prevalence of oesophageal cancer in Western Kenya: A review. *Infectious Agents and Cancer, 12*(1), 1–11. doi:10.118613027-017-0169-y

Kim, J.-O., & Mueller, C. W. (1978a). *Factor Analysis: Statistical Methods and Practical Issues.* SAGE Publications, Inc. doi:10.4135/9781412984256

Kim, J.-O., & Mueller, C. W. (1978b). *Introduction to Factor Analysis: What It Is and How To Do It* (1st ed.). SAGE Publications, Inc. doi:10.4135/9781412984652

Kim, S. Y., Sim, S., & Choi, H. G. (2017). High stress, lack of sleep, low school performance, and suicide attempts are associated with high energy drink intake in adolescents. *PLoS One, 12*(11), 1–12. doi:10.1371/journal.pone.0187759

Kim, W. B., Jerome, D., & Yeung, J. (2017). Diagnosis and management of psoriasis. *Canadian Family Physician Medecin de Famille Canadien, 63*(4), 278–285.

Kirchherr, J., Reike, D., & Hekkert, M. (2017). Conceptualizing the circular economy: An analysis of 114 definitions. *Resources, Conservation and Recycling, 127*, 221–232. doi:10.1016/j.resconrec.2017.09.005

Kisanga, P. S. (2015). *Perfomance of rice value chain in Kahama district, Tanzania* [Unpublished master dissertation]. Sokoine University of Agriculture. Morogoro, Tanzania.

Kissan Kerala. (n.d.). *About Kissan.* Retrieved from http://www.kissankerala.net/kissan/kissancontents/about.jsp

Klein, T., Calanca, P., Holzkamper, A., Lehmann, N., Roesch, A., & Fuhrer, J. (2012). Using farm accountancy data to calibrate a crop model for climate impact studies. *Agricultural Systems, 111*, 23–33. doi:10.1016/j.agsy.2012.05.001

Komarek, A., De Pinto, A., & Smith, V. (2020). A review of types of risks in agriculture: What we know and what we need to know. *Agricultural Systems, •••*, 178.

Kopnina, H. (2017). Sustainability: New strategic thinking for business. *Environment, Development and Sustainability, 19*(1), 27–43. doi:10.100710668-015-9723-1

Koprivica, A., Beljic, C., Vakanjac, B., Vakanjac, V. R., & Ilic, M. C. (2018). The Content of Toxic Metals in Agricultural Produce near a Coal Mine: Case Study KCB in Lazarevac, Serbia. *Minerals (Basel), 8*(4), 131. doi:10.3390/min8040131

Kosaka, S., Suda, K., Gunawan, B., Raksanagara, A., Watanabe, C., & Umezaki, M. (2018). Urban-rural difference in the determinants of dietary and energy intake patterns: A case study in west Java, Indonesia. *PLoS One, 13*(5), e0197626. Advance online publication. doi:10.1371/journal.pone.0197626 PMID:29768478

Kostov, P., Davidova, S., & Bailey, A. (2019). Comparative Efficiency of Family and Corporate Farms: Does Family Labour Matter? *Journal of Agricultural Economics, 70*(1), 101–115. doi:10.1111/1477-9552.12280

Krausmann, F., Haberl, H., Schulz, N. B., Erb, K. H., Darge, E., & Gaube, V. (2003). Land-use change and socio-economic metabolism in Austria - Part I: Driving forces of land-use change: 1950-1995. *Land Use Policy*, *20*(1), 1–20. doi:10.1016/S0264-8377(02)00048-0

Krauss, J. E., & Krishnan, A. (2021). Global decisions versus local realities: Sustainability standards, priorities and upgrading dynamics in agricultural global production networks. *Global Networks*, glob.12325. doi:10.1111/glob.12325

Kreuzwieser, J., & Gessler, A. (2010). Global climate change and tree nutrition: Influence of water availability. *Tree Physiology*, *30*(9), 1221–1234. doi:10.1093/treephys/tpq055 PMID:20581013

Krishnan, P., Ramakrishnan, B., Reddy, K. R., & Reddy, V. R. (2011). Chapter three—High-Temperature Effects on Rice Growth, Yield, and Grain Quality. In D. L. Sparks (Ed.), *Advances in Agronomy* (Vol. 111, pp. 87–206). Academic Press., doi:10.1016/B978-0-12-387689-8.00004-7

Krishnan, S., Laskowski, K., Shukla, V., & Merewitz, E. B. (2013). Mitigation of Drought Stress Damage by Exogenous Application of a Non-Protein Amino Acid γ– Aminobutyric Acid on Perennial Ryegrass. *Journal of the American Society for Horticultural Science*, *138*(5), 358–366. doi:10.21273/JASHS.138.5.358

Krizsan, S. J., Chagas, J. C., Pang, D., & Cabezas-Garcia, E. H. (2021). Sustainability aspects of milk production in Sweden. *Grass and Forage Science*, *76*(2), 205–214. doi:10.1111/gfs.12539

Kumar, R., & Gautam, H. (2014). Climate change and its impact on agricultural productivity in India. *Journal of Climatology & Weather Forecasting*, *2*(1), 1–4. doi:10.4172/2332-2594.1000109

Kuzmanova, M. (2016). Contemporary problems related to crisis management of Organizations. *Trakia Journal of Sciences*, *14*(3), 256–261. doi:10.15547/tjs.2016.03.010

Kvartiuk, V. (2015). Eastern Europe as source for skilled agricultural workers in Germany? *Berichte über Landwirtschaft*, *93*(3).

La Gra, J., Kitinoja, L. & Alpízar, K. (2016). *Metodología de evaluación de cadenas agroalimentarias para la identificación de problemas y proyectos: un primer paso para la disminución de pérdidas de alimentos*. IICA.

Lafargue, P., Rogerson, M., Parry, G.C., & Allainguillaume, J. (2021). Broken chocolate: Biomarkers as a method for delivering cocoa supply chain visibility. *Supply Chain Management*.

Lairez, J., Lopez-Ridaura, S., Jourdain, D., Falconnier, G. N., Lienhard, P., Striffler, B., Syfongxay, C., & Affholder, F. (2020). Context matters: Agronomic field monitoring and participatory research to identify criteria of farming system sustainability in South-East Asia. *Agricultural Systems*, *182*, 102830. doi:10.1016/j.agsy.2020.102830

Lai, S., Mastroluca, D., Matino, S., Panebianco, V., Vitarelli, A., Capotosto, L., Turinese, I., Marinelli, P., Rossetti, M., Galani, A., Baiocchi, P., D'Angelo, A. R., & Palange, P. (2017). Early Markers of Cardiovascular Risk in Autosomal Dominant Polycystic Kidney Disease. *Kidney & Blood Pressure Research*, *42*(6), 1290–1302. doi:10.1159/000486011

Lal, R. (2004). Soil carbon sequestration to mitigate climate change. *Geoderma*, *123*(1-2), 1–22. doi:10.1016/j.geoderma.2004.01.032

Lalwani, S. K., Nunes, B., Chicksand, D., & Boojihawon, D. K. (2018). Benchmarking self-declared social sustainability initiatives in cocoa sourcing. *Benchmarking*, *25*(9), 3986–4008. doi:10.1108/BIJ-07-2017-0186

Lambert, D. M., & Cooper, M. C. (2000). Issues in supply chain management. *Industrial Marketing Management*, *29*(1), 65–83. doi:10.1016/S0019-8501(99)00113-3

Lamine, C. & Chiffoleau, Y. (2012). Reconnecter agriculture e alimentation dans lês territoires: dynamiques et défis. *Pour, 3*(215-216), 85-92.

Lampayan, R. M., Rejesus, R. M., Singleton, G. R., & Bouman, B. A. M. (2015). Adoption and eco-nomics of alternate wetting and drying management for irrigated lowland rice. *Field Crops Research, 170*, 95–108.

Lampayan, R. M., Rejesus, R. M., Singleton, G. R., & Bouman, B. A. M. (2015). Adoption and economics of alternate wetting and drying management for irrigated lowland rice. *Field Crops Research, 170*, 95–108. doi:10.1016/j.fcr.2014.10.013

Lamy, E., Viegas, C., Rocha, A., Lucas, M. R., Tavares, S., Capela e Silva, F., Guedes, D., Laureati, M., Zian, Z., Salles Machado, A., Ellssel, P., Freyer, B., González-Rodrigo, E., Calzadilla, J., Majewski, E., Prazeres, I., Silva, V., Juračak, J., Platilová Vorlíčková, L., . . . Anzman-Frasca, S. (2020). Changes in food behavior during COVID-19 pandemic: a cross-country study about changes in eating habits, motivations and food-related. Academic Press.

Lancaster, A., & Torres, P. (2019). Investigating the Drivers of Farm Diversification Among US, Fruit and Vegetable Operations. *Sustainability, 2019*(11), 3380. doi:10.3390u11123380

Lan, H. Y. (2018). Research on the important impact of big data on agricultural economic management in the process of building smart agriculture. *Modern Business, 2021*(05), 40–42.

Lars, J. (2003). Hazards of heavy metal contamination. *British Medical Bulletin, 68*(1), 167–182. doi:10.1093/bmb/ldg032

Laskar, A. (2003). *Integrating GIS and Multicriteria Decision Making Techniques for Land Resource Planning* (MSc Thesis). University of Twente, Enschede, The Netherlands.

Latruffe, L., Balcombe, K., Davidova, S., & Zawalinska, K. (2004). Determinants of technical efficiency of crop and livestock farms in Poland. *Applied Economics, 36*(12), 1255–1263. doi:10.1080/0003684042000176793

Le Quéré, C., Jackson, R. B., Jones, M. W., Smith, A. J. P., Abernethy, S., Andrew, R. M., De-Gol, A. J., Willis, D. R., Shan, Y., Canadell, J. G., Friedlingstein, P., Creutzig, F., & Peters, G. P. (2020). Temporary reduction in daily global CO_2 emissions during the COVID-19 forced confinement. *Nature Climate Change, 10*(7), 647–653. doi:10.103841558-020-0797-x

Leal-Castañeda, E. J., Inchingolo, R., Cardenia, V., Hernandez-Becerra, J. A., Romani, S., Rodriguez-Estrada, M. T., & García Galindo, H. S. (2015). Effect of Microwave Heating on Phytosterol Oxidation. *Journal of Agricultural and Food Chemistry, 63*(22), 5539–5547. doi:10.1021/acs.jafc.5b00961

Leeuw, J., Njenga, M., Wagner, B., & Iiyama, M. (2014). An assessment of the resilience provided by tress in the drylands of Eastern Africa. World Agroforestry Centre.

León Bravo, V., Jaramillo Villacrés, M., & Silva, M.E. (2021). Analysing competing logics towards sustainable supplier management. *Supply Chain Management.*

Li, X. H., & Wang, Y. S. (2020). Problems and countermeasures of sustainable development of Agricultural economy in China. *Journal of Business Culture, 471*(18), 24-25.

Li,, X., & Jiang,, W., & Duan, D. (2020). Spatio-temporal analysis of irrigation water use coefficients in China. *Journal of Environmental Management*, 262. PMID:32090879

Liang, C., Li, D., Yuan, Z., Liao, Y., Nie, X., Huang, B., Wu, X., & Xie, Z. (2019). Assessing urban flood and drought risks under climate change, China. *Hydrological Processes, 33*(9), 1349–1361. doi:10.1002/hyp.13405

Li, G. X. (2019). Several issues to be considered in promoting the Revitalization of Rural Industry interpret the guidance on promoting the Revitalization of Rural Industry. *China Rural Science and Technology*, (8), 8–11.

Li, H., Zeng, E. Y., & Jing, Y. (2014). Mitigating pesticide pollution in china requires law enforcement, farmer training, and technological innovation. *Environmental Toxicology and Chemistry, 33*(5), 963–971. doi:10.1002/etc.2549 PMID:24753037

Li, J., Pu, L., Zhu, M., Zhang, J., Li, P., Dai, X., Xu, Y., & Liu, L. (2014). Evolution of soil properties following reclamation in coastal areas: A review. *Geoderma, 226*, 130–139. doi:10.1016/j.geoderma.2014.02.003

Liliana, Q. R. M. (2020). The cocoa agri-food chain in Venezuela: Towards territorial development. *Agroalimentaria, 26*(51), 213–238.

Lim, H. S., Lee, H. H., Byun, D. W., Lee, B., Enkhbold, T., & Kim, T. H. (2017). Serum Vitamin D Level Related to Coffee Consumption in Korean Young Adults Using the 5th Korea National Health and Nutrition Examination Survey. *Journal of Bone Metabolism, 24*(4), 229–233. doi:10.11005/jbm.2017.24.4.229

Lim, W. M., Yap, S. F., & Makkar, M. (2021). Home sharing in marketing and tourism at a tipping point: What do we know, how do we know, and where should we be heading? *Journal of Business Research, 122*, 534–566. doi:10.1016/j.jbusres.2020.08.051 PMID:33012896

Lin, M. (2018). Exploration and construction of rural e-commerce development model -- based on the investigation of E-commerce development in Wugong County. *Enterprise Reform and Management, 328*(11), 71-73.

Lindsey, R. (2020). *Climate Change: Atmospheric Carbon Dioxide*. Retrieved from https://www.climate.gov/news-features/understanding-climate/climate-change-atmospheric-carbon-dioxide

Lindsey, R., & Dahlman, L. (2021). *Climate Change: Global Temperature*. Retrieved from https://www.climate.gov/news-features/understanding-climate/climate-change-global-temperature

Linn, T., & Maenhout, B. (2019). Measuring the efficiency of rice production in Myanmar using data envelopment analysis. *Asian Journal of Agriculture and Development, 16*, 1-24.

Linquist, M. M., Adviento-Borbe, M. A. A., Chaney, R., Da Roda, E. F. F., & Van Kessel, C. (2014). Reducing greenhouse gas emissions, water use, and grain arsenic levels in rice systems. *Global Change Biology, 21*(1), 407–417. doi:10.1111/gcb.12701 PMID:25099317

Lioi, M. B., Scarfi, M. R., Santoro, A., Barbieri, R., Zeni, O., Di Berardino, D., & Ursini, M. V. (1998). Genotoxicity and oxidative stress induced by pesticide exposure in bovine lymphocyte cultures in vitro. *Mutation Research, 403*(1-2), 13–20. doi:10.1016/S0027-5107(98)00010-4

Liu, S. J. (2015). Key areas to improve productivity under the new normal economy. *The Trends of China's Industrial Economy,* (7), 7-9.

Liu, W. Y. (2020). Use Internet thinking to promote the transformation and upgrading of walnut industry. *Horticulture in the North,* (16).

Liu, Y., Hu, W., S Jetté-Nantel, & Tian, Z. (2014). The influence of labor price change on agricultural machinery usage in chinese agriculture. *Canadian Journal of Agricultural Economics/Revue canadienne d\"agroeconomie, 62*(2), 219-243.

Liu, D., Gong, Q. W., & Yang, W. J. (2019). The evolution and optimization path of China's cultivated land protection policy in the past 40 years of reform and opening up. *Zhejiang Land and Resources, 189*(01), 21–22.

Liu, G. M., Zhang, X. C., Wang, X. P., Shao, H. B., Yang, J. S., & Wang, X. P. (2017). Soil enzymes as indicators of saline soil fertility under various soil amendments. *Agriculture, Ecosystems & Environment, 237*, 274–279. doi:10.1016/j.agee.2017.01.004

Liu, W. Z., Zhang, W. H., Zheng, Z. H., Zou, J. X., Liu, X. X., Huang, S. H., You, W. J., He, Y., Zhang, J. Y., Wang, X. D., & Pan, B. X. (2020). Identification of a prefrontal cortex-to-amygdala pathway for chronic stress-induced anxiety. *Nature Communications*, *11*(1), 1–15. doi:10.103841467-020-15920-7

Liu, W., Oosterveer, P., & Spaargaren, G. (2016). Promoting sustainable consumption in China: A conceptual framework and research review. *Journal of Cleaner Production*, *134*, 13–21. doi:10.1016/j.jclepro.2015.10.124

Liu, Y. D. (2017). *Internet of Things and smart agriculture*. Hohai University.

Liu, Y. S. (2020). Efficiency analysis and countermeasures of agricultural digital transformation. *Economic Review (Kansas City, Mo.)*.

Liu, Y., Li, N., Zhang, Z. T., Huang, C. F., Chen, X., & Wang, F. (2020). Climate change effects on agricultural production: The regional and sectoral economic consequences in China. *Earth's Future*, *6*(5), 689–703. doi:10.1002/2017EF000768

Loly, J. F. R. B. (2014). *Análise de influência da utilização dos fatores da produção no aumento da produção de arroz no subdistrito de Maliana, Distrito de Bobonaro em Timor-Leste* [Unpublished master dissertation]. Universidade de Évora, Évora.

Lopes, J. M. C., & Gama, M. (2013). Comércio exterior competitive (4th ed.). São Paulo: Aduaneiras.

Lopes, C., Alberto, D., Luz, J. P., & Simões, M. P. (2018). As Políticas Agrícolas Aplicadas à Fruticultura em Produção Integrada na Beira Interior : O Estudo de Caso do Setor das Prunóideas na Cova da Beira [Agricultural Policies Applied to Integrated Fruit Production in Beira Interior: The Case Study of The Stone Fruits Sector in Cova da Beira]. *Revista Portuguesa de Estudos Regionais*, *49*, 98–101.

López, J., Villaseñor, J., López, M., Elvira, A., López, D., & Francisco, M. (2013). *El cultivo del Pistacho*. Edição Mundo-Prensa.

López, S., Rolando, D. G. F., Silva López, J. O., Rojas Briceño, N. B., Oliva, M., Terrones Murga, R. E., Iliquín Trigoso, D., Barboza Castillo, E., & Barrena Gurbillón, M. Á. (2020). Land Suitability for Coffee (*Coffea arabica*) Growing in Amazonas, Peru: Integrated Use of AHP, GIS and RS. *ISPRS International Journal of Geo-Information*, *9*(11), 673. doi:10.3390/ijgi9110673

Louro, M. J. S. (2011). *Modelação do Clima em Estufa Climatizada para Produção e Enraizamento de Estacas de Eucalyptus Globulus Labill* [Unpublished master thesis]. Instituto Politécnico de Viana do Castelo - Escola Superior Agrária de Ponte de Lima.

Louro, T., Simões, C., Castelo, P. M., Capela e Silva, F., Luis, H., Moreira, P., & Lamy, E. (2021a). How Individual Variations in the Perception of Basic Tastes and Astringency Relate with Dietary Intake and Preferences for Fruits and Vegetables. *Foods*, *10*(8), 1961. Advance online publication. doi:10.3390/foods10081961 PMID:34441738

Louro, T., Simões, C., Penetra, M. J., Carreira, L., Castelo, P. M., Luis, H., Moreira, P., & Lamy, E. (2021b). Relationship between mediterranean diet adherence and saliva composition. *Nutrients*, *13*(4), 1246. Advance online publication. doi:10.3390/nu13041246 PMID:33920127

Loviscek, V. (2021). Triple Bottom Line toward a Holistic Framework for Sustainability: A Systematic Review. *J. Contemporary Administrator*, *25*(3), e-200017. doi:10.1590/1982-7849rac2021200017.en

Lucas, M. R. (2006). Handbook of Consumer Behaviour. Universidade de Varsóvia, Polónia.

Ludwig, F., Scheltinga, C. T. V., Verhagen, J., Kruijt, B., Ierland, E. V., Dellink, R., Bruin, K. D., Bruin, K. D., & Kabat, P. (2007). *Climate change impacts on Developing Countries - EU Accountability. Policy Department Economic and Scientific Policy*. European Parliament. Retrieved from https://www.europarl.europa.eu/RegData/etudes/ etudes/join/2007/393511/IPOL-ENVI_ET(2007)393511_EN.pdf

Lu, Y., & de Vries, W. T. (2021). A Bibliometric and Visual Analysis of Rural Development Research. *Sustainability*, *13*(11), 6136. doi:10.3390u13116136

Lv, C., Huang, Y., Sun, W., Yu, L., & Zhu, J. (2020). Response of rice yield and yield components to elevated [CO2]: A synthesis of updated data from FACE experiments. *European Journal of Agronomy*, *112*, 125961. doi:10.1016/j.eja.2019.125961

Lyasnikov, N. (2019). Global challenges and threats of agrarian sector development in Russia. *Food Policy and Security*, 85 – 96.

Mabe, F. N., Danso-Abbeam, G., Azumah, S. B., Amoh Boateng, N., Mensah, K. B., & Boateng, E. (2020). Drivers of youth in cocoa value chain activities in Ghana. *Journal of Agribusiness in Developing and Emerging Economies*, *11*(4), 366–378. doi:10.1108/JADEE-10-2019-0177

MacMillan, A., & Turrentine, J. (2021). *Global Warming 101*. NRDC. Retrieved from https://www.nrdc.org/stories/global-warming-101

Madzaric, S., Ceglie, F. G., Depalo, L., Al Bitar, L., Mimiola, G., Tittarelli, F., & Burgio, G. (2018). Organic vs. Organic—Soil arthropods as bioindicators of ecological sustainability in greenhouse system experiment under Mediterranean conditions. *Bulletin of Entomological Research*, *108*(5), 625–635. doi:10.1017/S0007485317001158 PMID:29166981

Maffei, E. M. (2020). 5-Hydroxytryptophan (5-HTP): Natural Occurrence, Analysis, Biosynthesis, Biotechnology, Physiology and Toxicology. *International Journal of Molecular Sciences*, *22*(1), 1–25. doi:10.3390/ijms22010181

Magombeyi, M. T., & Odhiambo, N. M. (2017). Causal relationship between FDI and poverty reduction in South Africa. *Cogent Economic Finance*, *5*(1), 1357901. doi:10.1080/23322039.2017.1357901

Magrach, A., & Sanz, M. J. (2020). Environmental and social consequences of the increase in the demand for 'superfoods' world-wide. *People Nat.*, *2*(2), 267–278. doi:10.1002/pan3.10085

Maheshwari, M., & Chaudhary, S. (2017). Acute Atrial Fibrillation Complicating Organophosphorus Poisoning. *Heart Views*, *18*(3), 96–99. doi:10.4103/1995-705X.217856

Majeed, S., Akram, M., Latif, M., Ijaz, M., & Hussain, M. (2016). Mitigation of drought stress by foliar application of salicylic acid and potassium in mungbean (*Vigna radiata* L.). *Legume Research -. International Journal (Toronto, Ont.)*, *39*(2). Advance online publication. doi:10.18805/lr.v39i2.9532

Makosa, D. (2015). Constraints and opportunities to upgrading Uganda's rice markets: A value chain approach. *Journal of Development and Agricultural Economics*, *7*(12), 386–399. doi:10.5897/JDAE2015.0643

Malczewski, J. (2006). GIS-based multicriteria decision analysis: A survey of the literature. *International Journal of Geographical Information Science*, *20*(7), 703–726. doi:10.1080/13658810600661508

Malczewski, J., & Rinner, C. (2015). *Multicriteria Decision Analysis in Geographic Information Science*. Springer. doi:10.1007/978-3-540-74757-4

Malki, M., Bouchaou, L., Hirich, A., Ait Brahim, Y., & Choukr-Allah, R. (2017). Impact of agricultural practices on groundwater quality in intensive irrigated area of Chtouka-Massa, Morocco. *The Science of the Total Environment*, *574*, 760–770. doi:10.1016/j.scitotenv.2016.09.145 PMID:27664763

Mall, R. K., Gupta, A., & Sonkar, G. (2017). Effect of Climate Change on Agricultural Crops. In Current Developments in Biotechnology and Bioengineering: Crop Modification, Nutrition, and Food Production. doi:10.1016/B978-0-444-63661-4.00002-5

Manley, W., Foot, K., & Davis, A. (2019). Oxford Dictionary of Agriculture and Land Management (1st ed.). Academic Press.

Marabelli, C., Munarini, E., Lina, M., Mazza, R., Boffi, R., De Marco, C., Ruprecht, A., Angellotti, G., Veronese, C., Pozzi, P., Bruno, E., Gargano, G., Cavalleri, A., Garrone, G., & Berrino, F. (2017). A pilot study with early adolescents: Dealing with diet, tobacco and air pollution using practical experiences and biological markers. *Multidisciplinary Respiratory Medicine*, *12*(30), 1–9.

Marcus, A., & Fremeth, A. (2009). Green management matters regardless. *The Academy of Management Perspectives*, *23*(3), 17–26. doi:10.5465/amp.2009.43479261

Maris, C.-M. N. (2017). *Analysis of value chain in rice production systems in Ebonyi State, Nigeria* [Unpublished doctoral dissertation]. Federal University of Technology, Owerri, Nigeria.

Marques, F. C. (2017). *Metodologia TRIZ Aplicada ao Desenvolvimento de uma Estufa "Inteligente" Energeticamente Sustentável* [Unpublished master thesis]. Faculdade em Ciência e Tecnologia, Universidade Nova de Lisboa.

Marques, A., Cunha, J., De Meyer, A., & Navare, K. (2020). Contribution Towards a Comprehensive Methodology for Wood-Based Biomass Material Flow Analysis in a Circular Economy Setting. *Forests*, *11*(1), 21. doi:10.3390/f11010106

Marsden, T., & Morley, A. (2014). Current food questions and their scholarly challenges: creating and framing a sustainable food paradigma. In T. Marsden & A. Morley (Eds.), Sustainable Food Systems: Building a New Paradigm. Routledge. doi:10.4324/9780203083499-9

Marsden, T., Banks, J., & Bristow, G. (2000). Food supply chain approaches: Exploring their role in rural development. *Sociologia Ruralis*, *40*(4), 424–438. doi:10.1111/1467-9523.00158

Marshall, R., Lag-Brotons, A. J., Inam, E. J., Herbert, B. M. J., Hurst, L., & Semple, K. T. (2020). From Bioenergy By-products to Alternative Fertilisers: Pursuing a Circular Economy. In RSC Green Chemistry (pp. 289-314). Royal Society of Chemistry.

Martinho, V. J. P. D. (2015). Output Impacts of the Single Payment Scheme in Portugal: A Regression with Spatial Effects. *Outlook on Agriculture*, *44*(2), 109–118. doi:10.5367/oa.2015.0203

Martinho, V. J. P. D. (2017). Efficiency, total factor productivity and returns to scale in a sustainable perspective: An analysis in the European Union at farm and regional level. *Land Use Policy*, *68*, 232–245. doi:10.1016/j.landusepol.2017.07.040

Martinho, V. J. P. D. (2020a). Agricultural Entrepreneurship in the European Union: Contributions for a Sustainable Development. *Applied Sciences (Basel, Switzerland)*, *10*(6), 2080. doi:10.3390/app10062080

Martinho, V. J. P. D. (2020b). Comparative analysis of energy costs on farms in the European Union: A nonparametric approach. *Energy*, *195*, 116953. doi:10.1016/j.energy.2020.116953

Martinho, V. J. P. D. (2020c). Relationships between agricultural energy and farming indicators. *Renewable & Sustainable Energy Reviews*, *132*, 110096. doi:10.1016/j.rser.2020.110096

Martinho, V. J. P. D. (2021). Direct and indirect energy consumption in farming: Impacts from fertilizer use. *Energy*, *236*, 121504. doi:10.1016/j.energy.2021.121504

Martinsson, E., & Hansson, H. (2021). Adjusting eco-efficiency to greenhouse gas emissions targets at farm level? The case of Swedish dairy farms. *Journal of Environmental Management*, *287*, 112313. doi:10.1016/j.jenvman.2021.112313 PMID:33714041

Matissek, R., Reinecke, J., Von Hagen, O., & Manning, S. (2012). *Sustainability in the Cocoa Sector-Review*. Challenges and Approaches.

Matkovski, B., Kalas, B., Zekic, S., & Jeremic, M. (2019). Agri-food competitiveness in South East Europe. *Outlook on Agriculture*, *48*(4), 326–335. doi:10.1177/0030727019854770

Matopoulos, A., Vlachopoulou, M., Manthou, V., & Manos, B. (2007). A conceptual framework for supply chain collaboration: Empirical evidence from the agri-food industry. *Supply Chain Management*, *12*(3), 177–186. doi:10.1108/13598540710742491

Matran, I. M., & Dumitrascu, D. L. (2018). Legătura între alimente, factorii psihosociali şi suferinţă. *Revista Medicală Română*, *LXV*(1), 24–3.

Maugeri, A., Barchitta, M., Fiore, V., Rosta, G., Favara, G., La Mastra, C., La Rosa, M. C., San Lio, R. M., & Agodi, A. (2019). Determinants of adherence to the mediterranean diet: Findings from a cross-sectional study in women from Southern Italy. *International Journal of Environmental Research and Public Health*, *16*(16), 2963. Advance online publication. doi:10.3390/ijerph16162963 PMID:31426512

Ma, W., & Abdulai, A. (2019). IPM adoption, cooperative membership and farm economic performance: Insight from apple farmers in China. *China Agricultural Economic Review*, *11*(2), 218–236. doi:10.1108/CAER-12-2017-0251

Ma, W., & Wang, X. (2020). Internet use, sustainable agricultural practices and rural incomes: Evidence from china. *The Australian Journal of Agricultural and Resource Economics*, *64*(4), 1087–1112. doi:10.1111/1467-8489.12390

Mayén, A. L., Marques-Vidal, P., Paccaud, F., Bovet, P., & Stringhini, S. (2014). Socioeconomic determinants of dietary patterns in low- and middle-income countries: A systematic review. *The American Journal of Clinical Nutrition*, *100*(6), 1520–1531. Advance online publication. doi:10.3945/ajcn.114.089029 PMID:25411287

McDonough, W., & Braungart, M. (2002). Design for the triple top line: New tools for sustainable commerce. *Corporate Environmental Strategy*, *9*(3), 251–258. doi:10.1016/S1066-7938(02)00069-6

McLoughlin, K., & Meehan, J. (2021). The institutional logic of the sustainable organisation: The case of a chocolate supply network. *International Journal of Operations & Production Management*, *41*(3), 251–274. doi:10.1108/IJOPM-11-2020-0773

McMichael, A. J., Lendrum, D. H. C., Corvalan, C. F., Ebi, K. L., Githeko, A. K., Scheraga, J. D., & Woodward, A. (2003). *Climate change and human health*. World Health Organization. Retrieved from https://www.who.int/globalchange /publications/ climch ange .pdf?ua=1

McNamara, K. T., & Weiss, C. (2005). Farm household income and on and off-farm diversification. *Journal of Agricultural and Applied Economics*, *37*(1), 37–48. doi:10.1017/S1074070800007082

Medda, N., Kumar De, S., & Maiti, S. (2021). Different mechanisms of arsenic related signaling in cellular proliferation, apoptosis and neo-plastic transformation. *Ecotoxicology and Environmental Safety*, *208*, 1–9. doi:10.1016/j.ecoenv.2020.111752

MEDWATERICE Homepage. (n.d.). https://www.medwaterice.org/

Melchor-Ferrer, E. (2020). Determinants of labour productivity growth in Spanish and Portuguese regions: A spatial shift-share approach. *The Annals of Regional Science*, *65*(1), 45–65. doi:10.100700168-019-00973-4

Mendoza, M. L. B., Cedeño, M. R., & Cossío, N. S. (2020). Prospective of the circular economy in the agri-food chain of fine organic aromatic cocoa in the province of manabi. *Revista de la Facultad de Agronomía, 37*(1), 95–110.

Meng, L. (2009). *Study on the influence of RMB appreciation on the export trade of agricultural products in China* [Doctoral dissertation]. Hunan Agricultural University.

Mentzer, J. T., DeWitt, W., Keebler, J. S., Min, S., Nix, N. W., Smith, C. D., & Zacharia, Z. G. (2001). Defining Supply Chain Management. *Journal of Business Logistics, 22*(2), 1–25. doi:10.1002/j.2158-1592.2001.tb00001.x

Merkulova, T., Kononova, K., & Titomir, O. (2017). Development trends of circular economy: Case study of Ukraine. *Rivista di Studi sulla Sostenibilità*, (2), 51–66. doi:10.3280/RISS2017-002005

Meul, M., Nevens, F., Reheul, D., & Hofman, G. (2007). Energy use efficiency of specialised dairy, arable and pig farms in Flanders. *Agriculture, Ecosystems & Environment, 119*(1–2), 135–144. doi:10.1016/j.agee.2006.07.002

Meyfroidt, P., Schierhorn, F., Prishchepov, A. V., Mueller, D., & Kuemmerle, T. (2016). Drivers, constraints and trade-offs associated with recultivating abandoned cropland in Russia, Ukraine and Kazakhstan. *Global Environmental Change, 37*, 1–15. doi:10.1016/j.gloenvcha.2016.01.003

Middendorp, R. S., Boever, O., Rueda, X., & Lambin, E. F. (2020). Improving smallholder livelihoods and ecosystems through direct trade relations: High-quality cocoa producers in Ecuador. *Business Strategy and Development, 3*(2), 165–184. doi:10.1002/bsd2.86

Midingoyi, S. G., Kassie, M., Muriithi, B., Diiro, G., & Ekesi, S. (2019). Do farmers and the environment benefit from adopting integrated pest management practices? Evidence from Kenya. *Journal of Agricultural Economics, 70*(2), 452–470. doi:10.1111/1477-9552.12306

Mielke, E., Diaz Anadon, L., & Narayanamurti, V. (2010). *Water consumption of energy resource extraction, processing, and conversion. Discussion paper for Energy Technology Innovation and Policy.* Harvard Kennedy School.

Mighty, M. A. (2015). Site suitability and the analytic hierarchy process: How GIS analysis can improve the competitive advantage of the Jamaican coffee industry. *Applied Geography (Sevenoaks, England), 58*, 84–93. doi:10.1016/j.apgeog.2015.01.010

Milanovic, T., Popovic, V., Vuckovic, S., Rakascan, N., Popovic, S., & Petkovic, Z. (2020). Analysis of soybean production and biogas yield to improve eco-marketing and circular economy. *Ekonomika Poljoprivreda-Economics of Agriculture, 67*(1), 141–156. doi:10.5937/ekoPolj2001141M

Ministry of Home Affairs, Government of India. (2011). *Disaster Management in India.* Retrieved from file:///C:/Users/dell/AppData/Local/ Temp/disaster_ management _in_india.pdf

Ministry of Statistics and Programme Implementation and The World Food Programme. (2019). *Food and Nutrition Security Analysis, India, 2019.* Retrieved from http://mospi.nic.in/ sites/default/files/publication_reports/document%281% 29.pdf

Mishra, A., & El-Osta, H. (2002). Risk Management Through Enterprise Diversification: A Farm-Level Analysis. *American Agricultural Economics Association.*

Mistretta, A., Marventano, S., Antoci, M., Cagnetti, A., Giogianni, G., Nolfo, F., Rametta, S., Pecora, G., & Marranzano, M. (2017). Mediterranean diet adherence and body composition among Southern Italian adolescents. *Obesity Research & Clinical Practice, 11*(2), 215–226. Advance online publication. doi:10.1016/j.orcp.2016.05.007 PMID:27269367

Mithöfer, D., Roshetko, J. M., Donovan, J. A., Nathalie, E., Robiglio, V., Wau, D., Sonwa, D., & Blare, T. (2017). Unpacking 'sustainable' cocoa: Do sustainability standards, development projects and policies address producer concerns in Indonesia, Cameroon and Peru? *The International Journal of Biodiversity Science, Ecosystem Services & Management, 13*(1), 444–469. doi:10.1080/21513732.2018.1432691

Mkisan. (n.d.). *About KCC.* Retrieved from https://mkisan.gov.in/aboutkcc.aspx

Modica, G., Pollino, M., Lanucara, S., La Porta, L., Pellicone, G., Di Fazio, S., & Fichera, C. R. (2016). Land suitability evaluation for agro-forestry: Definition of a Web-Based Multi-Criteria Spatial Decision Support System (MC-SDSS): Preliminary results. In Computational Science and Its Applications - ICCSA 2016 (pp. 399-413). Academic Press.

Mohtadi, K., Msaad, R., Benalioua, N., Jafri, A., Meftah, H., Elkardi, Y., Lebrazi, H., Kettani, A., Derouiche, A., Taki, H., & Saïle, R. (2020). Sociodemographic and Lifestyle Factors Associated with Adherence to Mediterranean Diet in Representative Adult Population in Casablanca City, Morocco: A Cross-Sectional Study. *Journal of Nutrition and Metabolism, 3105271*, 1–9. Advance online publication. doi:10.1155/2020/3105271 PMID:32280541

Molina-Moreno, V., Leyva-Diaz, J. C., Llorens-Montes, F. J., & Cortes-Garcia, F. J. (2017). Design of Indicators of Circular Economy as Instruments for the Evaluation of Sustainability and Efficiency in Wastewater from Pig Farming Industry. *Water (Switzerland), 9*(9), 13. doi:10.3390/w9090653

Mondelaers, K., Verbeke, W., & Van Huylenbroeck, G. (2009). Importance of health and environment as quality traits in the buying decision of organic products. *British Food Journal, 111*(10), 1120–1139. doi:10.1108/00070700910992952

Montanarella, L., & Panagos, P. (2021). The relevance of sustainable soil management within the European Green Deal. *Land Use Policy, 100*, 104950. doi:10.1016/j.landusepol.2020.104950

Moreno-Cruz, J. B., & Smulders, S. (2017). Revisiting the economics of climate change: The role of geoengineering. *Research in Economics, 71*(2), 212–224. doi:10.1016/j.rie.2016.12.001

Moreno-Miranda, C., Molina, I., Miranda, Z., Moreno, R., & Moreno, P. (2020). The cocoa value chain in Ecuador: Proposal of strategies to support sustainability. *Bioagro-, 32*(3), 205–214.

Moretti, C. L., Mattos, L. M., Calbo, A. G., & Sargent, S. A. (2010). Climate changes and potential impacts on postharvest quality of fruit and vegetable crops: A review. *Food Research International, 43*(7), 1824–1832. doi:10.1016/j.foodres.2009.10.013

Mose, A. A., Mostaf, A., & David, J. B., & Lam-Son, P. T. (2018). Salt stress tolerance mechanisms and potential applications of legumes for sustainable reclamation of salt-degraded soils. *Land Degradation & Development, 29*, 3783–3794.

Moulton, A. (2015). *Greenhouse Governmentality: Discourses of Rural Development and the Negotiation of Farmer Subjectivity in Jamaica* [Unpublished master thesis]. East Carolina University.

Movilla-Pateiro, L., Mahou-Lago, X. M., Doval, M. I., & Simal-Gandara, J. (2021). Toward a sustainable metric and indicators for the goal of sustainability in agriculturaland food production. *Critical Reviews in Food Science and Nutrition, 61*(7), 1108–1129. doi:10.1080/10408398.2020.1754161 PMID:32319311

Mowat, A., & Collins, R. (2000). Consumer behaviour and fruit quality: Supply chain management in an emerging industry. *Supply Chain Management, 5*(1), 45–54. doi:10.1108/13598540010312963

Moyano Pesquera, P. B., Miranda Escolar, B., & Gordo Gomez, P. (2017). The involvement of economic and social stakeholders in the revitalization of rural municipalities. *Investigaciones Regionales, 38*, 91–115.

Mujcinovic, A., Nikolic, A., Tuna, E., Stamenkovska, I. J., Radovic, V., Flynn, P., & McCauley, V. (2021). Is It Possible to Tackle Youth Needs with Agricultural and Rural Development Policies? *Sustainability, 13*(15), 8410. doi:10.3390u13158410

Musaiger, A. O., Al-Hazzaa, H. M., Takruri, H. R., & Mokhatar, N. (2012). Change in nutrition and lifestyle in the eastern mediterranean region: Health impact. Journal of Nutrition and Metabolism. doi:10.1155/2012/436762

Muthayya, S., Sugimoto, J. D., Montgomery, S., & Maberly, G. F. (2014). An overview of global rice production, supply, trade, and consumption. *Annals of the New York Academy of Sciences*, *1324*(1), 7–14. doi:10.1111/nyas.12540 PMID:25224455

My, N. H. D., Demont, M., Loo, E. J. V., Guia, A., Rutsaert, P., Tuan, T. H., & Verbeke, W. (2018). What is the value of sustainably-produced rice? Consumer evidence from experimental auctions in Vietnam. *Food Policy*, *79*, 283–296. doi:10.1016/j.foodpol.2018.08.004

Nabhani, I., Daryanto, A., Yassin, M., & Rifin, A. (2015). Can Indonesia cocoa farmers get benefit on global value chain inclusion? A literature review. *Asian Social Science*, *11*(18), 288–294. doi:10.5539/ass.v11n18p288

Nakhooda, S., Watson, C., & Schalatek, L. (2016). The Global Climate Finance Architecture. *Climate Funds Update,* 1-5.

Nani, A., Murtaza, B., Khan, A. S., Khan, N. A., & Hichami, A. (2021). Antioxidant and anti-inflammatory potential of polyphenols contained in Mediterranean diet in obesity: Molecular mechanisms. *Molecules (Basel, Switzerland)*, *26*(4), 985. Advance online publication. doi:10.3390/molecules26040985 PMID:33673390

Narayanan, S. (2020). *How India's agrifood supply chains fared during the COVID-19 lockdown, from farm to fork.* International Food Policy Research Institute. Retrieved from https://www.ifpri.org/blog/how-indias-agrifood-supply-chains-fared-during-covid-19-lockdown-farm-fork

Narciso, A., Barzini, S. A., & Di Nuzzo, A. (2020). Discovering neverland: São tomé and príncipe and the development of the agricultural heritage of a multi-ethnic population. *Journal of Agriculture and Environment for International Development*, *114*(2), 63–84.

Nardone, A., Ronchi, B., Lacetera, N., Ranieri, M. S., & Bernabucci, U. (2010). Effects of climate changes on animal production and sustainability of livestock systems. *Livestock Science*, *130*(1–3), 57–69. doi:10.1016/j.livsci.2010.02.011

NCERT. (n.d.). *Contemporary India II. Agriculture.* Retrieved from https://ncert.nic.in/ncerts/l/jess104.pdf

Nda, M., Adnan, M. S., Ahmad, K. A., Usman, N., Razi, M. A. M., & Daud, Z. (2018). A Review on the Causes, Effects and Mitigation of Climate Changes on the Environmental Aspects. *International Journal of Integrated Engineering*, *10*(4), 169–175. doi:10.30880/ijie.2018.10.04.027

Ndirangu, S. N., Wilson, A., & Oyange, W. A. (2019). Analysis of Millers in Kenya's Rice Value Chain. *Journal of Agriculture and Veterinary Science*, Volume 12 (1). *Ser. III*, (January), 38–47.

Ndiritu, S. W., Kassie, M., & Shiferaw, B. (2014). Are there systematic gender differences in the adoption of sustainable agricultural intensification practices? Evidence from Kenya. *Food Policy*, *49*, 117–127. doi:10.1016/j.foodpol.2014.06.010

Nedoszytko, B., Reszka, E., Gutowska-Owsiak, D., Trzeciak, M., Lange, M., Jarczak, J., Niedoszytko, M., Jablonska, E., Strapagiel, J. D., Skokowski, J., Siekierzycka, A., Nowicki, R. J., Dobrucki, I. T., Zaryczańska, A., & Kalinowski, L. (2020). Genetic and Epigenetic Aspects of Atopic Dermatitis. *International Journal of Molecular Sciences*, *21*(18), 1–18. doi:10.3390/ijms21186484

Neill, A. O. (2020). *Distribution of the workforce across economic sectors in India 2020.* Statista. Retrieved from https://www.statista.com/statistics/271320/distribution-of-the-workforce-across-economic-sectors-in-india/

Nelson, G. C., Rosegrant, M. W., Koo, J., Robertson, R., Sulser, T., Zhu, T., Ringler, C., Msangi, S., Palazzo, A., Batka, M., Magalhaes, M., Santos, R. V., Ewing, M., & Lee, D. (2009). *Climate Change: Impact on agriculture and Costs of Adaptation.* Food Policy Report. International Food Policy Research Institute. Retrieved from http://ebrary.ifpri.org/utils/getfile/collection/p15738coll2/id/130648/filename/130821.pdf

Neto, J. B., Silva, J. R. I., Bezerra, C. W. F., Amorim, T. L., Hermínio, P. J., Morato, R. P., Albuquerque-Silva, M. M., & Simões, V. J. L. P. (2021). Impactos da seca e das mudanças climáticas no metabolismo das plantas: Uma revisão. *Research. Social Development, 10*(8), e16710817060–e16710817060. doi:10.33448/rsd-v10i8.17060

Nguyen, N. V., & Ferrero, A. (2012). Meeting the challenges of global rice production. *Paddy and Water Environment, 4*(1), 1–9. doi:10.100710333-005-0031-5

Nguyen, S. K., Vo, X. V., & Vo, T. M. (2020). Innovative strategies and corporate profitability: The positive resources dependence from political network. *Heliyon, 6*(4), 1–10. doi:10.1016/j.heliyon.2020.e03788 PMID:32373728

Nibbi, L., Chiaramonti, D., & Palchetti, E. (2019). Project BBChina: a new Master Program in three Chinese Universities on Bio-Based Circular Economy; from Fields to Bioenergy, Biofuel and Bioproducts. In J. Yan, H. X. Yang, H. Li, & X. Chen (Eds.), *Innovative Solutions for Energy Transitions* (Vol. 158, pp. 1261–1266). Elsevier Science Bv. doi:10.1016/j.egypro.2019.01.416

Nilsson, M., Griggs, D., & Visbeck, M. (2016). Policy: Map the interactions between Sustainable Development Goals. *Nature, 534*(7607), 320–322. doi:10.1038/534320a PMID:27306173

Nishi, M. (2020). Effects of Early-Life Stress on the Brain and Behaviors: Implications of Early Maternal Separation in Rodents. *International Journal of Molecular Sciences, 21*(9), 1–14. doi:10.3390/ijms21197212

Nkuba, J., Ndunguru, A., Madulu, R., Lwezaura, D., Kajiru, G., Babu, A., Chalamila, B., & Ley, G. (2016). Rice value chain analysis in Tanzania:Identification of constraints, opportunities and upgrading stategies. *African Crop Science Journal, 24*(1), 73–87.

Nogueira, L. R. T. (2014). *Ações voltadas para a Agricultura Familiar que Contribuem com o Desenvolvimento de Municípios do Sul de Minas Gerais* [Unpublished doctoral dissertation]. Minas Gerais. Brazil.

Nonci, N., Nonci, N., & Halim, H., & Muliani. (2019). Role of supply chain in cocoa blue print model performance in national development program. *Int. J. Supply Chain Managemet, 8*(6), 198–210.

Nørgaard, H., & Thuesen, A. A. (2021). Rural community development through competitions, prizes, and campaigns: The villagers' perspective. *Journal of Rural Studies, 87*, 465–473. doi:10.1016/j.jrurstud.2020.03.006

Norman, W., & MacDonald, C. (2004). Getting to the bottom of 'triple bottom line'. *Business Ethics Quarterly, 14*(02), 243–262. doi:10.5840/beq200414211

Notarnicola, B., Sala, S., Anton, A., McLaren, S. J., Saouter, E., & Sonesson, U. (2017). The role of life cycle assessment in supporting sustainable agri-food systems: A review of the challenges. *Journal of Cleaner Production, 140*(2), 399–409.

Nunes, M., Jordão, A., Paixão, J., Russo, A., Eugénio, R., Damásio, H., & Gonçalves, J. M. (2021). Experimentation on water saving in rice irrigation, in Beira Litoral (Portugal). *Proceedings of 15th Portuguese Water Congress.* www.aprh.pt/congressoagua2021/

Nunez, C. (2019). Sea level rise, explained. Reference. *National Geographic.* Retrieved from https://www.nationalgeographic.com/environment/global-warming/sea-level-rise/

Nurhadi, B., Sukri, N., Saputra, R. A., Wandhani, F. I., & Nurlita, A. I. (2020). Physical Characteristics of Amorphous and Crystalline Coconut Sugar Powder with the Addition of Tricalcium Phosphate (TCP) as an Anticaking Agent. *International Journal of Food Sciences, 2020*, 1–10. doi:10.1155/2020/5320173

Obura, J., Ombok, B. O., & Omuga, G. (2017). Analysis of Rice Supply Chain in Kenya. *International Journal of Managerial Studies and Research, 5*(8), 12-17.

Odijie, M. E. (2018). Sustainability winners and losers in business-biased cocoa sustainability programmes in West Africa. *International Journal of Agricultural Sustainability, 16*(2), 214–227. doi:10.1080/14735903.2018.1445408

Odokonyero, T., Barungi, M. & Mbowa, S. (2016). *Supply of Improved Rice Seed in Eastern Uganda: The Gap and Required Investment.* Series Policy Briefs 71, Economic Policy Research Centre (EPRC).

OECD. (2009). *Managing Risks in Agriculture: a Holistic Approach.* Organization for Economic Co-operation and Development.

OECD. (2009). *The Impact of the Global Crisis on SME and Entrepreneurship Financing and Policy Responses*. OECD Publishing.

OECD. (2019). *Risks that Matter: Main Findings from the 2018 OECD Risks that Matter Survey.* https://www.oecd.org/social/risks-that-matter.htm

OECD/FAO-Organization for Economic Cooperation and Development/Food and Agriculture Organization. (2019). *Agricultural Outlook 2019-2028.* OECD Publishing.

Ogunniyi, M. B., & Igberi, C. O. (2014). The impact of foreign direct investment on poverty reduction in Nigeria. *Journal of Economics and Sustainable Development, 5*(14), 73–89.

Oh-e, I., Saitoh, K., & Kuroda, T. (2007). Effects of High Temperature on Growth, Yield and Dry-Matter Production of Rice Grown in the Paddy Field. *Plant Production Science, 10*(4), 412–422. doi:10.1626/pps.10.412

Oh, T. K., Song, I. A., Park, Y. M., Hwang, J. W., Jeon, Y. T., Do, S. H., Yoon, Y. E., Ahn, S., & Lee, J. S. (2017). Prevalence and risk factors for postoperative stress-related cardiomyopathy in adults. *PLoS One, 12*(12), 1–11. doi:10.1371/journal.pone.0190065

Ojo, T. O., Ogundeji, A. A., Babu, S. C., & Alimi, T. (2019). *Estimating financing gaps in rice production in southwestern Nigeria.* IFPRI Discussion Paper 1818. Washington, DC: International Food Policy Research Institute (IFPRI).

Okpiaifo, G., Durand-Morat, A., West, G. H., Nalley, L. L., Nayga, R. M. Jr, & Wailes, E. J. (2020). Consumers' preferences for sustainable rice practices in Nigeria. *Global Food Security, 24*, 100345. doi:10.1016/j.gfs.2019.100345

Oliveira, F., Reis, P., Oliveira, S., Russo, A., Damásio, H., Eugénio, R., Paixão, J., Melo, F., Nunes, M., Teixeira, M., Bahcevandziev, K., & Gonçalves, J. (2019). Sustainability of Rice Production Systems: Agro-Economic Analysis of Baixo Mondego and Lis Irrigation Districts, Portugal. In *X International Scientific Agriculture Symposium "Agrosym 2019"* (pp. 1210-1215). Faculty of Agriculture East Sarajevo.

Olivier, J. G. J., & Peters, J. A. H. W. (2020). *Trends in Global CO_2 and Total Greenhouse Gas Emissions 2019 Report.* PBL Netherlands Environmental Assessment Agency. Retrieved from https://www.pbl.nl/sites/default/files/downloads/pbl-2020-trends-in-global-co2-and-total-greenhouse-gas-emissions-2019-report_4068.pdf

Ollivier de Leth, D., & Ros-Tonen, M. A. F. (2021). Creating shared value through an inclusive development lens: A case study of a CSV strategy in Ghana's cocoa sector. *Journal of Business Ethics.* Advance online publication. doi:10.100710551-021-04808-1

Olper, A., Raimondi, V., Cavicchioli, D., & Vigani, M. (2014). Do CAP payments reduce farm labour migration? A panel data analysis across EU regions. *European Review of Agriculture Economics*, *41*(5), 843–873. doi:10.1093/erae/jbu002

Omoare, O. A., & Oyeleke, O. W. (2017). Assessment of factors affecting rice (oryza spp.) value chain (RVC) in Ogun and Niger states, Nigeria. *Global Journal of Agricultural Research*, *5*(4), 43–59.

Online, T. (2021). *Hi-tech solutions to help farmers combat climate crisis*. https://www.timesonline.lk/features/Hi-tech-solutions-to-help-farmers-combat-climate-crisis/5-1135291

Opoku, R. A., & Akorli, P. A. K. (2009). The preference gap: Ghanaian consumers' attitudes toward local and imported products. *African Journal of Business Management*, *3*(8), 350–357.

Orozco-Aguilar, L., López-Sampson, A., Leandro-Muñoz, M. E., Robiglio, V., Reyes, M., Bordeaux, M., Sepúlveda, N., & Somarriba, E. (2021). Elucidating pathways and discourses linking cocoa cultivation to deforestation, reforestation, and tree cover change in Nicaragua and Peru. *Frontiers in Sustainable Food Systems*, *5*, 5. doi:10.3389/fsufs.2021.635779

Othman, S. N., Othman, Z., Yaacob, N. A., & Abhamid, K. (2016). Sustainable rice production and its impact on the rice value chain: A case study of rural paddy farm in Kedah. *Proceedings of the International Conference on Applied Science and Tecnhology 2016 (ICAST'16). Project Sustainable agriculture and value chain.* 10.1063/1.4960925

Ouellet, G., Lapen, D. R., Topp, E., Sawada, M., & Edwards, M. (2008). A heuristic model to predict earthworm biomass in agroecosystems based on selected management and soil properties. *Applied Soil Ecology*, *39*(1), 35–45. doi:10.1016/j.apsoil.2007.11.003

Owonubi, S. J., Linganiso, L. Z., Motaung, T. E., & Songca, S. P. (2019). Wood and its byproducts: Cascading utilization for biomass (re)generation. In "Waste-to-Profit" (W-t-P): Circular Economy in the Construction Industry for a Sustainable Future (Vol. 2, pp. 293-311). Nova Science Publishers, Inc.

Paladino, O., Massabo, M., & Gandoglia, E. (2020). Assessment of Nitrate Hazards in Umbria Region (Italy) Using Field Datasets: Good Agriculture Practices and Farms Sustainability. *Sustainability*, *12*(22), 9497. doi:10.3390u12229497

Pallaro, E., Subramanian, N., Abdulrahman, M. D., & Liu, C. (2015). Sustainable production and consumption in the automotive sector: Integrated review framework and research directions. *Sustainable Production & Consumption*, *4*(2), 47–61. doi:10.1016/j.spc.2015.07.002

Palmieri, N., Suardi, A., Alfano, V., & Pari, L. (2020). Circular Economy Model: Insights from a Case Study in South Italy. *Sustainability (Switzerland)*, *12*(8), 11. doi:10.3390u12083466

Panagiotakos, D. B., Tzima, N., Pitsavos, C., Chrysohoou, C., Zampelas, A., Toussoulis, D., & Stefanadis, C. (2007). The association between adherence to the mediterranean diet and fasting indices of glucose homoeostasis: The attica study. *Journal of the American College of Nutrition*, *26*(1), 32–38. Advance online publication. doi:10.1080/07315724.2007.10719583 PMID:17353581

Panpluem, N., Mustafa, A., Huang, X., Wang, S., & Yin, C. (2019). Measuring the Technical Efficiency of Certified Organic Rice Producing Farms in Yasothon Province: Northeast Thailand. *Sustainability*, *11*(24), 6974. doi:10.3390u11246974

Pan, S. Y., Du, M. A., Huang, I. T., Liu, I. H., Chang, E. E., & Chiang, P. C. (2015). Strategies on implementation of waste-to-energy (WTE) supply chain for circular economy system: A review. *Journal of Cleaner Production*, *108*, 409–421. doi:10.1016/j.jclepro.2015.06.124

Paraskevas, A. (2006). Crisis management or crisis response system? A complexity science approach to organizational crises. *Management Decision*, *44*(7), 892–907. doi:10.1108/00251740610680587

Parija, P., & Mishra, P. (2011). *India Allows Wheat Exports for the First Time in Four Years*. Bloomberg. Retrieved from https://www.bloomberg.com/news/articles/2011-09-08/india-allows-wheat-exports-for-the-first-time-in-four-years-1-

Pathak, H. (2015). Greenhouse Gas Emission from Indian Agriculture: Trends, Drivers and Mitigation Strategies. *Proceedings of the Indian National Science Academy*, *81*(5), 1133–1149. doi:10.16943/ptinsa/2015/v81i5/48333

Pathera, A. K., Riar, C. S., Yadav, S., & Sharma, D. P. (2017). Effect of Dietary Fiber Enrichment and Different Cooking Methods on Quality of Chicken Nuggets. *Han-gug Chugsan Sigpum Hag-hoeji*, *37*(3), 410–417. doi:10.5851/kosfa.2017.37.3.410

Paul, J. (2019). Marketing in emerging markets: A review, theoretical synthesis and extension. *Int. J Emerging Markets*, *15*(3), 446–468. doi:10.1108/IJOEM-04-2017-0130

Paul, J., & Benito, G. R. G. (2018). A review of research on outward foreign direct investment from emerging countries, including China: What do we know, how do we know and where should we be heading? *Asia Pacific Business Review*, *24*(1), 90–115. doi:10.1080/13602381.2017.1357316

Paul, J., & Criado, A. R. (2020). The art of writing literature review: What do we know and what do we need to know? *International Business Review*, *29*(4), 101717. doi:10.1016/j.ibusrev.2020.101717

Paulsson-Do, U., Edlund, B., Stenhammar, C., & Westerling, R. (2017). Psychosocial vulnerability underlying four common unhealthy behaviours in 15-16-year-old Swedish adolescents: A cross-sectional study. *BMC Psychology*, *5*(1), 1–11. doi:10.118640359-017-0209-9

Pearson, K. (1896). VII. Mathematical contributions to the theory of evolution.—III. Regression, heredity, and panmixia. *Philosophical Transactions of the Royal Society of London. Series A, Containing Papers of a Mathematical or Physical Character*, *187*, 253–318. doi:10.1098/rsta.1896.0007

Pearson, K., & Filon, L. N. G. (1898). VII. Mathematical contributions to the theory of evolution.—IV. On the probable errors of frequency constants and on the influence of random selection on variation and correlation. *Philosophical Transactions of the Royal Society of London. Series A, Containing Papers of a Mathematical or Physical Character*, *191*, 229–311. doi:10.1098/rsta.1898.0007

Perdigones, A., Benedicto, S., & García, J. L. (2015). *Automatização da estufa: a evolução da tecnologia*. Universidade Técnica de Madrid.

Perdomo, J. A., Capó-Bauçà, S., Carmo-Silva, E., & Galmés, J. (2017). Rubisco and Rubisco Activase Play an Important Role in the Biochemical Limitations of Photosynthesis in Rice, Wheat, and Maize under High Temperature and Water Deficit. *Frontiers in Plant Science*, *8*, 490. doi:10.3389/fpls.2017.00490 PMID:28450871

Perez-Araluce, R., Martinez-Gonzalez, M. A., Fernández-Lázaro, C. I., Bes-Rastrollo, M., Gea, A., & Carlos, S. (2021). Mediterranean diet and the risk of COVID-19 in the 'Seguimiento Universidad de Navarra' cohort. *Clinical Nutrition (Edinburgh, Lothian)*. Advance online publication. doi:10.1016/j.clnu.2021.04.001 PMID:33934925

Pestana, M. H., & Gagueiro, J. (2014). Data Analysis for Social Sciences: The Complementarity of SPSS [Análise de dados para as Ciências Sociais: A complementaridade do SPSS] (6th ed.). Sílabos Eds.

Petersm, E. M. J., Michenko, A., Kupfer, J., Kummer, W., Wiegand, S., Niemeier, V., Potekaev, N., Lvov, A., & Gieler, U. (2014). Mental stress in atopic dermatitis—neuronal plasticity and the cholinergic system are affected in atopic dermatitis and in response to acute experimental mental stress in a randomized controlled pilot study. *PLoS One*, *9*(12), 1–17.

Petrikov, A. V. (2020). Adaptation of the agri-food sector to post-pandemic reality. Free Economic Society of Russia, 99-105.

Pfeifer, D., Rešetar, J., Gajdoš Kljusurić, J., Panjkota Krbavčić, I., Vranešić Bender, D., Rodríguez-Pérez, C., Ruíz-López, M. D., & Šatalić, Z. (2021). Cooking at Home and Adherence to the Mediterranean Diet During the COVID-19 Confinement: The Experience From the Croatian COVIDiet Study. *Frontiers in Nutrition*, *8*, 617721. Advance online publication. doi:10.3389/fnut.2021.617721 PMID:33869262

Phua, M.-H., & Minowa, M. (2005). A GIS-based multi-criteria decision making approach to forest conservation planning at a landscape scale: A case study in the Kinabalu Area, Sabah, Malaysia. *Landscape and Urban Planning*, *71*(2-4), 207–222. doi:10.1016/j.landurbplan.2004.03.004

Pinho, I., Rodrigues, S., Franchini, B., & Graça, P. (2016). Padrão Alimentar Mediterrânico: Promotor De Saúde. *Direção-Geral Da Saúde Programa Nacional Para a Promoção Da Alimentação Saudável.*

Pivonello, R., Ferrigno, R., De Martino, M. C., Simeoli, C., Di Paola, N., Pivonello, C., Barba, L., Negri, M., De Angelis, C., & Colao, A. (2020). Medical Treatment of Cushing's Disease: An Overview of the Current and Recent Clinical Trials. *Frontiers in Endocrinology*, *11*, 1–29. doi:10.3389/fendo.2020.00648

Poczta, W., & Pawlak, K. (2011). Potential competitiveness and competitive position of the Polish agri-food sector on the Single European Market. *Berichte über Landwirtschaft*, *89*(1), 134–169.

Poelman, M. P., Gillebaart, M., Schlinkert, C., Dijkstra, S. C., Derksen, E., Mensink, F., Hermans, R. C. J., Aardening, P., de Ridder, D., & de Vet, E. (2021). Eating behavior and food purchases during the COVID-19 lockdown: A cross-sectional study among adults in the Netherlands. *Appetite*, *157*, 105002. Advance online publication. doi:10.1016/j.appet.2020.105002 PMID:33068668

Pohjolainen, P., Vinnari, M., & Jokinen, P. (2015). Consumers' perceived barriers to following a plant-based diet. *British Food Journal*, *117*(3), 1150–1167. Advance online publication. doi:10.1108/BFJ-09-2013-0252

Polit, D., & Hungler, B. (1995). *Fundamentação de Pesquisa de Enfermagem.* Artes Médicas.

Porter, M. E. (1985). *Competitive Advantage. Creating and Sustaining Superior Performance.* Free Press.

Pradhan, P., Costa, L., Rybski, D., Lucht, W., & Kropp, J. P. (2017). A Systematic Study of Sustainable Development Goal (SDG) Interactions. *Earth's Future*, *5*(11), 1169–1179. doi:10.1002/2017EF000632

Pradhan, P., Reusser, D. E., & Kropp, J. P. (2013). Embodied greenhouse gas emissions in diets. *PLoS One*, *8*(5), e62228. doi:10.1371/journal.pone.0062228 PMID:23700408

Pradhan, R., Singh, N., & Singh, R. P. (2019). Onset of summer monsoon in Northeast India is preceded by enhanced transpiration. *Scientific Reports*, *9*(1), 18646. doi:10.103841598-019-55186-8 PMID:31819130

Prasar Bharati. (2016). *National Programme, Farm and Home Unit.* Retrieved from https://prasarbharati.gov.in/all-india-radio-2/#National-Programs

Prasar Bharati. (2021). *About.* Retrieved from https://prasarbharati.gov.in/dd-kisan-homepage/about-dd-kisan/

Prazeres, I., Lucas, M. R., & Marta-Costa, A. (2021). Cocoa Markets and Value Chains: Dynamics and Challenges for Sao Tome and Principe Organic Smallholders. *Int. J. Innovation and Economic Development*, *7*(2), 64–77. doi:10.18775/ijied.1849-7551-7020.2015.72.2005

Predieri, S., Sinesio, F., Monteleone, E., Spinelli, S., Cianciabella, M., Daniele, G. M., Dinnella, C., Gasperi, F., Endrizzi, I., Torri, L., Toschi, T. G., Bendini, A., Pagliarini, E., Cattaneo, C., Di Monaco, R., Vitaglione, P., Condelli, N., & Laureati, M. (2020). Gender, age, geographical area, food neophobia and their relationships with the adherence to the mediterranean diet: New insights from a large population cross-sectional study. *Nutrients*, *12*(6), 1778. Advance online publication. doi:10.3390/nu12061778 PMID:32549267

Pretty, J., & Bharucha, Z. P. (2014). Sustainable intensification in agricultural systems. *Annals of Botany*, *114*(8), 1571–1596. doi:10.1093/aob/mcu205 PMID:25351192

Priyadarshini, P., & Abhilash, P. C. (2020a). Circular economy practices within energy and waste management sectors of India: A meta-analysis. *Bioresource Technology*, *304*, 11. doi:10.1016/j.biortech.2020.123018 PMID:32087547

Priyadarshini, P., & Abhilash, P. C. (2020b). Fostering sustainable land restoration through circular economy-governed transitions. *Restoration Ecology*, *28*(4), 719–723. doi:10.1111/rec.13181

Prowse, M. (2013). A history of tobacco production and marketing in Malawi, 1890 – 2010. *Journal of Eastern African Studies: the Journal of the British Institute in Eastern Africa*, *7*(4), 691–712. doi:10.1080/17531055.2013.805077

Puelles, J. A., Gómez, M., & Puelles, M. (2011). *Marcas de distribuidor, concepto, evolución, protagonistas y adaptación a los ciclos económicos*. Ediciones Pirámide.

Puia, A., & Leucuta, D. C. (2017). Children's lifestyle behaviors in relation to anthropometric indices: A family practice study. *Clujul Medical*, *90*(4), 385–391.

Qiang, L., Huang, J., Luo, R., & Liu, C. (2013). China's labor transition and the future of china's rural wages and employment. *China & World Economy*, (03), 4–24.

Quinta-Nova, L., & Roque, N. (2018). An integrated agroforestal suitability model using a GIS-Based Multicriteria Analysis Method: A case study of Portugal. The Eurasia Proceedings of Science, Technology. *Engineering & Mathematics*, *3*, 11–20.

Rakotovao, N. H., Chevallier, T., Chapuis-Lardy, L., Deffontaines, S., Mathé, S., Ramarofidy, M. A., Lepage, A., Masso, C., Albrech, A., & Razafimbelo, T. M. (2021). Impacts on greenhouse gas balance and rural economy after agroecology development in Itasy Madagascar. *Journal of Cleaner Production*, *291*, 125220. doi:10.1016/j.jclepro.2020.125220

Rampersad, G. C., Hordacre, A.-L., & Spoehr, J. (2020). Driving innovation in supply chains: An examination of advanced manufacturing and food industries. *Journal of Business and Industrial Marketing*, *35*(5), 835–847. doi:10.1108/JBIM-03-2019-0101

Randhawa, K., Wilden, R., & Hohberger, J. (2016). A bibliometric review of open innovation: Setting a research agenda. *Journal of Product Innovation Management*, *33*(6), 750–772. doi:10.1111/jpim.12312

Ranuzzi, A., & Srivastava, R. (2012). *Impact of Climate Change on Agriculture and Food Security*. ICRIER Policy Series, 16. Retrieved from http://www.icrier.org/pdf/ Policy_Series_No_16.pdf

Rashidi, K., Noorizadeh, A., Kannan, D., & Cullinane, K. (2020). Applying the triple bottom line in sustainable supplier selection: A meta-review of the state-of-the-art. *Journal of Cleaner Production*, *269*, 122001. doi:10.1016/j.jclepro.2020.122001

Rather, I. A., Koh, W. Y., Paek, W. K., & Lim, J. (2013). The Sources of Chemical Contaminants in Food and Their Health Implications. *Frontiers in Pharmacology*, *8*, 1–8.

Ray, D., Mueller, N., West, P., & Foley, J. (2013). Yield trends are insuficiente to double global crop production by 2050. *PLoS One*, *8*(6), e66428. doi:10.1371/journal.pone.0066428 PMID:23840465

Raza, A., Razzaq, A., Mehmood, S. S., Zou, X., Zhang, X., Lv, Y., & Xu, J. (2019). Impact of Climate Change on Crops Adaptation and Strategies to Tackle Its Outcome: A Review. *Plants (Basel, Switzerland)*, *8*(2), 34. doi:10.3390/plants8020034 PMID:30704089

Razafinjoelina, S. M. (2017). *Rice Value Chain Development and Policy Sequencing in Southern Africa* [Unpublished bachelor dissertation]. University of Arkansas, Fayetteville.

Reardon, T., Timmer, C. P., & Minten, B. (2012). Supermarket revolution in Asia and emerging development strategies to include small farmers. *Proceedings of the National Academy of Sciences of the United States of America, 109*(31), 12332–12337. doi:10.1073/pnas.1003160108 PMID:21135250

Reba, M. L., & Massey, J. H. (2020), Surface irrigation in the Lower Mississippi River Basin: Trends and Innovations. Trans. of the ASABE, 63, 1305–1314.

Rebouças, P. M., Dias, I. F., Alves, M. A., & Barbosa Filho, J. A. D. (2015). Radiação solar e temperatura do ar em ambiente protegido. *Revista Agrogeoambiental, Porto Alegre, 7*(2), 115–125.

Recanati, F., Marveggio, D., & Dotelli, G. (2017). From beans to bar: A life cycle assessment towards sustainable chocolate supplychain. *The Science of the Total Environment, 613614*, 1013–1023. PMID:28946374

Rerkasem, B. (2017). The Rice Value Chain: a Case Study of Thai Rice. *Journal of Social Sciences and Humanities, 4*(1).

Richardson, J. (2004). Accounting for sustainability: measuring quantities or enhancing qualities. In A. Henriques & J. Richardson (Eds.), *The Triple Bottom Line: Does it All Add Up?* (pp. 34–44). Routledge.

Ritchie, H. (2019). *Food production is responsible for one-quarter of the world's greenhouse gases emissions*. Our World in Data. Retrieved from https://ourworldindata.org/food-ghg-emissions

Ritchie, B. W. (2004). Chaos, crises and disasters: A strategic approach to crisis management in the tourism industry. *Tourism Management, 25*(6), 669–683. doi:10.1016/j.tourman.2003.09.004

Robertson, A., Tirado, C., Lobstein, T., Jermini, M., Knai, C., Jensen, J. H., Ferro-Luzzi, A., & James, W. P. (2004). Food and health in Europe: a new basis for action. WHO Regional Publications. European Series (Issue 96).

Robinson, G. M. (2009). Towards sustainable agriculture: Current debates. *Geography Compass, 3*(5), 1757–1773. doi:10.1111/j.1749-8198.2009.00268.x

Rodrigues, T. de S., Arge, L. W. P., Guedes, F. A. de F., Travassos-Lins, J., de Souza, A. P., Cocuron, J.-C., Buckeridge, M. S., Grossi-de-Sá, M. F., & Alves-Ferreira, M. (2021). *Integrative Analysis of Sorghum Bicolor Green Prop Roots Under Elevated CO_2 and Water Deficit Conditions* (p. 2021.03.12.435150). doi:10.1101/2021.03.12.435150

Rodriguez, J. M., Molnar, J. J., Fazio, R. A., Sydnor, E., & Lowe, M. J. (2009). Barriers to adoption of sustainable agriculture practices: Change agent perspectives. *Renewable Agriculture and Food Systems, 24*(1), 60–71. doi:10.1017/S1742170508002421

Rodriguez-Pose, A., & Tselios, V. (2009). Education and Income Inequality in the Regions of the European Union. *Journal of Regional Science, 49*(3), 411–437. doi:10.1111/j.1467-9787.2008.00602.x

Rodríguez-Tadeo, A., Patiño-Villena, B., González Martínez-La Cuesta, E., Urquídez-Romero, R., & Ros Berruezo, G. (2018). Food neophobia, mediterranean diet adherence and acceptance of healthy foods prepared in gastronomic workshops by spanish students. *Nutrición Hospitalaria, 35*(3). Advance online publication. doi:10.20960/nh.1337 PMID:29974774

Rojas-Downing, M. M., Nejadhashemi, A. P., Harrigan, T., & Woznicki, S. A. (2017). Climate change and livestock: Impacts, adaptation, and mitigation. In Climate Risk Management (Vol. 16). doi:10.1016/j.crm.2017.02.001

Romero-Meza, R., & Blanco-Vidal, C. (2011). Management of Crises: Lessons from Earthquake and Mining Crises in Chile. *Vikalpa*, 97–99.

Ronis, M. J. J., Pedersen, K. B., & Watt, J. (2018). Adverse Effects of Nutraceuticals and Dietary Supplements. *Annual Review of Pharmacology and Toxicology*, *58*(1), 583–601. doi:10.1146/annurev-pharmtox-010617-052844

Rosado-Serrano, A., Paul, J., & Dikova, D. (2018). International fran-chising: A literature review and research agenda. *Journal of Business Research*, *85*, 238–257. doi:10.1016/j.jbusres.2017.12.049

Rosenbauma, P. F., Weinstockb, R. S., Silverstonec, A. E., Sjödind, A., & Pavuk, M. (2017). Metabolic syndrome is associated with exposure to organochlorine pesticides in Anniston, AL, United State. *Environment International*, *108*, 11–21. doi:10.1016/j.envint.2017.07.017

Rosenthal, U., & Kouzmin, A. (1997). Crises and Crisis Management:Toward Comprehensive Government Decision Making. *Journal of Public Administration: Research and Theory*, *7*(2), 277–304. doi:10.1093/oxfordjournals.jpart.a024349

Ros-Tonen, M. A., Bitzer, V., Laven, A., Ollivier de Leth, D., Van Leynseele, Y., & Vos, A. (2019). Conceptualizing inclusiveness of smallholder value chain integration. *Current Opinion in Environmental Sustainability*, *41*, 10–17. doi:10.1016/j.cosust.2019.08.006

Rousset, L., & Halioua, B. (2018). Stress and psoriasis. *International Journal of Dermatology*, *57*(10), 1165–1172. doi:10.1111/ijd.14032

Rowan, N. J., & Galanakis, C. M. (2020). Unlocking challenges and opportunities presented by COVID-19 pandemic for cross-cutting disruption in agri-food and green deal innovations: Quo Vadis? *The Science of the Total Environment*, *748*, 1–15. doi:10.1016/j.scitotenv.2020.141362 PMID:32823223

Rowe, H., Withers, P. J. A., Baas, P., Chan, N. I., Doody, D., Holiman, J., Jacobs, B., Li, H., MacDonald, G. K., McDowell, R., Sharpley, A. N., Shen, J., Taheri, W., Wallenstein, M., & Weintraub, M. N. (2016). Integrating legacy soil phosphorus into sustainable nutrient management strategies for future food, bioenergy and water security. *Nutrient Cycling in Agroecosystems*, *104*(3), 393–412. doi:10.100710705-015-9726-1

Ruben, R., Van Boekel, M., Van Tilburg, A., & Trienekens, J. (2007). Governance for quality in tropical food chains. Wageningen: Wageningen Acadamic Publishers.

Ruben, R. (2017). Impact assessment of commodity standards: Towards inclusive value chains. *Enterprise Development & Microfinance*, *28*(1-2), 82–97. doi:10.3362/1755-1986.16-00020

Rueda, X., Garrett, R. D., & Lambin, E. F. (2017). Corporate investments in supply chain sustainability: Selecting instruments in the agri-food industry. *Journal of Cleaner Production*, *142*, 2480–2492. doi:10.1016/j.jclepro.2016.11.026

Rueda, X., Paz, A., Gibbs-Plessl, T., Leon, R., Moyano, B., & Lambin, E. F. (2018). Smallholders at a crossroad: Intensify or fall behind? Exploring alternative livelihood strategies in a globalized world. *Business Strategy and the Environment*, *27*(2), 215–229. doi:10.1002/bse.2011

Ru, F., Yin, A., Jin, J., Zhang, X., Yang, X., Zhang, M., & Gao, C. (2016). Prediction of cadmium enrichment in reclaimed coastal soils by classification and regression tree. *Estuarine, Coastal and Shelf Science*, *177*, 1–7. doi:10.1016/j.ecss.2016.04.018

Runkle, B. R. K., Suvoscarev, K., Reba, M. L., Reavis, C. W., Smith, S. F., Chiu, Y. L., & Fong, B. (2019). Methane emission reductions from the alternate wetting and drying of rice fields detected using the eddy covariance method. *Environmental Science & Technology*, *53*(2), 671–681. doi:10.1021/acs.est.8b05535 PMID:30566833

Ruwoldt, G. (2013). Food insecurity and climate change. *American Journal of Public Health*, *103*(1), e1. doi:10.2105/AJPH.2012.301082 PMID:23153137

RxList. (2020). *L-tryptophan*. https://www.rxlist.com/l-tryptophan/supplements.htm

Saaty, T. L. (1980). *The Analytic Hierarchy Process*. McGraw-Hill.

Saaty, T. L. (1993). The analytic hierarchy process: A 1993 overview. *Central European Journal of Operation Research and Economics*, *2*(2), 119–137.

Sabadas, M., Dirja, M., & Maxim, M. (2018). Study on Implemented and Future Projects for Sustainable Development in Aghiresu Area, Cluj County. *Scientific Papers-Series E-Land Reclamation Earth Observation & Surveying Environmental Engineering*, *7*, 113–116.

Sachs, J., Schmidt-Traub, G., Kroll, C., Durand-Delacre, D., & Teksoz, K. (2016). *SDG index and dashboard-global report*. Bertelsmann Stiftung and Sustainable Development Solutions Network.

Sadak, M. S. S. (2016). Mitigation of drought stress on Fenugreek plant by foliar application of trehalose. *International Journal of Chemtech Research*, *9*(2), 147–155.

Sadr, M. R., Reza, B., Fazlollah, M., Yazdi, S. N. M., & Anari, P. Y. (2016). Peganum Harmala (Aspand) Intoxication; a Case Report. *Emergency (Tehran, Iran)*, *4*(2), 106–107.

Sáez-Almendros, S., Obrador, B., Bach-Faig, A., & Serra-Majem, L. (2013). Environmental Footprints of Mediterranean versus Western Dietary Patterns: Beyond the Health Benefits of the Mediterranean Diet. *Environmental Health*, *12*(118), 1–8. doi:10.1186/1476-069X-12-118 PMID:24378069

Salas-Salvadó, J., Becerra-Tomás, N., García-Gavilán, J. F., Bulló, M., & Barrubés, L. (2018). Mediterranean Diet and Cardiovascular Disease Prevention: What Do We Know? *Progress in Cardiovascular Diseases*, *61*(1), 62–67. Advance online publication. doi:10.1016/j.pcad.2018.04.006 PMID:29678447

Salvioni, C., Henke, R., & Vanni, F. (2020). The Impact of Non-Agricultural Diversification on Financial Performance: Evidence from Family Farms in Italy. *Sustainability*, *12*(2), 486. doi:10.3390u12020486

Samborski, S. M., Gozdowski, D., Stepien, M., Walsh, O. S., & Leszczynska, E. (2016). On-farm evaluation of an active optical sensor performance for variable nitrogen application in winter wheat. *European Journal of Agronomy*, *74*, 56–67. doi:10.1016/j.eja.2015.11.020

Samson, E., Van der Werf, H. M. G., Dupraz, P., Ruas, J.-F., & Corson, M. S. (2012). Estimating environmental impacts of agricultural systems with LCA using data from the French Farm Accountancy Data Network (FADN). *Cahiers Agricultures*, *21*(4), 248–257. doi:10.1684/agr.2012.0581

Santangelo, C., Filesi, C., Varì, R., Scazzocchio, B., Filardi, T., Fogliano, V., D'Archivio, M., Giovannini, C., Lenzi, A., Morano, S., & Masella, R. (2016). Consumption of extra-virgin olive oil rich in phenolic compounds improves metabolic control in patients with type 2 diabetes mellitus: A possible involvement of reduced levels of circulating visfatin. *Journal of Endocrinological Investigation*, *39*(11), 1295–1301. Advance online publication. doi:10.100740618-016-0506-9 PMID:27344308

Santos, M. A. M. (2018). *Potencial e Viabilidade económica para Utilização de Estufas Para produção de Hortaliças em STP* [Unpublished master thesis]. Instituto Superior de Agronomia, Universidade de Lisboa.

Sarkar, S. (2018). *South and Southeast Asia most at risk by climate change*. Retrieved from https://www.thethirdpole.net/2018/12/05/south-asia-climate-change/

Savitz, A., & Weber, K. (2006). *The triple bottom line: How today's best-run organizations are achieving economic, social and environmental success -and how you can too*. Jossey Bass.

Sawada, Y., & Takasaki, Y. (2017). Natural Disaster, Poverty, and Development: An Introduction. *World Development*, *94*, 2–15. doi:10.1016/j.worlddev.2016.12.035

Sawinska, Z., Switek, S., Glowicka-Woloszyn, R., & Kowalczewski, P. L. (2020). Agricultural Practice in Poland Before and after Mandatory IPM Implementation by the European Union. *Sustainability*, *12*(3), 1107. doi:10.3390u12031107

Schaldach, R., Koch, J., Aus der Beek, T., Kynast, E., & Flörke, M. (2012). Current and future irrigation water requirements in pan-Europe: An integrated analysis of socioeconomic and climate scenarios. *Global and Planetary Change*, *94*, 33–45. doi:10.1016/j.gloplacha.2012.06.004

Schmidhuber, J., & Tubiello, F. (2007). Global Food Security under Climate Change. *Proceedings of the National Academy of Sciences of the United States of America*, *104*(50), 19703-19708. Retrieved from https://www.jstor.org/stable/25450779

Schmutz, U., Kneafsey, M., Kay, C. S., Doernberg, A., & Zasada, I. (2018). Sustainability impact assessments of different urban short food supply chains: Examples from London, UK. *Renewable Agriculture and Food Systems*, *33*(6), 518–529. doi:10.1017/S1742170517000564

Schnabel, L., Kesse-Guyot, E., Allès, B., Touvier, M., Srour, B., Hercberg, S., Buscail, C., & Julia, C. (2019). Association between Ultraprocessed Food Consumption and Risk of Mortality among Middle-aged Adults in France. *JAMA Internal Medicine*, *179*(4), 490. Advance online publication. doi:10.1001/jamainternmed.2018.7289 PMID:30742202

Schröder, H. (2007). Protective mechanisms of the Mediterranean diet in obesity and type 2 diabetes. Journal of Nutritional Biochemistry, 18(3). doi:10.1016/j.jnutbio.2006.05.006

Schroth, G., Garcia, E., Griscom, B. W., Teixeira, W. G., & Barros, L. P. (2016). Commodity production as restoration driver in the Brazilian amazon? Pasture re-agro-forestation with cocoa (theobroma cacao) in southern pará. *Sustainability Science*, *11*(2), 277–293. doi:10.100711625-015-0330-8 PMID:30174731

Scoones, I. (2009). The politics of global assessments: The case of the International Assessment of Agricultural Knowledge, Science and Technology for Development (IAASTD). *The Journal of Peasant Studies*, *36*(3), 547–571. doi:10.1080/03066150903155008

Scott, G. J., Donovan, J., & Higuchi, A. (2015). Costs, quality, and competition in the cocoa value chain in Peru: An exploratory assessment. *Custos e Agronegocio*, *11*(4), 324–358.

Sehgal, A., Sita, K., Siddique, K. H. M., Kumar, R., Bhogireddy, S., Varshney, R. K., HanumanthaRao, B., Nair, R. M., Prasad, P. V. V., & Nayyar, H. (1705). HanumanthaRao, B., Nair, R. M., Prasad, P. V. V., & Nayyar, H. (2018). Drought or/and Heat-Stress Effects on Seed Filling in Food Crops: Impacts on Functional Biochemistry, Seed Yields, and Nutritional Quality. *Frontiers in Plant Science*, *9*, 1705. Advance online publication. doi:10.3389/fpls.2018.01705 PMID:30542357

Shah, F., Huang, J., Cui, K., Nie, L., Shah, T., Chen, C., & Wang, K. (2011). Impact of high-temperature stress on rice plant and its traits related to tolerance. *The Journal of Agricultural Science*, *149*(5), 545–556. doi:10.1017/S0021859611000360

Shakina, E., & Barajas, A. (2020). "Innovate or Perish?": Companies under crisis. *European Research on Management and Business Economics*, 1–10.

Shao, H., Chu, L., Lu, H., Qi, W., Chen, X., Liu, J., Kuang, S., Tang, B., & Wong, V. (2019). Towards sustainable agriculture for the salt-affected soil. *Land Degradation & Development*, *30*(5), 574–579. doi:10.1002/ldr.3218

Sharma. M. (2021). *The Future of Indian Agriculture.* Down to Earth. Retrieved from https://www.downtoearth.org.in/blog/agriculture/the-future-of-indian-agriculture-75384

Sharma, V., Giri, S., & Rai, S. S. (2013, March). Supply chain management of rice in India: A rice processing company's perspective. *International Journal of Managing Value and Supply Chains*, *4*(1), 25–36. Advance online publication. doi:10.5121/ijmvsc.2013.4103

Sharma, Y. K., Mangla, S. K., Patil, P. P., & Uniyal, S. (2018). Analyzing sustainable food supply chain management challenges in India. In R. Mangey & D. J. Paulo (Eds.), *Soft computing techniques and applications in mechanical engineering* (pp. 162–180). IGI Global. doi:10.4018/978-1-5225-3035-0.ch008

Sherwood, J. (2020). The significance of biomass in a circular economy. *Bioresource Technology, 300*, 122755. Advance online publication. doi:10.1016/j.biortech.2020.122755 PMID:31956060

Shi, J., Yasuor, H., Yermiyahu, U., Zuo, Q., & Ben-Gal, A. (2014). Dynamic responses of wheat to drought and nitrogen stresses during re-watering cycles. *Agricultural Water Management, 146*, 163–172. doi:10.1016/j.agwat.2014.08.006

Sicher, R., Timlin, D., & Bailey, B. (2012). Responses of growth and primary metabolism of water-stressed barley roots to rehydration. *Journal of Plant Physiology, 169*(7), 686–695. doi:10.1016/j.jplph.2012.01.002 PMID:22285575

Silva, J. M. F. (2011). *Desenvolvimento dos recursos humanos no sector de produção de arroz em distrito de Baucau, Timor-Leste* [Unpublished master dissertation]. Universidade de Évora, Évora.

Silva, G. (2018). *Feeding the world in 2050 and beyond – Part 1: Productivity challenges*. Michigan State University.

Silver, R. (2005). The gap and policy between China's agricultural products processing technology and developed countries. *Scientific Management Research, 023*(006), 93–96.

Simões, M. P., Alberto, D., Dias, C., Duarte, A. C., Ferreira, D., Gavinhos, C., & Castanheira, I. (2015). O projeto + pêssego - inovação e desenvolvimento na cultura do pessegueiro na região da beira ["+Peach" Project: Innovation and development of peach production in Beira Interior Region of Portugal]. In III Jornadas Potencial Técnico e Científico do IPCB (pp. 25-27). Academic Press.

Singh, G. (2016). Climate Change and Food Security in India: Challenges and Opportunities. *Irrigation and Drainage, 65*, 5–10. doi:10.1002/ird.2038

Sintayehu, D. W. (2018). Impact of climate change on biodiversity and associated key ecosystem services in Africa: A systematic review. *Ecosystem Health and Sustainability, 4*(9), 225–239. Advance online publication. doi:10.1080/20964129.2018.1530054

Sjauw-Koen-Fa, A. R., Omta, S. W. F., & Blok, V. (2018). Constructing a multinationals' inclusive sourcing indicator for impacting farmer business models: Application in cocoa cases. *International Journal on Food System Dynamics, 9*(3), 207–225.

Sjöberg, L. (1998). World views, political attitudes and risk perception. *Risk–Health. Safety and Environment, 9*, 137–152.

Skot, J. (2016). *The State of Food and Agriculture: Climate Change, Agriculture and Food security*. Food and Agriculture Organization of United States.

Small Farmers' Agribusiness Consortium, Ministry of Agriculture and Farmers' Welfare, Government of India. (n.d.). *e-NAM Overview, Vision*. Retrieved from https://www.enam.gov.in/web/

Smith, P., Martino, D., Cai, Z., Gwary, D., Janzen, H., Kumar, P., McCarl, B., Ogle, S., O'Mara, F., Rice, C., Scholes, B., & Sirotenko, O. (2007). Agriculture. In Climate Change 2007: Mitigation. Contribution of Working Group III to the Fourth Assessment Report of the Intergovernmental Panel on Climate Change. Cambridge University Press.

Smith, D. L., & Almaraz, J. J. (2004). Climate change and crop production: Contributions, impacts, and adaptations. *Canadian Journal of Plant Pathology, 26*(3), 253–266. Advance online publication. doi:10.1080/07060660409507142

Soares, C. A., Viancelli, A., Michelon, W., Sbardelloto, M., Camargo, A. F., Vargas, G. D. L. P., Fongaro, G., & Treichel, H. (2020). Biogas yield prospection from swine manure and placenta in real-scale systems on circular economy approach. *Biocatalysis and Agricultural Biotechnology*, *25*, 101598. Advance online publication. doi:10.1016/j.bcab.2020.101598

Song, L. C. (1991). Analysis on the variation and influence of drought and flood during the active period of summer monsoon in China. *Disaster Science*, (03), 23–28.

Song, Y., Chen, Q., Ci, D., Shao, X., & Zhang, D. (2014). Effects of high temperature on photosynthesis and related gene expression in poplar. *BMC Plant Biology*, *14*(1), 111. doi:10.1186/1471-2229-14-111 PMID:24774695

Song, Y., Ci, D., Tian, M., & Zhang, D. (2014). Comparison of the physiological effects and transcriptome responses of *Populus simonii* under different abiotic stresses. *Plant Molecular Biology*, *86*(1–2), 139–156. doi:10.100711103-014-0218-5 PMID:25002226

Sonnino, R. (2013). Local Foodscapes: Place and Power in the Agri-food System. *Acta Agriculturæ Scandinavica. Section B, Soil and Plant Science*, *63*(1), 2–7. doi:10.1080/09064710.2013.800130

Sonwa, D. J., Weise, S. F., Schroth, G., Janssens, M. J. J., & Howard-Yana Shapiro. (2014). Plant diversity management in cocoa agroforestry systems in west and central Africa—Effects of markets and household needs. *Agroforestry Systems*, *88*(6), 1021–1034. doi:10.100710457-014-9714-5

Sørensen, J. F. (2018). The importance of place-based, internal resources for the population development in small rural communities. *Journal of Rural Studies*, *59*, 78–87. doi:10.1016/j.jrurstud.2018.01.011

Sottile, F., Massaglia, S., & Peano, C. (2020). Ecological and Economic Indicators for the Evaluation of Almond (*Prunus dulcis* L.) Orchard Renewal in Sicily. *Agriculture*, *10*(7), 301. doi:10.3390/agriculture10070301

Soubry, B., Sherren, K., & Thornton, T. F. (2020). Are we taking farmers seriously? A review of the literature on farmer perceptions and climate change, 2007–2018. *Journal of Rural Studies*, *74*, 210–222. doi:10.1016/j.jrurstud.2019.09.005

Soullier, G. (2017). *Modernization of domestic food chains in developing countries: What effects on small-scale farmers? The rice value chain in Senegal* [Unpublished doctoral dissertation]. Universidade de Montpellier.

Soumare, I. (2015). *Does Foreign Direct Investment Improve Welfare in North Africa?* Working Paper, North Africa Policy Series. Africa Development Bank.

Sousa, K. D., Casanoves, F., Sellare, J., Ospina, A., Suchini, J. G., Aguilar, A., & Mercado, L. (2018). How climate awareness influences farmers' adaptation decisions in Central America? *Journal of Rural Studies*, *64*, 11–19. doi:10.1016/j.jrurstud.2018.09.018

Spicka, J., & Smutka, L. (2014). The Technical Efficiency of Specialised Milk Farms: A Regional View. *TheScientificWorldJournal*, *985149*, 1–13. Advance online publication. doi:10.1155/2014/985149 PMID:25050408

Springmann, M., Clark, M., Mason-D'Croz, D., Wiebe, K., Bodirsky, B. L., Lassaletta, L., de Vries, W., Vermeulen, S. J., Herrero, M., Carlson, K. M., Jonell, M., Troell, M., DeClerck, F., Gordon, L. J., Zurayk, R., Scarborough, P., Rayner, M., Loken, B., Fanzo, J., ... Willett, W. (2018). Options for keeping the food system within environmental limits. *Nature*, *562*(7728), 519–525. Advance online publication. doi:10.103841586-018-0594-0 PMID:30305731

Spugnoli, P., Dainelli, R., D'Avino, L., Mazzoncini, M., & Lazzeri, L. (2012). Sustainability of sunflower cultivation for biodiesel production in Tuscany within the EU Renewable Energy Directive. *Biosystems Engineering*, *112*(1), 49–55. doi:10.1016/j.biosystemseng.2012.02.004

Srinivasa Rao, Ch., Prasad, R. S., & Mohapatra, T. (2019). *Climate Change and Indian Agriculture: Impacts, Coping Strategies, Programmes and Policy. Technical Bulletin/Policy Document 2019. Indian Council of Agricultural Research, Ministry of Agriculture and Farmers' Welfare and Ministry of Environment, Forestry and Climate Change*. Government of India.

SRP - Sustainable Rice Platform. (2020). *The SRP Assurance Scheme (Version 1.3)*. Sustainable Rice Platform.

Stata. (2021). *Stata: Software for Statistics and Data Science*. https://www.stata.com/

StataCorp. (2017a). *Stata 15 Base Reference Manual*. Stata Press.

StataCorp. (2017b). *Stata Statistical Software: Release 15*. StataCorp LLC.

Stenvinkel, P., Irving, G. F., Wegener, S., Eyre, S., Rippe, B., Segelmark, M., Barany, P., & Samuelsson, O. (2014). Phosphate additives in food a potential public health risk. High phosphate levels can lead to cardiovascular disease. *Lakartidningen*, *15*, 27–28.

Stephens, A. (2015). *Dictionary of Agriculture* (2nd ed.). Routledge.

Stern, N. (2018). Public economics as if time matters: Climate change and the dynamics of policy. *Journal of Public Economics*, *162*, 4–17. doi:10.1016/j.jpubeco.2018.03.006

Stevens, A. N. P. (2011). Factors Affecting Global Climate. *Nature Education Knowledge, 3*(10), 18. Retrieved from https://www.nature.com/scitable/knowledge/library/factors-affecting-global-climate-17079163/

Stiles, W. A. V., Styles, D., Chapman, S. P., Esteves, S., Bywater, A., Melville, L., Silkina, A., Lupatsch, I., Fuentes Grünewald, C., Lovitt, R., Chaloner, T., Bull, A., Morris, C., & Llewellyn, C. A. (2018). Using microalgae in the circular economy to valorise anaerobic digestate: Challenges and opportunities. *Bioresource Technology*, *267*, 732–742. doi:10.1016/j.biortech.2018.07.100 PMID:30076074

Stone, J., & Rahimifard, S. (2018). Resilience in agri-food supply chains: A critical analysis of the literature and synthesis of a novel framework. *Supply Chain Management*, *23*(3), 207–238. doi:10.1108/SCM-06-2017-0201

Straub, A. M., Gray, B. J., Ritchie, L. A., & Gill, D. A. (2020). Cultivating disaster resilience in rural Oklahoma: Community disenfranchisement and relational aspects of social capital. *Journal of Rural Studies*, *73*, 105–113. doi:10.1016/j.jrurstud.2019.12.010

Strelecek, F., Zdenek, R., & Lososova, J. (2007). Influence of farm milk prices in the EU 25 on profitability and production volume indicators. *Agricultural Economics-Zemedelska Ekonomika*, *53*(12), 545–557. doi:10.17221/1224-AGRICECON

Stryker, J. D. (2010). *Developing competitive rice value chains*. Second Africa Rice Congress, Bamako, Mali.

Stuart, T. (2015). *Food: How much does the world need?* World Economic Forum. Retrieved from https://www.weforum.org/agenda/2015/05/food-how-much-does-the-world-need/

Stubenrauch, J., Ekardt, F., Heyl, K., Garske, B., Schott, V. L., & Ober, S. (2021). How to legally overcome the distinction between organic and conventional farming—Governance approaches for sustainable farming on 100% of the land. *Sustainable Production and Consumption*, *28*, 716–725. doi:10.1016/j.spc.2021.06.006

Sturialea, L., & Scuderi, A. (2013). Evaluation of Social Media Actions for the Agrifood System. *Procedia Technology*, 200 – 208.

Sukhwani, V., Deshkar, S., & Shaw, R. (2020). COVID-19 Lockdown, Food Systems and Urban–Rural Partnership: Case of Nagpur, India. *International Journal of Environmental Research and Public Health*, *17*(16), 5710. doi:10.3390/ijerph17165710 PMID:32784648

Sun, S. F., Xin, X., Liu, M., Lu, Z. W., Qin, Y., Cai, H. L., & Ma, L. (2019). *China agricultural Economic Development Report and outlook*. China Economic Daily.

Sun, S. K., Yin, Y. L., Wu, P. T., Wang, Y. B., Luan, X. B., & Li, C. (2019). Geographical Evolution of Agricultural Production in China and Its Effects on Water Stress, Economy and the Environment: The Virtual Water Perspective. *Water Resources Research*, *55*(5), 4014–4029. doi:10.1029/2018WR023379

Sun, Y., Geng, Q., Du, Y., Yang, X., & Zhai, H. (2017). Induction of cyclic electron flow around photosystem I during heat stress in grape leaves. *Plant Science*, *256*, 65–71. doi:10.1016/j.plantsci.2016.12.004 PMID:28167040

Šupín, M., Loučanová, E., & Olšiaková, M. (2019). *Sustainable bioenergy policy for the period after 2020.* Paper presented at the 12th WoodEMA Annual International Scientific Conference - "Digitalisation and Circular Economy: Forestry and Forestry Based Industry Implications".

Suwannaporn, P., & Linnemann, A. (2008). Consumer Preferences and Buying Criteria in Rice: A Study to Identify Market Strategy for Thailand Jasmine Rice Export. *Journal of Food Products Marketing*, *14*(4), 33–53. doi:10.1080/10454440801986348

Suwa, R., Hakata, H., Hara, H., El-Shemy, H. A., Adu-Gyamfi, J. J., Nguyen, N. T., Kanai, S., Lightfoot, D. A., Mohapatra, P. K., & Fujita, K. (2010). High temperature effects on photosynthate partitioning and sugar metabolism during ear expansion in maize (*Zea mays* L.) genotypes. *Plant Physiology and Biochemistry*, *48*(2), 124–130. doi:10.1016/j.plaphy.2009.12.010 PMID:20106675

Swiss Re. (2020). *World economy set to lose up to 18% GDP from climate change if no action taken, reveals Swiss Re Institute's stress-test analysis*. Retrieved from https://www. swissre.com/media/news-releases/nr-20210422-economics-of-climate-change-risks.html

Székely, C., & Pálinkás, P. (2009). Agricultural Risk Management in the European Union and in the USA. *Studies in Agricultural Economics (Budapest)*, (109), 55–72.

Takele, A. (2010). *Analysis of rice profitability and marketing chain: The case of Fogera Woreda, South Gondar zone, Amhara National Regional State, Ethiopia* [Unpublished master dissertation]. Haramaya University.

Tambo, J. A., & Mockshell, J. (2018). Differential impacts of conservation agriculture technology options on household income in Sub-Saharan Africa. *Ecological Economics*, *151*, 95–105. doi:10.1016/j.ecolecon.2018.05.005

Tanaka, A., Diagne, M., & Saito, K. (2015). Causes of yield stagnation in irrigated lowland rice systems in the Senegal River Valley: Application of dichotomous decision tree analysis. *Field Crops Research*, *176*, 99–107. doi:10.1016/j.fcr.2015.02.020

Tanure, T. M., Miyajima, D. N., Magalhães, A. S., Domingues, E. P., & Carvalho, T. S. (2020). The Impacts of Climate Change on Agricultural Production, Land Use and Economy of the Legal Amazon Region between 2030 and 2049. *Economía*, *21*(1), 73–90. doi:10.1016/j.econ.2020.04.001

Taylor, D. H. (2005). Value chain analysis: An approach to supply chain improvement in agrifood chains. *International Journal of Physical Distribution & Logistics Management*, *35*(10), 744–761. doi:10.1108/09600030510634599

Taylor, K. E., Stouer, R. J., & Meehl, G. A. (2012). An Overview of CMIP5 and the Experiment Design. *Bulletin of the American Meteorological Society*, *93*(4), 485–498. doi:10.1175/BAMS-D-11-00094.1

Tedesco, I. (2017) A holistic approach to agricultural risk management for improving resilience. *2nd International Workshop on Modelling of Physical, Economic and Social Systems for Resilience Assessment*, 14-16.

Teklewold, H., Kassie, M., & Shiferaw, B. (2013). Adoption of multiple sustainable agricultural practices in rural Ethiopia. *Journal of Agricultural Economics*, *64*(3), 597–623. doi:10.1111/1477-9552.12011

Termeer, C. J. A. M., Drimie, S., Ingram, J., Pereira, L., & Whittingham, M. J. (2018). A diagnostic framework for food system governance arrangements: The case of South Africa. *NJAS Wageningen Journal of Life Sciences*, *84*(1), 85–93. Advance online publication. doi:10.1016/j.njas.2017.08.001

The Intergovernmental Panel on Climate Change. (2015). *Climate change 2014 Synthesis report*. IPCC.

The Royal Society. (2020). *The Basics of Climate Change*. Retrieved from https://royalsociety.org/topics-policy/projects/climate-change-evidence-causes/basics-of-climate-change/

The World Bank Report. (2013). *Turn Down the Heat: Climate Extremes, Regional Impacts, and the Case for Resilience*. Retrieved from https://www.worldbank.org/content/ dam/ Worldbank/document/Full_Report_Vol_2_Turn_Down_The_Heat_%20Climate_Extremes_Regional_Impacts_Case_for_Resilience_Print%20version_FINAL.pdf

The World Bank. (2020a). *Food Security and COVID-19*. Retrieved from https://www.worldbank. org/en/topic/agriculture/brief/food-security-and-covid-19

The World Bank. (2020b). *World Bank Announces $500 Million to Fight Locusts, Preserve Food Security and Protect Livelihoods*. Press Release. Retrieved from https://www.worldbank. org/ en/news/press-release/2020/05/21/world-bank-announces-500-million-to-fight-locusts-preserve-food-security-and-protect-livelihoods

The World Bank. (2021). *Climate Smart Agriculture*. Retrieved from https://www.worldbank. org/en/topic/climate-smart-agriculture

Theodoridis, X., Grammatikopoulou, M. G., Gkiouras, K., Papadopoulou, S. E., Agorastou, T., Gkika, I., Maraki, M. I., Dardavessis, T., & Chourdakis, M. (2018). Food insecurity and Mediterranean diet adherence among Greek university students. *Nutrition, Metabolism, and Cardiovascular Diseases*, *28*(5), 477–485. Advance online publication. doi:10.1016/j.numecd.2018.02.007 PMID:29655531

Thomas, I. A., Buckley, C., Kelly, E., Dillon, E., Lynch, J., Moran, B., Hennessy, T., & Murphy, P. N. C. (2020). Establishing nationally representative benchmarks of farm-gate nitrogen and phosphorus balances and use efficiencies on Irish farms to encourage improvements. *The Science of the Total Environment*, *720*, 137245. doi:10.1016/j.scitotenv.2020.137245 PMID:32325548

Thomassen, M. A., Dolman, M. A., van Calker, K. J., & de Boer, I. J. M. (2009). Relating life cycle assessment indicators to gross value added for Dutch dairy farms. *Ecological Economics*, *68*(8–9), 2278–2284. doi:10.1016/j.ecolecon.2009.02.011

Tilman, D., Balzer, C., Hill, J., & Befort, B. L. (2011). Global food demand and the sustainable intensification of agriculture. *Proceedings of the National Academy of Sciences of the United States of America*, *108*(50), 20260–20264. doi:10.1073/pnas.1116437108 PMID:22106295

Tingyu, L., Zhang, W., Cao, H., Hao, Y., Qingsong, Z., Siyang, R., Zitong, L., Yulong, Y., Wei, Q., Zhenling, C., Xuejun, L., Xiaotang, J., Oenema, O., de Vries, W., & Fusuo, Z. (2020). Region-specific nitrogen management indexes for sustainable cereal production in China. *Environmental Research Communications*, *2*(7), 075002. doi:10.1088/2515-7620/aba12d

Tinsley, R. (2012). *Rice Value Chain Analysis – Sokoto State Nigeria* (NIG 244). Winrock International Nigeria Farmer-to-Farmer Program, Consultant Volunteer Report.

Tipi, T., Yildiz, N., Nargeleçekenler, M., & Çetin, B. (2009). Measuring the technical efficiency and determinants of efficiency of rice (Oryza sativa) farms in Marmara region, Turkey. *New Zealand Journal of Crop and Horticultural Science*, *37*(2), 121–129. doi:10.1080/01140670909510257

Tisenkopfs, T., Adamsone-Fiskovica, A., Kilis, E. S., Grivins, M., Pinto-Correia, T., & Bjørkhau, H. (2020). Territorial fitting of small farms in Europe. *Global Food Security*, *26*, 1–9. doi:10.1016/j.gfs.2020.100425

Tittonell, P., Shepherd, K. D., Vanlauwe, B., & Giller, K. E. (2008). Unravelling the effects of soil and crop management on maize productivity in smallholder agricultural systems of western Kenya—An application of classification and regression tree analysis. *Agriculture, Ecosystems & Environment*, *123*(1–3), 137–150. doi:10.1016/j.agee.2007.05.005

Tiwari, P., Mishra, B. N., & Sangwan, N. S. (2014). Phytochemical and Pharmacological Properties of Gymnema sylvestre: An Important Medicinal Plant. *BioMed Research International*, *2014*, 1–18.

Tkemaladze, G. S., & Makhashvili, K. A. (2016). Climate changes and photosynthesis. *Annals of Agrarian Science*, *14*(2), 119–126. doi:10.1016/j.aasci.2016.05.012

TNAU Agritech Portal. (n.d.). *Agricultural Marketing in India.* Retrieved from https://agritech.tnau.ac.in/agricultural_marketing/agrimark_India.html

Todorovic, S., Papic, R., Ciaian, P., & Bogdanov, N. (2020). Technical efficiency of arable farms in Serbia: Do subsidies matter? *New Medit*, *19*(4), 81–97. doi:10.30682/nm2004f

Toledo, E., Salas-Salvado, J., Donat-Vargas, C., Buil-Cosiales, P., Estruch, R., Ros, E., Corella, D., Fito, M., Hu, F. B., Aros, F., Gomez-Gracia, E., Romaguera, D., Ortega-Calvo, M., Serra-Majem, L., Pinto, X., Schroder, H., Basora, J., Sorli, J. V., Bullo, M., ... Martinez-Gonzalez, M. A. (2015). Mediterranean diet and invasive breast cancer risk among women at high cardiovascular risk in the predimed trial a randomized clinical trial. *JAMA Internal Medicine*, *175*(11), 1752. Advance online publication. doi:10.1001/jamainternmed.2015.4838 PMID:26365989

Tong, T. Y. N., Imamura, F., Monsivais, P., Brage, S., Griffin, S. J., Wareham, N. J., & Forouhi, N. G. (2018). Dietary cost associated with adherence to the Mediterranean diet, and its variation by socio-economic factors in the UK Fenland Study. *British Journal of Nutrition*, *119*(6), 685–694. Advance online publication. doi:10.1017/S0007114517003993 PMID:29553031

Toni, J. (2019). *Curso de ZOPP e Método do Quadro Lógico*. Escola Nacional de Administração Pública.

Toop, T., Ward, S., Oldfield, T., Hull, M., Kirby, M., & Theodorou, M. K. (2017). AgroCycle – developing a circular economy in agriculture. *Energy Procedia*, *123*, 76–80. doi:10.1016/j.egypro.2017.07.269

Torraco, R. J. (2005). Writing integrative literature reviews: Guidelines and exemples. *HR Develop. RE:view*, *4*(3), 356–367.

Torres-Reyna, O. (n.d.). *Getting Started in Factor Analysis (using Stata)*. http://www.princeton.edu/~otorres/Stata/Factor

Tranfield, D., Denyer, D., & Smart, P. (2003). Towards a methodology for developing evidence-informed management knowledge by means of systematic review. *British Journal of Management*, *14*(3), 207–222. doi:10.1111/1467-8551.00375

Trenberth, K. E., Dai, A., van der Schrier, G., Jones, P. D., Barichivich, J., Briffa, K. R., & Sheffield, J. (2014). Global warming and changes in drought. *Nature Climate Change*, *4*(1), 17–22. doi:10.1038/nclimate2067

Trevor, R. W. & Lewis, I. (2015). *The Rice Value Chain in Tanzania. A report from the Southern Highlands Food Systems Programme*. FAO (Food and Agriculture Organization of the United Nations).

Trienekens, J. H. (2018). *A search for sustainability in cocoa supply chain business models* [Unpublished Bachelor thesis dissertation]. Wageningen University.

Trienekens, J. H. (2011). Agricultural value chains in developing countries: A framework for analysis. *The International Food and Agribusiness Management Review*, *14*(2), 51–81.

Tripathy, S. (2011). *Impact of TV programmes on agriculture and rural development.* www.Publishingindia.com

Trnkova, G., & Kroupova, Z. Z. (2020). Determinants of Persistent and Transient Technical Efficiency of Milk Production in Eu. *E+M. Ekonomie a Management*, *23*(4), 39–54. doi:10.15240/tul/001/2020-4-003

Tsakiraki, M., Grammatikopoulou, M. G., Stylianou, C., & Tsigga, M. (2010). Nutrition transition and health status of Cretan women: Evidence from two generations. *Public Health Nutrition*, *14*(5), 793–800. Advance online publication. doi:10.1017/S1368980010003010 PMID:21272421

Tseng, M. L. (2017). Using social media and qualitative and quantitative information scales to benchmark corporate sustainability. *Journal of Cleaner Production*, *142*, 727–738. doi:10.1016/j.jclepro.2016.05.062

Tseng, M. L., Chang, C. H., Lin, C. W. R., Wu, K. J., Chen, Q., Xia, L., & Xue, B. (2020). Future trends and guidance for the triple bottom line and sustainability: A data driven bibliometric analysis. *Environmental Science and Pollution Research International*, *27*(27), 33543–33567. doi:10.100711356-020-09284-0 PMID:32572746

Tseng, M. L., Chiu, A. S. F., Tan, R. R., & Siriban-Manalang, A. B. (2013). Sustainable consumption and production for Asia: Sustainability through green design and practice. *Journal of Cleaner Production*, *40*(2), 1–5. doi:10.1016/j.jclepro.2012.07.015

Tubiello, F. N., Donatelli, M., Rosenzweig, C., & Stockle, C. O. (2000). Effects of climate change and elevated CO_2 on cropping systems: Model predictions at two Italian locations. *European Journal of Agronomy*, *13*(2), 179–189. doi:10.1016/S1161-0301(00)00073-3

Tuong, T. P., & Bouman, B. A. M. (2003). Rice production in water-scarce environments. In Water Productivity: Limits and Opportunities for Improvement. CABI Pub. doi:10.1079/9780851996691.0053

Turati, F., Trichopoulos, D., Polesel, J., Bravi, F., Rossi, M., Talamini, R., Franceschi, S., Montella, M., Trichopoulou, A., La Vecchia, C., & Lagiou, P. (2014). Mediterranean diet and hepatocellular carcinoma. *Journal of Hepatology*, *60*(3), 606–611. Advance online publication. doi:10.1016/j.jhep.2013.10.034 PMID:24240052

Turck, D., Bresson, J. L., Burlingame, B., Dean, T., Fairweather-Tait, S., Heinonen, M., Ildico Hirsch-Ernst, K., Mangelsdorf, I., McArdle, J. M., Naska, A., Monika Neuhauser-Berthold, M., Nowicka, G., Pentieva, K., Sanz, Y., Siani, A., Sjodin, A., Stern, M., Tome, D., Vinceti, M., ... Loveren, H. (2018). Scientific opinion on the safety of xylo-oligosaccharides (XOS) as a novel food pursuant to Regulation (EU) 2015/2283. EFSA NDA Panel (EFSA Panel on Dietetic Products, Nutrition and Allergies). *EFSA Journal*, *16*(7), 5361–5381.

Uddin, S., Löw, M., Parvin, S., Fitzgerald, G. J., Tausz-Posch, S., Armstrong, R., O'Leary, G., & Tausz, M. (2018). Elevated [CO_2] mitigates the effect of surface drought by stimulating root growth to access sub-soil water. *PLoS One*, *13*(6), e0198928. doi:10.1371/journal.pone.0198928 PMID:29902235

Uddin, S., Löw, M., Parvin, S., Fitzgerald, G. J., Tausz-Posch, S., Armstrong, R., & Tausz, M. (2018). Yield of canola (*Brassica napus* L.) benefits more from elevated CO_2 when access to deeper soil water is improved. *Environmental and Experimental Botany*, *155*, 518–528. doi:10.1016/j.envexpbot.2018.07.017

UN. (n.d.). *The 17 Goals*. Retrieved September 26, 2021, from https://sdgs.un.org/goals

UNDP - The United Nations Development Program. (2016). *Reforço das capacidades das comunidades rurais para a adaptação aos efeitos às mudanças climáticas em São Tomé e Príncipe nos distritos de Cauê, Me-Zochi, Príncipe, Lembá, Cantagalo e Lobata (CMPLCL). Projeto Adaptação (CMPLCL) - PNUD/GEF (2015 - 2016)*. Programa Das Nações Unidas Para Desenvolvimento – STP.

UNESCO. (2010). *Evaluation of nominations for inscription in 2010 on the representative list of the intangible cultural heritage of humanity*. UNESCO ICH.

UNESCO. (n.d.). *Redesigning lifestyles, consumption and production through Education for Sustainable Development.* Retrieved September 26, 2021, from https://en.unesco.org/news/redesigning-lifestyles-consumption-and-production-through-education-sustainable-development

UNFCCC (United Nations Framework Convention on Climate Change). (2007). *Climate change: Impacts, vulnerability and adaptation in developing countries.* Retrieved from https://unfccc.int/resource/docs/publications/impacts.pdf

UNHCR. (2021). *Climate change and disaster displacement.* Retrieved from https://www. unhcr.org/climate-change-and-disasters.html

United Nations Development Programme. (2011). *Blending Climate Finance Through National Climate Funds.* Anvil Creatiosn Group.

US EPA. (2021). *Overview of Greenhouse Gases.* Retrieved from https://www. epa. gov/ ghgemissions/overview-greenhouse-gases

USDA, Foreign Agricultural Service. (2014). *India's Agricultural Exports Climb to Record High.* International Agricultural Trade Report. Retrieved from https://www.fas.usda.gov/data/india-s-agricultural-exports-climb-record-high

USDA/ERS-United States Department of Agriculture/Economic Research Service. (2019). *International Baseline Data:2019. International Long-Term Projections to 2028.* Author.

US-EPA - United States/Environmental Protection Agency. (2006). *Global Anthropogenic Non-CO2 Greenhouse Gas Emissions:1990-2020.* U.S. Environmental Protection Agency, Office of Atmospheric Programs, Climate Change Division.

USEPA. (2006). Global anthropogenic non-CO2 greenhouse gas emissions: 1990–2020. United States Environmental Protection Agency, EPA 430-R-06-003.

Vagias, W. (2006). *Likert-type scale response anchors.* Clemson International Institute for Management, Clemson University.

Valadez-Carmona, L., Cortez-García, R. M., Plazola-Jacinto, C. P., Necoechea-Mondragón, H., & Ortiz-Moreno, A. (2016). Effect of microwave drying and oven drying on the water activity, color, phenolic compounds content and antioxidant activity of coconut husk (Cocos nucifera L.). *Journal of Food Science and Technology, 53*(9), 3495–3501. doi:10.100713197-016-2324-7

Vallaster, C. (2017). Managing a Company Crisis through Strategic Corporate Social Responsibility: A Practice-Based Analysis. *Corporate Social Responsibility and Environmental Management, 24*(6), 509–523. doi:10.1002/csr.1424

van de Looij, Y., Larpin, C., Cabungcal, J. H., Sanches, E. F., Toulotte, A., Do, K. Q., & Sizonenko, S. V. (2019). Nutritional Intervention for Developmental Brain Damage: Effects of Lactoferrin Supplementation in Hypocaloric Induced Intrauterine Growth Restriction Rat Pups. *Frontiers in Endocrinology, 10*, 1–14. doi:10.3389/fendo.2019.00046

Van der Laan, H. L., H Dijkstra, T., & Tilburg, A. V. (1999). Agricultural marketing in tropical Africa: obstacles to systematic study. *Agricultural marketing in tropical Africa. Contributions from the Netherlands*, 1-17.

Van der Ploeg, J. D., Renting, H., Brunori, G., Knickel, K., Mannion, J., Marsden, T., De Roest, K., Sevilla-Guzmán, E., & Ventura, F. (2000). Rural development: From practices and policies towards theory. *Sociologia Ruralis, 40*(4), 391–408. doi:10.1111/1467-9523.00156

van Grinsven, H. J. M., Erisman, J. W., de Vries, W., & Westhoek, H. (2015). Potential of extensification of European agriculture for a more sustainable food system, focusing on nitrogen. *Environmental Research Letters, 10*(2), 025002. doi:10.1088/1748-9326/10/2/025002 van Huellen, S., & Abubakar, F. M. (2021). Potential for upgrading in financialised agri-food chains: The case of Ghanaian cocoa. *European Journal of Development Research, 33*(2), 227–252. doi:10.105741287-020-00351-3 PMID:33558793

Vanhamaki, S., Medkova, K., Malamakis, A., Kontogianni, S., Marisova, E., Dellago, D. H., & Moussiopoulos, N. (2019). Bio-based circular economy in European national and regional strategies. *International Journal of Sustainable Development and Planning, 14*(1), 31–43. doi:10.2495/SDP-V14-N1-31-43

Vardarlıer, P. (2016). Strategic approach to human resources management during crisis. *Procedia: Social and Behavioral Sciences, 235*, 463–472. doi:10.1016/j.sbspro.2016.11.057

Vaskalis, I., Skoulou, V., Stavropoulos, G., & Zabaniotou, A. (2019). Towards circular economy solutions for the management of rice processing residues to bioenergy via gasification. *Sustainability (Switzerland), 11*(22), 6433. Advance online publication. doi:10.3390u11226433

Vaz, J. (2018). *Plano de Gestão de Resíduos - 2018 – República Democrática de São Tomé e Príncipe*. Ecogestus Lda. doi:10.13140/RG.2.2.30218.62400

Vega-Quezada, C. A., & Blanco, M. (2017). Photobioreactors within the framework of circular economy and poverty eradication: Bioenergy production in the Ecuador-Peru border integration zone. In Photobioreactors: Advancements, Applications and Research (pp. 275-298). Nova Science Publishers, Inc.

Vega-Quezada, C., Blanco, M., & Romero, H. (2017). Synergies between agriculture and bioenergy in Latin American countries: A circular economy strategy for bioenergy production in Ecuador. *New Biotechnology, 39*, 81–89. doi:10.1016/j.nbt.2016.06.730 PMID:27287049

Velvizhi, G., Shanthakumar, S., Das, B., Pugazhendhi, A., Priya, T. S., Ashok, B., Nanthagopal, K., Vignesh, R., & Karthick, C. (2020). Biodegradable and non-biodegradable fraction of municipal solid waste for multifaceted applications through a closed loop integrated refinery platform: Paving a path towards circular economy. *The Science of the Total Environment, 731*, 138049. Advance online publication. doi:10.1016/j.scitotenv.2020.138049 PMID:32408201

Verbeke, W., Frewer, L. J., Scholderer, J., & De Brabander, H. F. (2007). Why consumers behave as they do with respect to food safety and risk information. *Analytica Chimica Acta, 586*(1-2), 2–7. doi:10.1016/j.aca.2006.07.065 PMID:17386689

Verberne, L., Bach-Faig, A., Buckland, G., & Serra-Majem, L. (2010). Association between the mediterranean diet and cancer risk: A review of observational studies. *Nutrition and Cancer, 62*(7), 860–870. Advance online publication. doi:10.1080/01635581.2010.509834 PMID:20924961

Verdoorn, P. J. (1949). Fattori che regolano lo sviluppo della produttivita del lavoro. *L'Industria, 1*, 45–53.

Via, M. (2012). The Malnutrition of Obesity: Micronutrient Deficiencies That Promote Diabetes. *ISRN Endocrinology, 2012*, 1–8. Advance online publication. doi:10.5402/2012/103472 PMID:22462011

Vogel, C., Mathé, S., Geitzenauer, M., Ndah, H. T., Sieber, S., Bonatti, M., & Lana, M. (2020). Stakeholders' perceptions on sustainability transition pathways of the cocoa value chain towards improved livelihood of small-scale farming households in Cameroon. *International Journal of Agricultural Sustainability, 18*(1), 55–69. doi:10.1080/14735903.2019.1696156

Vogeler, I., Beukes, P., Romera, A., & Cichota, R. (2012). Estimating nitrous oxide emissions from a dairy farm using a mechanistic, whole farm model and segregated emission factors for New Zealand. *Soil Research (Collingwood, Vic.), 50*(3), 188–194. doi:10.1071/SR12064

Vories, E., Stevens, W. G., Rhine, M., & Straatmann, Z. (2017). Investigating irrigation schedul-ing for rice using variable rate irrigation. *Agricultural Water Management, 179*, 314–323. doi:10.1016/j.agwat.2016.05.032

Waage, J., & Mumford, J. (2008). Agricultural biosecurity. *Philos Trans R Soc Lond B Biol Sci, 363*(1492), 863–876. doi:10.1098/rstb.2007.2188

Waage, J., Yap, C., Bell, S., Levy, C., Mace, G., Pegram, T., Unterhalter, E., Dasandi, N., Hudson, D., Kock, R., Mayhew, S., Marx, C., & Poole, N. (2015). Governing the UN Sustainable Development Goals: Interactions, infrastructures, and institutions. *The Lancet. Global Health, 3*(5), e251–ee252. doi:10.1016/S2214-109X(15)70112-9 PMID:25833663

Waas, T., Hugé, J., Verbruggen, A., & Wright, T. (2012). Sustainable Development: A Bird's Eye View. *Sustainability, 3*(10), 1637–1661. doi:10.3390u3101637

Wall, D. M., McDonagh, S., & Murphy, J. D. (2017). Cascading biomethane energy systems for sustainable green gas production in a circular economy. *Bioresource Technology, 243*, 1207–1215. doi:10.1016/j.biortech.2017.07.115 PMID:28803063

Wang, C., Ghadimi, P., Lim, M. K., & Tseng, M. (2018). A literature review of sustainable consumption and production: A comparative analysis in developed and developing economies. *Journal of Cleaner Production.* . doi:10.1016/j.jclepro.2018.09.172

Wang, F., Chen, W. K., & Chen, J. A. (2008). The Construction of New Countryside and the Development of Modern Agriculture in underdeveloped areas-- A Summary of the National Annual meeting of Young and Middle-aged Agricultural economists in 2007. *Agricultural Economic problems*, (1), 107-109.

Wang, G. R. (2017). *Study on benefit Evaluation and Popularization of High efficiency Water use Technology Model for main crops in Hebei Province* (Doctoral dissertation). Northeast Agricultural University.

Wang, X., Yu, J. F., Zhao, X. L., Huo, W. B., Qi, P., & Li, Y. H. (2020). *The invention relates to a management system based on agricultural data and production and an agricultural data monitoring terminal.* CN111352370A.

Wang, E., & Gao, Z. (2017). Chinese Consumer Quality Perception and Preference of Traditional Sustainable Rice Produced by the Integrated Rice–Fish System. *Sustainability, 9*(12), 2282. doi:10.3390u9122282

Wang, J., Liu, X., Zhang, X., Li, L., Lam, S. K., & Pan, G. (2019). Changes in plant C, N and P ratios under elevated [CO_2] and canopy warming in a rice-winter wheat rotation system. *Scientific Reports, 9*(1), 5424. doi:10.103841598-019-41944-1 PMID:30931987

Wang, J., Mendelsohn, R., Dinar, A., Huang, J., & Zhang, L. (2010). The impact of climate change on china's agriculture. *Agricultural Economics, 40*(3), 323–337. doi:10.1111/j.1574-0862.2009.00379.x

Wang, T. M., Wu, J. G., Kou, X. J., Oliver, C., Mou, P., & Ge, J. P. (2010). Ecologically asynchronous agricultural practice erodes sustainability of the loess plateau of china. *Ecological Applications, 20*(4), 1126–1135. doi:10.1890/09-0229.1 PMID:20597295

Wang, W. H. (2016). Soil control is in action. *China Rural Science and Technology*, (10), 52–55.

Wang, X., Yamauchi, F., & Huang, J. (2016). Rising wages, mechanization, and the substitution between capital and labor: Evidence from small scale farm system in china. *Agricultural Economics, 47*(3), 309–317. doi:10.1111/agec.12231

Wang, Y. K., Yang, X. N., Zhu, X., Xiao, X. R., Yang, X. W., Qin, H. B., Gonzalez, F. J., & Li, F. (2019). Role of Metabolic Activation in Elemicin-Induced Cellular Toxicity. *Journal of Agricultural and Food Chemistry, 67*(29), 8243–8252. doi:10.1021/acs.jafc.9b02137

Wang, Z. G. (2004). Measurement of rural marketization degree in China in 2002 and its influencing factors. *Annual conference of Chinese Society of Agricultural Economics.*

Wanowaa, Furaha, & Yan. (2020). Effects of Planting Field on Groundwater and Surface Water Pollution in China. *CLEAN - Soil Air Water*.

WCED. (1987). *Our Common Future. Brundtland Report* (Vol. 4). Oxford University Press.

Web, M. D. (2020). *AlgaOil. Side Effects and precautions*. Available at https://www.webmd.com/vitamins/ai/ingredientmono-1565/algal-oil

WebM. D. (2020). *Pectin*. https://www.webmd.com/vitamins/ai/ingredientmono-500/pectin

Wee, J. H., Min, C., Park, M. W., Park, I. S., Park, B., & Choi, H. G. (2020). Energy-drink consumption is associated with asthma, allergic rhinitis, and atopic dermatitis in Korean adolescents. *European Journal of Clinical Nutrition*, *75*(7), 1077–1087. doi:10.103841430-020-00812-2

Weerabahu, S., & Nanayakkara, J. (2014). A Strategic Framework for Managing Rice Supply Chains. *Conference: Wayamba International Conference, WinC 2014*.

Werle, S., Sobek, S., & Kaczor, Z. (2019). *The role of the thermal processes of waste biomass in the circular economy perspective*. Paper presented at the 32nd International Conference on Efficiency, Cost, Optimization, Simulation and Environmental Impact of Energy Systems, ECOS 2019.

Whiting, K. (2020). *Coronavirus isn't an outlier, it's part of our interconnected viral age*. World Economic Forum. Retrieved from https://www.weforum.org/agenda/2020/03/coronav irus-global-epidemics-health-pandemic-covid-19/

WHO. (2018). *Childhood Obesity Surveillance Initiative (COSI). Highlights 2015-17*. World Health Organization.

WHO. (2019). *Strategic Plan to Reduce Malnutrition in Africa Adopted*. WHO.

WHO. (2020a). *Naming the coronavirus disease (COVID-19) and the virus that causes it*. Retrieved from World Health Organization: https://www.who.int/emergencies/diseases/novel-coronavirus-2019/technical-guidance/naming-the-coronavirus-disease-(covid-2019)-and-the-virus-that-causes-it

WHO. (2020b). *Novel Coronavirus (2019-nCoV) Situation Report- 1*. Retrieved from World Health Organization: www.who.int/docs/default-source/coronaviruse/situation-reports/20200121-sitrep-1-2019-ncov.pdf?sfvrsn=20a99c10_4

Widmer, R. J., Flammer, A. J., Lerman, L. O., & Lerman, A. (2015). The Mediterranean diet, its components, and cardiovascular disease. *The American Journal of Medicine*, *128*(3), 229–238. Advance online publication. doi:10.1016/j.amjmed.2014.10.014 PMID:25447615

Wilk, E., Krowczyriska, M., & Zagajewski, B. (2019). Modelling the Spatial Distribution of Asbestos-Cement Products in Poland with the Use of the Random Forest Algorithm. *Sustainability*, *11*(16), 4355. doi:10.3390u11164355

Willett, W. C., Sacks, F., Trichopoulou, A., Drescher, G., Ferro-Luzzi, A., Helsing, E., & Trichopoulos, D. (1995). Mediterranean diet pyramid: A cultural model for healthy eating. *The American Journal of Clinical Nutrition*, *61*(6), 1402S–1406S. Advance online publication. doi:10.1093/ajcn/61.6.1402S PMID:7754995

Willy, D. K., Muyanga, M., Mbuvi, J., & Jayne, T. (2019). The effect of land use change on soil fertility parameters in densely populated areas of Kenya. *Geoderma*, *343*, 254–262. doi:10.1016/j.geoderma.2019.02.033 PMID:31160827

Wlodarczyk-Spiewak, K., & Korpysa, J. (2006). Changes of labour market as the main determinant of inequalities in the consumption structure in Polish households. *Transformations in Business & Economics*, *5*(2), 106–121.

Wojcik-Gront, E. (2018). Variables influencing yield-scaled Global Warming Potential and yield of winter wheat production. *Field Crops Research*, *227*, 19–29. doi:10.1016/j.fcr.2018.07.015

Wolfson, J. A., & Bleich, S. N. (2015). Is cooking at home associated with better diet quality or weight-loss intention? *Public Health Nutrition, 18*(8), 1397–1406. Advance online publication. doi:10.1017/S1368980014001943 PMID:25399031

World Bank Database. (2020). *Agriculture, forestry, and fishing, value added (% of GDP).* Retrieved from https://data.worldbank.org/indicator/NV.AGR.TOTL.ZS

World Bank Group. (2020). *Climate Change Knowledge Portal Country India.* Retrieved from https://climateknowledgeportal.worldbank.org/country/india

World Bank Group. (2021). *Climate smart agriculture.* https://www.worldbank.org/en/topic/climate-smart-agriculture

World Bank. (2018). *Turning Smallness into Uniqueness: Six Key Challenges to Unlock São Tomé and Príncipe Growth's Potential. Country Economic Memorandum for São Tomé and Príncipe Summary Report.* Author.

World Bank. (2019). *Rural Development: Sector Policy Paper.* http://documents.worldbank.org/curated/en/688121468168853933 /Rural-development-sector-policy-paper

World Bank. (2019). *World Bank Group Report – WBG (2019). The Obstacles And Opportunities for STP's Agriculture.* Value Chains.

World Food Summit. (1996). *Rome Declaration on World Food Security.* FAO.

World Meteorological Organization. (2021). *Climate change indicators and impacts worsened in 2020.* Retrieved from https://public.wmo.int/en/media/press-release/climate-change-indicators-and-impacts-worsened-2020

World Resource Institute. (2016). *Climate Watch.* https://www.wri.org/initiatives/climate-watch

World Water Development Report. (2020). *The United Nations world water development report 2020: water and climate change.* UNESCO World Water Assessment Programme. Retrieved from https://unesdoc.unesco.org/ark:/48223/pf0000372985.locale=en

Wossen, T., Berger, T., & Di Falco, S. (2015). Social capital, risk preference and adoption of improved farm land management practices in Ethiopia. *Agricultural Economics, 46*(1), 81–97. doi:10.1111/agec.12142

Wotlolan, D. L., Lowry, J. H., Wales, N. A., & Glencross, K. (2021). Land suitability evaluation for multiple crop agroforestry planning using GIS and multi-criteria decision analysis: A case study in Fiji. *Agroforestry Systems, 95*(8), 1519–1532. doi:10.100710457-021-00661-3

Wu, Y. N. (2021). *2020 rural online retail sales reached 1.79 trillion yuan. China's e-commerce market continues to lead the world.* https://www.chinairn.com/news/20210124/121901569.shtml

Wu, K. J., Liao, C. J., Tseng, M. L., & Chiu, K. S. (2016). Multi-attribute approach to sustainable supply chain management under uncertainty. *Industrial Management & Data Systems, 116*(4), 777–800. doi:10.1108/IMDS-08-2015-0327

Wu, K. J., Zhu, Y., Tseng, M. L., Lim, M. K., & Xue, B. (2018). Developing a hierarchical structure of the co-benefits of the triple bottom line under uncertainty. *Journal of Cleaner Production, 195*, 908–918. doi:10.1016/j.jclepro.2018.05.264

Xie, Y. K., & Xu, Y. M. (1988). Agricultural labor transfer and urbanization in China. *Journal of Renmin University of China*, (01), 80–86.

Xie, Z., Zhang, F., Lun, F., Gao, Y., Ao, J., & Zhou, J. (2020). Research on a diagnostic system of rural vitalization based on development elements in China. *Land Use Policy, 92*, 104421. doi:10.1016/j.landusepol.2019.104421

Xi, X. J. (2019). Research on the application of Internet of things production mode in intelligent agriculture. Scientific and technological. *Innovation*, (33), 77–78.

Xu, F. (2020). Analysis on the import trade structure of China's agricultural products-- ternary decomposition based on trade growth. *China's Circulation economy*, *34*(6), 96-104.

Xue, L., Weng, L., & Yu, H. (2018). Addressing policy challenges in implementing Sustainable Development Goals through an adaptive governance approach: A view from transitional China. *Sustainable Development*, *26*(2), 150–158. doi:10.1002d.1726

Xu, K., Yang, H. J., & Lu, Z. Y. (2021). The Prospect of the Development of Digital Agriculture during the 14th five-year Plan. *China Engineering Consulting*, (4), 4.

Xu, S. F., Zhang, Y. H., Wang, S., Pang, Z. Q., Fan, Y. G., Li, J. Y., Wang, Z. Y., & Guo, C. (2019). Lactoferrin ameliorates dopaminergic neurodegeneration and motor deficits in MPTP-treated mice. *Redox Biology*, *21*, 1–13. doi:10.1016/j.redox.2018.101090

Xu, X. B., Ma, Z., Chen, Y. Q., Gu, X. M., Liu, Q. Y., Wang, Y. T., Sun, M., & Chang, D. H. (2018). Circular economy pattern of livestock manure management in Longyou, China. *Journal of Material Cycles and Waste Management*, *20*(2), 1050–1062. doi:10.100710163-017-0667-4

Xu, Z. Q. (2015). Development trend of China's manufacturing informatization based on the Internet of Things era. *Economic Management (Abstracts)*, (8), 00283–00283.

Yadav, P. L., Han, S. H., & Rho, J. J. (2016). Impact of environmental performance on firm value for sustainable investment: Evidence from large US firms. *Business Strategy and the Environment*, *25*(6), 402–420. doi:10.1002/bse.1883

Yalew, S. G., van Griensven, A., & van der Zaag, P. (2016). AgriSuit: A web-based GIS-MCDA framework for agricultural land suitability assessment. *Computers and Electronics in Agriculture*, *128*, 1–8. doi:10.1016/j.compag.2016.08.008

Yamane, T. (1973). *Statistics: An Introductory Analysis* (3rd ed.). Harper and Row.

Yan, H. H., & Peng, S. Q. (2009). CREB signaling pathway mediates changes in cognitive function induced by organophosphorus pesticides in rats and its possible mechanism. *The Fifth National Conference of Chinese Society of Toxicology*.

Yang, H. P. (2018). Research on the Promotion of Agricultural products under the Mode of Internet + Agriculture-- taking native products as an example. *China Management Informatization*, *376*(10), 147-148.

Yang, J. (2014). *The impact of "aging" agricultural Labor force on Agricultural Production* (Dissertation). Northwest Normal University.

Yang, G., Ding, L., Huo, L., & Hu, Y. (2021). Coordinated development of agricultural water resources and the socio-economy in Shanxi province considering uncertainty. *Irrigation and Drainage*, *2021*(4), 1–10. doi:10.1002/ird.2581

Yan, W., Chen, C., & Chang, W. (2009). An investigation into sustainable product constructualization using a design knowledge hierarchy and Hopfield network. *Computers & Industrial Engineering*, *56*(4), 617–626. doi:10.1016/j.cie.2008.10.015

Yasir, S., & Schultz, K. (2020). Powerful Cyclone Lashes Mumbai as City Grapples With Coronavirus. *The New York Times*. Retrieved from https://www.nytimes.com/ 2020/06/03/world/asia/cyclone-nisarga-india-mumbai.html

Yazan, D. M., Cafagna, D., Fraccascia, L., Mes, M., Pontrandolfo, P., & Zijm, H. (2018). Economic sustainability of biogas production from animal manure: A regional circular economy model. *Management Research Review*, *41*(5), 605–624. doi:10.1108/MRR-02-2018-0053

Ye, Y., Jiang, S., Wen, T., & Tang, P. (2020). *The method and device of big data analysis in agriculture*. CN111582324A.

Ye, G. X. (2014). High-fidelity Agricultural Products E-commerce development analysis and research in Sichuan Province. *Sichuan Agriculture and Agricultural Machinery*, (4), 12–12.

Ye, T., Ye, L., Wang, S., Feng, Z. M., Dong, L. M., & Yin, Y. L. (2018). Building agricultural products traceability system to promote the development of smart agriculture. *High Technology and Industrialization*, *000*(005), 42–45.

Yong, H. N. (2008). *Supply Chain Model for Rice in Malaysia – Basics and Challenges*. ECER Regional Conference 2008 (ERC08), Kota Bharu, Kelantan, Malaysia.

Yuan, S., & Qi, W. (2019). The predicament and way out of promoting the value of rural land resources from the perspective of rural revitalization. *The 16th Shenyang Science Annual Meeting*.

Yu, C., Huang, X., Chen, H., Huang, G., Ni, S., Wright, J. S., Hall, J., Ciais, P., Zhang, J., Xiao, Y., Sun, Z., Wang, X., & Yu, L. (2018). Assessing the impacts of extreme agricultural droughts in China under climate and socioeconomic changes. *Earth's Future*.

Yue, D., Jiang, W., He, B., Zheng, C., & Kai, J. (2018). Change in intensity and frequency of extreme precipitation and its possible teleconnection with large-scale climate index over the China from 1960 to 2015. *Journal of Geophysical Research, D, Atmospheres*, *123*(4), 2068–2081. doi:10.1002/2017JD027078

Yu, S. Y., & Zhan, Y. P. (2006). The phenomenon of shortage of migrant workers and the regional transfer of labor-intensive industries in China. *Business Research*, (24), 105–107.

Yu, Y., Lian, X. Y., & Jiang, Y. H. (2017). Risk-based prioritization method for the classification of groundwater pesticide pollution from agricultural regions. *Integrated Environmental Assessment and Management*, *13*(6), 137. PMID:28577331

Zabaniotou, A., Rovas, D., Libutti, A., & Monteleone, M. (2015). Boosting circular economy and closing the loop in agriculture: Case study of a small-scale pyrolysis-biochar based system integrated in an olive farm in symbiosis with an olive mill. *Environmental Development*, *14*, 22–36. doi:10.1016/j.envdev.2014.12.002

Zaharia, R. M., & Zaharia, R. (2021). Triple Bottom Line. In D. Crowther & S. Seifi (Eds.), *The Palgrave Handbook of Corporate Social Responsibility*. Palgrave Macmillan.

Zbarsky, V. K., Trusova, N., Sokil, O. H., Pochernina, N., & Hrytsaienko, M. L. (2020). Social and Economic Determinants for the Development of Resource Potential of Small Forms of Agrarian Production in Ukraine. *Industrial Engineering and Management Systems*, *19*(1), 133–142. doi:10.7232/iems.2020.19.1.133

Zeng, S. Q. (2020). Information leads the Future-- an Analysis of the China Agriculture Outlook report (2020-2029). *Agricultural Products Market*, *939*(9), 8-10.

Zeng, Y. S. (2020). The long-term goal and strategic plan of agricultural and rural modernization. *Social Governance*, *55*(11), 22-24.

Zeweld, W., Van Huylenbroeck, G., Tesfay, G., & Speelman, S. (2017). Smallholder farmers' behavioural intentions towards sustainable agricultural practices. *Journal of Environmental Management*, *187*, 71–81. doi:10.1016/j.jenvman.2016.11.014 PMID:27883941

Zhai, Z., Martínez, J. F., Beltran, V., & Martínez, N. L. (2020). Decision support systems for agriculture 4.0: Survey and challenges. *Computers and Electronics in Agriculture*, *170*, 1–16. doi:10.1016/j.compag.2020.105256

Zhang, Y. Q. (2019). Environmental impact analysis of high-efficiency water-saving irrigation in groundwater over-exploitation area. *Henan Water Conservancy and South-to-North Water Transfer*, *338*(8), 24-35.

Zhang, D., & Managi, S. (2020). Financial development, natural disasters, and economics of the Pacific small island states. *Economic Analysis and Policy*, *66*, 168–181. doi:10.1016/j.eap.2020.04.003

Zhang, H. K. (2014). Imbalance of China's agricultural products international trade and development countermeasures. *Shopping Mall Modernization*, *000*(033), 2–2.

Zhang, J., Su, Y., Wu, J., & Liang, H. (2015). GIS based land suitability assessment for tobacco production using AHP and fuzzy set in Shandong province of China. *Computers and Electronics in Agriculture*, *114*, 202–211. doi:10.1016/j.compag.2015.04.004

Zhang, L. S., & Zhang, P. L. (2003). stimulating rural economy; Increasing rural income. *Construction of Small Towns*, *000*(004), 6–8.

Zhan, J. T., Xu, T., Zhang, Y. Y., Yang, X. L., Qu, Z. Q., & Tao, T. (2017). The Effects of Agricultural R&D on Chinese Agricultural Productivity Growth: New Evidence of Convergence and Implications for Agricultural R&D Policy. *Canadian Journal of Agricultural Economics*, *65*(3), 453–475. doi:10.1111/cjag.12137

Zhao, L., Shao, N., Dou, X. B., & Tao, W. Y. (2019). Analysis of China's agricultural products market in 2018 and prospects for 2019. *Agricultural Outlook*, *15*(1), 6-9: 17.

Zhao, C., Liu, B., Piao, S., Wang, X., Lobell, D. B., Huang, Y., Huang, M., Yao, Y., Bassu, S., Ciais, P., Durand, J. L., Elliott, J., Ewert, F., Janssens, I. A., Li, T., Lin, E., Liu, Q., Marte, P., Muller, C., ... Asseng, S. (2017). Temperature increase reduces global yields of major crops in four independent estimates. *Proceedings of the National Academy of Sciences of the United States of America*, *114*(35), 9326–9331. doi:10.1073/pnas.1701762114 PMID:28811375

Zhao, G. L. (2019). Study on the mechanism of enhancing the value of rural land resources. *Journal of Inner Mongolia University of Finance and Economics*, *17*(1), 8–12.

Zhao, H. J. (2018). How farmers can better participate in land consolidation. *Chinese and Foreign Entrepreneurs*, *596*(06), 213.

Zhao, H. M. (2018). Activating endogenous power to promote rural revitalization in Hebei Province. *Hainan Today*, *242*(09), 26.

Zheng, H., & Ma, W. (2021). The role of resource reallocation in promoting total factor productivity growth: Insights from China's agricultural sector. *Review of Development Economics*, *00*(4), 1–22. doi:10.1111/rode.12826

Zhiyan Consultation. (2020). *Analysis on the development trend of China's agricultural product output and agricultural product import and export trade in 2019*. https://www.chyxx.com/industry/202007/883979.html

Zhong, F. N., Zhu, J., & Xie, Z. Q. (2007). How Mobile Are Resources in Chinese Agriculture? — Implications for China's Agricultural Trade Policy. China & World Economy, 15(1), 43 - 61.

Zhou, D., & Lu, Y. (2014). Impact of aging of agricultural labor force on land use efficiency. *Tech Rich Guide*, (14), 52–52.

Zhu X. C. & J. Zhao. (2011). Subsidy policy of the purchase of machinery has accelerated progression of Chinese mechanization during the last seven years. *Chinese Agricultural News of Farmers' Daily*.

Zillman, J., & Sherwood, S. (2017). *The enhanced greenhouse effect*. Australian Academy of Science. Retrieved from https://www.science.org.au/curious/earth-environment/enhanced-greenhouse-effect

Zocca, R., Gaspar, P. P., Silva, P. D., Santos, F. D., Andrade, L. P., & Nunes, J. (2019). Decision-making computationally aided in the management of energy sources used in agrifood industries. *Energy Procedia*, *161*, 100–107. doi:10.1016/j.egypro.2019.02.063

Zou, L. (2009). *Effects of gradual and sudden heat stress on seed quality of Andean lupin, Lupinus mutabilis* [Master Degree]. University of Helsinki - Faculty of Agriculture and Forestry.

About the Contributors

Vítor João Pereira Domingues Martinho is Coordinator Professor with Habilitation at the Polytechnic Institute of Viseu, Portugal, and holds a Ph.D. in Economics from the University of Coimbra, Portugal. He was President of the Scientific Council, President of the Directive Council and President of the Agricultural Polytechnic School of Viseu, Portugal, from 2006 to 2012. He also was the President of the Direction of the Association of Forest Producers of Viseu, Portugal. He was an Erasmus student in the Faculty of Economics from the University of Verona, Italy, participated in various technical and scientific events nationally and internationally, has published several technical and scientific papers, is referee of various scientific and technical journals and participates in the evaluation of national and international projects.

* * *

Aashrika Ahuja is an experienced corporate and intellectual property rights lawyer, advisor, writer and author. With an experience of 8 years in the field of law, she is now well conversant with various aspects of the profession.She has handled legal matters in diverse fields including Civil, Criminal, Intellectual Property Rights, Sexual Harassment against Women and Children, Cyber laws, is a licensed Attorney and is eligible to appear and practice before various courts in India. She is well versed with corporate governance and with strong entrepreneurship skills, has successfully advised as well as worked with various corporates, organizations and institutions. She has an experience of working and handling clients all across the globe. She has a keen interest in research and writing and has authored research paper in the prestigious International Journal for Scientific and Engineering Research. While handling drafting, review of various kinds of legal Agreements, Contracts and other legal documents, she also provides legal opinion on diverse matters relating to Intellectual Property Rights. She is passionate about realizing and working towards United Nations 2030 Agenda for Sustainable Development Goals and with her strong advocacy skills has been actively working in the development field to leave no one behind.

José T. M. Aranha is a full professor of Geographical Information Systems and Remote Sensing (GIS – RS) in Agro-forestry applications at the Forestry Department in University of Trás-os-Montes e Alto Douro, Vila Real, Portugal. He earned his Ph.D. in GIS-RS (1998) from the Kingston University, England. His research interests are multidisciplinary, and include Forest Fire Hazard Indices, Forest Biomass calculation, Forest Inventory, Satellite Image processing and classification, Land use Land cover classification and mapping, Spatial analysis, Map Algebra modelling in Agro-forestry, Ecosystem Services modelling and Biological Struggle, Wildlife survey. He is the author (2021) of 1 Book; 5

Book chapters, 7 books published by UTAD, 25 papers in International Journals (ISI JCR), 6 papers in National Journals (Latindex, scoupus), 48 paper in International Congress Proceedings Books, 72 papers in National Conference Proceedings Books, 30 papers in dissemination and transmission of science. He is supervisor of several PhD and Msc students. Full member of Centre for the Research and Technology of Agro-Environmental and Biological Sciences – CITAB – https://www.citab.utad.pt/.

Kiril Bahcevandziev is professor of plant biotechnology Ph.D.

Oscar Bernardes holds a Ph.D. in management, since 2013. He worked as management consultant for several years, in several organizations (social, profitable, and governmental). He is Invited Professor at Polytechnic Institute of Porto (P.Porto), Portugal, where he has been lecturing courses in the following areas: entrepreneurship, innovation, marketing and general management. His research interests include entrepreneurship, gamification, innovation, and marketing.

Fernando Capela e Silva studied at the Universidade de Évora (Portugal) and Zaragoza (Spain) and holds a PhD in Biology. He is Associate Professor of Biology at the Department of Medical and Health Sciences, School of Health and Human Development, Universidade de Évora, and researcher at MED-Mediterranean Institute for Agriculture, Environment and Development, Universidade de Évora. His areas of interest are: oral tissues biology; determinants of food choices and their effects on the health of people and populations; food, health and society and the One Health approach. He has done research in the area of food choices and has participated in diverse projects. He is the author of several book chapters and academic scientific papers in peer-reviewed international journals and other publications.

Cristina Conceição holds a PhD in Agrarian Sciences, from the University of Evora, being Associated Professor of the Zootechnics Department of the same University and researcher at MED (Mediterranean Institute for Agriculture Environment and Development). Her areas of interest are traditional food products, with a major focus on dairy products. She has expertise in sensory analysis and is an active member of the Alentejo Gastronomy Confraternity, with particular interest in preserving food traditions. She has participated in different projects and is the author of several book chapters and academic scientific papers in peer-reviewed international journals and other publications.

Helena Maria de Paiva Martins Esteves Correia holds a Degree in Agronomic Engineering from the University of Trás-os-Montes e Alto Douro (UTAD), Postgraduate Course in Horticulture, FruitI-CULTURE and Viticulture (UTAD), Adjunct Professor at the Polytechnic Institute of Viseu, Portugal, since 2008, Director of the Department of Ecology and Sustainable Agriculture, 2010-2014. Director of the Degree Course in Agronomic Engineering at Agrarian School of Viseu (ESAV) 2017-2020. Vice-President of the ESAV, 2012-2016. Orientation of several final graduation works. Author of several publications in refereed journals and conference papers. Guiné R. P. F., Costa D. V. T. A., Correia PMR, Costa C. A., Correia H. E., Castro M., Guerra L. T., Seeds C., Coll C., Radics L., Arslan M., Soylu S., Tothova M., Toth P., Basile S. (2016) Designing training in organic farming on a multinational basis. The International Journal of Information and Learning Technology. 33(2): 99-114. https://doi.org/10.1108/IJILT-12-2015-0039 - Guiné, R., Costa, D., Correia, P., Costa, C.A., Correia, H.E., Castro, M., Guerra, L.T., Seeds, C., Coll, C., Radics, L., Arslan, M., Soylu, S., Tothova, M., Toth, P., Basile, S. (2017). Professional training in organic food production: a cross-country experience. The International Journal of Information and Learning Technology. 34(3): 259-273. https://doi.org/10.1108/IJILT-11-2016-0052. Guiné R. P. F., Gaião D., Costa D. V. T. A., Correia P. M. R., Guerra L. T., Correia H. E., Costa C. A. (2019). Bridges between family farming and organic farming: a study case of the Iberian Peninsula. De Gruyter 4: 727-73 Dias R. S., Costa D., Correia H. E., Costa C. A. (2021). Building Bio-districts or Eco-regions: participative processes. Agriculture, 11, 511. https://doi.org/10.3390/agriculture11060511 - Guiné R. P.F., Mesquita S., Oliveira J., Coelho C., Costa D. T., Correia P., Correia H. E., Dahle B., Oddie M., Karise R., Raimets R., Tourino L., Basile S., Buonomo E., Stefanic I., Costa C. A. (2021). Characterization of the beekeeping activity in seven European countries. Agronomy, 11, 2398. https://doi.org/10.3390/agronomy11122398 Collaborator of the Centre for Studies of Natural Resources, Environment and Society (CERNAS). Participation in various projects: COMPASS Programme "Further development of post-secondary education in the theme of ecological agriculture with special attention to the enhancement of transnationality, harmonising the conditions of the EU partners and associated countries" - Leonardo da Vinci. (2003-2006); Coordinator of the project PART.I.R.EC. Participation for the Improvement of Rural Economy (2009-2011). Leonardo da Vinci Programme, Transfer of Innovation (TOI); Participated in the project ECONEWFARMERS - Building a Future for New Farmers in Ecological Agriculture through Vocational Training of the Leonardo da Vinci Programme, Transfer of Innovation (TOI) action. 2014-16).

Daniela de Vasconcelos Teixeira Aguiar da Costa, is a PhD in Biotechnology. Master in Science and Technology Postharvest. Agricultural Engineering. Adjunct Professor at the Polytechnic Institute of Viseu, Portugal, since 2001. Research in CERNAS-IPV Research Centre. Responsible for the following subjects: Plant Physiology, Plant propagation, Botany and Plant Physiology, Forages and Pastures. As a teacher orientated the graduations projects of the students in the following areas: blueberry production and postharvest, Agriculture, Green Spaces and Plant Propagation. She was Vice President of the Assembly of Representatives of ESAV – IPV (2013 - 2018). Director of the Technological Specialization Course in Organic Farming at ESAV – IPV (2012 – 2015). Coordinator of the on-the-job training component of the Higher Professional Technical Course in Organic Farming at ESAV – IPV (2016-2018). Coordinator of the Superior Professional Technical Course in Organic Farming at ESAV – IPV (2018). Participation in the following projects: Organic Agriculture Project, Project AGRO No. 740 - "The Value of regional varieties of apple fruit through organic farming.", VARC Project - "recovery of agricultural waste from tanneries.", PART.I.REC: Participation for Improvement of Rural Economy, PTDC/AGR-CFL/114826. Potential woody species native to the energy production from biomass, CARDOP - Genetic Resources Carduus, Building a future for new farmers in ecological farming through vocational training", Aquasense project - Autonomous Crops Monitoring System and Intelligent Irrigation Control, Climcast - The new challenges for the chestnut tree in the context of climate change - PDR20202-101-032043, beeB Foster for beekeeping bridges through innovative and participate training. She is the author / co-author of 3 book chapters, more than 35 articles published in the books of national and international congresses in the area of postharvest physiology of fruits and vegetables, plant multiplication and organic farming. There are some publications and presentations in the area of postharvest physiology of fruits and plant propagation.

Andreia Dionisio is a Professor at the University of Évora, Management Department, specialized in statistics for business, data analysis, econometrics, and econophysics and is responsible for the Data Analysis courses. Her main areas of research have included the study of stock markets, global dependence, nonparametric methods, and fuzzy set analysis. Over the last 15 years, she has been involved in several research projects. She is co-author of more than 20 research articles (http://orcid.org/0000-0002-4289-9312). Presently is the Vice-Director of Center for Advanced Studies in Management and Economics of the University of Évora (CEFAGE).

Dora Ferreira is a Geographer with MSc in Urban and Regional Planning, and MSc in Agricultural Engineering. PhD Student at University of Extremadura in Sustainable Territorial Development doctoral program. Actually is researcher with grant of FCT and have several published works related with sustainability and rural innovation. Have special interest to investigate thematics, such as rural tourism, rural development, cultural landscape and sustainable agriculture.

José Manuel Gonçalves is Associate Professor with Habilitation in Agricultural Engineering.

Alpana Gupta is an Environmental Planner with bachelor's in Civil Engineering, having expertise in dealing with Urban & Environmental problems spatially, modeling assessment reports for the in-depth analysis, and the proposals of best practices that can be adopted for the effective implementation of the projects. Formerly as an Innovator, IIT Delhi, she has worked under the theme of "Combating Environmental Pollution" addressing the Global problem of plastic pollution & transforming plastic waste into useful products. She has also delivered projects as a Young professional (Consultant) for the Shyama Prasad Mukherji Rurban Mission, Panchayat and Rural Development Department, Govt. of Madhya Pradesh investigating infrastructure planning projects (check dams, construction of hostels, solar street light, Agro-park processing units, and pipe water supply, designed Toilets under Swachh Bharat Mission). Her experience of working as a Research Associate for the project "Livelihood and Community Mapping through GIS" for the School of Planning and Architecture, Bhopal under the Ministry of Tribal Affairs gave her deep-insights about interacting with communities and assessing them spatially. Her value proposition lies in holding hands-on experience of 2.6 years working in Government projects which makes her adaptable to the multi-dimensional field, perform inter-departmental coordination, Data collection & Analysis, Spatial Planning & Technical Role, Entrepreneurial Mind-Set, Drafting Government Proposal Reports and Research & Development Experience(+10 publications-Book/Book Chapter/Papers).

Pedro Damião Henriques holds a PhD in Agricultural Economics, from the University of Reading (1995), a McS in Agricultural from the Ohio State University (1987) and a B.A. in Agronomic Engineering from Universidade de Lisboa (1982). He is Researcher at the Mediterranean Institute of Agriculture Environment and Development (MED) of the University of Evora, since 2019. Her areas of interest are Agricultural Economics and Policies, Rural Development, Agriculture, Natural Resources and Environment. He has done extensive research in the area of agricultural economics and has participated in different national and international projects. He is author of several book chapters and academic scientific papers in peer-reviewed international journals and other publications.

Athar Hussain is working as Professor and Head Civil Engineering Department at NSUT West Campus, Delhi. Dr. Hussain completed his B.Tech. (Civil engineering) from Jamia Millia Islamia in 1998, M.Tech. from AMU Aligarh in 2001 and completed his Ph.D. in Environmental Engineering from IIT Roorkee in year 2007. Dr. Hussain Joined as Assistant Director in 2007 with Environment Management

Group at National Productivity Council, New Delhi an autonomous body under Ministry of Commerce & Industry and worked there till Oct, 2010. During the tenure he was involved in various national level and policy level projects sponsored from reputed government agencies and organizations such as MoEF, CPCB, SPCBs etc. in the field of environmental engineering. He has also worked as an Assistant Professor, civil engineering department in School of Engineering at Gautam Buddha University till December 2015. Thereafter he joined as Associate Professor through U.P.S.C. in Department of Civil Engineering at Ch. B. P. Govt. College of Engineering, New Delhi. His research area includes industrial wastewater treatment and solid waste management with special focus on anaerobic treatment of wastewater and solid waste with prime focus on renewable energy. He is supervising many Ph.D. Research Scholars in diversified areas of environmental engineering of which some has been awarded degree. Dr. Hussain has more than 15 years of experience in the area of teaching, research and consultancy. He has also guided more than 60 M.Tech dissertations. He has published more than 125 papers in international referred journals and international conferences. Dr. Hussain has published few books on water and wastewater treatment, IGI Global USA and is also serving as a reviewer for many International journals like Taylor and Francis, Inderscience Publishers, Elsevier Science etc. He is also an editorial board member of many journals some of which include- Elsevier science and Inderscience publications. He is also an Associate Member and Fellow member of The Institution of Engineers (IE) India, Member of Institution of valuers" (IV) N. Delhi, India and Member of Indian Society of Wind Engineering, India. Dr. Hussain has represented India many times through international conferences and also through training programmes held abroad being sponsored jointly by NPC, India and APO, Tokyo.

Badar Alam Iqbal, Ph.D, has been the DAAD, Fellow at German Institute of Economic Research; Berlin, Kiel Institute of World Economics; Kiel, South Asia Institute, University of Heidelberg; Germany; Ford Foundation, American Research Center USA; Fulbright Visiting Professor; US; Institute of Developing Economies; Tokyo; Former Fulbright scholar-in-Residence (SIR) 2004-05 at School of Business; Claflin University; South Carolina, USA. In 2016-17, the Fulbright Commission Washington has again nominated for Scholar-in-residence (SIR) at School of Business; Kentucky State University; USA. Now Dr. Iqbal is Emeritus Professor; External-Relation- Chair of Research; at Federic Bastiat Institute for African Research; Ghana. Adjunct Professor; Monarch Business School; Monarch University; Switzerland; Professor Extraordinarius; School of Transdisciplinary Research and Graduate Studies (STRGS), University of South Africa; Pretoria (South Africa).

António José Jordão is Engineer at the Regional Directorate of the Ministry of Agriculture.

Elsa Lamy holds a PhD in Zootechnics, from the University of Evora (2009) and a McS in Metabolic Diseases and Ingestive behaviour, from the Faculty of Medicine-University of Lisbon (2014). She is Assistant Researcher at the Mediterranean Institute of Agriculture Environment and Development (MED) of the University of Evora, since 2014. Her areas of interest are oral biology and salivary proteomics, for the understanding of factors influencing oral food perception and dietary habits. She has done extensive research in the area of salivary proteomics in health and disease, both in animals and humans. She is in the actual direction of the Portuguese High Schools Hub for promotion of the Mediterranean Diet and has participated in different national and international projects. Assistant Editor for 4 different peer-reviewed Journals indexed in Scopus and ISI, she edited 2 books and is author of several book chapters and academic scientific papers in peer-reviewed international journals and other publications.

Tuan Quoc Le attended Kasetsart University, Bangkok, Thailand for MSc & PhD in Food Science with many studies interdisciplinary with food processing, engineering, and health science. He has held several important roles in faculty as dean or acting dean, head of department or program director, international relations, and food technology program consultant at the different universities in Ho Chi Minh City, Vietnam. He also started up with the role of founder or co-founder for several sections of the food industry. He has interests in food R&D in rice and rice products, coffee and bakery technology, innovative drying technologies, microwave techniques, food post-harvest, accelerated food process, colloid systems in foods, health science, gastronomy, and food science.

Ana Bela M. Lopes is teacher English.

Maria de Fátima Martins Lorena de Oliveira is Associate Professor of Definitive Appointment at the Department of Human Social Sciences (DCSH) of the School of Agriculture (ESAC) of the Polytechnic Institute of Coimbra (IPC). He has a Degree in Agro-Industrial Engineering and a PhD in Agronomic Engineering from the Higher Institute of Agronomy, Technical University of Lisbon. Along with teaching has assumed, from the first hour, important administrative and technical-scientific positions, such as: Director of the Course of Higher Studies Specialized in Management and Agricultural Extension; Member of the Forest Resources Engineering Course Commission; Coordinator of the Commission of the Agricultural Engineering Course; Member of the Master of Environmental Management Committee. Throughout his academic and technical-scientific career, he has participated in various national and international training courses in a wide range of fields, from Common Agricultural Policy to international agri-food marketing and trade, through the development of differentiated products, always from the perspective of the agrarian economy. It should be mentioned that he coordinated and integrated the executive and technical-scientific committees of several relevant national and international events, in the context of agricultural policies, international agri-food trade and marketing and innovation and productive differentiation. He is a specialist in sustainable development, economics, agriculture and food security and develops work in projects within the Operational Group, in Prima project and ERASMUS + in the area of irrigation development, cereal logistics and development work on consumer behavior in organic farming and farmer development in different aspects. production to enhance its products.

Maria Raquel Lucas studied at the Universidade de Évora (Portugal) and Zaragoza (Spain) and holds a PhD in Management, Agricultural Management Speciality. She is Associate Professor of Management at the Universidade de Évora, visiting professor of Warsaw University and researcher at CEFAGE (Center for Advanced Studies and Training in Management and Economics). Her areas of interest are agribusiness and sustainability, agri-food marketing, consumer studies, value food chain and rural development. She has done extensive research in the area of management and marketing and has participated in diverse projects including transnational studies with Latin America, Asia and Africa. Referee for several peer-reviewed Journals indexed in Scopus and ISI, she edited 6 books and is author of several book chapters and academic scientific papers in peer-reviewed international journals and other publications. She belongs to the coordinating committee of the Phd programme of Agribusiness and Sustainability and is Director of two Msc programmes (Applied Economics and Management and Quality Management and Agri-food Marketing). Till know, she has been the supervisor of more than one hundred PhD and Msc students.

Junyi Lv is an independent researcher. Her research interests include FinTech, Chinese business and economy.

Ana Marta-Costa holds a PhD in Agri-social Sciences. She is Assistant Professor at the Department of Economics, Sociology and Management of the University of Trás-os-Montes e Alto Douro (UTAD) and researcher at the Centre for Transdisciplinary Development Studies (CETRAD). Her main research fields are agribusiness and sustainability, with emphasis on sustainability assessment, management and planning of the farming systems. She is author of more than 100 scientific publications in these fields, in several peer-reviewed international journals, books and conference proceedings. She has participated in several European and national funded research projects, related with the aforementioned research fields. She belongs to the coordinating committee of the Phd programme of Agribusiness and Sustainability.

Irina Mihaela Matran, Graduate in Food Technology Bachelor program, and Master in Food Quality and Safety Assurance at "Lucian Blaga" University from Sibiu, and also Master in Pharmacovigilence: Monitoring the safety of Medicines" at "Iuliu Hatieganu" University of Medicine and Pharmacy from Cluj-Napoca, Romania. PhD in Medicine with Doctoral Thesis entitled "Development of new products for the treatment of chronic digestive diseases". Co-inventor for Patent of Invention no. RO 132472-A0 - Food supplement for reducing gastroesophageal reflux, awarded with gold in the International Salons of Invention in 2019-2021 and Patent Application No. A/00588/14.08.2018 - "Composition and process for obtaining an assortment of flours from various cereals and functional products with the addition of sericin, as an unconventional ingredient", awarded gold, silver, and bronze at the International Salons of Invention in 2019-2020. Associate Professor and also student at the Faculty of Pharmacy within University of Medicine, Pharmacy, Science and Technology "George Emil Palade" from Targu Mures. Researcher in Food processing technology, member of 2 European funding projects, and legal project representative, all with European funding.

Monica Monica is a graduate in Food Technology Bachelor program, and Master in Food Quality and Safety Assurance at "Lucian Blaga" University from Sibiu, and also Master in Pharmacovigilence: Monitoring the safety of Medicines" at "Iuliu Hatieganu" University of Medicine and Pharmacy from Cluj-Napoca, Romania. PhD in Medicine with Doctoral Thesis entitled "Development of new products for the treatment of chronic digestive diseases". Co-inventor for Patent of Invention no. RO 132472-A0 - Food supplement for reducing gastroesophageal reflux, awarded with gold in the International Salons of Invention in 2019-2021 and Patent Application No. A/00588/14.08.2018 - "Composition and process for obtaining an assortment of flours from various cereals and functional products with the addition of sericin, as an unconventional ingredient", awarded gold, silver, and bronze at the International Salons of Invention in 2019-2020. Associate Professor and also student at the Faculty of Pharmacy within University of Medicine, Pharmacy, Science and Technology "George Emil Palade" from Targu Mures. Researcher in Food processing technology, member of 2 European funding projects, and legal project representative, all with European funding.

Kiakisiki Nascimento has a degree in police sciences and a Master in Applied Economics and Management, specialization in Agribusiness from Universidade de Évora and is a PhD student in Agribusiness and Sustainability programme at Universidade de Évora and Universidade de Trás-os-Montes e Alto Douro. He is a collaborating member of CEFAGE- UE and CETRAD-UTAD and employee of

PNUD - United Nations Development Programme in São Tomé and Príncipe. He participates, as a team member, in projects and develops his research in agribusiness and sustainability, namely, in the transition to circular economy of cocoa and palm oil value chains in São Tomé and Príncipe.

Manuel Nunes is Professor of agricultural engineering, specializing in hydraulics, irrigation and soils.

Sergio Oliveira is a graduate in Crops and Livestock Engineering and Master student in Crops and Livestock Engineering Professional in Agronomics.

Ibrahim Prazeres has a degree in Communication from Universidade Lusófona, a Master in Management, specialization in Marketing from Universidade de Évora and is a PhD student in Agribusiness and Sustainability programme at Universidade de Évora and Universidade de Trás-os-Montes e Alto Douro. He is a collaborating member of CEFAGE- UE and CETRAD-UTAD and responsible for several agricultural entrepreneurship projects in São Tomé and Príncipe, including agriculture field. He participates, as a team member, in projects and develops his research in agribusiness and sustainability, namely, in the sustainability and valorization of organic cocoa production in São Tomé and Príncipe and in the governance and performance of the respective value chain.

Luís Quinta-Nova is a PhD in Environmental Sciences, Master in Spatial Planning and Environmental Planning, and a degree in Biophysical Engineering. Professor at ESA-IPCB since 1996, and Adjunct Professor since 2009 in the scientific area of Planning and Management of Natural Resources, Environmental Impact Assessment and Landscape Ecology. His scientific activity includes areas such as the use of multicriteria analysis tools in the planning and management of rural and natural areas, the study of the influence of environmental factors on present and future species distribution. Also took part in 10 national and international research projects. He was author of 17 books and book chapters and 42 articles published in scientific journals and proceedings of scientific meetings with peer-review and presented 97 communications in scientific conferences. He also participated as staff member and scientific committee member in 26 conferences.

Cristovão Reis studied at the Universidade de Évora (Portugal) and holds a PhD in Management. He is Assistant Professor of Economics at the Universidade Nacional Timor Lorosa'e, visiting professor in several private universities in Timor-Leste. Specialized in value food chain, marketing, economics studies and consumer studies. He has been involved in several research projects, is supervisor of more than one hundred Bachelor's students and is author of 3 research articles.

Pedro Reis is a Master in Agricultural Economics and Rural Sociology, PhD in Agronomic Engineering from the University of Lisbon. Researcher at the National Institute of Agrarian and Veterinary Research, (INIAV), working in the area of agrarian systems, knowledge management and rural development. He has worked on various research, innovation and rural development projects, as coordinator or responsible for INIAV participation. From 2014 to 2019 he was a visiting assistant professor at the Instituto Superior de Agronomia, in Economics and Sociological Analysis. From 2008 to 2009, he was advisor to the Secretary of State for Rural Development and Forests. Since 2019 he is advisor of the Minister of Agriculture. From 1999 to 2008, he was assistant researcher at INIA, where he carried out planning and monitoring activities for agricultural technological development programs. President of the Scientific Council of INIAV, from 2012 to 2015, and of INRB, from 2011 to 2012.

António Terrão Russo is Senior Irrigation and Rural Development Expert.

Subir Sinha is an Indian scholar and a Lecture (State Aided College Teacher) in the Department of Journalism and Mass communication of Dum Dum Motijheel College (Affiliated under West Bengal State University). He was formerly attached with the Department of Journalism of Surendranath College (Affiliated under University of Calcutta) and Vijaygarh Jyotish Ray College (Affiliated under University of Calcutta)as a Guest Faculty . He was also remained attached with Indra Gandhi National Open University (IGNOU) under convergence scheme with Dum Dum Motijheel College as an Academic Counsellor of PGJMC for a short period. Among his interest- impact of media, role of new media, social media, communication and development communication are significant. He presented papers in various National and International seminars and conference and published various articles in various books and journals.

Anastasiia Sliusarenko has a Master's degree in Business Management. Main areas of research involve crisis management, innovation, technology and business strategy.

Oloiva Maria Sousa is a 3rd student of the PhD course in Agribusiness and Sustainability, a partnership between the University of Trás-os-Montes and Alto Douro (UTAD) and the University of Évora in Portugal.Her research interests are circular economy, bioenergy, sustainability and ecology.She has a Msc in health services management from UTAD, specializing in quality management system ISO 9001, EFQM, CHK, JCI, ACSA, in Health Units, a post- graduation in accounting and finance and a degree in economics from UTAD. She is a researcher at the Center for Advanced Studies and Training in Management and Economics (CEFAGE) at the University of Évora (UÉvora) whose lines of research are directed to the areas of Management and Taxation and, research member of the Center for Transdisciplinary Studies for Development (CETRAD) by UTAD affects the group of innovation, markets and organization, research lines in Economic Dynamics & Territorial Social Cohesion.

Monica Tarcea has a Licence degree in Medicine (1992), PhD in Medicine / Health (1999) and Habilitation degree in 2017, at the University of Medicine and Pharmacy Targu Mures, Romania. Professor and Head of Community Nutrition and Food Safety Department, also Coordinator of Bachelor program of Nutrition and Dietetics and Master program of Clinical and Community Nutrition, at the same university (from 2009 to 2019). University teacher since 1994, senior specialist in Hygiene and Food Safety, former Dean of Faculty of Medicine (2013-2016), and member of Hygiene Committee from Romanian Ministry of Health, Romanian Association of Hygiene, and Romanian Nutrition and Dietetics Association. Head of Environmental Health Department, at regional Public Health Center Targu Mures (with part time job) branch of Romanian Public Health Institute (Bucharest). Author of 10 hygiene and nutrition books, 7 manuals for students, 162 articles (24 in ISI journals), 95 scientific conferences participations and coordinator of 12 national conferences. Director and member in 8 national and 5 international projects. Research interests include Community Nutrition, Public Health, Food Safety, and Risk Behavior Management.

Arti Yadav is presently Post Doctoral Research Fellow at Indian Council of Social Science Research, India. She has also worked as an assistant professor at the Department of Commerce, Daulat Ram College, University of Delhi, India. She is a Ph.D. in Commerce from the Department of Commerce, Aligarh Muslim University, India and holds her master's degree from Delhi School of Economics, University of Delhi. She has also qualified UGC-NET/JRF. She has a number of research papers and chapters to her credit, published in well-known journals including Springer International Publishing, France forum, IGI global and ADBC listed journal.

Poshan (Sam) Yu is a Lecturer in Accounting and Finance in the International Cooperative Education Program of Soochow University (China). He is also an External Professor of FinTech and Finance at SKEMA Business School (China), a Visiting Professor at Krirk University (Thailand) and a Visiting Researcher at the Australian Studies Centre of Shanghai University (China). Sam leads FasterCapital (Dubai, UAE) as a Regional Partner (China) and serves as a Startup Mentor for AIC RAISE (Coimbatore, India). His research interests include financial technology, regulatory technology, public-private partnerships, mergers and acquisitions, private equity, venture capital, start-ups, intellectual property, art finance, and China's "One Belt One Road" policy.

Index

R

S

T

U

V

W

www.ingramcontent.com/pod-product-compliance
Lightning Source LLC
LaVergne TN
LVHW060359300926
849162LV00006B/1303